Lecture Notes in Computer Science 4549

Commenced Publication in 1973
Founding and Former Series Editors:
Gerhard Goos, Juris Hartmanis, and Jan van Leeuwen

James Aspnes   Christian Scheideler
Anish Arora   Samuel Madden (Eds.)

Distributed Computing in Sensor Systems

Third IEEE International Conference, DCOSS 2007
Santa Fe, NM, USA, June 18-20, 2007
Proceedings

 Springer

Volume Editors

James Aspnes
Yale University, Department of Computer Science
51 Prospect Street, New Haven, CT 06520-8285, USA
E-mail: aspnes@cs.yale.edu

Christian Scheideler
Technical University of Munich, Institute of Informatics
Boltzmannstr. 3, 85748 Garching, Germany
E-mail: scheideler@in.tum.de

Anish Arora
Ohio State University, Department of Computer Science and Engineering
395 Dreese Hall, Columbus, OH 43210-1277, USA
E-mail: anish@cse.ohio-state.edu

Samuel Madden
Massachusetts Institute of Technology
Computer Science and Artificial Intelligence Laboratory (CSAIL)
32 Vassar St., Cambridge, MA 02139, USA
E-mail: srmadden@mit.edu

Library of Congress Control Number: 2007928495

CR Subject Classification (1998): C.2.4, C.2, D.4.4, E.1, F.2.2, G.2.2, H.4

LNCS Sublibrary: SL 5 – Computer Communication Networks and
Telecommunications

ISSN   0302-9743
ISBN-10   3-540-73089-3 Springer Berlin Heidelberg New York
ISBN-13   978-3-540-73089-7 Springer Berlin Heidelberg New York

Springer is a part of Springer Science+Business Media

springer.com

© Springer-Verlag Berlin Heidelberg 2007
Printed in Germany

Typesetting: Camera-ready by author, data conversion by Scientific Publishing Services, Chennai, India
Printed on acid-free paper   SPIN: 12077670   06/3180   5 4 3 2 1 0

Message from the General Co-chairs

We are pleased to welcome you to the proceedings of DCOSS 2007, the IEEE International Conference on Distributed Computing in Sensor Systems, the third event in this annual conference series. The DCOSS meeting series covers the key aspects of distributed computing in sensor systems, such as high-level abstractions, computational models, systematic design methodologies, algorithms, tools and applications.

This meeting would not have been possible without the tireless effort of many volunteers. We are indebted to the DCOSS 2007 Program Chair, James Aspnes, for overseeing the review process and composing the technical program. We appreciate his leadership in putting together a strong and diverse Program Committee (PC) covering various aspects of this multidisciplinary research area.

We would like to thank the PC Vice Chairs, the members of the Program Committee, the external referees consulted by the PC, as well as all of the authors who submitted their work to DCOSS 2007.

Several volunteers contributed significantly to the realization of the meeting. We wish to thank the organizers of the three workshops that were collocated with DCOSS 2007. We would like to thank Wendi Heinzelman and Bhaskar Krishnamachari for their efforts in organizing the poster session. Special thanks go to Amol Bakshi for handling Web-based publicity and local arrangements, and Yang Yu for his assistance in putting together this proceedings volume.

We would like to especially thank Jose Rolim, DCOSS Steering Committee Chair, for inviting us to be the General Co-chairs. His invaluable input in shaping this conference series and his timely contributions in resolving meeting-related issues are gratefully acknowledged.

We wish to thank the keynote speakers, Richard M. Karp (University of California at Berkeley) and P.R. Kumar (University of Illinois, Urbana-Champaign). We deeply appreciate their participation in the meeting.

Finally, we would like to acknowledge the sponsorship by the IEEE Technical Committee on Parallel Processing, the IEEE Technical Committee on Distributed Processing, the TCS-Sensor Lab of the Centre Universitaire d. Informatique of the University of Geneva, Switzerland and the European Association for Theoretical Computer Science (EATCS).

The research area of sensor networks is rapidly evolving, influenced by fascinating advances in supporting technologies. We sincerely hope that this conference series will serve as a forum for researchers working in different, complementary aspects of this multidisciplinary field to exchange ideas and interact and cross-fertilize research in the algorithmic and foundational aspects, high-level approaches as well as more applied and technology-related issues focusing on tools and applications of wireless sensor networks.

June 2007

Sotiris Nikoletseas
Viktor K. Prasanna

Message from the Program Chair

This volume holds the proceedings of the 3rd Annual Conference on Distributed Computing on Sensor Systems (DCOSS 2007). The 27 papers in this volume where selected from 71 submissions in three tracks covering the areas of algorithms, applications, and systems. The conference continues in its mission to bring together researchers in all areas of sensor systems and ensure cross-pollination both between theory and practice and between the broader field of distributed computing and the specific issues arising in sensor networks and related systems.

Selecting the papers for this volume required the efforts of many people, including the members of the Program Committee and numerous outside referees. The work of the three Program Committee Vice Chairs—Christian Scheideler, for the Algorithms track; Anish Arora, for the Applications track; and Samuel Madden, for the Systems track—to make this process run smoothly was an invaluable contribution to its success. I am deeply in their debt for agreeing to take on this responsibility, as I am to the many members of the Program Committee and the many external referees who worked with them.

I would also like to thank the Proceedings Chair, Yang Yu, for his tireless work in assembling the conference proceedings, and the General Co-chairs Sotiris Nikoletseas and Viktor Prasanna for their guidance and countless efforts in organizing the conference.

June 2007   James Aspnes

Organization

General Chair

Sotiris Nikoletseas   University of Patras and CTI, Greece
Viktor K. Prasanna   University of Southern California, USA

Vice General Chair

Azzedine Boukerche   University of Ottawa, Canada

Program Chair

James Aspnes   Yale University, USA

Program Vice Chairs

Algorithms
Christian Scheideler   Technical University of Munich, Germany

Applications
Anish Arora   Ohio State University, USA

Systems
Samuel Madden   Massachusetts Institute of Technology, USA

Steering Committee Chair

Jose Rolim   University of Geneva, Switzerland

Steering Committee

Sajal Das   University of Texas at Arlington, USA
Josep Diaz   UPC Barcelona, Spain
Deborah Estrin   University of California, Los Angeles, USA
Phillip B. Gibbons   Intel Research, Pittsburgh, USA
Sotiris Nikoletseas   University of Patras and CTI, Greece
Christos Papadimitriou   University of California, Berkeley, USA
Kris Pister   University of California, Berkeley, and Dust, Inc., USA
Viktor Prasanna   University of Southern California, Los Angeles, USA

Poster Chair

Wendi Heinzelman   University of Rochester, USA
Bhaskar Krishnamachari   University of Southern California, USA

Proceedings Chair

Yang Yu   Motorola Labs, USA

Publicity Co-chairs

Amol Bakshi   University of Southern California, USA
Sanjay Jha   University of New South Wales, Australia
Christian Schindelhauer   University of Freiburg, Germany

Finance Chair

Germaine Gusthiot   University of Geneva, Switzerland

Sponsoring Organizations

IEEE Computer Society Technical Committee on Parallel Processing (TCPP)
IEEE Computer Society Technical Committee on Distributed Processing
    (TCDP)

Held in Cooperation with

ACM Special Interest Group on Computer Architecture (SIGARCH)
ACM Special Interest Group on Embedded Systems (SIGBED)
European Association for Theoretical Computer Science (EATCS)
IFIP WG 10.3

Program Committee

Tarek Abdelzaher   University of Illinois at Urbana-Champaign,
                                USA
Philippe Bonnet   University of Copenhagen, Denmark
Bogdan Chlebus   of Colorado at Denver, USA
Alfredo Cuzzocrea   University of Calabria, Italy
Uday Desai   Indian Institute of Technology at Bombay,
                                India
Amol Deshpande   University of Maryland, College Park, USA
Deborah Estrin   University of California at Los Angeles, USA
Andras Farago   University of Texas at Dallas, USA

Referees

Alexander Kroeller

Vinod Kulathumani

Santosh Kumar

Rafael Laufer

Xin Liu

Daniel Massaguer

Michele Nati

Melih Onus

Jeongyeup Paek

Joseph Paradiso

Sundeep Pattem

Roberto Petroccia

Mohammad Rahimi

Rik Sarkar

Tom Schoellhammer

Michael Segal

Mohamed Sharaf

Abhishek Sharma

Simone Silvestri

Primoz Skraba

Dongjin Son

Avinash Sridharan

Bin Tang

Bishal Thapa

Claudio Vicari

Yin Wang

Karen Weeks

Donglin Xia

Bo Xing

Guoliang Xue

Kiran Yedavalli

Xingbo Yu

Xianjin Zhu

Table of Contents

Distributed Coalition Formation in Visual Sensor Networks: A Virtual Vision Approach

Faisal Qureshi[1] and Demetri Terzopoulos[1,2]

[1] Dept. of Computer Science, University of Toronto, Toronto, ON, Canada
faisal@cs.toronto.edu
[2] Computer Science Dept., University of California, Los Angeles, CA, USA
dt@cs.ucla.edu

Abstract. We propose a distributed coalition formation strategy for collaborative sensing tasks in camera sensor networks. The proposed model supports task-dependent node selection and aggregation through an announcement/bidding/selection strategy. It resolves node assignment conflicts by solving an equivalent constraint satisfaction problem. Our technique is scalable, as it lacks any central controller, and it is robust to node failures and imperfect communication. Another unique aspect of our work is that we advocate visually and behaviorally realistic virtual environments as a simulation tool in support of research on large-scale camera sensor networks. Specifically, our visual sensor network comprises uncalibrated static and active simulated video surveillance cameras deployed in a virtual train station populated by autonomously self-animating pedestrians. The readily reconfigurable virtual cameras generate synthetic video feeds that emulate those generated by real surveillance cameras monitoring public spaces. Our simulation approach, which runs on high-end commodity PCs, has proven to be beneficial because this type of research would be difficult to carry out in the real world in view of the impediments to deploying and experimenting with an appropriately complex camera network in extensive public spaces.

Keywords: Camera sensor networks, Sensor coordination and control, Distributed coalition formation, Video surveillance.

1 Introduction

Camera sensor networks are becoming increasingly important to next generation applications in surveillance, in environment and disaster monitoring, and in the military. In contrast to current video surveillance systems, camera sensor networks are characterized by smart cameras, large network sizes, and ad hoc deployment.[1] These systems lie at the intersection of machine vision and sensor networks, raising issues in the two fields that must be addressed simultaneously. The effective visual coverage of extensive areas—public spaces, disaster zones, and battlefields—requires multiple cameras to collaborate towards common sensing goals. As the size of the camera network grows, it

---

[1] Smart cameras are self-contained vision systems, complete with image sensors, power circuitry, communication interfaces, and on-board processing capabilities [1].

J. Aspnes et al. (Eds.): DCOSS 2007, LNCS 4549, pp. 1–20, 2007.

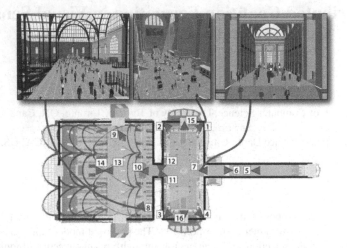

Fig. 1. Plan view of the virtual Penn Station environment with the roof not rendered, revealing the concourses and train tracks (left), the main waiting room (center), and the long shopping arcade (right). (The yellow rectangles indicate station pedestrian portals.) An example visual sensor network comprising 16 simulated active (pan-tilt-zoom) video surveillance cameras is shown.

becomes infeasible for human operators to monitor the multiple video streams and identify all events of possible interest, or even to control individual cameras in performing advanced surveillance tasks. Therefore, it is desirable to design camera sensor networks that are capable of performing visual surveillance tasks autonomously, or at least with minimal human intervention.

In this paper, we demonstrate a camera network model comprising *uncalibrated* passive and active simulated video cameras that with minimal operator assistance can perform persistent surveillance of a large virtual public space—a train station populated by autonomously self-animating virtual pedestrians (Fig. 1). Once a human operator or an automated visual behavior analysis routine monitoring the surveillance video feeds identified a pedestrian of interest, the cameras decide amongst themselves how best to observe the subject. For example, a subset of the active pan/tilt/zoom (PTZ) cameras can collaboratively track the pedestrian as (s)he weaves through the crowd. The problem of assigning cameras to persistently monitor pedestrians becomes challenging when there are multiple pedestrians of interest. To deal with the numerous possibilities, the cameras must be able to *reason* about the dynamic situation. To this end, we propose a distributed camera network control strategy that is capable of dynamic task-driven node aggregation through local decision making and inter-node communication.

1.1   Virtual Vision

Deploying a large-scale camera sensor network in the real world is a major undertaking whose cost can easily be prohibitive for most researchers interested in designing and experimenting with sensor networks. Moreover, privacy laws generally restrict the monitoring of people in public spaces for experimental purposes. As a means of

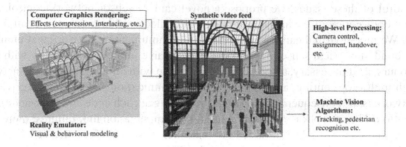

Fig. 2. Virtual vision paradigm (image from [2])

overcoming these impediments, we advocate the pursuit of camera sensor network research in the context of a unique synthesis of advanced computer graphics and vision simulation technologies. In particular, we demonstrate the design of simulated camera sensor network systems and meaningful experimentation with such systems within visually and behaviorally realistic virtual environments (Fig. 2).

Legal impediments and cost considerations aside, the use of realistic virtual environments in sensor network research offer significantly greater flexibility during the design and evaluation cycle, thus expediting the engineering process: The multiple virtual cameras, which generate synthetic video feeds that emulate those generated by real surveillance cameras monitoring public spaces, are very easily reconfigurable in the virtual space. The virtual world provides readily accessible ground-truth data for the purposes of visual sensor network algorithm validation. Experiments are perfectly repeatable in the virtual world, so we can readily modify algorithms and parameters and immediately determine their effect. The hard real-time constraints of the real world can easily be relaxed in the simulated world; i.e., simulation time can be prolonged relative to real, "wall-clock time", in principle permitting arbitrary amounts of computational processing to be carried out during each unit of simulated time. Finally, despite its sophistication, our simulator runs on high-end commodity PCs, thus obviating the need to grapple with special-purpose hardware and software.

1.2 Distributed Control in Camera Sensor Networks

Our work deals with distributed control in camera sensor networks and many of the characteristic challenges associated with sensor networks are relevant. Task-based sensor selection is a fundamental issue in sensor networks [3]. The selection process must take into account the information contribution of each node against its resource consumption or potential utility in other tasks. Another key issue in sensor networks is node organization, which has been proposed by researchers as a means to limit the communication to those nodes that are relevant to the task at hand. Distributed approaches for node selection or node organization are preferable to centralized approaches and offer what are perhaps the greatest advantages of networked sensing—robustness and scalability. Also, in a typical sensor network, each sensor node has local autonomy and can communicate with a small number of neighboring nodes that are within radio communication range.

Mindful of these issues, we propose a novel camera sensor network control strategy that does not require camera calibration, a detailed world model, or a central controller. We model virtual cameras as nodes in a communication network that emulates those found in physical sensor networks: 1) nodes can communicate directly with their neighbours, 2) if necessary, a node can communicate with another node in the network through multi-hop routing, and 3) unreliable communication. The overall behavior of the network is the consequence of the local processing at each node and inter-node communication. The network is robust to node and communication link failures; moreover, it is scalable due to the lack of a central controller. Visual surveillance tasks are performed by groups of one or more camera nodes. These groups, which are created on the fly, define the information sharing parameters and the extent of collaboration between nodes. During the lifetime of the surveillance task, a group evolves—i.e., old nodes leave the group and new nodes join it. One node in each group acts as the group leader and is responsible for group-level decision making. We also present a new constraint satisfaction problem formulation for resolving group interactions.

1.3   Overview

The contributions of this paper are twofold. We introduce a novel camera sensor network framework suitable for next generation visual surveillance applications. Furthermore, we demonstrate the advantages of developing and evaluating camera sensor networks within a sophisticated virtual reality simulation environment. The remainder of the paper is organized as follows: Section 2 covers relevant prior work. We explain the low-level vision emulation and behavior models for camera nodes in Section 3. Section 4 introduces the sensor network communication model. In Section 5, we demonstrate the application of this model in the context of visual surveillance. We present our results in Section 6 and our conclusions and future research directions in Section 7.

2   Related Work

As was argued in [4, 5], computer graphics and virtual reality technologies are rapidly presenting viable alternatives to the real world for developing sensory systems (see also [6]). Our camera network is deployed and tested within the virtual train station simulator that was developed in [2]. The simulator incorporates a large-scale environmental model (of the original Pennsylvania Station in New York City) with a sophisticated pedestrian animation system. The simulator can efficiently synthesize well over 1000 self-animating pedestrians performing a rich variety of activities in the extensive indoor urban environment. Like real humans, the synthetic pedestrians are fully autonomous. They perceive the virtual environment around them, analyze environmental situations, make decisions and behave naturally within the train station. Standard computer graphics techniques enable a photorealistic rendering of the busy urban scene with considerable geometric and photometric detail (Fig. 1).

The problem of forming sensor groups based on task requirements and resource availability has received much attention within the sensor networks community [3]. Reference [1] argues that task-based grouping in ad hoc camera networks is highly advantageous. Collaborative tracking, which subsumes the above issue, is considered an essential capability in many sensor networks [3]. Reference [7] introduces an information driven approach to collaborative tracking, which attempts to minimize the energy expenditure at each node by reducing inter-node communication. A node selects the next node by utilizing the information gain vs. energy expenditure tradeoff estimates for its neighbor nodes. In the context of camera networks, it is often difficult for a camera node to estimate the expected information gain by assigning another camera to the task without explicit geometric and camera-calibration knowledge, but such knowledge is tedious to obtain and maintain during the lifetime of the camera network. Therefore, our camera networks do without such knowledge; a node needs to communicate with nearby nodes in order to select new nodes.

Nodes comprising sensor networks are usually untethered sensing units with limited onboard power reserves. Consequently, a crucial concern in sensor networks is the energy expenditure at each node, which determines the life-span of a sensor network [8]. Node communications have large power requirements; therefore, sensor network control strategies attempt to minimize the inter-node communication [9,7]. Presently, we do not address this issue. However, the communication protocol that we propose limits the communication to the active nodes and their neighbors.

Little attention has been paid in computer vision to the problem of controlling active cameras to provide visual coverage of an extensive public area, such as a train station or an airport [10,11]. Previous work on camera networks in computer vision has dealt with issues related to low-level and mid-level computer vision, namely segmentation, tracking, and identification of moving objects [12], and camera network calibration [13]. Our approach does not require calibration; however, we assume that the cameras can identify a pedestrian with reasonable accuracy. To this end, we employ color-based pedestrian appearance models.

IrisNet is a sensor network architecture tailored towards high-capability multimedia sensors connected via high-capacity communication channels [14]. It takes a *centralized* view of the network and models it as a distributed database, allowing efficient access to sensor readings. We consider this work to be orthogonal to ours. SensEye is a recent sensor-network inspired multi-camera systems [15]. It demonstrates the benefits of a multi-tiered network—each tier defines a set of sensing capabilities and corresponds to a single class of smart camera sensors—over single-tiered networks in terms of low-latencies and energy efficiency. However, SensEye does not deal with the distributed camera control issues that we address.

Our node grouping strategy is inspired by the ContractNet distributed problem solving protocol [16] and realizes group formation via inter-node negotiation. Unlike Mallett's [1] approach to node grouping where groups are defined implicitly via membership nodes, our approach defines groups explicitly through group leaders. This simplifies reasoning about groups; e.g., Mallett's approach requires specialized nodes for group termination. Our strategy handles group leader failures through group merging and group leader demotion operations.

Resolving group-group interactions requires sensor assignment to various tasks, which shares many features with Multi-Robot Task Allocation (MRTA) problems studied by the multi-agent systems community [17]. Specifically, according to the taxonomy provided in [17], our sensor assignment formulation belongs to the single-task robots (ST), multi-robot tasks (MR), instantaneous assignment (IA) category. ST-MR-IA problems are significantly more difficult than single robot task MTRA problems. Task-based robot grouping arise naturally in ST-MR-IA problems, which are sometimes referred to as *coalition formation*. ST-MR-IA problems are extensively studied and can be reduced to a Set Partitioning Problem (SPP), which is strongly NP-hard [18]. However, many heuristics-based set partitioning algorithms exist that produce good results on large SPPs [19]. Fortunately, the sizes of MRTA problems, and by extension SPPs, encountered in our camera sensor network setting are small due to the spatial/locality constraints inherent to the camera sensors.

We model sensor assignments as a Constraint Satisfaction Problem (CSP), which we solve using "centralized" backtracking. Each sensor assignment that passes the hard constraints is assigned a weight, and the assignment with the highest weight is selected. We have intentionally avoided distributed constraint optimization techniques, such as [20] and [21], due to their explosive communication requirements even for small sized problems. Additionally, it is not obvious how they handle node and communication failures. Our strategy lies somewhere between purely distributed and fully centralized schemes for sensor assignments—sensor assignment is distributed at the level of the network, whereas it is centralized at the level of a group.

3   Camera Nodes

Each virtual camera node in the sensor network is able to perform low-level visual processing and is an active sensor with a repertoire of camera behaviors. The next two sections describe each of these aspects of a camera node.

3.1   Local Vision Routines

Each camera has its own suite of visual routines for pedestrian recognition, identification, and tracking, which we dub "Local Vision Routines" (LVRs). The LVRs are computer vision algorithms that directly operate upon the synthetic video acquired by the virtual cameras. LVRs do not have access to any 3D information available from the virtual world, and they mimic the performance of a state-of-the-art pedestrian segmentation and tracking module (Fig. 3(a)). In particular, pedestrian tracking can fail due to occlusions, poor segmentation, bad lighting, or crowding. Tracking sometimes locks on the wrong pedestrian, especially if the scene contains multiple pedestrians with similar visual appearance; i.e., wearing similar clothes. Our imaging model emulates artifacts that are of interest to camera network researchers, such as video compression and interlacing. It also models camera jitter and imperfect color response.

We employ appearance-based models to track pedestrians. Pedestrians are segmented to construct unique and robust color-based signatures (appearance models), which are then matched across the subsequent frames. Color-based signatures have found widespread use in tracking applications [22], but they are sensitive to illumination changes. However, this shortcoming can be mitigated by operating in HSV space instead of RGB space. Furthermore, zooming can drastically change the appearance of a pedestrian, thereby confounding conventional appearance-based schemes. We employ a

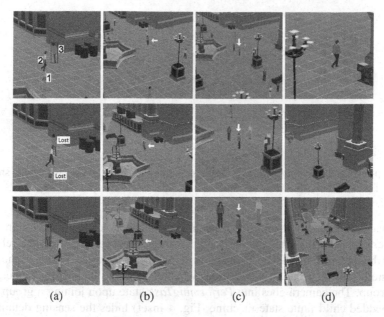

Fig. 3. (a) The LVRs are programmed to track Pedestrians 1 and 3. Pedestrian 3 is tracked successfully; however, track is lost of Pedestrian 1 who blends in with the background. The tracking routine loses Pedestrian 3 when she is occluded by Pedestrian 2, but it regains track of Pedestrian 3 when Pedestrian 2 moves out of the way. (b) Tracking while fixating on a pedestrian. (c) Tracking while zooming in on a pedestrian. (d) Camera returns to its default settings upon losing the pedestrian; it is now ready for another task.

modified *color-indexing* scheme [23] to tackle this problem. Thus, a distinctive characteristic of our pedestrian tracking routine is its ability to operate over a range of camera zoom settings. It is important to note that we do not assume camera calibration. See [24] for more details.

3.2 Camera Node Behaviors

Each camera node is an autonomous agent capable of communicating with nearby nodes. The LVRs determine the sensing capabilities of a camera node, whose overall behavior is determined by the LVR (bottom-up) and the current task (top-down). We model the camera controller as an augmented hierarchical finite state machine (Fig. 4).

In its default state, *Idle*, the camera node is not involved in any task. A camera node transitions into the *ComputingRelevance* state upon receiving a *queryrelevance* message from a nearby node. Using the description of the task that is contained within the *queryrelevance* message and by employing the LVRs, the camera node can compute its relevance to the task. For example, a camera can use visual search to find a pedestrian that matches the appearance-based signature passed by the querying node. The relevance encodes the expectation of how successful a camera node will be at a particular sensing task. The camera returns to the *Idle* state when it fails to compute

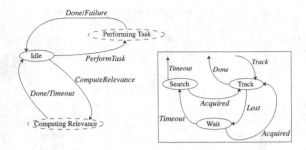

Fig. 4. Top-level camera controller. Dashed states contain the child finite state machine shown in the inset.

the relevance because it cannot find a pedestrian that matches the description. When the camera successfully finds the desired pedestrian, however, it returns the relevance value to the querying node. The querying node passes the relevance value to the leader (leader node) of the group, which decides whether or not to include the camera node in the group. The camera goes into *PerformingTask* state upon joining a group where the embedded child finite state machine (Fig. 4 inset) hides the sensing details from the top-level controller and enables the node to handle short-duration sensing (tracking) failures. Built-in timers allow the camera node to transition into the default state instead of hanging in some state waiting for a message from another node, which might never arrive due to a communication error or node failure.

Each camera can *fixate* and *zoom* in on an object of interest. Fixation and zooming routines are image driven and do not require any 3D information, such as camera calibration or a global frame of reference. We discovered that traditional Proportional Derivative (PD) controllers generate unsteady control signals, resulting in jittery camera motion. The noisy nature of tracking forces the PD controller to try continuously to minimize the error metric without ever succeeding, so the camera keeps servoing. Hence, we model the fixation and zooming routines as dual-state controllers. The states are used to activate/deactivate the PD controllers. In the *act* state the PD controller tries to minimize the error signal; whereas, in the *maintain* state the PD controller ignores the error signal altogether and does nothing.

The *fixate* routine brings the region of interest—e.g., a pedestrian's bounding box—into the center of the image by tilting the camera about its local x and y axes (Fig. 3(b)). The *zoom* routine controls the FOV of the camera such that the region of interest occupies the desired percentage of the image (Fig. 3(c)). See [24] for the details.

A camera node returns to its default stance after finishing a task using the *reset* routine, which is a PD controller that attempts to minimize the error between the current zoom/tilt settings and the default zoom/tilt settings (Fig. 3(d)).

4   Sensor Network Model

We now explain the sensor network communication scheme that enables task-specific coalition formation. The idea is as follows: A human operator presents a particular

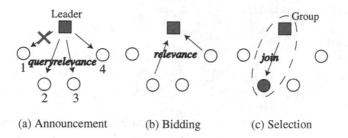

(a) Announcement   (b) Bidding   (c) Selection

Fig. 5. Task auction supports coalition formation. The red cross indicates a lost message.

sensing request to one of the nodes. In response to this request, relevant nodes self-organize into a group with the aim of fulfilling the sensing task. The *group*, which formalizes the collaboration between member nodes, is a dynamic arrangement that evolves throughout the lifetime of the task. At any given time, multiple groups might be active, each performing its respective task. Group formation is determined by the local computation at each node and the communication between the nodes. Specifically, we employ the ContractNet protocol, which models auctions (announcement, bidding, and selection) for group formation [16] (see Fig. 5). The local computation at each node involves choosing an appropriate bid for the announced sensing task.

From the standpoint of user interaction, we have identified two kinds of sensing queries: 1) where the queried sensor itself can measure the phenomenon of interest—e.g., when a human operator selects a pedestrian to be tracked within a particular video feed—and 2) when the queried node might not be able to perform the required sensing and needs to route the query to other nodes. For instance, an operator can request the network to count the number of pedestrians wearing green shirts. To date we have experimented only with the first kind of queries, which are sufficient for setting up collaborative tracking tasks; however, this is by no means a limitation of the proposed communication model.

4.1 Coalition Formation

Node grouping commences when a node n receives a sensing query. In response to the query, the node sets up a named task and creates a single-node group. Initially, as node n is the only node in the group, it is chosen as the leader node. To recruit new nodes for the current task, node n begins by sending *queryrelevance* messages to its neighboring nodes, N_n. This is akin to auctioning the task in the hope of finding suitable nodes. A subset N' of N_n respond by sending their relevance values for the current task (*relevance* message). This is the bidding phase. Upon receiving the relevance values, node n selects a subset M of N' to include in the group, and sends *join* messages to the chosen nodes. This is the selection phase. When there is no resource contention between groups (tasks)—e.g., when only one task is active, or when multiple tasks that do not require the same nodes for successful operation are active—the selection process is relatively straightforward; node n picks those nodes from N' that have the highest relevance values. On the other hand, a conflict resolution mechanism is required when multiple groups vie for the same nodes; we present a scheme to handle this situation in

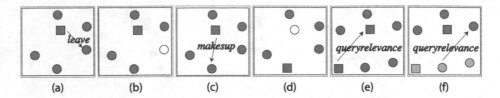

Fig. 6. (a)-(b) A node leaves a group after receiving a leave message from the group leader. (c)-(d) Old group leader selects a new group leader and leaves the group. (e) A leader node detects another leader performing the same task; leader/supervisor demotion commences. (f) Conflict detection between two resources.

the next section. A node that is not already part of any group can join the group upon receiving a *join* message from the leader of that group. After receiving the *join* message, a subset M' of M elect to join the group.

For multinode groups, if a group leader decides to recruit more nodes for the task at hand, it instructs group nodes to broadcast task requirements. This is accomplished via sending *queryrelevance* to group nodes. The leader node is responsible for group-level decisions, so member nodes forward to the group leader all the group-related messages, such as the *relevance* messages from potential candidates for group membership. During the lifetime of a group, group nodes broadcast *status* messages at regular intervals. Group leaders use *status* messages to update the relevance information of the group nodes. When a leader node receives a *status* message from another node performing the same task, the leader node includes that node into its group. The leader node uses the most recent relevance values to decide when to drop a member node. A group leader also removes a node from the group if it has not received a *status* message from the node in some preset time limit.[2] Similarly, a group node can choose to stop performing the task when it detects that its relevance value is below a certain threshold. When a leader detects that its own relevance value for the current task is below the predefined threshold, it selects a new leader from amongst the member nodes. The group vanishes when the last node leaves the group.

4.2 Conflict Resolution

A conflict resolution mechanism is needed when multiple groups require the same resources (Fig. 6(f)). The problem of assigning sensors to the contending groups can be treated as a Constraint Satisfaction Problem (CSP) [25]. Formally, a CSP consists of a set of variables $\{v_1, v_2, v_3, \cdots, v_k\}$, a set of allowed values $\mathrm{Dom}[v_i]$ for each variable v_i (called the domain of v_i), and a set of constraints $\{C_1, C_2, C_3, \cdots, C_m\}$. The solution to the CSP is a set $\{v_i \leftarrow a_i | a_i \in \mathrm{Dom}[v_i]\}$, where the a_is satisfy all the constraints.

We treat each group g as a variable, whose domain consists of the non-empty subsets of the set of sensors with relevance values (with respect to g) greater than a predefined

---

[2] The relevance value of a group node decays over time in the absence of new *status* messages from that node. Thus, we can conveniently model node dependent timeouts; i.e., the time duration during which at least one *status* message must be received by the node in question.

threshold. The constraints restrict the assignment of a sensor to multiple groups. Assume, for example, a group g and a set of nodes $\{n_1, n_2, n_3\}$ with relevance values $\{r_1, r_2, r_3\}$, respectively. If r_3 is less than the predefined threshold, the set of nodes that will be considered for assignment to g is $\{n_1, n_2\}$, and the domain of g is the set $\{\{n_1\}, \{n_2\}, \{n_1, n_2\}\}$. We define a constraint C_{ij} as $a_i \cap a_j = \{\Phi\}$, where a_i and a_j are sensor assignments to groups g_i and g_j, respectively; k groups give rise to $k!/2!(k-2)!$ constraints.

We can then define a CSP as $P = (G, D, C)$, where $G = \{g_1, g_2, \cdots, g_k\}$ is the set of groups (variables) with non-empty domains, $S = \{\text{Dom}[g_i] | i \in [1, k]\}$ is the set of domains for each group, and $C = \{C_{ij} | i, j \in [1, k], i \neq j\}$ is the set of constraints. To solve P, we employ *backtracking* to search systematically through the space of possibilities. We find all solutions, rank these solutions according to the relevance values for sensors (with respect to each group), and select the best solution to find the optimal assignments. The solution ranking procedure can easily incorporate other relevant concerns such as a preference for sensors that are positioned orthogonal to each other with respect to the pedestrian so as to increase the position estimate accuracy or using sensors that are within one hop distance of each other. When P has no solution, When P has no solution, we solve smaller CSPs by relaxing the node requirements for each task.

A node initiates the conflict resolution procedure upon identifying a group-group conflict; e.g., when it intercepts a *queryrelevance* message from multiple groups, or when it already belongs to a group and it receives a *queryrelevance* message from another group. The conflict resolution procedure begins by *centralizing* the CSP in one of the leader nodes that uses *backtracking* to solve the problem.[3] The result is then conveyed to the other leader nodes.

CSPs have been studied extensively in the computer science literature and there exist more powerful variants of the basic backtracking method; however, we employ the naive backtracking approach in the interest of simplicity and because it can easily cope with the size of problems encountered in the current setting. A key feature of our conflict resolution scheme is centralization, which requires that all the relevant information be gathered at the node that is responsible for solving the CSP. For smaller CSPs, the cost of centralization is easily offset by the speed and ease of solving the CSP.

Solving the CSP. Any solution of the above CSP P is a valid sensor node assignment; however, some solutions are better than others as not all nodes are equally suitable for any given sensing task. The node relevance value with respect to a group quantifies the suitability of the node to the task performed by that group, and we can view the quality of a solution as a function of the quality of sensor assignments to different groups. In a restrictive setting, we can define the quality of a solution to be the sum of the quality of sensor assignments to individual groups.

When it is possible to compare the quality of a partial solution to that of a full solution, we can store the currently best result and backtrack whenever the current partial solution is of poorer quality. Using this strategy, we can guarantee an optimal solution

---

[3] Leader node where centralization occurs is selected using a strategy similar to that used for group merging (Fig. 7).

Table 1. Finding an optimal sensor node assignment. The problem is to assign three sensors each to two groups. The average number of relevant nodes for each group is 12 and 16. *AllSolu* finds all solutions, ranks them, and picks the best one, whereas *BestSolu* computes the optimal solution by storing the currently best solution and backtracking when partial assignment yields a poorer solution. As expected, the *BestSolu* solver outperforms the *AllSolu*solver.

Test cases	1	2	3	4
Number of groups	2	2	2	2
Number of sensors per group	3	3	3	3
Average number of relevant sensors	12	12	16	16
Average domain size	220	220	560	560
Number of solutions	29290	9	221347	17
Nodes explored	29511	175	221908	401
Number of Backtracks	48620	36520	314160	215040
Solver used	AllSolu	BestSolu	AllSolu	BestSolu

under the assumption that the quality of solutions increase monotonically as values are assigned to more variables. For example, compare test cases 1 and 2 in Table 1. The goal was to assign 3 sensors each to the two groups. Optimal assignments were found in both cases; however, *BestSolu*, which employs backtracking based on the quality of the partial solution, visited only 175 nodes to find the optimal solution, as opposed to *AllSolu*, which visited 29290 nodes. The *AllSolu* solver enumerates every solution to find the optimal sensor assignment. The same trend is observed in columns 3 and 4 in the table. The *BestSolu* solver clearly outperforms the *AllSolu* solver in finding the optimal node assignment. Of course, when operating under time/resource constraints, we can always choose the first solution or pick the best solution after a predetermined number of nodes have been explored.

4.3 Node Failures and Communication Errors

The purposed communication model takes into consideration node and communication failures. Communication failures are perceived as sensor failures; for example, when a node is expecting a message from another node, and the message never arrives, the first node concludes that the second node is malfunctioning. A node failure is assumed when the leader node does not receive a *status* from the node during some predefined interval, and the leader node removes the problem node from the group. On the other hand, when a member node does not receive any message (*status* or *queryrelevance*) from the leader node during a predefined interval, it assumes that the leader node has experienced a failure and selects itself to be the leader of the group. An actual or perceived leader node failure can therefore give rise to multiple single-node groups performing the same task. Multiple groups assigned to the same task are merged by demoting all of the leader nodes of the constituent groups, except one. Consider, for example, a group comprising three nodes a, b, and c, with node a being the leader node. When a fails, b and c form two single-node groups and continue to perform the sensing task. In due course, nodes b

---

Assumptions: Nodes n and m are two leader nodes performing Task 1.

Case 1: Node n receives a *queryrelevance* or *status* message from node m.

if Node n is not involved in demotion negotiations with another node **then** send *demote* message to node m after a random interval.

Case 2: Node n receives a *demote* message from node m.

a) **if** Node n has not sent a *demote* message to another node **then** demote node n and send *demoteack* message to node m.

b) **if** Node n has sent a *demote* message to node m **then** send *demoteretry* message to node m and send a *demote* message to node m after a random interval.

c) **if** Node n has sent a *demote* message to another node **then** send a *demotenack* message to node m.

Case 3: Node n receives a *demotenack* message from node m.

Terminate demotion negotiations with node m.

Case 4: Node n receives a *demoteack* message from node m.

Add m to node n's group.

Case 5: Node n receives a *demoteretry* message from node m.

Send a *demote* message to node m after a random interval.

Fig. 7. Group merging via leader demotion

and c discover each other—e.g., when b intercepts a *queryrelevance* or a *status* message from c—and they form a new group comprising b and c, demoting node c in the process. Thus, our proposed communication model is able to handle node failures.

Demotion is either carried out based upon the unique ID assigned to each node— among the conflicting nodes, the one with the highest ID is selected to be the group leader—or, when unique node IDs are not guaranteed, demotion can be carried out via the process shown in Fig. 7. The following observations suggest that our leader demotion strategy is correct; i.e., only a single leader node survives the demotion negotiations and every other leader node is demoted.

– **Observation 1:** The demotion process between two leader nodes either succeeds or fails. It succeeds when one of the two nodes is demoted. Demotion between two nodes is based on the contention management scheme that was first introduced in the ALOHA network protocol [26]. The ALOHA network protocol was developed in the late 60s and it is a precursor to the widely used Ethernet protocol. In its basic version, the ALOHA protocol states

• if you have data to send, send it.

• if there is a collision, resend after a random interval.

We point the interested reader to [27] for the details. What is important here is to note that eventually one of the two leader nodes will be demoted; i.e., the demotion process between two nodes will eventually succeed.

– **Observation 2:** The demotion process between more than two nodes involves repeated (distributed and parallel) application of the demotion process between two nodes.

5 Video Surveillance

We now consider how a sensor network of dynamic cameras may be used in the context of video surveillance. A human operator spots one or more suspicious pedestrians in one of the video feeds and, for example, requests the network to "observe this pedestrian," "zoom in on this pedestrian," or "observe the entire group." The successful execution and completion of these tasks requires intelligent allocation and scheduling of the available cameras. In particular, the network must decide which cameras should track the pedestrian and for how long.

A detailed world model that includes the location of cameras, their fields of view, pedestrian motion prediction models, occlusion models, and pedestrian movement pathways may allow (in some sense) *optimal* allocation and scheduling of cameras; however, it is cumbersome and in most cases infeasible to acquire such a world model. Our approach does not require such a knowledge base. We assume only that a pedestrian can be identified by different cameras with reasonable accuracy and that the camera network topology is known *a priori*. A direct consequence of this approach is that the network can easily be modified through removal, addition, or replacement of camera nodes.

5.1 Computing Camera Node Relevance

The accuracy with which individual camera nodes are able to compute their relevance to the task at hand determines the overall performance of the network. Our scheme for computing the relevance of a camera to a video surveillance task encodes the intuitive observations that 1) a camera that is currently free should be chosen for the task, 2) a camera with better tracking performance with respect to the task at hand should be chosen, 3) the turn and zoom limits of cameras should be taken into account when assigning a camera to a task; i.e., a camera that has more leeway in terms of turning and zooming might be able to follow a pedestrian for a longer time, and 4) it is better to avoid unnecessary reassignments of cameras to different tasks, as that might degrade the performance of the underlying computer vision routines.

Upon receiving a task request, a camera node returns to the leader node a relevance metric—a list of attribute-value pairs describing its relevance to the current task along multiple dimensions (Fig. 8). The leader node converts this metric into a scalar relevance value r as follows:

$$
r = \begin{cases} \exp\left(-\frac{(c-1)^2}{2\sigma_c{}^2} - \frac{(\theta-\hat{\theta})^2}{2\sigma_\theta{}^2} - \frac{(\alpha-\hat{\alpha})^2}{2\sigma_\alpha{}^2} - \frac{(\beta-\hat{\beta})^2}{2\sigma_\beta{}^2} \right) \\ \qquad\qquad\qquad\qquad \text{when } s = \text{free} \\ \frac{t}{t+\gamma} \quad \text{when } s = \text{busy} \end{cases}
\tag{1}
$$

where $\hat{\theta} = (\theta_{\min} + \theta_{\max})/2$, $\hat{\alpha} = (\alpha_{\min} + \alpha_{\max})/2$, and $\hat{\beta} = (\beta_{\min} + \beta_{\max})/2$, and where $\theta_{\min}$ and $\theta_{\max}$ are extremal field of view settings, $\alpha_{\min}$ and $\alpha_{\max}$ are extremal rotation angles around the x-axis (up-down), and $\beta_{\min}$ and $\beta_{\max}$ are extremal rotation angles around the y-axis (left-right). Here, $0.3 \leq \sigma_c \leq 0.33$, $\sigma_\theta = (\theta_{\max} - \theta_{\min})/6$, $\sigma_\alpha = (\alpha_{\max} - \alpha_{\min})/6$, and $\sigma_\beta = (\beta_{\max} - \beta_{\min})/6$. The value of γ is chosen empirically (for our experiments we have selected γ to be 1000).

Status	= $s \in$	{busy, free}
Quality	= $c \in$	$[0, 1]$
Fov	= $\theta \in$	$[\theta_{min}, \theta_{max}]$ degrees
XTurn	= $\alpha \in$	$[\alpha_{min}, \alpha_{max}]$ degrees
YTurn	= $\beta \in$	$[\beta_{min}, \beta_{max}]$ degrees
Time	= $t \in$	$[0, \infty)$ seconds
Task	= $a \in$	$\{a_i | i = 1, 2, \cdots\}$

Fig. 8. The relevance metric returned by a camera node relative to a new task request. The leader node converts the metric into a scalar value representing the relevance of the node for the particular surveillance task.

The computed relevance values are used by the node selection scheme described above to assign cameras to various tasks. The leader node gives preference to the nodes that are currently free, so the nodes that are part of another group are selected only when an insufficient number of free nodes are available for the current task.

5.2 Surveillance Tasks

We have implemented an interface that presents the operator a display of the synthetic video feeds from multiple virtual surveillance cameras. The operator can select a pedestrian in any video feed and instruct the camera network to perform one of the following tasks: 1) follow the pedestrian, 2) capture a high-resolution snapshot, or 3) zoom-in and follow the pedestrian. The network then automatically assigns cameras to fulfill the task requirements. The operator can also initiate multiple tasks, in which case either cameras that are not currently occupied are chosen for the new task or some currently occupied cameras are reassigned to the new task.

6   Results

To date, we have tested our visual sensor network system with up to 16 stationary and pan-tilt-zoom cameras, and we have populated the virtual Penn Station environment with up to 100 pedestrians. The sensor network correctly assigned cameras in most of the cases. Some of the problems that we encountered are related to pedestrian identification and tracking. As we increase the number of virtual pedestrians in the train station, the identification and tracking module has increasing difficulty following the correct pedestrian, so the probability increases that the surveillance task fails (and the cameras just return to their default settings).

For the example shown in Fig. 9, we placed 16 active PTZ cameras in the train station, as shown in Fig. 1. The operator selects the pedestrian with the red shirt in Camera 7 (Fig. 9(e)) and initiates the "follow" task. Camera 7 forms the task group and begins tracking the pedestrian. Subsequently, Camera 7 recruits Camera 6, which in turn recruits Cameras 2 and 3 to track the pedestrian. Camera 6 becomes the leader of the group when Camera 7 loses track of the pedestrian and leaves the group. Subsequently, Camera 6 experiences a tracking failure, sets Camera 3 as the group leader, and leaves the group. Cameras 2 and 3 track the pedestrian during her stay in the main waiting room, where she also visits a vending machine. When the pedestrian starts walking towards the concourse, Cameras 10 and 11 take over the group from Cameras 2 and 3.

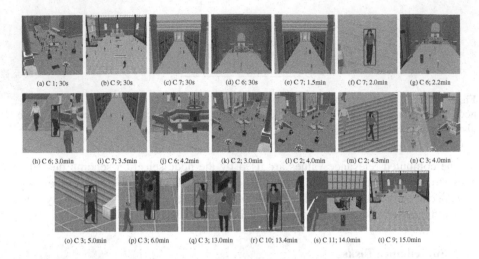

(a) C 1; 30s   (b) C 9; 30s   (c) C 7; 30s   (d) C 6; 30s   (e) C 7; 1.5min   (f) C 7; 2.0min   (g) C 6; 2.2min

(h) C 6; 3.0min   (i) C 7; 3.5min   (j) C 6; 4.2min   (k) C 2; 3.0min   (l) C 2; 4.0min   (m) C 2; 4.3min   (n) C 3; 4.0min

(o) C 3; 5.0min   (p) C 3; 6.0min   (q) C 3; 13.0min   (r) C 10; 13.4min   (s) C 11; 14.0min   (t) C 9; 15.0min

Fig. 9. A pedestrian is successively tracked by Cameras 7, 6, 2, 3, 10, and 9 (see Fig. 1) as she makes her way through the station to the concourse. (a-d) Cameras observing the station. (e) The operator selects a pedestrian in the video feed from Camera 7. (f) Camera 7 has zoomed in on the pedestrian, (g) Camera 6, which is recruited by Camera 7, acquires the pedestrian. (h) Camera 6 zooms in on the pedestrian. (i) Camera 7 reverts to its default mode after losing track of the pedestrian and is now ready for another task (j) Camera 6 has lost track of the pedestrian. (k) Camera 2. (l) Camera 2, which is recruited by Camera 6, acquires the pedestrian. (m) Camera 2 tracking the pedestrian. (n) Camera 3 is recruited by Camera 6; Camera 3 has acquired the pedestrian. (o) Camera 3 zooming in on the pedestrian. (p) Pedestrian is at the vending machine. (q) Pedestrian is walking towards the concourse. (r) Camera 10 is recruited by Camera 3; Camera 10 is tracking the pedestrian. (s) Camera 11 is recruited by Camera 10. (t) Camera 9 is recruited by Camera 10.

(a)   (b)   (c)   (d)   (e)

Fig. 10. "Follow" sequence. (a) The operator selects a pedestrian in Camera 1 (upper row). (b) and (c) Camera 1 and Camera 2 (lower row) are tracking the pedestrian. (d) Camera 2 loses track. (e) Camera 1 is still tracking; Camera 2 has returned to its default settings.

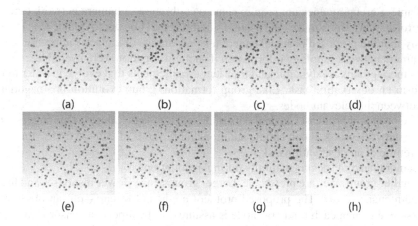

Fig. 11. Group merging and leader failure. Blue nodes are idle. Red nodes are following the targets shown as pink cones. Square nodes represent group leaders and black nodes indicate node failures.

Cameras 2 and 3 leave the group and return to their default modes. Later Camera 11, which is now acting as the group's leader, recruits Camera 9, which tracks the pedestrian as she enters the concourse.

Fig. 10 illustrates a "follow" task sequence. An operator selects the pedestrian with the green shirt in Camera 1 (top row). Camera 1 forms a group with Camera 2 (bottom row) to follow and zoom in on the pedestrian. At some point, Camera 2 loses the pedestrian (due to occlusion), and it invokes a search routine, but it fails to reacquire the pedestrian. Camera 1, however, is still tracking the pedestrian. Camera 2 leaves the group and returns to its default settings.

Fig. 11 presents a simulation of larger sensor networks outside our virtual vision simulator. It shows a sensor network of 50 nodes placed randomly in a 25 m^2 area. The nodes that are within 5 m of each other can directly communicate with each other. Each node can communicate with another node in the network through multi-hop routing. Fig. 11(a)–(e) show group merging. When the leader of the group fails (Fig. 11(f)), multiple member nodes assume leadership (Fig. 11(g)). These nodes negotiate each other to select a single leader (Fig. 11(h)).

6.1 Discussion

Given the above results, we make the following observations about the proposed scheme:

- The proposed protocol successfully forms camera groups to carry out various observation tasks. Cameras that belong to a single group collaborate with each other for the purposes of carrying out the observation task. Currently we support a few observation tasks that are of interest to the visual surveillance community. These are 1) taking snapshots of a pedestrian, 2) closely observing a pedestrian during his/her stay in the designated region, and 3) following a pedestrian across multiple cameras.

- Camera grouping does not require camera calibration or camera network topology information, which makes our system suitable for ad hoc deployment. This is not to say that the proposed protocol cannot take advantage of camera calibration and/or camera network topology information if such information were available.
- Camera groups are dynamic and transient arrangements that evolve in order to perform an observation task. Like group formation, group evolution is a negotiation between the relevant nodes.
- The proposed protocol can deal with node and message failures. This suggests that the network protocol can handle addition and removal of camera nodes during the lifetime of an observation task.
- Camera hand off occurs naturally during negotiations.
- Smaller group sizes are preferable. Larger groups have slower responses and higher maintenance costs. The proposed protocol might fail to carry out an observation task even when each (camera) node is assumed to be a perfect sensor if the group evolution cannot keep up with a fast changing observation task.
- Assuming that each (camera) node is a perfect sensor, the proposed protocol still might fail to carry out an observation task if a large fraction of nodes fail or a significant fraction of messages are lost.
- Camera node aggregation is fully distributed and lacks a central controller, so it is scalable. Sensor assignment in the presence of conflicts, however, is centralized over the involved groups. Therefore, our scheme lies somewhere between a fully distributed and a fully centralized system. In the interest of scalability, group sizes should be kept small.

7   Conclusion

We envision future video surveillance systems to be networks of stationary and active cameras capable of providing perceptive coverage of extensive environments with minimal reliance on human operators. Such systems will require not only robust, low-level vision routines, but also novel sensor network methodologies. The work presented in this paper is a step toward the realization of these new sensor networks and our initial results are promising.

A unique and, in our view, important aspect of our work is that we have developed and demonstrated our prototype video surveillance system in a realistic virtual train station environment populated by lifelike, autonomously self-animating virtual pedestrians. Our sophisticated sensor network simulator should continue to facilitate our ability to design large-scale networks and experiment with them on commodity personal computers.

The overall behavior of our prototype sensor network is governed by local decision making at each node and communication between the nodes. Our approach is new insofar as it does not require camera calibration, a detailed world model, or a central controller. We have intentionally avoided multi-camera tracking schemes that assume prior camera network calibration which, we believe, is an unrealistic goal for a large-scale camera network consisting of heterogeneous cameras. Similarly, our approach does not expect a detailed world model, which is generally hard to acquire. We expect the proposed approach to be robust and scalable.

We are currently pursuing a *Cognitive Modeling* [28] approach to node organization and camera scheduling. We are also investigating scalability and node failure issues. Moreover, we are constructing more elaborate scenarios involving multiple cameras situated in different locations within the train station, with which we would like to study the performance of the network when it is required to follow multiple pedestrians during their prolonged stay in the train station.

Acknowledgments

The research reported herein was supported in part by a grant from the Defense Advanced Research Projects Agency (DARPA) of the Department of Defense. We thank Tom Strat, formerly of DARPA, for his generous support and encouragement. We also thank Wei Shao and Mauricio Plaza-Villegas for their invaluable contributions to the implementation of the Penn Station simulator. Deborah Estrin provided helpful advice and pointers into the sensor networks literature.

References

1. Mallett, J.: The Role of Groups in Smart Camera Networks. PhD thesis, Program of Media Arts and Sciences, School of Architecture, Massachusetts Institute of Technology (Feb. 2006)
2. Shao, W., Terzopoulos, D.: Autonomous pedestrians. In: Proc. ACM SIGGRAPH / Eurographics Symposium on Computer Animation, Los Angeles, CA (July 2005) 19–28
3. Zhao, F., Liu, J., Liu, J., Guibas, L., Reich, J.: Collaborative signal and information processing: An information directed approach. Proceedings of the IEEE **91**(8) (2003) 1199–1209
4. Terzopoulos, D., Rabie, T.: Animat vision: Active vision in artificial animals. Videre: Journal of Computer Vision Research **1**(1) (September 1997) 2–19
5. Terzopoulos, D.: Perceptive agents and systems in virtual reality. In: Proc. 10th ACM Symposium on Virtual Reality Software and Technology, Osaka, Japan (October 2003) 1–3
6. Santuari, A., Lanz, O., Brunelli, R.: Synthetic movies for computer vision applications. In: Proc. 3rd IASTED International Conference: Visualization, Imaging, and Image Processing (VIIP 2003). Number 1, Spain (September 2003) 1–6
7. Zhao, F., Shin, J., Reich, J.: Information-driven dynamic sensor collaboration for tracking applications. In: IEEE Signal Processing Magazine. Volume 19. (March 2002) 61–72
8. Bhardwaj, M., Chandrakasan, A., Garnett, T.: Upper bounds on the lifetime of sensor networks. In: IEEE International Conference on Communications. Number 26 (2001) 785 – 790
9. Chang, J.H., Tassiulas, L.: Energy conserving routing in wireless adhoc networks. In: Proceedings of the IEEE Conference on Computer Communications (INFOCOM), Tel Aviv, Israel (March 2000) 22–31
10. Collins, R., Lipton, A., Fujiyoshi, H., Kanade, T.: Algorithms for cooperative multisensor surveillance. Proceedings of the IEEE **89**(10) (October 2001) 1456–1477
11. Costello, C.J., Diehl, C.P., Banerjee, A., Fisher, H.: Scheduling an active camera to observe people. In: Proc. 2nd ACM International Workshop on Video Surveillance and Sensor Networks, New York, NY, ACM Press (2004) 39–45
12. Collins, R., Amidi, O., Kanade, T.: An active camera system for acquiring multi-view video. In: Proc. International Conference on Image Processing, Rochester, NY, USA (September 2002) 517–520

13. Devarajan, D., Radke, R.J., Chung, H.: Distributed metric calibration of ad hoc camera networks. ACM Transactions on Sensor Networks 2(3) (2006) 380–403
14. Campbell, J., Gibbons, P.B., Nath, S., Pillai, P., Seshan, S., Sukthankar, R.: Irisnet: An internet-scale architecture for multimedia sensors. In: Proc. of the 13th annual ACM international conference on Multimedia (MULTIMEDIA '05), New York, NY, USA, ACM Press (2005) 81–88
15. Kulkarni, P., Ganesan, D., Shenoy, P., Lu, Q.: Senseye: a multi-tier camera sensor network. In: MULTIMEDIA '05: Proc. of the 13th annual ACM international conference on Multimedia, New York, NY, USA, ACM Press (2005) 229–238
16. Smith, R.G.: The contract net protocol: High-level communication and control in a distributed problem solver. IEEE Transctions on Computers C-29(12) (Dec 1980) 1104–1113
17. Gerkey, B., Matari, M.: A formal analysis and taxonomy of task allocation in multi-robot systems. International Journal of Robotics Research 23(9) (2004) 939–954
18. Garey, M.R., Johnson, D.S.: "strong" npcompleteness results: Motivation, examples, and implications. Journal of the ACM 25(3) (1978) 499–508
19. Atamturk, A., Nemhauser, G., Savelsbergh, M.: A combined lagrangian, linear programming and implication heuristic for large-scale set partitioning problems. Journal of Heuristics 1 (1995) 247–259
20. Modi, P.J., Shen, W.S., Tambe, M., Yokoo, M.: Adopt: asynchronous distributed constraint optimization with quality guarantees. Artificial Intelligence 161(1–2) (Mar 2006) 149–180 Elsevier.
21. Yokoo, M.: Distributed Constraint Satisfaction: Foundations of Cooperation in Multi-agent Systems. Springer-Verlag, Berlin, Germany (2001)
22. Comaniciu, D., Ramesh, V., Meer, P.: Real-time tracking of non-rigid objects using mean shift. In: Proc. of the 2000 IEEE Conference on Computer Vision and Pattern Recognition (CVPR 00). Volume 2., Hilton Head Island, South Carolina, USA (2000) 142–151
23. Swain, M.J., Ballard, D.H.: Color indexing. International Journal of Computer Vision 7(1) (Nov 1991) 11–32
24. Qureshi, F.Z.: Intelligent Perception in Virtual Camera Networks and Space Robotics. PhD thesis, Department of Computer Science, University of Toronto (January 2007)
25. Pearson, J.K., Jeavons, P.G.: A survey of tractable constraint satisfaction problems. Technical Report CSD-TR-97-15, Royal Holloway, University of London (July 1997)
26. Kuo, F.F.: The aloha system. ACM SIGCOMM Computer Communication Review 25(1) (1995) 41–44 Special twenty-fifth anniversary issue. Highlights from 25 years of the Computer Communication Review.
27. Murthy, C., Manoj, B.: Ad Hoc Wireless Networks Architectures and Protocols. Prentice Hall (2004)
28. Qureshi, F., Terzopoulos, D., Jaseiobedzki, P.: Cognitive vision for autonomous satellite rendezvous and docking. In: Proc. IAPR Conference on Machine Vision Applications, Tsukuba Science City, Japan (May 2005) 314–319

Efficient and Distributed Access Control for Sensor Networks

Donggang Liu

iSec laboratory, CSE Department
The University of Texas at Arlington
dliu@uta.edu

Abstract. Sensor networks are often used to sense the physical world and provide observations for various uses. In hostile environments, it is critical to control the network access to ensure the integrity, availability, and at times confidentiality of the sensor data. This paper develops efficient methods for distributed access control in sensor networks. The paper starts with a *baseline approach*, which provides a more flexible and efficient way to enforce access control when compared with previous solutions. This paper then extends the baseline approach to enable *privilege delegation*, which allows a user to delegate its privilege to other users without using a trusted server, and *broadcast query*, which allows a user to access the network at a large scale efficiently. The privilege delegation and broadcast query are very useful in practice; none of the current solutions can achieve these two properties.

1 Introduction

The primary purpose of deploying a sensor network is to monitor the physical world and provide observations for various uses. In hostile environments, it is critical to control the access to the sensor nodes (e.g., reading sensor data), especially when there are many users in the system. For example, a sensor network may be deployed to monitor the activities in a battlefield and provide information for soldiers and commanders to make decisions during military operations. Every soldier or commander will be a potential system user. In applications like this, different users may have different access privileges. A soldier may be only allowed to read the sensor data, while a commander may be able to re-configure the network or update the internal states of the sensor nodes. The application will be compromised if the access control is not properly enforced.

However, enforcing access control in sensor networks is particularly challenging. First, the resource constraints on sensor nodes often make it undesirable to implement expensive algorithms. For example, the commonly used MICAz platform uses an 8-bits ATmega128 CPU that operates at 7.7MHz [1]. It only has 128KB ROM for programming code and 4KB RAM for buffers and variables. Second, sensor nodes are usually deployed unattended and may be compromised after deployment [2]. Hence, any security protocol has to be resilient to node compromises.

J. Aspnes et al. (Eds.): DCOSS 2007, LNCS 4549, pp. 21–35, 2007.
© Springer-Verlag Berlin Heidelberg 2007

A number of techniques have been developed recently to achieve access control in sensor networks. These include the *Least-Privilege* scheme [3] and the *Wang-Li* scheme [4]. However, the former can only be used for a specific type of access control where the access is limited to the data at a pre-determined physical path in the field. The latter is based on Elliptic Curve Cryptography (ECC), which has been shown to be feasible for sensor nodes [5]. However, the DoS attacks on signature verification were not properly addressed, making this scheme less attractive. In addition, this scheme can only access one sensor node at a time, while a significant part of actual access requests are targeted to many sensor nodes via broadcast.

In this paper, we start with a baseline approach to deal with the situation where every access request is targeted to a specific sensor node. Compared with the previous schemes, our baseline approach has a number of advantages. First, it provides a generic access control method and can be easily used to enforce any type of access control policies. Second, it only uses symmetric key cryptography, which is much more efficient than public key cryptography. Third, our approach only involves a single message for every request from the user to a sensor node, while the scheme in [4] needs to employ a more expensive challenge-responsive protocol for accessing a local sensor node and a much more complicated cooperative protocol for accessing a remote sensor node. Finally, our approach provides strong security guarantee against compromised sensor nodes or users. Indeed, no matter how many sensor nodes or users are compromised, a benign sensor node will not grant the attacker any access that is beyond the simple union of the privileges of the compromised users.

In addition to the baseline approach, we also develop techniques that enable *privilege delegation* and *broadcast query*. The idea of privilege delegation allows a user to delegate its access privilege to other users without a trusted third party. The idea of broadcast query allows a user to efficiently access the sensor nodes at a large scale. These two additional functionalities can be very useful in many sensor operations, making our access control approach much more appealing than the previous solutions.

This paper is organized as follows. The next section gives the system model and assumptions. Section 3 presents the proposed techniques for access control in sensor networks. Section 4 shows the evaluation of the proposed techniques. Section 5 reviews some related work on sensor network security. The last section summarizes this paper and discusses future research directions.

2 System Models and Assumptions

In this paper, we consider wireless sensor networks that consist of a large number of resource-constrained sensor nodes, many system users, and a trusted server such as the base station. The sensor nodes are used to sense conditions in their local surroundings and report their observations to system users based on various query commands. The system users (e.g., soldiers) use *access devices* such as

PDAs and Laptops to access the sensor data. The trusted server is used to bootstrap the keying materials for access devices to enforce access control policy.

Dataset Table: For every sensor node, the sensor data that are accessible for system users can be viewed as a *dataset table*. Every record in this table is a snapshot of the values of different *attributes* at a particular time. An attribute is defined as a specific type of sensor data in the network. Examples of attributes include the sensor readings (e.g., temperature) and the sensor states (e.g., the active sensor). Table 1 shows an example of the dataset table of a sensor node at time t_n. This node is able to sense the temperature and humidity of its local environments. It may activate one or both sensors, depending on the value of the third attribute, the active sensor.

Let $\mathcal{A} = \{a_1, a_2, ..., a_m\}$ be the set of attributes in the system. Any value d in the dataset table can be located based on the attribute name $a \in \mathcal{A}$ and the time t. For simplicity, let $d_{a,t}$ be the value of the attribute a at time t, $A(d)$ be the attribute name related to data d, and $T(d)$ be the time related to data d. Clearly, $A(d_{a,t}) = a$ and $T(d_{a,t}) = t$.

Table 1. Dataset table at time t_n

Time	Temp.	Hum.	Active Sensor
t_1	80	N/A	temp. sensor
t_2	82	N/A	temp. sensor
t_3	N/A	45	hum. sensor
⋮	⋮	⋮	⋮
t_n	N/A	50	hum. sensor

Access Model: A system user can perform two possible access operations on a particular data value in the table, *read* and *write*. For example, the sensor readings such as temperature and humidity are usually read only, while the attributes related to the system states can be modified by a user with some proper privilege. Based on the dataset table, we define a *capability matrix* for a given user. A two-bit value at column i and row j in this matrix indicates the user's access privilege to d_{a_i,t_j} in the dataset table. This value can be 00 (no access), 01 (read-only), 10 (write-only), or 11 (read and write).

Note that the capability matrix can be very large. Fortunately, most sensor applications use a simple way to specify a user's capability. For example, the policy "user U can only read the humidity readings collected from time t_1 to time t_2" can be encoded as "$(A(d) = humidity) \wedge (t_1 \leq T(d) \leq t_2) \wedge (O = 01)$", where O denotes the operation performed by the user. Hence, we can usually encode the capability matrix into a short *capability string*. We will not discuss the details of the encoding and decoding schemes for capability strings since they are application-specific and orthogonal to our approaches.

A user's capability string can be used to define its access privilege. However, we might have other requirements for the user's access to the sensor data. For example, the user may only be able to access the data at a rate of no more than one packet per 30 seconds. Hence, *we define a user's access privilege as a* **constraint**. The user's capability string can be also viewed as a constraint. A constraint can be built from multiple constraints by simply using "AND" ($\wedge$) or "OR" ($\vee$) operator. For example, "$(A(d) = humidity \wedge O = 01) \wedge (R \geq 30)$"

means that the user can only read the sensed humidity at a rate of no more than one packet per 30 seconds. In this constraint, the first part is the capability string of the user, while the second part ($R \geq 30$) is an additional constraint on the packet rate.

A constraint can usually be written in an appropriate formal language. However, our method is independent from how we write the constraints. Hence, we simply discuss it informally rather than proposing a formal language to define constraints. As a result, we denote the access privilege of a user u as a constraint C_u. The access will be granted to u only when the queried sensor data meets the constraint C_u. This access model is simple yet flexible for a sensor network. We believe that it is sufficient for most sensing applications.

Attack Model: An attacker can launch a wide range of attacks against the network. For example, he can simply perform a denial-of-service (DoS) attack to jam the wireless channel and disable the network operation. Such DoS attack is easy to mount and common to the protocols in every sensor network. There are no effective ways to stop the attacker from doing such attack. However, in this paper, *we are more interested in the attacks whose goal is to access the valuable sensor data that he is not supposed to access.* We assume an attacker can eavesdrop, modify, forge, replay or block any network traffic. We assume that it is computationally infeasible to break the underlying cryptographic primitives such as encryption, decryption, and hash. We assume that the attacker is able to compromise a few sensor nodes and learn the key materials stored on the compromised sensor nodes. We also assume that some users may be compromised. Once a user is compromised, all the secrets assigned to the user will be disclosed.

3 Efficient and Distributed Access Control

This section presents the proposed method for access control as well as the detailed analysis. We start with a baseline approach and then develop techniques to enable *privilege delegation* and *broadcast query*.

3.1 The Baseline Approach

The main focus of the baseline approach is to provide efficient access control when a user queries only one node at a time. With this technique, it will be also practical to access a few nodes at a time as long as the query packet has space for additional information such as MACs. We assume that the user knows the ID of the sensor that has the requested data. We also assume a routing protocol that can be used to deliver the query to such node.

It is clearly possible that the query message gets lost on the way to the queried node due to the unreliable channel or malicious attacks. For example, an attacker can always jam the channel and interrupt the delivery of any message. There are no effective methods to prevent an attacker from mounting such DoS attacks. Therefore, our focus in this paper is to stop unauthorized user access to the sensor networks.

The main idea of our baseline approach is to map the access privilege of a user into a cryptographic key. This key is used to prove that the user does have the access privilege he claims. As a result, achieving confidentiality and integrity using this key also achieves the access control.

- *Initialization:* Before deployment, every node i shares a key K_i with a trusted server. Every user u will contact this server for the keying materials for access control. The server first determines the constraint C_u based on the user's credentials, depending on the application. With the constraint C_u, the user u gets pre-loaded a hash key $K_{u,i} = H(u||C_u||K_i)$ for every node i, where H is a one-way hash function and "$||$" denotes the concatenation of two bit strings. Clearly, for a network of n nodes, a user will need to store n hash keys. We believe that this storage overhead will not be a problem for users who have powerful computing devices such as PDAs or Laptops. In fact, even for a resource-poor sensor node such as TelosB motes [1], it has 1MB flash memory to store the keying materials for a network of 128,000 nodes if every hash image is 8-byte long.
- *Access Query:* Let $\{\cdot\}_K$ be the message authentication operation using K. When a user u wants to access a sensor node i, it constructs an appropriate query command $Q(u)$ and send the following message to node i.

$$u \rightarrow i : \{C_u, Q(u)\}_{K_{u,i}}$$

Once node i receives such message, it can re-construct the hash key $K_{u,i}$ from C_u since it has the key K_i. With this hash key, it can check the authenticity of the query message. If the message is authenticated, it is certain that user u does have the access privilege defined by C_u and the query $Q(u)$ does come from u. If $Q(u)$ also meets the constraint C_u, the access to i's data defined by $Q(u)$ will be granted to u.

Security: The focus of our security study is *the access to the data at non-compromised nodes.* For a benign user u, we can see that a benign node i will provide access to this user only when u does make a corresponding request. The reason is that every request from user u to node i has to be authenticated by the key $K_{u,i}$, which is only known by user u, node i and the trusted server.

We then study the access capability of an adversary who compromised a set of system users U_c. Assume that an adversary issues a query command Q_c to access the sensor data. The following theorem tells us that no matter how many sensor nodes are compromised, the collusion of compromised users and sensor nodes will not give the adversary any additional privilege that is beyond the simple union of privileges of compromised users.

Theorem 1. *To access the sensor data at benign sensor nodes, the query Q_c from the attacker must meet the constraint $\bigvee_{v \in U_c} (C_v)$, where U_c is the set of compromised users.*

Proof. According to the initialization step, every keying material at a user is generated using a one-way hash function. Hence, no matter how many sensor nodes or system users are compromised, an adversary is unable to access the communication between a benign user and a benign sensor node. Hence, the adversary must issue Q_c as a user $v \in U_c$ since otherwise Q_c will be immediately rejected because of the failure in the authentication. When $v \in U_c$, if the query Q_c doesn't meet the constraint C_v, Q_c will also be rejected by any benign sensor node i. Hence, the query Q_c must meet $\bigvee_{v \in U_c}(C_v)$ in order to gain access to a benign sensor node.

Overheads: In the baseline approach, a sensor node only needs to store a single key, while a system user needs to store n keys for a network of n nodes. However, as we mentioned, this is usually not a problem for the users who have powerful and resourceful platforms such as PDAs and Laptops. To access the data at a sensor node, a user only needs to send a single message. For every query message, the user only needs to perform one efficient hash operation, while the target node only needs to do two efficient hash operations, one for generating the authentication key and the other for authentication.

Comparison: The baseline approach has many advantages over the previous schemes in [3, 4]. First, it provides a generic access control method and can be easily used to enforce any type of access policies. While the technique in [3] can only achieve a specific type of access control where the system can only force the user to access the network at a pre-determined physical path in the field. Second, our approach only involves a few efficient one-way hash operations. This is much more efficient than the ECC-based signature scheme in [4]. Third, our approach only involves a single message for each query, while the schemes in [4] need three messages for a local sensor node and a lot more messages for a remote sensor node. In their method, a certain number of local sensor nodes have to collaborate together to commit the query message for a remote sensor node. This complicates the protocol and increases the overheads significantly. Finally, our approach is more resilient to the compromise of sensor nodes. Indeed, no matter how many sensor nodes or users are compromised, the control of the access from any user to a benign sensor node can still be properly enforced. In contrast, the collusion of a small number of sensor nodes will allow the attacker to access any sensor node for the technique in [4].

3.2   Enabling Privilege Delegation

In many cases, a user may want to directly delegate its privilege to other users in the system without going through a trusted server. This can be very useful when the trusted server is not available for bootstrapping the keying materials. For example, in a battlefield, the trusted server may stay in the military base, while an officer may want to temporarily delegate its privilege to a soldier during the military operation in the field. In this case, it is often not practical to contact the trusted server to bootstrap the keying materials for the soldier.

Fortunately, privilege delegation can be easily implemented in our baseline approach. The overall process is similar to the bootstrap in the baseline approach.

Basically, whenever a user wants to delegate its privilege to other users, it can act as a trusted server since it shares a unique key with every sensor node. A user with a delegated privilege can further delegate its privilege to other users, forming a *delegation tree*. For the sake of presentation, when a user u delegates its privilege to user v, we call u as v's *parent user* and v as u's *child user*. Figure 1 shows an example of delegation trees.

For every user u in the delegation tree, it knows the IDs as well as the constraints of the users who are on the path from u to the root (the trusted server). This information is pre-loaded to u when the trusted server or a user bootstraps u. For the sake of presentation, we use $S(u)$ and $I(u)$ to denote the set of constraints and IDs of the users on the path from u to the root in the tree respectively. For example, in Figure 1, we have $S(7) = \{C_7, C_6, C_2\}$ and $I(7) = \{7, 6, 2\}$.

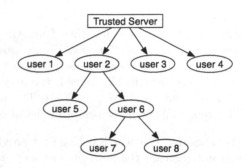

Fig. 1. An example of delegation tree

- *Delegation:* Assume a user u wants to delegate its privilege to another user v. u first determines the constraint C_v for user v. User u will then assign a hash key $K_{v,i} = H(v\|C_v\|K_{u,i})$ to user v for every sensor node i. User u also pre-loads $I(v) = I(u) \cup \{v\}$ and $S(v) = S(u) \cup \{C_v\}$ to v. The actual access privilege of user v will be determined by $S(v)$. For example, in Figure 1, the privilege of user 6 is $C_2 \wedge C_6$.
- *Access Query:* When user v needs to access node i, it sends the following message to i.

$$v \to i : \{I(v), S(v), Q(v)\}_{K_{v,i}}$$

After the node i receives such message, it can easily reconstruct $K_{v,i}$ based on $I(v)$ and $S(v)$. When the message is verified, it is certain that user v does have the access privilege defined by $S(v)$ and the query $Q(v)$ does come from v as long as the users on the path from v to the root of the delegation tree are still benign. Finally, node i will check if $Q(v)$ meets all the constraints in $S(v)$. If yes, the access is granted.

Privilege delegation may cause problems for *stateful* constraints. A user's constraint is said to be stateful if it requires the node to maintain certain history information for this user. A potential security problem for these constraints is that a malicious user can easily bypass them by privilege delegation. As one example, assume a malicious user can only collect data at a rate of no more than one packet per 30 seconds. To bypass this constraint, this user can simply create another user and delegate its privilege to the newly created user to initiate another information flow with up to one packet per 30 seconds.

To address this problem, we propose an *upward updating* approach. In this approach, when node i receives an authenticated query from user u, it will consider it as the valid query for every user $v \in I(u)$ and check with every $v \in I(u)$ to see if it meets all the constraints in $S(v)$. If not, the query command will be rejected; otherwise, the query will be accepted, and the sensor node i will update its state for every user $v \in I(u)$ based on the query as well as the response to such query.

Security: The delegation of a user's privilege is similar to the process of bootstrapping the keying materials from the trusted server. Due to the one way property of the hash function H, we can see that an adversary who compromised a set U_c of users can only access the sensor data that are accessible for at least one user in U_c. Indeed, the access of adversary is still limited by the following constraint: $\bigvee_{u \in U_c}(S(u))$. The proof will be the same as the proof of Theorem 1. We will then skip the detail of the proof for simplicity.

Overheads: It is worth noting that providing the privilege delegation property will not increase the storage overhead at sensor nodes. The only difference is that the states of different users are now organized as a tree structure. As for the computational overhead, a user will need to generate n hash values to delegate its access privilege. A sensor node needs to perform a number of hash operations to generate the key to authenticate the query and protect the sensor data. This number depends on the distance from the corresponding user to the root of the delegation tree. Fortunately, the height of the delegation tree will not be very large in most cases. As for the communication overhead, we can clearly see that allowing privilege delegation will not introduce much more communication cost. A user only needs to send a single message, which has the similar format as the query message in the baseline approach, to access the data at a sensor node.

3.3   Enabling Efficient Broadcast Query

The baseline approach can only support the queries that are targeted to one sensor node at a time. In practice, however, a system user may want to access the data at many sensor nodes at a time. We can certainly use the baseline approach to query these nodes one by one. However, in this case, the communication overhead will increase linearly with the number of queried nodes. Hence, it is particularly desirable to enable *broadcast query*, where many sensor nodes can be queried at the same time efficiently.

A broadcast authentication method will be needed for broadcast query. Two possible candidates are μTESLA [6] and the ECC-based signature scheme [5]. μTESLA is based on symmetric cryptography and is believed to be more efficient than the ECC-based scheme. However, it has some undesirable features such as the need for time synchronization, the authentication delay, and the update of key chains. These limit the application of μTESLA in real-world scenarios. On the other hand, the performance of ECC-based signature schemes have been and will continue to be optimized by researchers. We thus use ECC-based signature schemes for broadcast authentication in this paper. However, a critical problem that has to be addressed before making the ECC-based signature scheme a reality is the DoS attacks on signature verification. Hence, the main challenge is *how to thwart the DoS attacks against the signature verification.*

Note that broadcast is usually done through a network-wide flooding. Many energy efficient flood mechanisms could be used for this purpose [7,8,9]. Our main observation is that every flooding message from a sensor node only has a small number of receivers due to the limited neighbor size. This actually allows us to *weakly* authenticate the broadcast message before verifying the digital signature to thwart the DoS attacks. Since our weak authentication method is independent from the broadcast protocol, we simply assume a broadcast protocol and will not discuss how it is achieved. In the following discussion, we assume that every two neighbor sensor nodes u and v share a pairwise key $k_{u,v}$. Many existing pairwise key establishment protocols could be used for this purpose [10, 11, 12, 13]. Note that we use the lower-case "k" for the pairwise key between two sensor nodes and the upper-case "K" for the shared key between a user and a sensor node.

- *Initialization:* The initialization step is similar to the baseline approach. However, we have additional keying materials for sensor nodes and users. Specifically, the public key of the trusted server will be pre-loaded to every sensor node. Every user u also has a pair of private and public keys $(K_e(u), K_d(u))$. The public key $K_d(u)$ and C_u are either signed by the trusted server or another user who is willing to delegate its privilege to u. Let $Cert(u)$ be the resulting signature. In the end, user u will get pre-loaded with $Cert(v)$ for every $v \in I(u)$, where $I(u)$ is the set of user IDs on the path from u to the root of the delegation tree.

After deployment, every node i first discovers a set of neighbors $N(i)$ and exchange it with every node $j \in N(i)$. In the end, we want to make sure that a sensor node i has the neighbor list $N(i)$ and also knows its position $P_{j,i}$ in the set $N(j)$ for every $j \in N(i)$. Such position information is used to identify the values for weakly authenticating the broadcast message.

When a user u needs to query a large number of sensor nodes, it will first generate a broadcast message M_B that includes the constraint C_u, its public key $K_d(u)$, the certificate $Cert(u)$, the query command $Q(u)$, and a signature on $Q(u)$ using its private key. User u then sends M_B to one or a few randomly picked nearby nodes to start the broadcast query process.

- *Broadcast:* When a sensor node i gets an authenticated copy of M_B, it will check to see if it needs to re-broadcast the message based on the broadcast protocol. If yes, it will compute a set of *committing values* (one for each neighbor node) to *weakly authenticate* the broadcast message. Specifically, for each neighbor node j, node i computes $H(M_B||k_{i,j})$ and uses the most significant l bits of this hash as the $P_{i,j}$-th committing value for weak authentication. Let W be the set of these committing values. The final broadcast message will be $\{M_B, W\}$.

- *Authentication:* When a sensor node j receives the message $\{M_B, W\}$ from node i for the first time, it computes $H(M_B \| k_{i,j})$, extracts the most significant l bits, and then compares it with the value at the position $P_{i,j}$ in W. If it finds a different value, the packet will be ignored; otherwise, node j starts to perform two signature verifications on M_B, one for the certificate and the other for the signature signed by node i. Once both signatures are verified, it is certain that the privilege and the query of user u are correct. Node j will then grant access to u if the query meets the constraint. Similar to the baseline approach, the access will be protected by the key $K_{u,j}$. Note that when privilege delegation is enabled, node j will need to receive all the certificates of the users on the path from u to the root. This can be provided by u when it is necessary.

Note that a compromised or malicious node can forge the broadcast queries that can always pass the weak authentication and trigger the expensive signature verification at other sensor nodes. To deal with this problem, we develop an *anomaly suppression* method to make sure that a compromised sensor node cannot generate a large impact on the signature verification at other sensor nodes. The main observation is that *the fraction of forged broadcast messages that can pass weak authentication will be very small when the sender is a benign node*. Hence, if a sensor node observes a large fraction of forged messages passing weak authentication, the sender is very likely to be compromised.

- *Anomaly Suppression:* We keep track of the number x of forged broadcast messages that passed the weak authentication at node i during the previous M forged broadcast messages from node j. If $x > m$, node i will directly consider the message as forged and stop any further verification; otherwise, it will verify the certificate and signature once the broadcast message passes the weak authentication check. Clearly, the fraction of unnecessary signature verifications is bounded by $\frac{m}{M}$.

Security Analysis: We study how well our approach performs under DoS attacks. Assume that node i receives a forged broadcast message $\{M_B', W'\}$ from the adversary who impersonates node j. Note that the committing value for node i is generated using the key $k_{i,j}$ that is only known by nodes i and j. The probability that W' includes a correct guess of this value can be estimated by $\frac{1}{2^l}$. Clearly, our method can effectively reduce the impact of DoS attacks on signature verification even for a small l. For example, when $l = 8$, node i only needs to verify one out of 256 forged messages on average.

When node j is a compromised sensor node, the problem becomes much more complicated since node j can always forge broadcast messages with correct committing values. Fortunately, such impact is bounded by the threshold m and the window size M. Intuitively, setting a smaller m or a larger M brings more resilience against DoS attacks but generates more false suppressions for benign sensor nodes. Hence, we study the *false suppression rate*, which is *the probability that the number of forged messages that passed the weak authentication in a window of M fake messages exceeds m when the sender is actually benign*. The false suppression is caused by the messages forged by the adversary who impersonates the sender. This can be viewed as a binomial distribution $B(M, \frac{1}{2^l})$. The false suppression rate can thus be estimated by

$$P_{false-sup} = 1 - \sum_{i=0}^{m} \frac{M!}{i!(M-i)!}(\frac{1}{2^l})^i(1 - \frac{1}{2^l})^{M-i}$$

Figure 2 shows the false suppression rate under different settings of l, m, and M. We can clearly see that the false suppression rate is very low for a reasonable setting of l, m and M.

Overheads: Compared with the baseline approach, the additional storage overhead at sensor nodes mainly comes from the public keys of the trusted server and the system users. The additional computation overhead at sensor nodes mainly comes from the signature verification. Since the DoS attacks against signature verification can be greatly reduced by our weak authentication method, we believe that supporting broadcast query using ECC-based signature schemes is practical for the current generation of sensor networks.

Our weak authentication scheme needs additional space for committing values at every broadcast message. Let b denote the average number of neighbor nodes for every sensor node in the network. The average overhead is about $b \times l$ bits for every broadcast message from a sensor node. We argue that this additional communication overhead is usually affordable. First, in many applications, most sensor nodes only have a small number of neighbors in their radio ranges. It is unlikely to see a very large b in practice. Second, the value b can be further reduced in many flooding methods

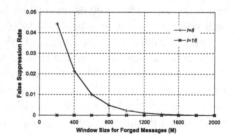

Fig. 2. False suppression rate under different settings of l and M. Assume $m = 0.01M$.

[7, 8, 9] where every broadcast has the receiver size that is much smaller than the neighbor size. The reason is that two neighbor nodes usually share a large number of common neighbor nodes, and it is not necessary to have redundant transmission. Third, in cases where the committing values cannot fit into one packet, we can simply re-broadcast the same message multiple times with different set of committing values every time.

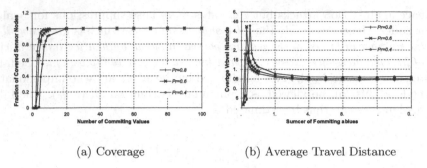

(a) Coverage   (b) Average Travel Distance

Fig. 3. Impact on broadcast protocol. Assume channel loss rate $P_l = 0.2$.

4   Simulation Evaluation

The main focus of our simulation experiments is the security as well as the performance of the broadcast query. The simulation study will focus on two aspects of the proposed weak authentication method, *its impact on the broadcast protocol* and *its performance in dealing with the DoS attacks against the signature verification*.

Many energy efficient flooding mechanisms could be used for broadcast [7,8,9]. However, most of them are developed for wireless ad hoc networks and do not work well in sensor networks due to the limited bandwidth and the lossy channel. Instead, we use a naive broadcast protocol in our evaluation. In this method, when a sensor node receives an authentic broadcast message for the first time, it will re-broadcast it at a probability of P_r.

In our simulation, we randomly deploy 5,093 sensor nodes in a field of size 1000×1000 square meters. Every two sensor nodes can talk to each other if they are no more than 50 meters away. Thus, there are 40 neighbor nodes on average for every node. We assume there are 40 bytes available in each packet for committing values. The broadcast will always start from the center of the field. Every node will only pick b neighbors for re-broadcasting. Figure 3 shows the *coverage* (the fraction of sensor nodes that receives the broadcast message) and the *average travel distance* (the average number of hops a broadcast message travels to reach a sensor node) under different settings. The coverage affects the effectiveness of the broadcast, and the average travel distance affects the latency of the broadcast. From the figure, we can see that increasing b does improve the coverage and reduce the latency. However, after certain point (e.g., 20), increasing b will not generate much benefit. Thus, as long as b is large than certain value such as 20, we will not see big differences in terms of the performance of the broadcast. As a result, the length l of a committing value can be from 8 bits to 16 bits given 40 bytes space. The simulation result in the next part will further show that this is enough for us to effectively defend against the DoS attacks. Note that when b is very small, we will see a small average travel distance. The reason is that broadcast stops quickly after a few hops due to a small b.

We now show that the DoS attacks against signature verification can be mitigated significantly using our approach. We conducted two set of experiments. In the first set of experiments, the attacker impersonates a benign node; in the second set of experiments, the attacker launches the DoS attacks through a compromised sensor node. Clearly, in the second case, the attacker can always pass the weak authentication if he wants.

Figure 4 shows the performance of our protocol in dealing the DoS attacks on signature verification. The *suppression point* in the figure is defined as the point (i.e., the index of the fake broadcast message) where a sensor node starts to suppress the broadcast message from a given neighbor node. For example, a suppression point of 500 means that the sensor node starts to suppress the broadcast message after receiving the 500-th fake message. As discussed before, this usually happens when *the neighbor node is malicious* or *the attacker has impersonated this neighbor node for a long period of time*. From the figure, we can see that impersonating a benign node will only generate a small impact on sensor nodes, while launching attacks through a compromised node actually reveals the identity of the compromised sensor node and will be suppressed. In conclusion, our weak authentication approach effectively thwarts the DoS attacks against signature verification.

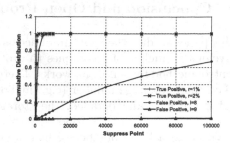

Fig. 4. Performance under DoS attacks. r is the fraction of forged messages with correct committing values (generated by a compromised node).

5   Related Work

This section reviews current research studies on sensor network security. To establish pairwise keys between sensor nodes, researchers have developed many key pre-distribution techniques [10, 11, 12, 13]. μTESLA protocol was developed to provide broadcast authentication for sensor networks [6]. DoS attacks in sensor networks have been studied in [14]. Attacks on routing protocols and counter measures were studied in [15]. Wormhole attacks have been identified as a major threat to sensor networks [16]. A typical sensor network has many supporting services such as data aggregation. These services have to be protected properly. Several techniques were developed to protect in-network processing [17, 18]. Security issues in localization and time synchronization has been studied in [19, 20]. This paper considers a critical security service, access control, in sensor networks. This is a useful service that is complementary to the above studies.

6   Conclusion and Open Problems

This paper studies the problem of access control in sensor networks. The paper develops a number of techniques to provide a practical and distributed access control method for sensor networks. There are a number of open problems. First, when there are huge number (e.g., millions) of sensor nodes and the capability string are long and detailed, the storage overhead at a user could be a big problem. Second, in the proposed technique, a malicious user can delegate its privilege to a large number of fake users and exhaust the memory at sensor nodes. It is interesting to study ideas to address these issues.

Acknowledgment. The authors would like to thank the anonymous reviewers for their valuable comments.

References

1. Crossbow Technology Inc.: Wireless sensor networks. Accessed in February 2006 http://www.xbow.com/Products/Wireless_Sensor_Networks.htm
2. Hartung, C., Balasalle, J., Han, R.: Node compromise in sensor networks: The need for secure systems. Technical Report CU-CS-990-05, U. Colorado at Boulder (January 2005)
3. Zhang, W., Song, H., Zhu, S., Cao, G.: Least privilege and privilege deprivation: Towards tolerating mobile sink compromises in wireless sensor networks. In: Proceedings of ACM Mobihoc'05 (2005)
4. Wang, H., Li, Q.: Distributed user access control in sensor networks. In: Gibbons, P.B., Abdelzaher, T., Aspnes, J., Rao, R. (eds.) DCOSS 2006. LNCS, vol. 4026, pp. 305–320. Springer, Heidelberg (2006)
5. Gura, N., Patel, A., Wander, A.: Comparing elliptic curve cryptography and rsa on 8-bit CPUs. In: Joye, M., Quisquater, J.-J. (eds.) CHES 2004. LNCS, vol. 3156, pp. 119–132. Springer, Heidelberg (2004)
6. Perrig, A., Szewczyk, R., Wen, V., Culler, D., Tygar, D.: SPINS: Security protocols for sensor networks. In: Proceedings of Seventh Annual International Conference on Mobile Computing and Networks (July 2001)
7. Lim, H., Kim, C.: Multicast tree construction and flooding in wireless ad hoc networks. In: Proceedings of ACM Modeling, Analysis, and Simulation of Wireless and Mobile Systems (2000)
8. Peng, W., Lu, X.: On the reduction of broadcast redundancy in mobile ad hoc networks. In: Proceedings of ACM International Symposium on Mobile and Ad Hoc Networking and Computing (2000)
9. Wu, J., Dai, F.: Broadcasting in ad hoc networks based on self-pruning. In: Proceedings of INFOCOM (2003)
10. Eschenauer, L., Gligor, V.D.: A key-management scheme for distributed sensor networks. In: Proceedings of the 9th ACM Conference on Computer and Communications Security, pp. 41–47 (November 2002)
11. Chan, H., Perrig, A., Song, D.: Random key predistribution schemes for sensor networks. In: IEEE Symposium on Research in Security and Privacy, pp. 197–213 (2003)

12. Liu, D., Ning, P.: Establishing pairwise keys in distributed sensor networks. In: Proceedings of 10th ACM Conference on Computer and Communications Security (CCS'03). pp. 52–61 (October2003)
13. Du, W., Deng, J., Han, Y.S., Varshney, P.: A pairwise key pre-distribution scheme for wireless sensor networks. In: Proceedings of 10th ACM Conference on Computer and Communications Security (CCS'03) pp. 42–51(October 2003)
14. Wood, A.D., Stankovic, J.A.: Denial of service in sensor networks. IEEE Computer 35(10), 54–62 (2002)
15. Karlof, C., Wagner, D.: Secure routing in wireless sensor networks: Attacks and countermeasures. In: Proceedings of 1st IEEE International Workshop on Sensor Network Protocols and Applications (May 2003)
16. Hu, Y., Perrig, A., Johnson, D.: Packet leashes: A defense against wormhole attacks in wireless ad hoc networks. In: Proceedings of INFOCOM 2003 (April 2003)
17. Du, W., Deng, J., Han, Y.S., Varshney, P.K.: A witness-based approach for data fusion assurance in wireless sensor networks. In: Proceedings of IEEE Global Communications Conference (GLOBECOM 03) (December 2003)
18. Przydatek, B., Song, D., Perrig, A.: SIA: Secure information aggregation in sensor networks. In: Proceedings of the First ACM Conference on Embedded Networked Sensor Systems (SenSys '03) (November 2003)
19. Liu, D., Ning, P., Du, W.: Attack-resistant location estimation in wireless sensor networks. In: Proceedings of the Fourth International Conference on Information Processing in Sensor Networks (IPSN '05) (April 2005)
20. Sun, K., Ning, P., Wang, C.: Fault-tolerant cluster-wise clock synchronization for wireless sensor networks. IEEE Transactions on Dependable and Secure (TDSC) 2(1), 177–189 (2005)

Optimizing End to End Routing Performance in Wireless Sensor Networks

Chen Wang, Guokai Zeng, and Li Xiao

Department of Computer Science and Engineering
Michigan State University, East Lansing, MI 48824
{wangchen,zengguok,lxiao}@cse.msu.edu

Abstract. The geographic routing is an ideal approach to realize point-to-point routing in wireless sensor networks because packets can be delivered by only maintaining a small set of neighbors' physical positions. The geographic routing assumes that a packet can be moved closer to the destination in the network topology if it is moved geographically closer to the destination in the physical space. This assumption, however, only holds in an ideal model where uniformly distributed nodes communicate with neighbors through wireless channels with perfect reception. Because this model oversimplifies the spatial complexity of a wireless sensor network, the geographic routing may often lead a packet to the local minimum or low quality route. Unlike the geographic forwarding, the ETX-embedding proposed in this paper can accurately encode both a network's topological structure and channel quality to small size nodes' virtual coordinates, which makes it possible for greedy forwarding to guide a packet along an optimal routing path. Our performance evaluation based on both the MICA2 sensor platform and TOSSIM simulator shows that the greedy forwarding based on ETX-embedding outperforms previous geographic routing approaches.

1 Introduction

The geographic routing is a promising approach to realize point-to-point routing in a wireless sensor network comprising large number of randomly deployed resource-constrained sensors. By greedily forwarding a packet to the next neighbor which is geographically closer to the destination, the geographic routing can gradually move the packet towards the destination and finally deliver the packet through consecutive hop by hop forwarding. Because the simple greedy forwarding only requires each intermediate node to maintain a small set of neighbors' physical positions, the geographic routing incurs small storage and communication overhead, which makes it preferable to the shortest path routing and the on-demand routing for resource-constrained wireless sensor networks.

The shortest path routing requires each node to maintain a per-destination state routing table whose size is proportional to the total number of nodes in a network. The shortest path routing cannot scale to a large size wireless sensor network because the large size routing table is unaffordable for individual nodes

J. Aspnes et al. (Eds.): DCOSS 2007, LNCS 4549, pp. 36–49, 2007.

with limited storage capacity. The on-demand routing such as AODV [1] builds and maintains a route only when necessary which reduces the size of routing table because intermediate nodes only keep routing states to currently active destinations. However, AODV frequently initiates broadcast messages to discover the routing paths and therefore incurs massive communication overhead.

The geographic routing assumes that a packet can be moved closer to the destination in the network topology when it is moved geographically closer to the destination in the Euclidean space. This assumption is based on the observation that the topological structure of a wireless sensor network can be approximated by its geographic structure. Because a wireless node can only communicate with its neighbors within the maximum radio transmission range, pairwise nodes may have a short communication path in the network topology if they are geographically closer to each other in the Euclidean space, i.e. the hop count distance between pairwise nodes is proportional to their Euclidean distance. This observation is correct in an ideal wireless sensor network model where uniformly distributed nodes communicate with connected neighbors through wireless channels with perfect reception. Nevertheless, this ideal model often oversimplifies the spatial complexity of a realistic wireless sensor network which may have *complicated topological structure* and *irregular wireless communication channels*.

Due to the discrepancy between a wireless sensor network's complex spatial characteristics and its oversimplified geographic description, the geographic routing may fail to deliver a packet or forward a packet along a suboptimal routing path. For example, a packet may be trapped in a local minimum where none of the neighbors is closer to the destination. A packet may also be forwarded along a route consisting of long distance hops with low quality wireless channels. Numerous approaches have been proposed to recover the geographic routing from local minimum [2] or find the proper forwarding advance without sacrificing channel quality [3] [4] [5]. Constrained by the inaccurate geographic model on a network topology, these workaround solutions cannot guarantee a packet to be efficiently forwarded along the optimal routing path.

In this paper, we try to improve the end-to-end routing performance of the greedy forwarding by improving the expression accuracy of a wireless sensor network. Unlike the simple geographic model where the communication route is approximated by the geographic path, we embed a wireless sensor network into a Euclidean space where nodes' virtual distance is equal to the number of expected transmissions for a packet to be successfully delivered between the pairwise nodes. Because the virtual distance directly reflects the end-to-end communication channel quality, the greedy forwarding can guide a packet along the optimal routing path which has the shortest virtual distance.

In the following discussion, we first introduce the spatial complexity of wireless sensor networks. We further show how the wireless sensor networks' complex spatial characteristics can be efficiently encoded to nodes' virtual coordinates to support optimal greedy forwarding. The performance of our proposed approach is evaluated on the MICA2 sensor platform and TOSSIM simulator.

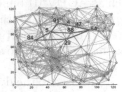
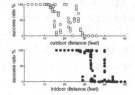
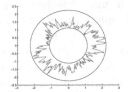

Fig. 1. Long distance radio links of geographic routing

Fig. 2. Packet reception between pairwise wireless nodes

Fig. 3. Irregular radio transmission pattern

2   Spatial Complexity of Wireless Channels

The geographic routing uses a simple connectivity model to describe the wireless channels between pairwise nodes, i.e. pairwise nodes have the perfect reception channel if they are within the maximum transmission range of radio signals. In a realistic wireless sensor network, neighboring nodes are often connected through unreliable wireless channels where packets may be lost due to the transmission error of radio signals. It is normal that packet loss rate is increased with the transmission range because the radio signals attenuate during their transmission, which leads to low signal to noise ratio (SNR). Since the geographic routing greedily selects the next hop which is closest to the destination and therefore furthest to the sender, the geographic routing tends to include long distance hops in the routing path which are often unreliable and have high packet loss rate. An example is shown in Fig 1. Based on the greedy forwarding policy of the geographic routing, a packet is forwarded along the routing path $84 \to 29 \to 54$, which may have higher packet loss rate and lower throughput than the routing path of $84 \to 5 \to 91 \to 4 \to 47 \to 54$ consisting of more while shorter intermediate links.

Several approaches [4] [3] [6] have been proposed to balance the forwarding distance and radio link quality, which can be divided into two categories:

1. Define a threshold to exclude low quality radio links.
2. Define a new metric which can be maximized under the constraints of both forwarding distance and radio link quality.

The complex radio signal transmission pattern, however, makes it difficult to improve the performance of the geographic routing through these two strategies.

For the first strategy, it is difficult to determine a proper threshold value which can maximize the end-to-end routing performance. Fig. 2 shows the packet reception between pairwise nodes we measured on the MICA2 sensor platform, which demonstrates that the perfect radio channel with 100% reception only exists between transceivers within a short distance. If the threshold is aggressively set to only include links with 100% packet reception, we may have a disconnected network or a routing path comprising excessive intermediate nodes, which increases both the processing cost and delay. We use an example to further explain how the threshold values affect the end-to-end routing performance. As shown in Fig. 1, the routing path $84 \rightarrow 5 \rightarrow 88 \rightarrow 54$ selected by the 85% threshold outperforms the routing path $84 \rightarrow 5 \rightarrow 91 \rightarrow 4 \rightarrow 47 \rightarrow 54$ selected by the 100% threshold, because the former uses less intermediate nodes in the packet forwarding with slightly inferior links. Because a proper threshold to determine the optimal routing path may vary from different pairwise nodes, the geographic routing with a constant threshold cannot provide a universal solution to find the optimal routing path between any pairwise nodes.

Instead of simply excluding low quality radio links below a certain constant threshold, the second strategy selects radio links by optimizing the forwarding advance and quality of radio links simultaneously. For example, the energy-efficient forwarding [3] chooses the next hop which can maximize the product of the packet reception rate (PRR) and the distance traversed towards destination. This strategy can achieve good routing performance when i) nodes are uniformly distributed; ii) the packet reception rate of the wireless channels can be explicitly modeled by the transmission distance. However, in an obstructed environment, radio signals have complex transmission patterns because the signal strength may be either strengthened or weakened due to multipathing or shading such that the packet reception rate is less correlated to the transmission distance and difficult to model. Fig. 2 shows pairwise transceivers have different packet reception rate in outdoor and indoor environments. The radio signal transmission may also vary in different directions as shown in Fig. 3.

Even if the geographic routing can find the optimal tradeoff between the forwarding advance and the link quality for individual hops, it may fail to find the path with the optimal end-to-end routing performance. Because both the advance distances and PRR are local metrics, the greedy forwarding may lead to a local minimum and fail to find the global optimal path, which often happens in a network topology with complex spatial characteristics.

3   Greedy Forwarding Based on ETX-Distance

In a wireless sensor network with complex spatial characteristics including irregular wireless channels and concave network topology, the routing cost between pairwise sensors cannot be estimated simply from their Euclidean distance. Consequently, the greedy forwarding based on geographic distance comparison may lead a packet along low quality links or towards a local minimum. Instead of using the geographic distance to indirectly approximate the routing cost, we argue that the optimal end-to-end routing performance of the greedy forwarding can only be achieved by comparing neighboring nodes based on the routing metric with the following properties: i) the metric should reflect the underlying wireless channel quality between neighboring nodes; ii) the metric should also reflect the end-to-end channel quality between pairwise nodes.

3.1   Underlying Wireless Channel Evaluation

The communication quality and cost of a wireless channel can be evaluated by various metrics such as packet reception ratio, transmission delay, and throughput. A detailed comparison of three link-quality metrics - expected transmission count (ETX) [7], per-hop round trip time (RTT) [8], and per-hop packet pair delay - has been conducted in [9], which concludes that the ETX metric has the best performance in a static wireless sensor network. In this paper, we show that ETX is an ideal metric to define the virtual distance to support the greedy forwarding to achieve optimal end-to-end routing performance.

To route data over unreliable wireless channels, hop-by-hop recovery is usually preferred over end-to-end recovery [10]. The hop-by-hop recovery is realized by acknowledging received packets and retransmitting loss packets. For neighboring nodes i and j in a route path, the receiver j will send back an acknowledgment to sender i when a packet is correctly delivered; and the sender i will retransmit a packet if it has not received the acknowledgment within a certain time period after a packet transmission. Assume the packet loss rate from node i to j is P_{ij} and the packet loss rate from node j to node i is P_{ji}. The probability of a successful packet transmission is $(1 - P_{ij})(1 - P_{ji})$, and the expected number of retransmissions defined as the expected transmission count metric in [7] between node i and j is:

$$ETX(i,j) = 1/(1 - P_{ij})(1 - P_{ji}).$$

Assume that a pairwise node p_1 and p_n has the routing path l comprising intermediate nodes $p_2, p_3, \ldots, p_{n-1}$; we have the expected transmission count of the routing path l as:

$$ETX(l) = \sum_{i=1}^{n-1} ETX(p_i, p_{i+1})$$

It has been proposed in [7] to incorporate the ETX into the on-demand routing such as DSR to find the optimal routing path between pairwise wireless nodes. In the combined approach, the source broadcasts route probing messages to an entire network. The routing paths can be discovered when the destination sends back the response messages along the reversed paths of the probing message. Among all the paths connecting the source and the destination, the optimal routing path can be determined with the minimal ETX.

In this paper, we propose to combine the ETX metric with the greedy forwarding such that the optimal routing path can be found without reliance on the frequently broadcast route probing messages.

3.2   Virtual Distance Based Greedy Forwarding

We define the ETX virtual distance between pairwise nodes $\mathbf{x}_i$ and $\mathbf{x}_j$ as the minimal ETX among all the routing paths connecting $\mathbf{x}_i$ and $\mathbf{x}_j$, i.e.

$$\delta(\mathbf{x}_i, \mathbf{x}_j) = \min_{l_i \in L} ETX(l_i),$$

where L is the set of routing paths connecting nodes $\mathbf{x}_i$ and $\mathbf{x}_j$. In this section, we assume that ETX distance between pairwise nodes in a wireless sensor network can be inferred from their virtual coordinates. How to acquire nodes' virtual coordinates will be discussed in detail in the coming section.

Based on the comparison of the ETX distances between neighboring nodes, the greedy forwarding can determine the next hop as follows: suppose a packet need to be forwarded to the destination $\mathbf{x}_k$. Let node $\mathbf{x}_i$ be the intermediate node with the routing packet and set N define all the neighbors of node $\mathbf{x}_i$. The next forwarding hop can be selected from the neighbor set N as:

$$\widehat{\mathbf{x}}_j = \arg \min_{\mathbf{x}_j \in N} (\delta(\mathbf{x}_i, \mathbf{x}_j) + \delta(\mathbf{x}_j, \mathbf{x}_k)), \tag{1}$$

i.e. the packet is greedily forwarded to the next hop $\mathbf{x}_j$ which minimizes the summary of the ETX distances $\delta(\mathbf{x}_i, \mathbf{x}_j)$ and $\delta(\mathbf{x}_j, \mathbf{x}_k)$.

Because the ETX distance directly reflects the length of a communication path between pairwise nodes in a wireless sensor network, the greedy forwarding based on ETX distance comparison can guide a packet towards the correct direction and deliver the packet through consecutive hop by hop forwarding.

4   ETX Embedding

The greedy forwarding can achieve optimal end-to-end routing performance based on ETX distance comparison between neighboring nodes. A simple solution of realizing ETX distance comparison is to assign each node with a virtual coordinate which contains ETX distances to all the nodes in the network:

$$\mathbf{x}_i = [x_{i1}, x_{i2}, \ldots, x_{iN}],$$

where x_{ij} is the ETX distance to node $\mathbf{x}_j$. The optimal routing performance, however, is achieved at the cost of $O(N)$ states maintained by individual nodes, which is unaffordable to resource-constrained wireless sensors. In order to achieve high routing performance on the basis of small size routing states, we seek an optimal routing design for wireless sensor networks which can compress the $O(N)$ routing states into constant size routing states while accurately preserving the ETX distances between pairwise nodes. We generalize the routing states compression problem as embedding a N-dimensional ETX distance metric space into a M-dimensional Euclidean space which has the following properties: i) M is a small constant $(M \ll N)$ which does not scale with network size N; ii) ETX distances between pairwise nodes can be precisely inferred from Euclidean distances in the M-dimensional embedded space.

An embedding problem can be formalized as follows. Let (X, δ) define a metric space. Here X is a set and δ is a metric which defines a distance function between elements in X. For elements $\mathbf{x}_i, \mathbf{x}_j, \mathbf{x}_k \in X$, the distance function δ satisfies (1) symmetry: $\delta(\mathbf{x}_i, \mathbf{x}_j) = \delta(\mathbf{x}_j, \mathbf{x}_i)$; (2) positive definiteness: $\delta(\mathbf{x}_i, \mathbf{x}_j) \geq 0$ and $\delta(\mathbf{x}_i, \mathbf{x}_j) = 0$ iff $i = j$; (3) triangular inequality: $\delta(\mathbf{x}_i, \mathbf{x}_k) + \delta(\mathbf{x}_k, \mathbf{x}_j) \geq \delta(\mathbf{x}_i, \mathbf{x}_j)$.

Let (P, d) define the Euclidean space. Here P is a set of points mapped from elements in the set X and d is a metric which defines the function of 2-norm Euclidean distances between pairwise points in P. For $\mathbf{p}_i, \mathbf{p}_j \in P$, we have $d(\mathbf{p}_i, \mathbf{p}_j) = |\mathbf{p}_i - \mathbf{p}_j|$.

Definition 1. *An embedding of metric space (X, δ) into an Euclidean space (P, d) is a mapping $\phi : X \to P$ such that*

1. $\boldsymbol{p}_i = \phi(\boldsymbol{x}_i)$;
2. $\delta(\boldsymbol{x}_i, \boldsymbol{x}_j) = d(\phi(\boldsymbol{x}_i), \phi(\boldsymbol{x}_j)) = d(\boldsymbol{p}_i, \boldsymbol{p}_j) = |\boldsymbol{p}_i - \boldsymbol{p}_j|$

It can be easily verified that the ETX distance defines a distance metric which satisfies the properties of symmetry, positive definiteness and triangular inequality. Consequently, we can embed a wireless sensor network defined by the ETX distance function into an Euclidean space. The ETX embedding can be intuitively explained as given ETX distances between pairwise nodes in a network topology, finding nodes' coordinates in a M-dimensional Euclidean space such that the ETX distances can be inferred from the Euclidean distance of the mapped space. The objective of the embedding is to find the minimal M such that $\delta(\mathbf{x}_i, \mathbf{x}_j) = d(\phi(\mathbf{x}_i), \phi(\mathbf{x}_j))$, i.e. to embed a ETX distance metric space into the lowest dimensional Euclidean space in which ETX distances between pairwise nodes can still be precisely preserved.

Instead of an exact embedding, a network topology can be approximately embedded into an Euclidean space by relaxing condition 2 as $\delta(\mathbf{x}_i, \mathbf{x}_j) \approx d(\phi(\mathbf{x}_i), \phi(\mathbf{x}_j))$. Here, we slightly sacrifice the embedding accuracy to achieve lower dimensionality of the embedded Euclidean space and therefore smaller routing states. Consequently, we are seeking nodes' coordinates to minimize differences between ETX distances of the network topology and Euclidean distances of the embedded space:

$$\min \sum_{\mathbf{x}_i, \mathbf{x}_j \in X} (\delta(\mathbf{x}_i, \mathbf{x}_j) - d(\phi(\mathbf{x}_i), \phi(\mathbf{x}_j)))^2.$$

We use the multidimensional scaling (MDS) [11] to embed a N-dimensional ETX distance metric space to a M-dimensional Euclidean space. MDS has shown great success in psychology and pattern recognition in discovering meaningful low-dimensional structures hidden in high-dimensional observations [12]. In this paper, we show MDS is effective to reduce N-dimensional per-destination state to low dimensional virtual coordinates from which ETX distances can be accurately recovered. We use a short deduction below to show how a wireless sensor network can be embedded to a low-dimensional Euclidean space through MDS. More details on MDS can be found in [11]. For simplicity, we use short notation δ_{ij} to refer to the ETX distance $\delta(\mathbf{x}_i, \mathbf{x}_j)$ and d_{ij} to refer to the Euclidean distance $d(\mathbf{x}_i, \mathbf{x}_j)$

According to the embedding defined in Definition 1, we have

$$\delta_{ij}^2 = |\mathbf{p}_i - \mathbf{p}_j|^2 = (\mathbf{p}_i - \mathbf{p}_j)^t (\mathbf{p}_i - \mathbf{p}_j) = \mathbf{p}_i^t \mathbf{p}_i + \mathbf{p}_j^t \mathbf{p}_j - 2\mathbf{p}_i^t \mathbf{p}_j$$

By shifting matrix P to the center, nodes' coordinates $\mathbf{p}_i$ and $\mathbf{p}_j$ can be expressed as the function of ETX distance δ_{ij} and we have

$$P^t P = F,$$

where $P = [\mathbf{p}_1^t; \mathbf{p}_2^t; \ldots \mathbf{p}_N^t]$ and $F = [f(\delta_{ij}^2)]$. Because F is symmetric, it can be decomposed through singular value decomposition (SVD) as:

$$P^t P = F = V \Sigma V^t$$
$$P^t = V \Sigma^{1/2}.$$

Here, Σ is a diagonal matrix with the rank-ordered set of singular values $\sigma_1 \geq \sigma_2 \geq \ldots \geq \sigma_r \geq \sigma_{r+1} = \sigma_{r+2} = \ldots \sigma_N = 0$. Let $\sigma_i^{1/2} \equiv \lambda_i$. We have

$$[\mathbf{p}_1, \mathbf{p}_2, \ldots, \mathbf{p}_N] = [\lambda_1 \mathbf{v}_1, \lambda_2 \mathbf{v}_2, \ldots, \lambda_N \mathbf{v}_N]^t. \tag{2}$$

From Eqn (2), we have node i's N-dimensional coordinate

$$\mathbf{p}_i = [\lambda_1 \mathbf{v}_{1i}, \lambda_2 \mathbf{v}_{2i}, \ldots, \lambda_N \mathbf{v}_{Ni}]^t$$

such that

$$\delta_{ij} = |\mathbf{p}_i - \mathbf{p}_j|.$$
$$\text{Let } \mathbf{p}_i' = [\lambda_1 \mathbf{v}_{1i}, \lambda_2 \mathbf{v}_{2i}, \ldots, \lambda_m \mathbf{v}_{mi}]^t, m < r,$$

where $\mathbf{p}_i$ defines the m-dimensional virtual coordinates of node i. The difference between the ETX distance and the Euclidean distance estimated from m-dimensional coordinates is:

$$\delta_{ij}^2 - |\mathbf{p}_i' - \mathbf{p}_j'|^2 = |\mathbf{p}_i - \mathbf{p}_j|^2 - |\mathbf{p}_i' - \mathbf{p}_j'|^2$$
$$= \sum_{m+1 \leq k \leq r} \lambda_k^2 (\mathbf{v}_{ki} - \mathbf{v}_{kj})^2 \tag{3}$$

Based on Eqn (3), we have $|\mathbf{p}_i' - \mathbf{p}_j'| \to \delta_{ij}$ when $m \to r$. This reflects the tradeoff between the accuracy of network expression and the size of network expression. Moreover, if rank-ordered singular value λ_i quickly converges to zeros after the first several most important ones, we have $|\mathbf{p}_i' - \mathbf{p}_j'| \approx \delta_{ij}$ for a relative small m. This is the case for network embedding and is verified in our performance evaluations. Here, MDS not only reveals the inherent tradeoff between the accuracy and size of a network expression, but also provide a viable approach to efficiently approximate a network topology.

5   Sample a Wireless Sensor Network with Beacons

In order to embed a network to a low-dimensional Euclidean space through MDS, we need to measure ETX distances between all pairwise nodes of a network, which involves massive communication messages. In this section, we show

how to achieve the ETX-embedding by sampling the network through a small number of beacons in a distributed fashion. Instead of measuring ETX-distance between any pair of nodes, we only measure the ETX-distances from a node to a set of reference points defined as beacons. Each beacon broadcasts a beacon message with a transmission counter initialized by zero. The transmission counter is increased after each transmission (including retransmission). By finding out the smallest transmission counter among all the received beacon messages, a node can infer its ETX distance to the sampling beacon.

Based on the knowledge of ETX-distances to the set of beacons, a wireless sensor network can be embedded into a low dimensional Euclidean space through the two steps below:

1. Use MDS to embed all the beacons to the low dimensional Euclidean space based on the ETX-distances between all pairwise beacons. We have

$$K_M = [\mathbf{k}_1, \mathbf{k}_2, \ldots, \mathbf{k}_M],$$

where $\mathbf{k}_i$ is beacon i's virtual coordinate in the embedded space.
2. Node i's virtual coordinate in the embedded space can be inferred from its ETX-distances to the set of beacons using the least square fitting:

$$\widehat{\mathbf{p}}_i = P(\delta, K_M) = \arg\min_{\mathbf{p}_i} \sum_{j=1}^{M}(\delta_{ij} - |\mathbf{p}_i - \mathbf{k}_j|)^2.$$

The ETX-embedding described above can be intuitively explained as follows. First, all the beacons are embedded into the low dimensional Euclidean space based on the full measurement between any pair of beacons. After that, the rest nodes can be embedded according to their relative ETX-distances to the beacons. Accurate embedding can be achieved when sufficient beacons are uniformly distributed such that a network's spatial characteristics can be fully represented by sampling beacons.

6   Performance Evaluation

We evaluate the ETX distance based greedy forwarding through a small scale experiment based on MICA2 platform and a large scale simulation based on TOSSIM/TYTHON [13]. TOSSIM is a bit level simulator which shares the same TinyOS code with the MICA2 platform. We first evaluate the packet transmission in multihop forwarding on the MICA2 platform. After that, we compare the greedy forwarding based on ETX distance with the geographic routing in the TOSSIM simulator. The two metrics below are used in our performance evaluation.

1. **Number of transmissions per packet**: The number of packet transmissions per packet is the total number of transmissions, including retransmissions required for a packet to be forwarded from the source to the destination.

2. **Packet failure rate**: A packet routing may fail due to the unreliable wireless transmission channels. The packet failure rate is the percentage of failed routing among all the routing tests.

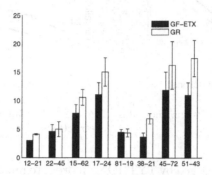

Fig. 4. Number of transmissions between pairwise nodes

6.1   Evaluate the ETX-Embedding in TOSSIM

TOSSIM is a bit level simulator which can accurately simulate the radio transmission channels between wireless sensors. We use the lossy radio model which assumes each bit of the transmission packet has the probability of p to be flipped. The probability of bit flipping is measured from the real experiment. When a packet is received, the correctness is verified/recovered by FEC code. A packet will be dropped if it cannot be recovered by the FEC verifying code. We use the TYTHON to control the simulated nodes in TOSSIM, which act as the two laptops attached to the source and the destination. The pairwise source and destination are randomly selected from 100 nodes deployed in a 125 by 125 feet square area. The testing packet is injected to the simulated network through TYTHON which is forwarded along different routing paths computed from different routing algorithms. The average number of packet transmissions and the

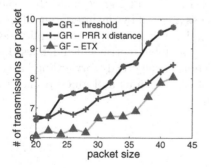

Fig. 5. Number of transmissions under different packet sizes

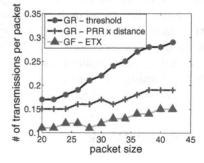

Fig. 6. Packet failure rate under different packet sizes

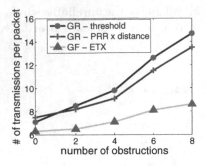
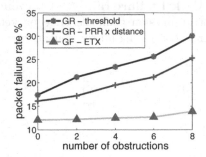

Fig. 7. Number of transmissions under different obstructions

Fig. 8. Packet failure rate under different obstructions

packet delivery ratio is logged by TYTHON. In order to minimize the interference between radio links, we inject the testing packets in a sequential order with a time interval between consecutive packets.

We first evaluate the greedy forwarding on 8 pairwise nodes randomly selected. We use both greedy forwarding based on ETX distance (GF-ETX) and the geographic routing (GR) to forward packets between each pairwise nodes and record the average number of transmissions. Fig. 4 shows that the GF-ETX uses less number of transmissions to deliver packets from a source to a destination.

We further compare the GF-ETX, the geographic routing with threshold (GR-threshold) and the geographic routing based on the product of the packet reception rate and the forwarding distance (GR-PRR x distance). Fig. 5 and Fig. 6 show that both the number of packet transmissions and the packet failure rate are increased with the increment of packet size. Moreover, the geographic routings are more susceptible to the packet size and have high average number of packet transmissions for larger size packets. This is because both the routing approaches tend to select low quality routing links with high bit flipping error and multiple bit errors may happen in long length packets which cannot be recovered by FEC code.

We compare the GF-ETX with the geographic routing in complex deployed environment by adding obstructions into the deployed area, which increases the spatial complexity of the network topology. Fig. 7 and Fig. 8 show that the greedy forwarding based on ETX distance is less affected by the interference from obstructions and can always learn the optimal routing paths in a complicated network topology. Because the ETX distance is a global metric which defines the end-to-end channel quality between pairwise nodes, the greedy forwarding based on ETX distance can foresee the affection of obstructions and guide packet to bypass the obstruction along an optimal route.

7   Related Work

Packet routing is a fundamental problem in wireless networks, which has been studied for decades. Geographic routing has been proposed to route packets based on nodes' geographic positions [14] [15] [2] [16] [6] [17] [18] [4] [19] [20] [21]. In geographic routing, a packet is greedily forwarded to the next hop which is geographically closer to the destination. Nevertheless, a packet may be trapped in a local minimum where none of the neighbors are closer to the destination. Numerous schemes have been proposed to recover the geographic forwarding from the local minimum [2] [22] [23] by exploring the geographic characteristics of a wireless sensor network or relying on small range flooding. The recovery schemes, however, often have higher computation complexity than the simple greedy forwarding. In contrast, the ETX embedding proposed in this paper directly improves the routing success rate of greedy forwarding with no extra overhead.

Until recently, it has been proposed to construct nodes' coordinates from hop count distances instead of geographic distance to support the greedy forwarding [24] [25] [26] [27] [28] [23] [29] [30]. The hop count distance based greedy forwarding outperforms the geographic routing because i) it does not require nodes' physical positions; ii) it can find a route close to the shortest path. The hop count distance based greedy forwarding, however, suffers the same problem as the geographic routing in that the quality of underlying wireless channels are not reflected in the routing decision. Both approaches tend to select a route with fewer intermediate forwarding nodes and therefore longer distance hops.

Several approaches [3] [4] [5] [31] have been proposed to balance the forwarding distance and radio link quality, which either defines a threshold to exclude low quality radio links; or defines a new metric which can be maximized under the constraints of both forwarding distance and radio link quality. As we discussed before, the greedy forwarding based on defined local metrics may fail to find the optimal end-to-end routing path because the local metrics combined by the forwarding advance and the link quality between immediate neighbors can not reflect the global communication channel qualities.

The optimal routing metrics in a wireless network have been investigated in [7] [8] [9] [32] [33], which propose the ETX and RTT instead of the shortest path as the routing metric. The proposed metrics are incorporated into the on-demand routing such as DSR to discover the optimal routing path.

Different from previous work, the ETX embedding proposed in this paper embeds a wireless sensor network into a Euclidean space to support the greedy forwarding. The ETX distance based greedy forwarding finds the route without relying on frequently broadcast route probing messages. Moreover, it can find the optimal routing path because the ETX distance is a global metric which defines end-to-end channel quality.

8 Conclusion

We propose the ETX-embedding for greedy forwarding to achieve optimal end-to-end routing performance in wireless sensor networks. We show that MDS can embed a wireless sensor network into a low dimensional Euclidean space where ETX distances are accurately preserved. Based on the ETX distance comparison from nodes' small size virtual coordinates, the simple greedy forwarding can forward a packet along the optimal routing path without relying on broadcast route probing messages. Consequently, the optimal end-to-end routing performance is achieved with small overhead in terms of storage and communication costs, which makes the ETX distance based greedy forwarding an ideal routing approach for resource-constrained wireless sensor networks deployed in complicated environment.

Acknowledgment. This work was supported in part by the US National Science Foundation under grants CCF-0514078, CNS-0549006, and CNS 0551464.

References

1. Perkins, C., Royer, E.: Ad-hoc on-demand distance vector (AODV) routing. Internet Draft (1998)
2. Karp, B., Kung, H.T.: GPSR: Greedy Perimeter Stateless Routing for Wireless Networks. In: MobiCom (2000)
3. Seada, K., Zuniga, M., Helmy, A., Krishnamachari, B.: Energy-efficient forwarding strategies for geographic routing in lossy wireless sensor networks. In: SenSys (2004)
4. Lee, S., Bhattacharjee, B., Banerjee, S.: Efficient geographic routing in multihop wireless networks. In: MobiHoc (2005)
5. Wu, S., Candan, K.S.: GPER: Geographic power efficient routing in sensor networks. In: ICNP (2004)
6. Wu, S., Candan, K.S.: Gper: Geographic power efficient routing in sensor networks. In: ICNP (2004)
7. Couto, D.S.J.D., Aguayo, D., Bicket, J., Morris, R.: A high-throughput path metric for multi-hop wireless routing. In: MobiCom (2003)
8. Adya, A., Bahl, P., Padhye, J., Wolman, A., Zhou, L.: A multi-radio unification protocol for ieee 802.11 wireless networks. In: Broadnets (2004)
9. Draves, R., Padhye, J., Zill, B.: Comparison of routing metrics for static multi-hop wireless networks. In: SIGCOMM (2004)
10. Stann, F., Heidemann, J.: RMST: Reliable data transport in sensor networks. In: SNPA (2003)
11. Cox, T.F., Cox, M.A.A.: Multidimensional Scaling. Chapman and Hall (1994)
12. Tenenbaum, J.B., de Silva, V., Langford, J.C.: A global geometric framework for nonlinear dimensionality reduction. Science (2000)
13. Levis, P., Lee, N., Culler, M.W.D.: Tossim: Accurate and scalable simulation of entire tinyos applications. In: SenSys (2003)
14. Bose, P., Morin, P., Stojmenovic, I., Urrutia, J.: Routing with guaranteed delivery in ad hoc wireless networks. In: ACM Int. Workshop on Discrete Algorithms and Methods for Mobile Computing and Communications (1999)
15. Ko, Y., Vaidya, N.H.: Location-aided routing (LAR) in mobile ad hoc networks. In: MobiCom (1998)
16. Kim, Y.-J., Govindan, R., Karp, B., Shenker, S.: Geographic routing made practical. In: NSDI (2005)
17. Seada, K., Helmy, A., Govindan, R.: On the effect of localization errors on geographic face routing in sensor networks. In: IPSN (2004)

18. Kuhn, F., Wattenhofer, R., Zhang, Y., Zollinger, A.: Geometric adhoc routing: Of theory and practice. In: PODC (2003)
19. Xing, G., Lu, C., Pless, R., Huang, Q.: On greedy geographic routing algorithms in sensing covered networks. In: MobiHoc (2004)
20. Bruck, J., Gao, J., Jiang, A.: Localization and routing in sensor networks by local angle information. In: MobiHoc (2005)
21. Melodia, T., Pompili, D., Akyildiz, I.: On the interdependence of distributed topology control and geographical routing in ad hoc and sensor networks. JSAC (2005)
22. Fang, Q., Gao, J., Guibas, L.J.: Locating and bypassing routing holes in sensor networks. In: InfoCom (2004)
23. Fonseca, R., Ratnasamy, S., Zhao, J., Ee, C.T., Culler, D., Shenker, S., Stoica, I.: Beacon-vector routing: Scalable point-to-point routing in wireless sensor networks. In: NSDI (2005)
24. Newsome, J., Song, D.: Gem: Graph embedding for routing and data-centric storage in sensor networks without geographic information. In: SenSys (2003)
25. Rao, A., Ratnasamy, S., Papadimitriou, C., Shenker, S., Stoica, I.: Geographic routing without location information. In: MobiCom (2003)
26. Leong, B., Mitra, S., Liskov, B.: Path vector face routing: Geographic routing with local face information. In: ICNP (2005)
27. Lim, M., Greenhalgh, A., ChesterField, J., Crowcroft, J.: Landmark guided forwarding. In: ICNP (2005)
28. Zhao, Y., Li, B., Zhang, Q., Chen, Y., Zhu, W.: Efficient hop id based routing for sparse ad hoc networks. In: ICNP (2005)
29. Cao, Q., Abdelzaher, T.F.: A scalable logical coordinates framework for routing in wireless sensor networks. In: RTSS (2004)
30. Caruso, A., Chessa, S., De, S., Urpi, A.: GPS free coordinate assignment and routing in wireless sensor networks. In: InfoCom (2005)
31. Zhang, H., Arora, A., Sinha, P.: Learn on the fly: Data-driven link estimation and routing in sensor network backbones. In: InfoCom (2006)
32. Woo, A., Tong, T., Culler, D.: Taming the underlying challenges of reliable multihop routing in sensor networks. In: SenSys (2003)
33. Zhao, J., Govindan, R.: Understanding packet delivery performance in dense wireless sensor networks. In: SenSys (2003)

Improving Event-to-Sink Throughput in Wireless Sensor Networks

Chen Wang and Li Xiao

Department of Computer Science and Engineering
Michigan State University, East Lansing, MI 48824
{wangchen, lxiao}@cse.msu.edu

Abstract. Disaster relief is an important application of sensor networks, in which bursting data needs to be collected in a short period to the sink through a multi-hop wireless network. In general, packets containing the reported data have few correlations among each other, such that meaningful information can be inferred from partially received packets. For better understanding of monitored events, it is more important to capture the total number of unique reports rather than to reliably deliver each individual packet. Therefore, in the case of monitoring disaster filed with bursting data, communication channel throughput has higher priority than the channel reliability. Under this circumstance, we revisit the sensor network transport protocols, which use hop-by-hop recovery to provide reliable data transmission over unreliable links. We found that the complex recovery mechanism, while assuring high reliable individual packet delivery, reduces channel throughput when measured data is reported at high rate. To provide optimal data transport in a sensor network with bursting data generation, we propose a light weight sink centric transport protocol, which maximizes the channel throughput by minimizing the interference of packet recovery process. We implement the sink centric transport protocol in TinyOS and evaluate its performance in the TOSSIM simulator. The comparison shows that our proposed approach outperforms the hop-by-hop recovery approach in terms of event reporting throughput and transmission costs.

1 Introduction

Wireless sensor networks have broad applications in habitat monitoring, battle field surveillance and disaster relief. While sensors normally operate under light load, they may be suddenly activated by abrupt events such as enemy attack or fire spread. In such cases, large volume of data may be generated and transmitted to sink within a short period, which demands high throughput transmission channels to transfer the bursting data in a timely fashion. However, the lossy nature of wireless communication, together with the channel contention incurred by multi-hop forwarding, makes it a challenging task to achieve the high throughput data transfer in sensor networks.

Because sensors are equipped with low-power radio transceivers, radio signals have low signal-to-noise ratio when transmitted to long distance. Consequently,

J. Aspnes et al. (Eds.): DCOSS 2007, LNCS 4549, pp. 50–63, 2007.

the packet can be easily corrupted and have high lossy rate during the long range transmission[1]. To achieve reliable packet delivery, sensors are usually densely deployed such that packets can be forwarded between pairwise sensors within short range[2]. However, channel interference can easily happen in a dense sensor network when a large volume of generated packets are quickly forwarded along the multi-hop paths towards the center of the sink. In order to efficiently collect data over the unreliable and crowded wireless channels, a suit of transport protocols have been developed to recover the lossy data and mitigate congested channels.

The hop-by-hop recovery has been proposed to retransmit a lost packet when the acknowledgment has not been received within a certain period [3]. The reliable data transfer is necessary for critical messages that cannot tolerate packet loss. However, this timer based retransmission mechanism may degrade channel throughput because packet acknowledging consumes extra bandwidth and intensifies channel contention. On the other hand, data collection in many sensor network applications are usually loss tolerant[4][5]. For example, in fire monitoring, it is unnecessary to collect all the measured temperature to the sink while the fire status can still be well understood. Instead of reliably delivering all measured data, it is more important to capture the total number of unique reports in a short period, such that the fire spread and development can be learned in a timely fashion. Under this background, the best effort data transmission mechanism[5], in which each intermediate node immediately forwards the received packet to the next hop, may be more effective to deliver bursting data than the stop-and-wait retransmission mechanism.

The best effort mechanism, however, is not energy efficient because the transmission energy is wasted when packets are lost in the middle of multi-hop paths. In this paper, we seek an optimal transport protocol design for sensor networks, which can well balance the channel throughput and energy consumption. We believe that bursting data can be efficiently collected by a transport protocol which has the following design principles: i) the generated data has the highest priority to access the wireless channel which forms a continuous upstream towards the sink; ii) the channel bandwidth consumed by control messages should be minimized; iii) the data reporting rate is properly set to fully utilize the channel throughput while not incurring intensive channel contention and congestion.

Following the principles above, we propose the sink centric transport protocol, where the major function of packet recovery and congestion control is achieved in the sink. In the sink centric transport protocol, packet loss is detected by the sink from the discontinuous packet sequence numbers, and the sequence numbers of lost packets are sent back towards sources through virtual backward channels. In order to maintain the continuous data stream forwarding from sources to the sink, the backward channel is created by appending missing sequence numbers to the packets forwarded to the sink, which can be overheard by the previous node due to the broadcast nature of wireless communication. As a result, the missing sequence numbers can be relayed backward to the source. If the missing sequence numbers hit the local buffer of intermediate nodes, the lost packets will be re-entered into the the transmission queue. Here, the lost packets are pulled back to the sink through the backward channel, which does not incur extra message transmission and has minimum impact on the forward data stream.

It is possible that the local buffer of intermediate nodes are overflowed by a high data reporting rate, such that the missing sequence numbers miss all the intermediate buffers and reach the sources. We regard this as the symptom of channel congestion and lower the data reporting rate accordingly. Here, the backward channel is reused for congestion control.

The sink centric transport protocol is a semi-design between the hop-by-hop recovery and the end-to-end retransmission. By deferring the packet loss to the sink, the resources of intermediate nodes are devoted to forward data stream. Meanwhile, energy is saved by recovering lost packets in the middle of the forwarding paths. The sink centric transport protocol tries to maximize the channel throughput of sensor networks through the following strategies.

1. Like the best effort mechanism, each intermediate node immediately forwards a received packet to the next hop. As a result, the continuous forward data stream can be maintained.
2. The retransmission control messages are hitchhiked into the forward data stream, which has minimal interference on the upstream data forwarding.
3. Because the retransmission decision is deferred to the sink, it is possible for the sink to intelligently ask for critical lost packets based on the collected data from multiple sources.
4. Moreover, because the congestion control is integrated into the packet recovery function by sharing the same messaging bytes and operation logic, the overhead of the transport protocol is minimized.

In the following text of Section II, we compare the packet reliability and event throughput and discuss how the event throughput is limited by current transport protocols. We detail our sink centric transport protocol in Section III, which is evaluate in Section IV. We summarize previous work in Section V and conclude this paper in Section VI.

2   Preliminary

Because data collection in sensor networks is often loss tolerant, the necessity of developing reliable transport protocols for upstream data has been argued in [5][4]. In the following discussion, we compare the difference between event throughput and packet reliability. After that, we illustrate that current sensor network transport protocols, which aim to optimize the packet reliability, may fail to maximize the event throughput.

2.1   Packet Reliability v.s. Event Throughput

The packet reliability R of a multi-hop forwarding path can be defined as below $R = N_r/N_s$, where N_s is the number of packets sent by the source node and N_r

is the number of packets received by the sink. The event throughput E is defined as $E = N_\tau/\tau$, where N_τ is the number of packets received by the sink within the time period τ. The importance of packet reliability R and event throughput E vary in different sensor applications. In general, packet reliability is critical in control message transmission, such as sending reprogramming packets from the sink to sensors, where any packet loss will break the reprogrammed functionality. The packet reliability R is also important for rare event monitoring, where a small packet loss may jeopardize the application fidelity. On the other hand, the event throughput E is more important in sensor applications that generate bursting data within a short period. It is critical for the sink to collect the bursting data as soon as possible, while partial data loss is tolerable.

2.2   Hop-by-Hop Recovery Based on Time Out Mechanism

In hop-by-hop recovery based on time out mechanism[3], each intermediate node waits for the acknowledgment from the next hop after sending out a packet. The retransmission is initiated when the acknowledgment has not been received within a certain period. The hop-by-hop recovery based on time out mechanism reduces the event throughput because of several reasons: i) the stop-and-wait nature interrupts the continuous forward data stream from sources to the sink; ii) the acknowledgment messages consume extra bandwidth and intensify channel contention; iii) the acknowledgment may be lost in unreliable channels, which leads to false packet loss detection and unnecessary packet retransmissions.

It is notable that the implicit acknowledgment based on overhearing cannot be directly applied in the hop-by-hop recovery based on time out mechanism. For instance, assume that node P_i sends packet d to node P_{i+1} and waits for the implicit acknowledgment by overhearing node P_{i+1} to transmit packet d. Node P_{i+1}, however, will not immediately forward packet d since it is at the tail of the queue. Consequently, we have the overall acknowledgment delay $D_k = D_t + D_q$, where D_t is the transmission delay and D_q is the queue waiting delay. Here, D_q is not a constant, which makes it difficult to set a proper waiting time for the time out mechanism to wait for acknowledgments.

2.3   Hop-by-Hop Recovery Based on Out-of-Sequence Mechanism

In hop-by-hop recovery based on out-of-sequence mechanism[6], each intermediate node monitors the sequence numbers of forwarded packets. When the sequence numbers of received packets are not in the continuous incremental order, the packet loss is detected and retransmission request containing the missing sequences is sent back to the previous hop, which resends the lost packets based on the missing sequences. The retransmission request itself may be lost due to the unreliable nature of wireless channels. In this case, the receiver will initiate another retransmission request if the lost packets have not received within a certain period. The process is repeatedly applied until the packet sequence order is restored.

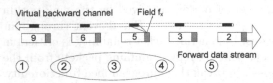

Fig. 1. Virtual backward channel is formed by utilizing the wireless broadcast nature. Node P_2 can overhear packet d_5 transmitted by node P_3 to P_4. Therefore, the missing sequence in f_x is relayed backward to P_2 by the forward data stream.

Fig. 2. Three hand circular queue consists of buffer units linked by double pointers. The grey region contains sent packets that may be recovered later.

The out-of-sequence mechanism outperforms the time out mechanism in that i) it does not involve stop-and-wait operations; and ii) it eliminates the unnecessary retransmission since lost packets are retransmitted only when requested. However, the hop-by-hop recovery based on out-of-sequence mechanism relies on the continuous in-order sequences, which limits the event throughput of communication channels. For example, when node P_k receives packet sequences $[...i-1, i+1, i+2, i+3, ...]$, it detects that packet d_i is lost based on the missing sequence i. Node P_k will hold all the subsequent packets $d_{i+1}, d_{i+2}, ...$ until packet d_i is restored from the previous hop. Here node P_k cannot send out all the subsequent packets before the packet d_i is sent out because the sequence order must be strictly maintained in the packet forwarding. Otherwise, the next hop P_{k+1} will falsely detect that packet i is lost in the transmission from node P_k to node P_{k+1}, which incurs unnecessary packet retransmission requests.

3   Sink Centric Transport Protocol

The analysis above shows that the complicated hop-by-hop recovery, based on either the time-out mechanism or the out-of-sequence mechanism, is not an optimal design to maximize the event throughput because i) the continuous forward data stream is interrupted; and ii) the acknowledgments or retransmission requests consume extra bandwidth and intensify channel contention. In order to provide high event throughput for bursting data, we propose the sink centric transport protocol to optimize the event throughput by deferring the packet loss detection to the sink, which has the following advantages:

1. It uses the out-of-sequence mechanism for packet loss detect and therefore inherits its strength which eliminates stop-and-wait operations and unnecessary retransmissions;
2. It does not need to maintain the continuous in-order sequence numbers as the hop-by-hop recovery. Because loss detection is deferred to the last hop of the sink, intermediate nodes do not have to main strict sequence order

for loss detection. As a result, the continuous forward data stream can be achieved by the sink centric approach based on out-of-sequence mechanism.

We detail the packet loss detection and notification, packet retransmission, and congestion control in the following discussion.

3.1 Packet Loss Detection and Notification

We use the out-of-sequence mechanism to detect lost packets in the sink. When a source node sends a serial of packets to the sink, it labels each packet with a sequence number, which is increased by one after each sending. As a result, the source injects a serial of packets to the communication channel, where each packet is labeled with a sequence number associated with its transmission order. All the packets are queued and forwarded by intermediate nodes. Some of the packets are dropped due to the lossy wireless channels, and the rest are delivered to the sink. As a result, the sink receives a serial of packets with discontinuous, while still in-order, sequences. Here, "in-order" means a packet with smaller sequence must be sent by the source earlier. An example is shown in Fig. 1, where transmitted packets are labeled with $9, 6, 5, 3, 2$, which are not continuous but still in-order.

Based on the discontinuous and in-order sequences, the sink can detect lost packets by recording the largest sequences l that it has received. When the sink receives a new packet with sequence number r, it computes the sequence gap between r and l as $g = r - l$. If $g = 1$, this is the normal status that packets are continuously received. We simply set the value of the largest sequence number l as r. If $g > 1$, this means packets with sequences between l and r are lost, and all the sequences $[l+1, l+2, ...r-1]$ will be entered into the circular buffer recording missing sequences. If $g < 1$, which means the received packet is recovered by the retransmission mechanism, and we clear the corresponding field in the missing sequence buffer. In the above process, the sink detects lost packets based on the sequence continuality.

The missing sequence is sent back towards the source through the *virtual backward channel*, which is constructed by overhearing the forward data stream. We illustrate how missing sequences are sent back along the backward channel in Fig. 1. When intermediate nodes forward packets from the source to the sink, the extra field f_k in the forwarded packets are allocated to the backward channel. When node P_i forwards packet d to the next hop P_{i+1}, node P_{i-1} can overhear packet d due to the broadcast nature of wireless channels. Consequently, node P_{i-1} can retrieve the information in filed f_x, which is specially reserved for the backward channel. In our design, node P_i sets the value of field f_x as the missing sequences such that the missing sequences can be overheard by previous hop P_{i-1} when packets are being transmitted from node P_i to P_{i+1}. Missing sequences can be transmitted from the sink to sources via the continuous backward relay of intermediate nodes.

One important issue of the backward channel is how to inject missing subsequences from the sink. Because the sink does not forward received packets any more, the last hop prior to the sink cannot overhear missing sequences hitchhiked to the forward data stream. To solve this problem with a uniform solution, we force the sink to forward received packets to the next virtual destination such that the last hop can work as other intermediate nodes to overhear packets transmitted by the next hop. When the sink forwards a received packet to the next virtual destination, the fields of f_x are set with missing sequences extracted from the circular buffer, which is populated by the packet loss detection process. In order to inject all the missing sequences into the backward channel, the pointer *head* is maintained for the missing sequence circular buffer. The field of f_x is set by the value pointed by *head* which is increased automatically. The missing sequences will be recycled after all sequences in the buffer have been scanned by *head*.

3.2 Packet Retransmission

When intermediate nodes receive missing sequences through the backward channel, they will check the local transmission buffer. If the lost packet can be found, the retransmission is initiated by re-entering the lost packet to the transmission queue. In order to minimize the interference of packet retransmission to the forward data stream, we implement the packet retransmission mechanism through a *three-hand double linked queue*.

As shown in Fig. 2, the three-hand double linked queue consists of buffer units connected by double links. Pointer *head* points to the buffer containing the sending packet. Pointer *tail* points to the buffer ready to receive a packet. Here, the black region between *head* and *tail* works as the normal transmission queue, which follows the FIFO rule to receive packets at tail and send packet at head. When a packet is sent out, pointer *head* is moved to the next buffer unit and the buffer of the sent packet is moved to the grey region accordingly, which is between pointer *head* and *sent*.

When a missing sequence is received by intermediate node P_i, it will scan the grey region containing sent packets. If the lost packet is found, the buffer containing the lost packet will re-enter to the black region of the transmission queue by relinking the buffer unit to the black region. Otherwise, node P_i will continue to scan the black region of the transmission queue, if the lost packet is found, which means that the lost packet has been recovered by the previous retransmission request, node P_i will not start the buffer relink. For both cases, node P_i will drop the missing sequence and not relay it backward to the source any more. This is achieved by setting the field of f_x to zero in the backward channel. If the missing sequence can not be matched in both the black region and the gray region, the sequence number will be continuously relayed backward to the source through the backward channel.

The process above illustrates how lost packets are recovered in the sink centric approach. After a packet is sent out and leaves the transmission region, it enters the temporal buffer of the grey region instead of being discarded. If a missing sequence matches the packet temporally kept in the grey region, the packet will be re-linked into the black region of the transmission queue.

It is notable that the operation of missing sequence matching and the packet forwarding can be finished concurrently. Similar to the best effort mechanism, each intermediate node receives packets into the tail of the transmission queue and sends out packets at the head immediately when the communication channel is available. Consequently, the continuous forward data stream is maintained, which maximizes the utilization of the communication channel throughput. Here, the packet retransmission is achieved by simply re-linking the lost packets back to the transmission queue and therefore minimizes the interference to the forward data stream.

4   Performance Evaluation

We implement the sink centric transport protocol on TinyOS and compare the proposed protocol with the best effort approach and the hop-by-hop recovery approach based on time out mechanism in TOSSIM [7]. TOSSIM is a bit-level simulator that accurately simulates radio communications in sensor networks. TOSSIM re-implements the hardware platform of the TinyOS, such that the same code that runs on sensor network hardware can directly run in TOSSIM. TOSSIM models network channels with a directed graph. Each edge (u, v) is described by its bit error rate e_{uv}. When a bit is sent from node u to node v, it may be flipped with the probability of e_{uv}. The asymmetric links of wireless channels can be modeled with different e_{uv} and e_{vu} ($e_{uv} \neq e_{vu}$).

Based on the bit error model, TOSSIM simulates the TinyOS network stack with high fidelity, which includes the CSMA media access protocol and packet error detection/correction with a full packet CRC. Before sender u transmits a packet, it enters the CSMA mode to listen for an idle channel. After the CSMA model, the sender transmits a serial of start symbols to synchronize the receiver, which is followed by the transmission packet. During the transmission, each bit has the error probability to be flipped between zero and one. The incorrect bytes in a packet may be detected and corrected by the CRC code appended to the end of the packet. If the packet cannot be recovered, it will be discarded by the receiver. Because TOSSIM simulates the radio communication in sensor networks in high fidelity, where channel contention and packet corruption are accurately described, it provide a good basis for us to compare the sink centric approach, the best effort approach, and the hop-by-hop recovery approach based on time out mechanism.

In our evaluation, we use TYTHON[8], a scripting environment to interact and observe the simulated sensor nodes. In our simulation, a group of sensors $P_1, P_2, ..., P_n$ are arranged in a line. We use TYTHON to inject radio packets to the sensor network, which are received by the first node P_1, and thereafter forwarded by node $P_2, P_3...$ in sequence. When packets reach the destination, they will be transmitted back to the TYTHON script through the serial port. Here, we implement the sink function in the combination of the destination and the TYTHON script. When a packet is received from the serial port, its sequence number r will be checked against the largest sequence number l ever received. If r is larger than l but not increased by one, all the missing sequences will be entered into the circular buffer, which will be sent back towards the source along the backward channel.

In our simulation, we vary the packet inject rate at the source to test the event throughput of different transport protocols. The comparison is also conducted under different channel qualities. We achieve this by varying the interval between forwarding sensors. The bit error rates are assigned to the wireless channels between pairwise sensors according to their distances, which is based on empirical data from the real experiment. The number of received packets received by the sink are recorded by the TYTHON script. By monitoring the internal variables of each sensor node, the total number of packet transmissions (including retransmissions) can be read and logged by the TYTHON scripts. Based on all the observed data, we evaluate the three transport protocols in terms of transmission quality and cost. The evaluation metrics are defined as below.

- **Event throughput:** The number of unique packets received by the sink per unit time.
- **Packet reliability:** The number of unique packets received by the sink divided by the number packets sent by the source.
- **Transmission cost:** The number of transmitted (including retransmitted) bits per packet which is successfully delivered to the sink.
- **Inject rate:** The number of packets sent by the source per unit time.

In the following simulation, we compare the three transport protocols under different conditions by varying the packet inject rate, the interval between forwarding sensors, and the hop distance between the source and the sink.

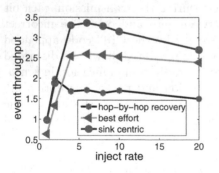
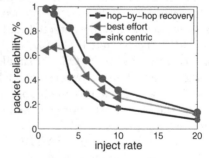

Fig. 3. Event throughput comparison under inject rate

Fig. 4. Packet reliability comparison under different inject rate

4.1   Impact of Packet Inject Rate

We first evaluate the maximum event throughput that a transport protocol can achieve under a certain channel quality. In this evaluation, we use the TYTHON script to inject packets, which are forwarded by 6 sensor nodes before reaching the destination. We increase the packet inject rate from 2 packet/second to 20 packets/second. The event throughput measured at the sink is shown in Fig. 3, which illustrates that the event throughput can be increased by increasing packet inject rate. However, after a certain critical point, the event throughput is decreased when packet inject rate is further increased. This demonstrates that a maximum event throughput exists for each transport protocol. When the packet inject rate exceeds the maximum event throughput, the actual event throughput is reduced due to buffer overflow and channel contention.

We also observe that the maximum of event throughput of the best effort approach is higher than that of the hop-by-hop recovery approach. This happens because the timer-based hop-by-hop recovery approach uses the stop-and-wait mechanism to retransmit lost packets, which interrupts the continuous forward data stream and therefore lowers the maximum event throughput a channel can offer. Meanwhile, the extra acknowledge messages intensify channel contention, which further reduce the maximum event throughput.

Fig. 3 also shows that among all the three approaches, the sink centric transport protocol has the highest maximum event throughput. This is because the sink centric approach maintains the continuous forward data stream and recovers lost packets with minimum overhead. Since no extra acknowledge messages are incurred to compete for channels, and lost packets are retransmitted by simple buffer relinking, the interference of recovery process on the forward data stream is minimized. Consequently, extra event throughput can be obtained by the light weight recovery mechanism of the sink centric approach.

We further compare the packet reliability under different packet inject rates. Fig. 4 shows that the packet reliability decreases with the increased packet inject rates. This is because the increased packet inject rate intensifies channel contention which leads to packet dropping, and therefore the number of successfully delivered packets is reduced. Fig. 4 also shows that at the low packet inject rate, the hop-by-hop recovery approach outperforms the other two approaches with the highest packet reliability. However, the packet reliability of the hop-by-hop recovery approach drops faster when the packet inject rate is increased, and the sink centric approach has the best performance under high packet inject rates. This illustrates again that channel contention and interference have severe impact on data transmission when wireless channels are heavily loaded. Reducing the channel contention not only improves the event throughput but also improves the packet reliability.

Fig. 5 shows the transmission cost comparison under different packet inject rates. The transmission costs increase with the increased packet inject rates. This happens because more packets are dropped in the middle of the forwarding path when the packet inject rates are increased. As a result, the average number of transmissions per successful packet is increased. Fig. 5 also shows that the hop-by-hop recovery incurs higher packet transmission cost than the best effort approach and the sink centric approach. The extra communication costs

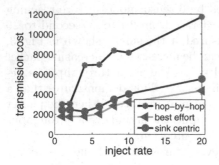
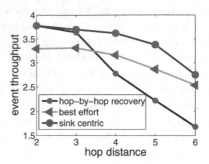

Fig. 5. Transmission cost comparison under different inject rate

Fig. 6. Event throughput under different hop distances

of the hop-by-hop approach are incurred by the acknowledge messages and their intensified channel contention.

Based on the comparison above, we conclude that if we aim to improve the event throughput, instead of the high reliable data transmission at low transfer speed, the light weight sink centric transport protocol is preferable over the hop-by-hop recovery approach based on time-out mechanism. The former can provide higher maximum event throughput with lower transmission cost.

4.2   Impact of the Hop Distances

We discuss the impact of the hop distances between the source and the sink on the event throughput and communication costs. In the evaluation we fix the packet inject rate at 4 packets/second. Along the forwarding path with uniformly distributed nodes $P_1, P_2, ...P_n$, we select different destination P_d ($3 \leq d \leq 7$) for each test. As a result, the number of forwarding hops varies from 2 to 6 and the interval distance between neighbors in each test is the same.

Fig. 6 shows that the event throughput of all the three approaches decreases with the increased lengths of the forwarding paths, which is resulted from two reasons: i) packets are dropped with higher probability when they have to be forwarded through more unreliable links; ii) channel contention and interference are intensified when more forwarding hops are involved in the packet transmission. Fig. 6 also shows that both the sink centric approach and the hop-by-hop recovery approach have high event throughput when the forwarding path is short. However, the event throughput of the hop-by-hop recovery approach decreases fast with the increased number of forwarding hops. This happens because of two reasons: i) channel contention incurred by the acknowledgments becomes more intense when more intermediate nodes are involved in the packet forwarding; ii) lost acknowledgments incur unnecessary retransmissions that further intensify the channel contention and interference.

5   Related Work

Packet delivery performance in sensor networks has been intensively studied in [2][1], which shows that packet loss is common when they are transmitted in the unreliable links between pairwise nodes equipped with low-power radio transceivers. To successfully deliver packets over lossy and collisional wireless channels, a suit of protocols have been developed to properly allocate communication channels and recovery lost packets.

To allocate the shared wireless channels among neighboring nodes, a variety of media access protocols have been proposed[9] [10][11][12][13], which aim to solve the channel contention between neighboring nodes. However, media access protocols neither reduce packet loss due to buffer overflow incurred by channel congestion, nor recover lost packets corrupted by the channel interference. Consequently, it is necessary to develop the transport protocol in sensor networks for reliable data transmission, which is achieved by congestion control and packet recovery.

Channel congestion can easily happen in sensor networks due to their unsymmetrical communication model, where a large volume of data packets are converged from multiple sources to the sink. This makes the wireless channels close to the sink become heavily loaded. To mitigate channel congestion in sensor networks, several transport protocols focusing on the congestion control have been proposed[14][15][16][17][18]. Different from these approaches, our proposed sink centric approach focuses on efficiently recovering lost packets with minimal interference on the forward data stream. In the sink centric approach, the congestion control message and logic are integrated into the packet recovery process and therefore the overhead incurred by the control messages of transport protocols is minimized.

To reliably transmit packets over lossy wireless channels, a group of transport protocols have been proposed to recover lost packets based on the retransmission mechanism. Among the proposed approaches, the RMST[3] and RBC[19] approaches use time out mechanism to detect lost packets. As we discussed before, the stop-and-wait nature of the time out mechanism reduces channel throughput. On the other hand, the out-of-sequence mechanism is used by the PSFQ[4], GARUDA[20], and the lazy loss detection [6] to detect lost packets in a hop-by-hop fashion. The out-of-sequence mechanism uses the strict continuous in-order sequence numbers to detect lost packets, which may incur packet delay and reduce channel throughput. This happens because packets cannot be forwarded by an intermediate node if any packets with lower sequence numbers have not been recovered. Otherwise, unnecessary packets retransmission requests will be falsely invoked. The above recovery approaches ensure high data transmission reliability through complex mechanisms to detect and retransmit lost packets, which is critical to applications where packet loss is intolerable. For example, the PSFQ and GARUDA approaches are specially designed for download stream transmission, such as delivering reprogramming packets, from the sink to sensors. Different from the approaches above, the sink centric approach aims to improve event throughput instead of packet reliability, which maximizes channel throughput by minimizing interference from packet recovery process.

The sink centric protocol proposed in this paper is designed for upstream data forwarding from sensors to the sink, which are loss tolerant[5][4]. To improve the event reliability instead of the packet reliability, ESRT[5] has been proposed to dynamically adjust event report rate in the source. If the number of packets received by the sink within a time period is less than the threshold, the source will aggressively increase the event report rate such that more packets can arrive in the sink. ESRT uses the best effort mechanism to transmit packets and uses sink broadcast to adjust source report rate. However, low data arrival rate at the sink incurred by channel congestion may be falsely judged by the ESRT as by packet loss. As a result, ESRT will increase the source data reporting rate, which further intensify channel congestion. the sink centric approach improves the ESRT approach in that i) it is more energy efficient since lost packets are recovered from the middle of forwarding paths; ii) control messages are sent through virtual backward channel instead of broadcast; iii) it can distinguish the reasons of low data arrival rate at the sink between channel congestion and packet loss by integrating congestion detection into the lost packet recovery process.

6   Conclusion

We propose the sink centric transport protocol to improve event throughput in sensor networks. The sink centric transport protocol is specially designed for applications where bursting data is generated within a short time. Aiming at improving the event throughput instead of reliably transmitting each individual packets, the sink centric approach optimizes the sensor network transport protocol as follows: i) it maintains the continuous upstream data forwarding by minimizing overhead incurred by control messages; ii) the virtual backward channel is created by utilizing the broadcast nature of wireless channels; iii) the packet recovery and congestion control are integrated into the same logic, which further minimizes the overhead incurred by control messages. We prove the effectiveness of the sink centric protocol through both theoretical analysis and experimental comparison conducted in a TOSSIM environment, which shows that the sink centric transport protocol outperforms the hop-by-hop recovery and the best effort approach when events are reported with high frequency through a dense sensor network.

Acknowledgment. This work was supported in part by the US National Science Foundation under grants CCF-0514078, CNS-0549006, and CNS 0551464.

References

1. Woo, A., Tong, T., Culler, D.: Taming the underlying challenges of reliable multi-hop routing in sensor networks. In: SenSys (2003)
2. Zhao, J., Govindan, R.: Understanding packet delivery performance in dense wireless sensor networks. In: SenSys (2003)

3. Stann, F., Heidemann, J.: RMST: Reliable data transport in sensor networks. In: SNPA (2003)
4. Wan, C.-Y., Campbell, A.T., Krishnamurthy, L.: PSFQ: a reliable transport protocol for wireless sensor networks. In: WSNA (2002)
5. Sankarasubramaniam, Y., Akan, O.B., Akyildiz, I.F.: ESRT: event-to-sink reliable transport in wireless sensor networks. In: MobiHoc (2003)
6. Cao, Q., He, T., Fang, L., Abdelzaher, T., Stankovic, J., Son, S.: Efficiency centric communication model for wireless sensor networks. In: INFOCOM (2006)
7. Levis, P., Lee, N., Culler, M.W.D.: Tossim: Accurate and scalable simulation of entire tinyos applications. In: SenSys (2003)
8. Demmer, M., Levis, P.: Tython: Scripting tossim (2004) in http://www.tinyos.net/tinyos-1.x/doc/tython/manual.html
9. Woo, A., Culler, D.E.: A transmission control scheme for media access in sensor networks. In: MobiCom, pp. 221–235 (2001)
10. Ye, W., Heidemann, J., Estrin, D.: An energy-efficient mac protocol for wireless sensor networks. In: INFOCOM (2002)
11. van Dam, T., Langendoen, K.: An adaptive energy-efficient mac protocol for wireless sensor networks. In: SenSys (2003)
12. Polastre, J., Hill, J., Culler, D.: Versatile low power media access for wireless sensor networks. In: SenSys (2004)
13. Rhee, I., Warrier, A., Aia, M., Min, J.: Z-MAC: a hybrid mac for wireless sensor networks. In: SenSys (2005)
14. Hull, B., Jamieson, K., Balakrishnan, H.: Mitigating congestion in wireless sensor networks. In: SenSys (2004)
15. Wan, C.-Y., Eisenman, S.B., Campbell, A.T.: CODA: congestion detection and avoidance in sensor networks. In: SenSys (2003)
16. Wan, C.-Y., Eisenman, S.B., Campbell, A.T., Crowcroft, J.: Siphon: overload traffic management using multi-radio virtual sinks in sensor networks. In: SenSys (2005)
17. Ee, C.T., Bajcsy, R.: Congestion control and fairness for many-to-one routing in sensor networks. In: SenSys (2004)
18. Wang, C., Sohraby, K., Lawrence, V., Li, B., Hu, Y.: Priority-based congestion control in wireless sensor networks. In: Sensor Networks, Ubiquitous, and Trustworthy Computing (2006)
19. Zhang, H., Arora, A., ri Choi, Y., Gouda, M. G.: Reliable bursty convergecast in wireless sensor networks. In: MobiHoc (2005)
20. Park, S.-J., Vedantham, R., Sivakumar, R., Akyildiz, I.F.: A scalable approach for reliable downstream data delivery in wireless sensor networks. In: MobiHoc (2004)

Dwarf: Delay-aWAre Robust Forwarding for Energy-Constrained Wireless Sensor Networks

Mario Strasser[1], Andreas Meier[1], Koen Langendoen[2], and Philipp Blum[3]

[1] ETH Zurich, Switzerland (equally contributing authors)
[2] Delft University of Technology, The Netherlands
[3] Siemens Building Technologies, Switzerland

Abstract. With the field of wireless sensor networks rapidly maturing, the focus shifts from "easy" deployments, like remote monitoring, to more difficult domains where applications impose strict, real-time constraints on performance. One such class of applications is safety critical systems, like fire and burglar alarms, where events detected by sensor nodes have to be reported reliably and timely to a sink node. A complicating factor is that systems must operate for years without manual intervention, which puts very strong demands on the energy efficiency of protocols running on current sensor-node platforms.

Since we are not aware of a solution that meets all requirements of safety-critical systems, i.e. provides reliable data delivery *and* low latency *and* low energy consumption, we present Dwarf, an energy-efficient, robust and dependable forwarding algorithm. The core idea is to use unicast-based partial flooding along with a delay-aware node selection strategy. Our analysis and extensive simulations of real-world scenarios show that Dwarf tolerates large fractions of link and node failures, yet is energy efficient enough to allow for an operational lifetime of several years.

1 Introduction

The state of the art in Wireless Sensor Networks (WSN) is rapidly changing. From the Smart Dust vision in 1999 [4], through the early Great Duck Island experiment with first-generation hardware in 2002 [9], to a host of (pilot) deployments in operation today [1,11,17]. Although the application domains vary, these deployments typically fall into the class of remote monitoring, where rather soft constraints on performance (e.g., latency, throughput, and lifetime) allow for straightforward engineering solutions. The experience gained with these pilots is being incorporated into a second-generation software that is better tuned, more robust, and offers the potential for enlarging the scope to more demanding applications.

Using WSN technology for implementing safety-critical applications such as fire and burglar alarm systems is a big challenge because of the real-time constraints imposed by their users. Typically, alarms detected by sensor nodes have to be reported reliably and within a few seconds to at least one sink node, even in

J. Aspnes et al. (Eds.): DCOSS 2007, LNCS 4549, pp. 64–81, 2007.

case that some of the nodes and communication links fail. Additionally, safety-critical applications are required to observe the status of the network and report node failures within a specified time. A complicating factor is that maintenance costs have to be very low to make an application economically feasible. This requires energy-efficient operation of the sensor network, because batteries should not be replaced more often than once every two to three years.

With current generation sensor node hardware built out of COTS components, the radio consumes the most power, and the above lifetime requirement translates into a duty cycle of well below 1%. This, in turn, limits the data rate to about 1 message per second and leaves little room for re-transmissions in a multi-hop scenario. Multi-path routing proposed for ad-hoc networks is an alternative way of handling link and node failures, but solutions either compromise on latency or rely on broadcast for efficiency. A broadcast-based approach, however, is consuming way too much energy as it implies that nodes listen to *all* neighboring traffic including regular status-update messages for monitoring system integrity. Furthermore, broadcast causes channel contention problems and introduces synchronization overhead. The need to avoid broadcast is also a reason that standard data gathering algorithms for WSN networks, like TAG [8] and Synopsis Diffusion [12], cannot be used.

To jointly address the three fundamental requirements associated with safety-critical applications (reliable data delivery, low latency, low energy consumption) we advocate an integrated approach that cuts across the individual MAC, routing, and transport layers, to arrive at a working solution for commodity sensor nodes in use today. To this end, we present Dwarf, a Delay-aWAre Robust Forwarding algorithm that is based on the following observations and assumptions:

a) One of the most robust, yet simple forwarding algorithms is flooding because it ensures that a message will eventually reach its destination as long as the network remains connected.
b) Traditional flooding is very expensive (with regard to energy consumption and transfer costs) and does not consider the message delivery time at all.
c) Nodes duty-cycle their radio to increase network lifetime and spend most of their time in sleep mode. Also, to minimize overheads and reduce protocol complexity, nodes do not synchronize globally (as in TDMA-based systems [5,7]) and wake up independently of each other.
d) The node-to-sink notification time is determined by the, relatively long, sleep periods of the destination nodes along the path. What is more, the transfer time of a message is much smaller than the time between two wake-ups (10 ms vs. 1000 ms in our alarm-system scenario, see Section 3).

The fundamental idea of Dwarf is to perform a unicast-based partial flooding towards the sink in combination with a (greedy) delay-aware node selection strategy to overcome the drawbacks mentioned above. More precisely, the number of neighbors k to which an alarm is forwarded determines the degree of introduced redundancy, thus making the algorithm more robust at the expense of an increase in the number of messages and the associated complexity in handling peak loads (e.g., collisions). The selection of the destination nodes according to their wake-up time and relative position aims at reducing the overall alarm notification time. That is, neighbors that wake up first and are closer to a sink are favored over nodes that wake up later or are not on the shortest path towards the sink. In order to maintain system integrity, status messages are exchanged between neighboring nodes on a regular basis. This enables to detect (temporary) link failures as well as (permanent) node failures, which must be reported to the sink so operators can take appropriate action (e.g., replace batteries). The status messages are also used to account for clock drifts in individual nodes and keep an up-to-date view on neighboring nodes' wake-up times.

Summarized, the main contributions of this paper are threefold:

1. It presents a novel, integrated algorithm (Dwarf) for the robust and timely delivery of alarm messages at the sink node in a energy-constrained multi-hop sensor network.
2. It provides a theoretical analysis of fundamental performance guarantees that Dwarf can achieve.
3. It shows through a set of detailed simulations that the Dwarf algorithm meets the requirements of an alarm system taken in a real-world scenario.

The remainder of this paper is organized as follows. Section 2 discusses related work, followed by a list of requirements and assumptions in Section 3, which set the boundary for the actual Dwarf algorithm presented in Section 4. The evaluation in Section 5 includes a formal analysis as well as an extensive set of simulations. Finally, Section 6 concludes the paper.

2   Related Work

The need for energy-efficient operation is at the core of WSN research and has received considerable attention. At the MAC layer, the excess channel capacity can be exploited by duty-cycling the radio; at the routing layer, the redundancy in the number of nodes can be exploited by rotating on/off duties. The latter approach, however, is not an option in an alarm system where all nodes are essential. WSN-specific MAC protocols generally save energy at the expense of an increase in (multi-hop) latency, but differ in their exact trade-off [6]. The class of Low-Power Listening protocols, such as B-MAC [13] or more sophisticated approaches like SCP-MAC [21] and WiseMAC [2] fit best because of the low power consumption doing idle listening and explicit control of the length of the interval between wake-ups. SCP-MAC is based on a global synchronization of the wake-up slots and hence introduces overhearing of the regular status-update messages. WiseMAC on the other hand is based on asynchronous wake-up slots, naturally minimizing message overhearing. Dwarf will be working with a slightly enhanced version of WiseMAC that is discussed in detail in Section 3.2.

Another corner stone of WSN research is the need to handle errors in the wireless channel (short-term packet loss, long-term link failures) and the possibilities of node failures. Surprisingly little research has been done on providing reliable end-to-end message delivery. The transport protocols that do address link failures (e.g., ESRT [15], RMST [16], and PSFQ [18]) include techniques like retransmissions and path diversity to overcome the errors on individual links, but do not provide end-to-end reliability because of the high costs and long (multi-hop) latencies involved. At best, advanced protocols like MMSPEED [3] provide probabilistic bounds on end-to-end delivery ratios[1]. Therefore, an application should always be prepared to tolerate (residual) packet loss.

An effective, but expensive, approach to handle communication errors is to use flooding. Quite often network-wide redundancy is not needed and partial flooding suffices. For example, the GRAB protocol [20] uses a credit mechanism to specify how many additional hops may be made to reach the destination, effectively creating a "wide-path". GRAB requires the set-up of a gradient field towards the destination, hence, is only applicable to a few, popular destinations like the sink(s) in an alarm system. The DFRF framework [10] generalizes this idea and allows for easy creation of tailor-made partial-flooding protocols. Unfortunately, DFRF does not integrate well with the MAC layer below making it difficult to control latency and energy consumption.

An issue specific to safety-critical systems is the importance of detecting failed nodes, which compromise the integrity of the system. A straightforward solution is to make use of heart-beat style failure detectors where nodes periodically send out a message notifying neighbors of their status. Wang and Kuo extend this idea to a two-phase gossiping protocol suited for ad-hoc networks [19]. Although very robust, information propagates slowly and at high cost. Recent work by Rost and Balakrishnan shows that more-advanced failure detectors help in reducing the message overhead [14], but latency remains an issue.

The lack of an integrated protocol that provides fast and robust delivery of alarm messages in a multi-hop WSN, combined with a low-overhead failure detector prompted us to design Dwarf.

3   Requirements and Assumptions

The design of the Dwarf forwarding and failure detection algorithms were driven, on the one hand, by the requirements from the safety-critical application that it should support and, on the other hand, by the functionality and configurability that the MAC layer below provides. As always in a design process, the boundary conditions were not crystal clear and subject to change. Hence, we had to make some assumptions, which are also detailed in this section.

3.1   Alarm-System Scenario

The concrete application scenario for which Dwarf was designed is a distributed indoor wireless alarm system. Each sensor node consists of a micro controller (ATMEL ATMega128), a communication unit (CC1000 transceiver), a power supply (2 AA batteries) and a sensor for detecting a specific alarm condition.

---

[1] MMSPEED in fact provides QoS guarantees on reliability and latency, but ignores energy consumption, rendering it unsuitable for our purposes.

All nodes are manually deployed at fixed locations in a building as with ordinary, wired sensors. In addition, there is at least one mains-powered sink node that is connected to a central control station. Domain specific regulations require that an alarm raised by a sensor is reported at the control station (sink) **within 10 seconds**, which leaves little room for per-hop delays in typical office buildings with long corridors and one control station per floor.

A second domain specific requirement is that failing nodes must be reported **within 5 minutes** at the control station. Since link errors caused by environmental interference are much more likely to occur than a node running out of energy or failing for some other reason, we assume that unreachable nodes, although technically still alive, must also be reported. This requires a periodic status observation of the nodes, which must send out at least one message per 5 minutes. The control station needs to be positively informed about the aliveness of each node, necessitating a collective, multi-hop forwarding scheme.

A failed node must be replaced, which is a costly operation due to the need of calling in a qualified technician asserting the integrity of the complete alarm system. Thus, it makes sense to replace the batteries of all nodes as soon as the first one runs out of energy. However, to reduce operational costs, such a grand replacement procedure should not occur more often than every two to three years. This consideration requires Dwarf to minimize and to equalize the power consumption for all sensor nodes.

3.2   MAC Protocol

A first requirement on the MAC protocol is that the effective duty-cycle must be well below 1 %, which follows from the minimum lifetime (2 years), the power consumption of the target radio in use (15 mA) and the battery capacity (2800 mAh). A further constraint follows from the maximum end-to-end latency of 10 s and the assumption that topologies with a depth of at least 5 hops must be supported, together bounding the maximum wake-up interval (T_w) to at most 2 s (careful staggering wake-up periods a la DMAC [7] would allow for even longer intervals, but runs the risk of excessive delays in the case of link errors).

A 1 % duty cycle and a 2 s wake-up interval allow for an active period of about 20 ms, which is enough to perform a carrier sense operation taking approximately 2.5 ms on a CC1000 radio [13]. Recall that Dwarf is based on partial flooding to overcome link and node failures, and needs to contact several nodes per hop. To avoid accumulating delays in doing so, it is essential that a MAC protocol provides an interface which allows for querying the wake-up schedule of neighboring nodes. Note further that, although most MAC protocols do not provide such functionality, only minor modifications are required to enhance them.

To allow for easy deployment and a high resilience to errors, a MAC protocol that requires no time synchronization between nodes is strongly preferred. This limits the choice to the class of low-power listening protocols, in which a sender pretends each message with a preamble that is slightly longer than T_w to ensure that the intended receiver will sense a busy channel and listen in on the complete transmission. These long preambles increase latency, but a sophisticated protocol like WiseMAC, which learns the wake-up schedules of its neighbors, can still use ordinary (short) preambles in most cases; the exact length of the preamble depends on the clock drift and the time passed since the last message was exchanged with the intended receiver. Thus, in order to use WiseMAC effectively, Dwarf should ensure that neighboring nodes are periodically contacted.

3.3   Definitions

Throughout this paper, we represent the sensor network by the graph $G := (V, E)$ consisting of the set of sink nodes $S \subset V$, the set of sensor nodes $V \setminus S$, and the set of edges E. All communication links are considered to be bidirectional and two nodes $u, v \in V$ can directly communicate with each other (i.e., are neighbors) if and only if $\{u, v\} \in E$. Furthermore, all sensor nodes are organized in rings according to their distance to the nearest sink; nodes with the same distance are said to be in the same ring:

Definition 1. *Let $d(u, v)$ be the distance (i.e., the length of the shortest path) between two nodes u and v. A node u is said to be in the i-th ring, or alternatively to be on level $l(u) = i$ with respect to the set of sink nodes S if and only if $\min\{d(u, s) : s \in S\} = i$. The set $R_i := \{u : u \in V \land l(u) = i\}$ contains all nodes on level i and the maximal level is denoted by $L := \max\{l(u) : u \in V\}$.*

Based on Definition 1, the neighbors of a node are divided into parents, peers, and children (see Figure 1(a)):

Definition 2. *We denote the set $N_u^- := \{v : \{u, v\} \in E \land l(v) = l(u) - 1\}$ as the parents of a node u, the set $N_u^0 := \{v : \{u, v\} \in E \land l(v) = l(u)\}$ as its peers, and the set $N_u^+ := \{v : \{u, v\} \in E \land l(v) = l(u) + 1\}$ as its children, respectively.*

As already mentioned, we assume that all nodes but the sinks sleep most of the time in order to save energy, and only wake up periodically. The wake-up period of node u is denoted by $T_w(u)$, its wake-up times by $\tau_{u,i}$.

Definition 3. *Let $\tau_{u,i}$, $u \in V \setminus S$ and $i \in \{0, 1, 2, \ldots\}$ be the wake-up times of node u, such that $\tau_{u,i+1} = \tau_{u,i} + T_w(u)$ and $0 < T_w(u) < \infty$. The duration until the upcoming wake-up time relative to the current time t is denoted by $\tau(u, t) := \min\{\tau_{u,i} : \tau_{u,i} > t\} - t$. Sink nodes are assumed to be always listening hence, $\forall s \in S : \tau(s, u) := 0$.*

Finally, we set $T_w := \max\{T_w(u) : u \in V \setminus S\}$, assume that there exists a constant upper bound $t_m \in O(1)$ on the transmission time of any message m, and use k to denote the (constant) upper bound on the number of neighbors to which a message is forwarded.

4   Algorithms

In this section, we present the proposed Dwarf algorithm which consists of two main tasks: alarm forwarding and node or link status observation.

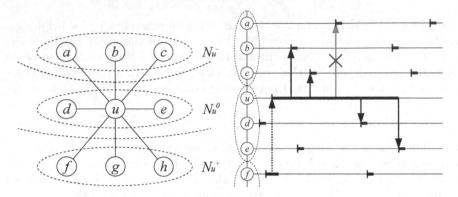

Fig. 1. On the left, the local neighborhood of node u with parents $N_u^- := \{a, b, c\}$, peers $N_u^0 := \{d, e\}$ and children $N_u^+ := \{f, g, h\}$ is depicted. A corresponding forwarding example with $k = 4$ and a failing transmission to node a is shown on the right.

4.1 Alarm Forwarding

When a node detects an alarm, it creates an appropriate alarm message m and calls the function *forwardAlarm()* which, in turn, forwards it to $\min(k, |N_u^-| + |N_u^0|)$ parents and peers (see Algorithm 1). To this end, a set of parent candidates $\widetilde{N}_{u,m}^-$ as well as a set of peer candidates $\widetilde{N}_{u,m}^0$ is maintained for each message. The actual selection of the nodes to which a message is forwarded is performed by the function *getNextHop()*. The function determines the parent candidate that wakes up next and, after removing it from the candidate set, returns it as the current destination. If there are no more parents to chose from (i.e., $\widetilde{N}_{u,m}^- = \emptyset$), the peer that wakes up next is returned instead. Once both sets are empty, they are reinitialized with $N_u^- \setminus \widehat{N}$ and $N_u^0 \setminus \widehat{N}$, respectively, with $\widehat{N}$ being the set of neighbors that successfully received the message (or forwarded it in the first place). In order to keep the induced traffic as low as possible and because there would be no additional gain otherwise, a node on level one aborts the forwarding process as soon as it has successfully delivered the message to at least one sink.

Upon reception of an alarm message m, a node first verifies that the message has not already been forwarded (i.e., $m \notin H$). New messages are appended[2] to the message history H and forwarded in the same manner as a newly generated alarm using the function *forwardAlarm()*.

Should an alarm message be dropped because of a send, receive, or transmission failure, it is retransmitted up to r_a times, resulting in maximal $k + r_a$ transmissions per message m. For each retransmission, however, a new destination is selected with *getNextHop()*, thus ensuring that retransmissions are also forwarded in the fastest way possible.

A forwarding example for the scenario depicted in Figure 1(a) and $k = 4$ is presented in Figure 1(b). After receiving an alarm message, node u forwards the

---

[2] Of course not a whole message m has to be stored but only an unique identifier such as the tuple (alarm originator, sequence number).

Algorithm 1. Alarm forwarding for node u

```
 1: var H ← ∅
 2:
 3: function INITCANDIDATES(m)
 4:     Ñ⁻_{u,m} ← N⁻_u \ N̂_m
 5:     Ñ⁰_{u,m} ← N⁰_u \ N̂_m
 6: end function
 7:
 8: function GETNEXTHOP(m)
 9:     if Ñ⁻_{u,m} = ∅ and Ñ⁰_{u,m} = ∅ then
10:         INITCANDIDATES(m)
11:     end if
12:     if Ñ⁻_{u,m} ≠ ∅ then
13:         select v ∈ Ñ⁻_{u,m} such that
14:             τ(v) = min{τ(w) : w ∈ Ñ⁻_{u,m}}
15:         Ñ⁻_{u,m} ← Ñ⁻_{u,m} \ {v}
16:         return v
17:     else if Ñ⁰_{u,m} ≠ ∅ then
18:         select v ∈ Ñ⁰_{u,m} such that
19:             τ(v) = min{τ(w) : w ∈ Ñ⁰_{u,m}}
20:         Ñ⁰_{u,m} ← Ñ⁰_{u,m} \ {v}
21:         return v
22:     else
23:         return ⊥
24:     end if
25: end function
26:
27: function FORWARDALARM(m)
28:     H ← H ∪ {m}
29:     INITCANDIDATES(m)
30:     r_m ← 0
31:     i_m ← 1
32:     v ← GETNEXTHOP(m)
33:     send alarm message m to node v
34: end function

35: upon acknowledgment of alarm message m
        sent to w
36:     if i_m < min(k, |N⁻_u| + |N⁰_u|) and w ∉ S
        then
37:         N̂_m ← N̂_m ∪ {w}
38:         i_m ← i_m + 1
39:         v ← GETNEXTHOP(m)
40:         send alarm message m to node v
41:     end if
42: end upon
43:
44: upon drop of alarm message m sent to w
45:     if r_m < r_a then
46:         r_m ← r_m + 1
47:         v ← GETNEXTHOP(m)
48:         send alarm message m to node v
49:     else if i_m < min(k, |N⁻_u| + |N⁰_u|) then
50:         i_m ← i_m + 1
51:         v ← GETNEXTHOP(m)
52:         send alarm message m to node v
53:     end if
54: end upon
55:
56: upon reception of alarm message m from v
57:     if m ∉ H then
58:         N̂_m ← {v}
59:         FORWARDALARM(m)
60:     else if v ∉ N̂_m then
61:         N̂_m ← N̂_m ∪ {v}
62:     end if
63: end upon
64:
65: upon detection of an alarm
66:     create alarm message m
67:     N̂_m ← {}
68:     FORWARDALARM(m)
69: end upon
```

message to the parents b, and c as well as to peers d and e, assuming that the transmission to parent a failed.

4.2 Node Status Observation

The purpose of the status messages is twofold: On the one hand, they are required in order to detect node or link failures, and on the other hand, they keep the mutual knowledge of neighboring nodes regarding their wake-up times up to date. Therefore, nodes send a status message to a peer as well as to a parent in a round robin fashion every interval T_s (see Algorithm 2). In the scenario depicted in Figure 1(a), for instance, node u would first send its status message to a and d, after interval T_s to b and e, then to c and d etc. Each status message contains a list of nodes which are known to be up and running; or put differently, present a node's (limited) view of the network. Whenever a node receives a status message, the included node status list X' is merged with its own list X (i.e., $X := X \,\dot\cup\, X'$); the list is cleared every interval T_s once it has been sent to a parent. Should a message to a peer (parent) be dropped, the next peer (parent) is chosen and the message retransmitted up to r_s times.

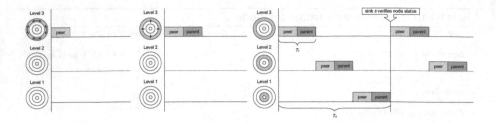

Fig. 2. Each node periodically sends a status message to first a peer and then a parent, containing a list of all running nodes it knows about. When a node receives such a status message it appends the mentioned nodes to its own status list. Nodes in the outer rings send first such that the sink will eventually receive the complete status of the network.

By having the nodes in the outer rings of the network send first (see Figure 2), node states are disseminated — and thereby gradually updated — in form of waves towards the sink which eventually receives a complete list of all running nodes every interval $T_s{}^3$. Therefore, the status-update interval T_s is divided into L sub-intervals of length $T_r := T_s/L$, which, in turn, are halved into a peer and a parent slot. Furthermore, each node in the i-th ring is associated with i-th sub-interval and sends its status message to the selected peer and parent in the corresponding peer or parent slot, respectively. By scheduling the peer slot before the parent slot, some additional reliability is introduced as a node's status list is now forwarded by one of its peers as well. The exact sending time within a slot is independently and uniform-randomly chosen by each node and, if possible, rounded to the nearest regular wake-up time. It might be worth mentioning that, due to the relatively large delay between two subsequent parent slots within a wave, a loose time synchronization (i.e., in the order of seconds) is sufficient for the proposed algorithm.

4.3 Startup

The required knowledge of a node u consists of: (i) its level $l(u)$, (ii) its one-hop neighbors and their levels (i.e., N_u^-, N_u^0, and N_u^+), and (iii) the starting time for the status waves. The information regarding level and neighbors can easily be obtained as part of an enhanced neighborhood discovery algorithm that works as follows:

1. Initially, the level of all nodes but the sinks is set to infinite and they have no information about their neighborhood.
2. The sinks initiate the algorithm by broadcasting their ID and level.
3. A node u which receives a message from a neighbor v with $l(v) < l(u) - 1$ sets its own level to $l(v) + 1$, updates the parent, peer, and children and (re)broadcasts its new level.

---

[3] If there is more than one sink, each of them receives only a partial list. However, we assume that all sink nodes are connected with each other, either directly or via a central control station, and thus can easily obtain all partial lists.

Algorithm 2. Status message exchange for node u

1: **var** next_parent $\leftarrow 0$	28: $r_m \leftarrow 0$
2: **var** next_peer $\leftarrow 0$	29: $v \leftarrow$ NEXTNODETOPROBE()
3: **var** $X \leftarrow \{u\}$	30: create status message m
4: **var** $t_u \leftarrow$ uniform randomly out of $[0, T_r)$	31: **send** status message m to node v
5: **start** peer timer for $\Delta t = t_u + (L - l(u))T_r$	32: **start** peer timer for $\Delta t = T_s$
6: **start** parent timer for $\Delta t = t_u + ((L - l(u)) +$	33: **end upon**
$\frac{1}{2})T_r$	34:
7:	35: **upon timeout** of parent timer
8: **function** NEXTNODETOPROBE	36: use_parent $\leftarrow$ true
9: **if** use_parent $=$ true **then**	37: $r_m \leftarrow 0$
10: **if** next_parent $< |N_u^-| - 1$ **then**	38: $v \leftarrow$ NEXTNODETOPROBE()
11: next_parent $\leftarrow$ next_parent $+ 1$	39: create status message m
12: **else**	40: **send** status message m to node v
13: next_parent $\leftarrow 0$	41: $X \leftarrow \{u\}$
14: **end if**	42: **start** parent timer for $\Delta t = T_s$
15: **return** $N_u^-[$next_parent$]$	43: **end upon**
16: **else**	44:
17: **if** next_peer $< |N_u^0| - 1$ **then**	45: **upon drop** of status message m
18: next_peer $\leftarrow$ next_peer $+ 1$	46: **if** $r_m < r_s$ **then**
19: **else**	47: $r_m \leftarrow r_m + 1$
20: next_peer $\leftarrow 0$	48: $v \leftarrow$ NEXTNODETOPROBE()
21: **end if**	49: **send** message m to node v
22: **return** $N_u^0[$next_peer$]$	50: **end if**
23: **end if**	51: **end upon**
24: **end function**	52:
25:	53: **upon reception** of status message m
26: **upon timeout** of peer timer	54: merge received status X' with X
27: use_parent $\leftarrow$ false	55: **end upon**

An accurate propagation of the starting time of the status waves could be achieved by flooding a relative starting time which is gradually updated by subtracting the (approximated) transfer times.

5 Evaluation

In this section, we provide an analytical evaluation of the proposed algorithms and present findings of simulations for scenarios derived from real-world data.

5.1 Analytical

In the following, we present proofs for the maximal number of link and node failures that can be tolerated, show an upper bound on the required hop count, and prove that the proposed destination selection algorithm is only a constant factor worse than the optimal solution.

Theorem 1. *The proposed alarm forwarding algorithm (Algorithm 1) can tolerate up to $t_l := \min(k, \delta) - 1$ link failures with $\delta = \min\{|N_u^-| + |N_u^0| : u \in V\}$ and k being the number of forwarding destinations.*

Proof. Let us assume that u is the first node on level $i > 0$ to receive or create an alarm which cannot be forwarded to a node on level $i - 1$. Consequently, all links to u's parents must be broken. In addition, for $\geq \min(k, |N_u^0|)$ of u's peers either the links from u to them or from them to their parents must be broken.

Per definition, each node has at least one parent. Thus, in total $\geq |N_u^-| + \min(k, |N_u^0|) \geq \min(k, \delta) > t_l$ links must be down, contradicting the assumption that there are at most t_l link failures. The threshold is tight for $k \geq \delta$ as a node u with $|N_u^-| + |N_u^0| = \delta$, which exists per definition, can be isolated from all its parents and peers if we allow $\geq \delta = t_l + 1$ link failures.

Theorem 2. *If at most t_l links fail (see Theorem 1), an alarm initiated by node u on level $l(u)$ will reach the nearest sink after at most $l(u) + \min(t_l, l(u)) \leq 2l(u)$ hops.*

Proof. In order to extend the number of required hops by one, the links to all parents of a node v must be broken. However, as a message is forwarded to $\min(k, |N_v^0|)$ peers, at least $\min(k, |N_v^0|) - (t_l - |N_v^-|) = \min(k + |N_v^-|, |N_v^0| + |N_v^-|) - t_l \geq \min(k, \delta) - t_l = 1$ of them are able to forward it to one of their parents. As a result, the number of required hops can be extended by only one for each level and requires that at least one link is down. The maximal number of additional hops is thus $\min(t_l, l(u))$.

Definition 4. *For a node u, we denote by $|N_u^Z|$ the maximal number of peers such that: (i) for each peer there exists a path to a node on level $l(u) - 1$ that is not a parent of u; (ii) on each path, all but the last node are on level $l(u)$; and (iii) all paths are mutually node-disjoint. More precisely, $|N_u^Z| = \max\{|a| \mid a \subseteq N_u^0 \wedge \forall v \in a : \exists$ path $(v, v_1, v_2, v_3, \ldots, v_m)$ such that $\forall v_i, 1 \leq i < m : v_i \in R_{l(u)}$ and $v_m \in R_{l(u)-1} \setminus N_u^- \wedge \forall v, w \in a, v \neq w : \forall i, j : v_i \neq w_j\}$*

Theorem 3. *The proposed alarm forwarding algorithm (Algorithm 1) can tolerate up to $t_p := \min(k, \gamma) - 1$ node failures with $\gamma = \min\{|N_u^-| + |N_u^Z| : u \in V\}$ and k being the number of forwarding destinations.*

Proof. Let us assume that u is the first node on level $i > 0$ to receive or create an alarm which cannot be forwarded to a node on level $i - 1$. Consequently, all parents of u must have failed. In addition, for $\geq \min(k, |N_u^Z|)$ of u's peers, they, a node on the corresponding node disjoint path, or the corresponding node in the next level must have failed. Thus, in total $\geq |N_u^-| + \min(k, |N_u^Z|) \geq \min(k, \gamma) > t_p$ nodes must be down, contradicting the assumption that there are at most t_p node failures. The threshold is tight for $k \geq \gamma$ as a node u with $|N_u^-| + |N_u^Z| = \gamma$, which exists per definition, can be isolated from all nodes in the next level if we allow $\geq \gamma = t_p + 1$ node failures.

Theorem 4. *If at most t_p nodes fail (see Theorem 3), an alarm initiated by node u on level $l(u)$ will reach the nearest sink after at most $l(u) + \beta t_p$ hops with $\beta = \max\{|N_u^+| : u \in V\}$.*

Proof. Given that at most t_p nodes fail, there exists a path $(u, u_1, u_2, u_3, \ldots, u_m)$ which connects a node u with a node u_m in the next lower level. A node v that fails has at most $|N_v^+|$ children and thus can prevent at most $|N_v^+|$ nodes on level i from forwarding an alarm to level $i - 1$. Consequently, after at least $|N_v^+|$ hops in the same level a node with a different parent is reached. As a result, each failed node v can extend the number of required hops by at most $|N_v^+| \leq \beta$ and the maximal number of additional hops is bounded by βt_p.

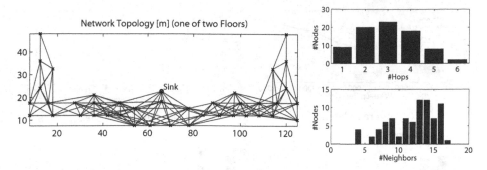

Fig. 3. 80 sensor nodes are positioned according to a real world, but wired deployment; connectivity is based on measured path-loss coefficients

Theorem 5. *The presented (greedy) destination selection algorithm selects a route that is at most $1 + \frac{T_w}{t_m} \in O(1)$ times slower than the optimal route.*

Proof. If there are no link failures, an alarm m initiated by node u on level $l(u)$ will reach the nearest sink in time $t_b = l(u)t_m$ in the best case and in time $t_w = l(u)(T_w + t_m)$ in the worst case. Thus, if the algorithm prefers parent v over w because $\tau(v, t_0) = T < \tau(w, t_0) = T + \varepsilon$ we get a worst case ratio of

$$c = \frac{T + (l(u) - 1)(T_w + t_m)}{T + \varepsilon + (l(u) - 1)t_m} \leq \frac{T_w + t_m}{t_m} = 1 + \frac{T_w}{t_m}$$

5.2   Simulation

To evaluate Dwarf under appropriate and realistic conditions, we conducted a set of measurements in an existing (wired) real-world alarm system located in a large historic public building comprising 80 sensor nodes and one central control station (sink). In particular, the complete 80x80 path-loss matrix was recorded, capturing the link quality between any pair of nodes. In addition, we enhanced GloMoSim such that it can be directed to work with this recorded data. During simulation, a signal-to-interference-plus-noise-rate (SINR) model is used to compute the bit-error rate for each transmission individually. The packet-error rates are then calculated based on the packet length not assuming any bit-error correction. Figure 3 shows the topology and some basic characteristics in the case without interference; links are shown if the bit-error rate is below 0.1%. The realistic channel model is complemented with the original GloMoSim implementation of the WiseMAC protocol, enhanced by its authors to include an API for querying the wake-up times of contacted neighbors. This functionality was needed to implement Dwarf's parent selection for rapid alarm forwarding.

In the remainder of this section, we present the results of our simulations with respect to alarm notification time, message complexity, energy consumption, and robustness against link failures.

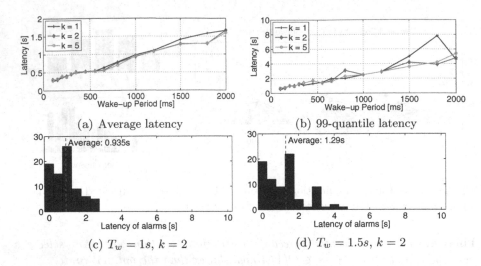

(a) Average latency

(b) 99-quantile latency

(c) $T_w = 1s$, $k = 2$

(d) $T_w = 1.5s$, $k = 2$

Fig. 4. Fire-alarm performance without link failures

Alarm Notification Time. Figure 4(b) presents the 99-quantile of the latency for different T_w and k in a first experiment without any node and link failures. Note that, since no absolute guarantees can be given in any wireless system, we report the worst case latency of all simulations except the 1% pathological cases. The first observation that can be made is that the maximum (6-hop) notification delay increases linearly with T_w due to the wake-up period dominating the actual message transfer time ($T_w \gg t_m$). In addition to the 99-quantile, the latency distribution of all alarm messages for $k = 2$ and $T_w = \{1s, 1.5s\}$ is depicted in Figures 5.2 and 5.2. This latency distribution shows that the average (3-hop) latency is just $0.94s$ for $T_w = 1s$ and $1.29s$ for $T_w = 1.5s$, which is even smaller than the nodes' wake-up time. This is a consequence of Dwarf's forwarding scheme that selects the next hop according to its level *and* wake-up time. The average hop delay is therefore much smaller than T_w. In contrast to the 99-quantile, the average latency increases less than linear with T_w. This behavior can be explained by the fact that an always listening sink has a much bigger impact on the average latency than on the 99-quantile. Since the outliers become more distinct with an increasing wake-up time, we will use $T_w = 1s$ from now on as the default setting for the remaining experiments.

Figure 4(b) also shows the impact of the parameter k. One can observe that the smaller k, the more outliers occur. This is especially severe for $k = 1$ where such a distinct outlier can be noticed at $T_w = 1.8s$. Choosing $k > 1$ prevents them to occur, since a local jam can be bypassed on another route. Even though $k = 5$ seems to be more stable in terms of small outliers, $k = 2$ already delivers the messages well within the required $10s$.

Table 1. The number of messages Dwarf injects into the network as a function of trigger level and k. For comparison, a network-wide flood generates 479 messages.

Level	k=1	k=2	k=3	k=4	k=5
1	1.00	1.00	1.00	1.00	1.00
2	2.00	4.23	7.51	11.46	19.84
3	3.01	9.25	18.30	27.19	35.81
4	4.02	15.97	26.73	35.51	40.95
5	5.03	22.69	34.56	43.20	48.40
6	6.07	30.68	42.14	49.23	53.29

Message Complexity. The parameter k has not only an impact on the latency but also on the number of propagated alarm messages. Table 1 shows the average number of totally generated messages per alarm, depending on k and the level on which the alarm was triggered. For $k = 1$ the number of messages is just slightly larger than the level, showing that messages can usually be forwarded to parents. For $k = 2$ the number of messages increases in the order of 2^{level} up to the 4th level. For $k = 5$, on the other hand, the increase in the number of messages is far below 5^{level}, mainly because messages are not sent backwards to children and due to the topology of real deployments that enforce frequent unifications of alarm messages routed on different paths. As a result, increasing k from 2 to 5 for a level-6 alarm will not even double the number of generated messages. Furthermore, an alarm is considered to be a very rare event, allowing a certain message overhead in order to ensure robust operation.

Energy Consumption. There are two main sources of energy consumption: First, the radio must be turned on regularly in order to check for a possible alarm message and second, sending and receiving status messages. The energy consumption of the former increases linearly with the wake-up frequency $1/T_w$, while the latter depends on the total number of sent messages. This number, in turn, heavily depends on the update interval T_s, which was chosen to be 150 seconds in order to ensure that node failures are reported in time ($2T_s \leq 5min$). Figure 5(a) depicts this partitioning of the maximal energy consumption for the usual idle state where no alarms are generated. It shows that the energy consumption of status messages is constant for $T_w \gtrsim 500ms$, but significantly increases for shorter wake-up times due to a more frequent overhearing of status messages. As already mentioned, the targeted duty cycle is required to be below 1%, which can be achieved with $T_w \gtrsim 500ms$. However, having a wake-up time of about $1s$ provides some additional flexibility and accounts for additional maintenance tasks as well as network initialization. Finally, Figure 5(b) shows that Dwarf provides the desired equalized energy consumption of all nodes; the maximum duty cycle is only about 25% higher than the average.

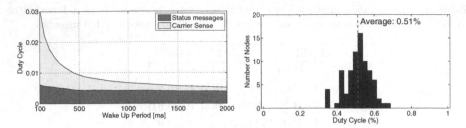

Fig. 5. Worst-case energy consumption for status message exchange and carrier sense. As shown on the right, the node's energy consumption well-balanced.

Robustness Analysis. So far we considered a benign communication environment without any failing links. In order to analyze Dwarf's performance in a harsher environment, we added random but static link failures while ensuring that the network remains connected. Furthermore, we triggered the alarm messages synchronous to the link failures, in order to avoid that Dwarf adapts to the limited communication environment, and, as discussed before, chose T_w being equal to $1s$.

Figure 6(b) presents the 99-quantile of the latency with up to 40% link failures and different k's. Setting $k = 1$ clearly does not provide a lot of robustness. This can also be seen in Figure 6(c) showing the fraction of messages that are not delivered at all. For instance, there is a single alarm that is not reported with only 7.5% link failures. This can happen if the communication of a node towards the sink is blocked and no redundant messages are sent. In contrast to the failure-less case in Figure 4, having $k = 2$ or $k = 5$ makes a difference in the alarm performance. Especially the 99-quantile shows fewer outliers with bigger k's. The unreported alarms, on the other hand, show a similar trend for different $k \geq 2$. The explanation for this is that each $k > 2$ provides enough redundancy to find a way to the sink, except when the way towards the sink is completely blocked at the alarm-triggering node (i.e., neither parents nor peers are available).

Figure 6(d) shows the distribution of the latency for 30% link failures with $T_w = 1s$ and $k = 2$. Compared to the case without link failures (cf. Figure 4), the alarms take about 50% longer to reach the sink, but still get there on time, except for one outlier. We determined that such outliers are caused by the increase in the number of status messages in response to the link failures, which effectively blocks the channel for alarm messages. There is little we can do about this because ongoing transmissions cannot be aborted.

The impact of link failures on the average latency is shown in Figure 6(a). The main observation that can be made is that $k = 2$ shows a better average performance than $k = 1$, since alarm messages are bypassed on other routes.

Not only the alarm messages are affected by link failures, but also the status messages. Their performance in combination with link failures is shown in Figure 7. By design, Dwarf does not report any false negatives that is, Dwarf never reports nodes being alive which actually have failed. In contrast, the number of false positives (alive nodes reported missing) suffers severely from the

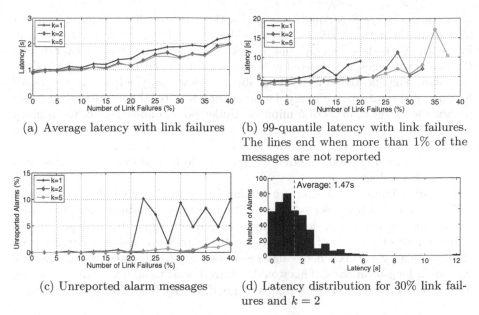

(a) Average latency with link failures

(b) 99-quantile latency with link failures. The lines end when more than 1% of the messages are not reported

(c) Unreported alarm messages

(d) Latency distribution for 30% link failures and $k = 2$

Fig. 6. Fire-alarm performance with link failures, based on 400 triggered alarm messages per sample point.

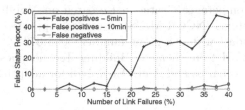

Fig. 7. Status message performance with link failures

bad communication environment as shown in Figure 7. The performance can be increased significantly, however, when the reporting time for node failures is increased and the results of several status monitoring intervals can be combined, as shown in Figure 7 for a doubled reporting time of 10min. Alternatively, T_s could be decreased, resulting in increased power consumption.

6   Conclusions

In this paper we presented Dwarf, an energy-efficient, robust and dependable forwarding algorithm for the accurate notification of alarm messages in safety-critical WSN applications. The fundamental idea of Dwarf is to perform a unicast-based partial flooding in combination with a (greedy) delay-aware node selection strategy. Our evaluation, based on a real-world scenario, shows that alarm messages are dependably reported at the sink, even if a substantial number of links in the network fail. On average alarms are delivered over multiple hops in less than a node's wake-up time T_w. For a T_w in the order of 1s, over 99% of the alarm messages are reported well within the required 10s, even if 30% of the links fail. The effective duty-cycle is always below 1%, yielding an operational lifetime of several years. Finally, Dwarf manages to reliably report failed nodes within the target interval of 5 minutes. Under poor conditions, i.e. when many links fail, alarm and status messages actually interfere showing that application scenarios should always be regarded as a whole. We firmly believe that alarm reporting, failure detection, and duty-cycling should be addressed in an integrated way, as Dwarf does.

Possible enhancements that are left for future work are to also consider the children of a weakly connected node (with a certain probability) and to change the number of forwarding destinations k (dynamically) per node, depending on either a node's local view of the network or the history of a message. Furthermore, the link-quality between nodes has been assumed to be static but is likely to vary on a long-term basis. This would require Dwarf to adapt its ring structure accordingly in order to ensure that only good-quality links are being used.

Acknowledgments

The work presented in this paper was supported by CTI grant number 8222.1 and the National Competence Center in Research on Mobile Information and Communication Systems (NCCR-MICS), a center supported by the Swiss National Science Foundation under grant number 5005-67322. We would like to thank Bernhard Imfeld from Siemens Building Technologies for providing us with the real-world scenarios and Amre El-Hoiydi for adapting WiseMAC.

References

1. Culler, D., Estrin, D., Srivastava, M. (eds.): Special issue IEEE Computer on Wireless Sensor Networks (August 2004)
2. El-Hoiydi, A., Decotignie, J.-D.: WiseMAC: An ultra low power MAC protocol for multi-hop wireless sensor networks. In: Nikoletseas, S.E., Rolim, J.D.P. (eds.) ALGOSENSORS 2004. LNCS, vol. 3121, pp. 18–31. Springer, Heidelberg (2004)
3. Felemban, E., Lee, C.-G., Ekici, E.: MMSPEED: Multipath Multi-SPEED protocol for QoS guarantee of reliability and timeliness in wireless sensor networks. IEEE Trans. on Mobile Computing 5(6), 738–754 (2006)
4. Kahn, J., Katz, R., Pister, K.: Next Century Challenges: Mobile Networking for Smart Dust. In: 5th ACM/IEEE Conf. on Mobile Computing and Networks (MobiCom '99), Seatle, WA pp. 271–278(August 1999)
5. Keshavarzian, A., Lee, H., Venkatraman, L.: Wakeup scheduling in wireless sensor networks. In: 7th ACM symposium on Mobile ad hoc networking and computing (MobiHoc) , Florence, Italy pp. 322–333 (2006)
6. Langendoen, K., Halkes, G.: Energy-efficient medium access control. In: Zurawski, R. (ed.) Embedded Systems Handbook, pp. 34.1–34.29 CRC press, Boca Raton, USA (2005)

7. Lu, G., Krishnamachari, B., Raghavendra, C.: An adaptive energy-efficient and low-latency MAC for data gathering in sensor networks. In: Algorithms for Wireless, Mobile, Ad Hoc and Sensor Networks (WMAN), Santa Fe, NM, (April 2004)
8. Madden, S., Franklin, M., Hellerstein, J., Hong, W.: TAG: a tiny aggregation service for ad-hoc sensor networks. ACM SIGOPS Operating Systems Review, vol. 36(SI) pp. 131–146 (2002)
9. Mainwaring, A., Polastre, J., Szewczyk, R., Culler, D., Anderson, J.: Wireless sensor networks for habitat monitoring. In: ACM Workshop on Wireless Sensor Networks and Application (WSNA), Atlanta, GA pp. 88–97 (September 2002)
10. Maroti, M.: Directed flood-routing framework for wireless sensor networks. In: 5th ACM/IFIP/USENIX Conf. on Middleware, pp. 99–114 (2004)
11. Marrón, P.J., Voigt, T., Rohner, C., Ahlgren, B. (eds.): 2nd ACM Workshop on Real-World Wireless Sensor Networks (REALWSN), Uppsala, Sweden (June 2006)
12. Nath, S., Gibbons, P., Seshan, S., Anderson, Z.: Synopsis diffusion for robust aggregation in sensor networks. In: 2nd ACM Conf. on Embedded Networked Sensor Systems, Baltimore, MD pp. 250–262 (November 2004)
13. Polastre, J., Hill, J., Culler, D.: Versatile low power media access for wireless sensor networks. In: 2nd ACM Conf. on Embedded Networked Sensor Systems, Baltimore, MD pp. 95–107 (November 2004)
14. Rost, S., Balakrishnan, H.: Memento: A health monitoring system for wireless sensor networks. In: IEEE SECON, Reston, VA (September 2006)
15. Sankarasubramaniam, Y., Akan, O., Akyildiz, I.: ESRT: Event-to-sink reliable transport in wireless sensor networks. In: 4th ACM Symposium on Mobile Ad Hoc Networking & Computing (MobiHoc), pp. 177–188 (June 2003)
16. Stann, F., Heidemann, J.: RMST: Reliable data transport in sensor networks. In: First IEEE Workshop on Sensor Net Protocols and Applications, Anchorage, AK, pp. 102–112 (April 2003)
17. Voigt, T., Rohner, C. (eds.): Workshop on Real-World Wireless Sensor Networks (REALWSN), Stockholm, Sweden, (June 2005)
18. Wan, C.-Y., Campbell, A., Krishnamurthy, L.: PSFQ: A reliable transport protocol for wireless sensor networks. In: ACM Workshop on Wireless Sensor Networks and Application (WSNA), Atlanta, GA pp. 1–11 (September 2002)
19. Wang, S.-C., Kuo, S.-Y.: Communication strategies for heartbeat-style failure detectors in wireless ad hoc networks. In: Conf. od Dependable Systems and Networks, San Francisco, CA, pp. 361–370 (June 2003)
20. Ye, F., Zhong, G., Lu, S., Zhang, L.: GRAdient Broadcast: A robust data delivery protocol for large scale sensor networks. Wireless Networks 11(3), 285–298 (2005)
21. Ye, W., Silva, F., Heidemann, J.: Ultra-low duty cycle mac with scheduled channel polling. In: 4th ACM Conf. on Embedded Networked Sensor Systems (SenSys 2006), Boulder, CO pp. 321–334 (November 2006)

Localization for Anchoritic Sensor Networks[*]

Yuliy Baryshnikov[1] and Jian Tan[2]

[1] Bell Laboratories, Murray Hill, 07974 NJ
[2] Department of Electrical Engineering, Columbia University
New York, NY 10027

Abstract. We introduce a class of *anchoritic* sensor networks, where communications between sensor nodes are undesirable or infeasible due to, e.g., harsh environments, energy constraints, or security considerations. Instead, we assume that the sensors buffer the measurements over the lifetime and report them directly to a sink without necessarily requiring communications. Upon retrieval of the reports, all sensor data measurements will be available to a central entity for post processing.

Our algorithm is based on the further assumption that some of the data fields that are being observed by the sensors can be modeled as a local (i.e. having decaying spatial correlations) stochastic process; if not, then choose an auxiliary field, e.g., carefully engineered random signals intentionally generated by arranged devices, "cloud shadows" cast on the ground, or animal heat. The sensor nodes record the measurements, or a function of the measurements, e.g., "1" when the measured signal is above a threshold, and "0" otherwise. These time-stamped sequences are ultimately transferred to the sink. The localization problem is then approached by analyzing the correlations between these sequences at pairs of nodes.

As for applications, we discuss the localization scheme for large-scaled sensor networks deployed on the seabed and study a two-tiered architecture that organizes deaf sensors with local masters.

1 Introduction

The "coordinate-free" localization problem in sensor networks has attracted significant attention in the literature; see, e.g., the survey [9] and the references therein. The problem is to determine positions of the nodes in the network without absolute reference information, like GPS or direction/distance information relative to some known beacons. Coordinate-free localization problem is therefore to determine the *absolute positions* of the nodes using only the *local information*, e.g., the internode distances or relative directions. This local-to-global localization problem presents a serious research challenge and the amount of work on it is rapidly growing [16,10,4,17,7,14,12,13,18,11,1]. However, in the majority of the publications, solving the localization problem assumes extensive internode communications, i.e., bidirectional exchanges of signals used to infer the pairwise

---

[*] This work is supported by DARPA SToMP grant and Bell Labs SIP program.

J. Aspnes et al. (Eds.): DCOSS 2007, LNCS 4549, pp. 82–95, 2007.
© Springer-Verlag Berlin Heidelberg 2007

distances, or globally received calibrating signals serving as a system of beacons (the "poor man's GPS").

As for localization schemes that rely on internode communications, people use one or a combination of RSS(received signal strength), ToA (time of arrival) and AoA (angle of arrival) data to reconstruct the mutual positions of the nodes, and consequently to determine their absolute positions. The beacons with known positions provide absolute reference points for the remaining sensor nodes.

Methods using centrally controlled global signals can also be applied to calibrate locations. For example, the Spotlight system [20] with the aid of steerable laser rays that sweep over the monitored terrain locate the sensors without equipping them with specialized ranging hardware. However, this approach requires precise knowledge on the trajectory of the ray.

In this paper, we study networks that are subject to severe communication constraints. Particularly, we do not allow either *internode communications* or any *centrally structured signals* for the localization. We decouple the localization problem into two steps: the first stage is to recover the internode mutual positions, and the second stage is to reconstruct the global positions of the nodes by the information obtained from the first stage.

Traditionally, the localization problem, in the absence of centrally steered signals, implicitly assumes that the relative position information is obtained from the signal exchange between the nodes, and that a subsequent processing of the exchanged signals is necessary. However, as we argue below, the requirement that the sensors are able to regularly emit signals and to process information is undesirable in some applications, due to, e.g., harsh environments, energy constraints, or security reasons. Assuming this for an instant, we ask:

> *Can the localization problem, in particular the first stage of recovering the inter node distances, be solved under the conditions that the sensors do not have the ability to exchange signals, and that no global signals are applicable for reference?*

Clearly, the sensors have to be able to gather some measurements and eventually report them to a sink/processing entity; a sensor unable to do even that much can be removed without any detrimental effect for the network operation.

We refer to sensor networks that are deprived of the ability to chat and lack centrally controlled global signals as *anchoritic* sensor networks. More precisely, chatting is a two-way process involving listen and talk, while for anchoritic sensors, either they can not listen or they can not talk, or even neither.

It is perhaps counterintuitive that the localization problem for anchoritic sensor networks can be solved. Before presenting our approach, we need, however, answer the following natural questions:

1. When and where are anchoritic sensor networks necessary?
2. If no communications are allowed, how can the measurements collected by each individual sensors be transferred to the sink/center?

These questions are addressed in the next subsection. In the last subsection of the introduction we describe related works. The rest of this paper is organized as follows: after describing our approach to the localization problem for anchoritic sensor networks and some engineering ramifications, we study several random field models in section 3. These models are candidates for recovering the inter node distances that are inspired by possible real applications, for which we conduct some simulation experiments.

1.1 Motivation

Scenarios that one has to resort to the anchoritic assumption are much more widespread than one might suppose.

First of all, when the sensor networks are immersed into harsh environments where the communications between the sensor nodes are difficult or infeasible, standard approaches that use inter node communication signals to infer the pairwise distances no longer applies. For example, consider a large-scaled sensor network deployed on the seabed. Using sophisticated techniques such as SONAR may not be a good solution because of the latency of acoustic signals, the effective data rates [19] and the size of the devices. Hence one might look into the possibility of keeping the sensors silent over the whole life.

Also, the cost of the devices for locating and communication scales with the number of sensors. To equip a single sensor only increases innocuous amount of cost, but this results in a prohibitive expense for large-scaled sensor networks. Hence, the idea of only choosing *mute or deaf* sensors that are cheap, or organizing them with some expensive and powerful sensors, might force the developers to adopt the *anchoritic* requirements.

Another situation that relates to security is in adversary environments. Transmitting signals may reveal the presence of the sensors, and therefore makes them vulnerable to suppression and manipulation. Similarly, one would not deploy any globally structured signals for localization purposes as an adversary could generate noises or worse, emulate the system signals to compromise the localization completely.

While there are further scenarios where anchortic networks could be necessary, these situations — physical constraints, cost considerations, security requirements — seem to cover most of them.

Now, if the sensors in the network are silent, how can they report the data to the sink? There are again several scenarios.

First, the sensors (or their data storage units) can be indirectly collected after their mission is completed. As for the sensor networks deployed on the seabed, buoyant sensors can be attached to heavy ballasts and sink themselves. After a period of time during which the sensors perform their measurements, the ballasts are released and the sensors emerge to the surface where their measurements can be collected, for example, by radio. Clearly, the original positions of the sensors cannot be reliably estimated just by their locations on the sea surface.

Another case involves sensors that may only be able to pass their measurements to a collection node once, perhaps after the moving collection node approaches them close enough. For example, in some adversary environments sensors should not reveal themselves except being activated. Here, the sensors are equipped with devices capable of talking. However, for the reasons stated above, the communication should be made on demand and kept short. Hence, the sensors transmit just once in their operation cycle, transferring to the center (which can be, e.g., a mobile agent that passes by) all the information they gathered. After that, the sensors either wait for another transmission cycle, or even are compromised.

In either case, the central entity has to recover the original positions of the sensors depending only on the individual measurements from each sensor, and this information is oblivious of the positions and the existence of all the other sensors in the network.

1.2 Related Work

As we mentioned above, most of the publications on the "coordinate-free" localization problem follow the path of reconstructing locations from the proximity data. These approaches typically assume the distances being given by RSS data, or by the connectivity patters of the network formed by the sensors. Then some geometric properties are used, followed possibly by iterative adjustment and fine tuning. The works following to some degree this direction are, e.g., [16,10,4,17,7,12,13,18,11,1].

Some deviations from this scheme are also considered in literature. For example, the work [2,14] assumes a lack of communications between the nodes, yet relies on several anchors which can communicate with significant parts of the network. The distances to these anchors are then used for the localization. Similarly, [15] assumes a system of beacons having known positions and sending acoustic signals used for the localization. A somewhat more complicated approach mixing internode chatter and beacons is used in [6]. The localization system [20] achieves high accuracy in recovering coordinates of the nodes without requiring internode chatter. However, some ranging signals (steerable laser rays sweeping over the terrain populated with sensors) are necessary.

In the existing literature, the one most close to our techniques appears to be the ingenious SLAT (simultaneous localization and tracking) proposal [5]. There, the authors consider a network of cameras tracking a moving object by recovering their own positions and then the trajectory of the tracked object. While conceptually not completely disjoint from our model of random walkers (see Section 3.3), the approach of [5] relies heavily on the uniqueness of the moving object and on the far range of their sensing devices. Introducing many targets seems to require a major overhaul of the approach used there, which might lead to a statistical procedure of distinguishing multiple targets, and thus to techniques close to ours.

On a conceptual level, the correlations between the measurements have been used in the sensor/ad-hoc networks, most notably to develop coding schemes. It has been proposed to use correlations in the measurements to improve the network throughput. Similarly, using correlated signals in ad-hoc wireless MIMO networks can improve the transmission rates. Here, however, we do not try to filter the noise out of the signals, but rather to use the noise (insofar it admits some decaying correlation functions) for the localization.

2 Our Approach

We approach the problem of recovering the mutual distances between the nodes in an anchoritic sensor network by exploiting the time-space correlation structure of some random field observed by the sensors. This random field can be what the sensors are tasked with measuring, or some auxiliary signals randomly generated by the sensors intentionally (no need for any coordination) .

2.1 Description

Our algorithm is based on the assumption that the data observed by the sensors can be modeled as a locally isotropic stationary random field (for a precise definition, see the discussion before Theorem 1); if not, then choose another auxiliary field that satisfies the condition, e.g., hydroacoustic noise near the seabed, cloud shadows cast on the ground, or even carefully engineered random radio signals. The sensor nodes record the measurements, or a function of the measurements. For example, simply record "1" when the measured signal is above a threshold; otherwise record "0". In the end, the Boolean sequences as well as the data measurements will be transferred to the sink. The crucial intuition behind this approach is that the correlations between these Boolean sequences at different sensors decrease with the mutual distances. Then, by analyzing the sequence correlations at pairs of nodes, we can approach the localization problem indirectly. Note that, though each sensor has to reserve an extra space for the Boolean sequence, the overhead is negligible since even 250 bytes contains 2000 bits.

More precisely, we assume that a random field $\{\xi(z,t)\}_{z\in\mathbb{R}^2, t\in\mathbb{R}}$ is measured by the sensors $\mathbf{N} = \{1, 2, \ldots, N\}$ at synchronized instants $t_o, t_1, \ldots, t_T$. The position of sensor i is denoted as $z_i = (x_i, y_i)$, and it records "1" when the measured signal $\xi(z_i, t)$ is above a threshold, otherwise record "0" (this can be generalized to multivalued records) at time t.

Definition 1. *For a subset $I = \{i_1, i_2, \ldots, i_S\} \subset \mathbf{N}$, define empirical instantaneous correlation functions to be*

$$\kappa^T(I) = \frac{1}{T} \sum_{j=1}^{T} \xi_{i_1}(t_j)\xi_{i_2}(t_j) \cdots \xi_{i_S}(t_j), \tag{1}$$

where $\xi_i(t_j)$ is the record of the field ξ by sensor i at time t_j.

Remark 1. Here the requirement for synchronization is not tight, since even when clocks drift over a long period of time, a small time lag between the measurements will not affect the spatial-temporal correlation too much.

Remark 2. In this paper, we are only interested in analyzing the correlation at pairs of nodes ($S = 2$). For $S \geq 3$, it contains more information on the relative positions of the sensors (e.g., $S = 3$ forms a triangle). We refrain from this generalization in this work.

2.2 Theoretic Framework

Assume that N independent nodes are selected uniformly in a plane region $A \subset \mathbb{R}^2$. The positions of the first $B < N$ nodes (called local *beacons*) are assumed to be known, and the positions of the rest are to be determined.

If the mutual distances $d_{ij} = |z_i - z_j|$ are known for all pairs of sensors, it is possible to reconstruct the whole configuration $\mathcal{Z} = \{z_1, \ldots, z_N\}$ up to an isometry of the plane preserving the positions of beacons (that is up to a rotation if $B = 1$ or up to an axis symmetry if $B = 2$). Obviously, the positions and the number of beacons affect the reconstruction. In an anchoritic network, the sensors do not know their mutual distances, and we resort to the measurements of (empirical) correlations/cumulants as a proxy for the internode distances to construct a proximity graph.

Empirical Correlations and Cumulants. When the field ξ is stationary and ergodic with respect to time t, the empirical correlation function, in the limit of $T \to \infty$, converges (e.g., see Theorem 9.6 of [8]) to its expected value $\kappa(I) = \mathbb{E}\xi_{i_1}\xi_{i_1} \cdots \xi_{i_s}$. In many cases, when the sensors form spatially separated clusters, the correlation function also clusters correspondingly $\kappa(I) \approx \prod_{I_i \subset I} \kappa(I_i)$, and the approximation is good when the distances between the clusters $\{I_i\}$ are large. Cumulants $\{c_l\}_{l \geq 1}$ are defined by $\sum_l \frac{c_l s^l}{l!} = \log(\sum_l \frac{\kappa_l s^l}{l!})$, where $\kappa_l = \kappa(\{1, \ldots, l\})$ are the correlations.

We focus on the pairwise cumulants, which reduce to the standard statistical correlation

$$c(i, j) = \mathbb{E}\xi_i\xi_j - \mathbb{E}\xi_i\mathbb{E}\xi_j$$

for sensor i and j. The notation $c(z_i, z_j) \equiv c_2(i, j)$ is used to indicate the exact locations of sensor i and j.

The cumulants described above capture the instantaneous spatial dependencies of the random field. One can also exploit the spatial and temporal dependencies simultaneous. A general way to do so is to extend the dimension of the random field ξ to a new field $\tilde{\xi}_\Delta(z, t) = \{\xi(z, s)\}_{s \in [t-\Delta, t+\Delta]}, \Delta > 0$ with the value at point z and time t being the trajectory of ξ at point z between $t - \Delta$ and $t + \Delta$. In this case, the 2-point correlation functions involve a kernel function $K(u, v)$ and are given by

$$\tilde{c}(i, j) = \int_{[-\Delta, \Delta]^2} \xi_i(u)K(u, v)\xi_j(v)dudv.$$

Proximity Graph. To solve the localization problem, we need to construct the *proximity graph* Γ_N that connects each node i to k_N nodes with the *largest* empirical mutual cumulants. The resulting graph Γ_N can approximate the corresponding k_N-nearest neighbor graph that is built on the Euclidean space with each node being connected to k_N nearest nodes under the following two assumptions.

A: Convergence of empirical cumulants: $c^{(T)}(z_i, z_j) \to c(z_i, z_j)$ for all $i, j \in \mathbf{N}$ as $T \to \infty$, and,

B: Asymptotic isotropy: for all $x, y \in A$, the set $S_x(\delta) \triangleq \{y : c(x,y) \geq c(x) - \delta\}$
satisfies, as $\delta \to 0$,

$$\frac{S_x(\delta)\sqrt{\pi}}{\sqrt{\mathrm{Area}(S_x(\delta))}} \to \text{a unit circle.}$$

Theorem 1. *Under the assumptions A and B, if $k_N = \log N^c, c > 1$, then, for any $\epsilon > 0$,*

$$\lim_{N \to \infty} \lim_{T \to \infty} \mathbb{P}\left[\left|h_{ij}^{(N)}\sqrt{\frac{k_N}{\pi N}} - d_{ij}\right| < \epsilon, 1 \leq i, j \leq N\right] = 1,$$

where $h_{ij}^{(N)}$ is the hop distance between node i and j in Γ_N and d_{ij} is the Euclidean distance between z_i and z_j.

The proof of this theorem is presented in Section 6. From this theorem, we see that knowing the cumulants is enough to reconstruct, with an arbitrary precision, the positions of all the nodes in the network, assuming that the network is large enough.

In the next section we study three different models that could be applied in an anchoritic sensor network.

3   Random Field Models

The sensors are assumed to be scattered uniformly in an open area. For the case when the sensors are not uniformly positioned, one may have to introduce local masters/sinks. We will discuss the ramifications in Section 4.

3.1   Boolean Model

This model imitates a random field with shadow/light patterns. To model the shadow patterns generated by "clouds", we will apply the widely used *Boolean model* (see e.g. [3]). The model is specified by a Poisson point process $\mathcal{P}$ and a class of bounded sets $\mathcal{B}$ where $B \in \mathcal{B}$ is assumed to be a circle with a random radius R. Given the pair $(\mathcal{P}, \mathcal{B})$, the random set C is

$$C = \bigcup_{Z_\alpha \in \mathcal{P}} (Z_\alpha + B_\alpha),$$

where $\{B_\alpha\}$ are *iid* realizations of the sets from $\mathcal{B}$. At each time t, the sensors located in the field observe

$$\xi(z,t) = \begin{cases} 1 \text{ if } & z \in C; \\ 0 \text{ otherwise.} \end{cases}$$

3.2 Large Clouds

A variant of the Boolean model deals with the unbounded shapes, the "large clouds". Here the clouds are represented as the parallel strips of random widths. More precisely, we consider the random set C to be bounded by a family of parallel lines, which are orthogonal to a direction that is chosen uniformly from the unit circle with crossing points forming a Poisson point process (one can check that this definition is independent of the choice of the origin in the plane).

In other words, $z \in C$ if, for some i,

$$x_{2i} \leq \langle z, e \rangle \leq x_{2i+1},$$

where $\langle , \rangle$ denote Euclidean scalar product, e is a random vector chosen uniformly from $\{|e| = 1\}$ and $\{x_k\}_{-\infty}^{\infty}$ is a Poisson point process with constant intensity.

3.3 Random Walkers

This model describes the random field generated by some independent random walkers in the area A. At time t a sensor at position z records $\xi(z, t) = 1$ if there is a walker within a distance less than r from itself; otherwise records $\xi(z, t) = 0$. We will use a spatial-temporal correlation function defined by

$$\sum_s \mathbb{E}\xi(z, t)\xi(z', t + s), |s| < \Delta,$$

which sums up the cumulants of two points over the interval Δ. The precise expression for the cumulants in this model is a polynomial in Gaussian functions.

4 Engineering Ramifications

The idea of anchoritic sensors can be extended to situations where some powerful sensors form a class of local masters and other sensors only report measurements to the nearest masters. The network architecture is depicted as in Figure 1.

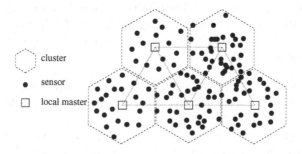

Fig. 1. Combine deaf sensors and local masters

The local masters can exchange information and, if necessary, can even be carefully engineered to generate random radio or acoustic signals such that we obtain a desired cumulant function. This can easily build an auxiliary random field that may be used to locate other sensors. The sensors in a cluster periodically report their data to the local master. Using the correlation values, the local master can determine the positions of the sensors very precisely, possibly by using the inverse of the already known cumulant function.

The distinguishing feature of this structure is that it does not require globally coordinated signals from a central entity.

5   Experimental Results

In this section, based on the random field models described in the preceding section, we present simulation experiments on constructing the proximity graph in anchoritic networks. In all these simulations, $N = 1000$ sensors are chosen independently at random in the unit square A and are represented as little squares in the plots. The simulation is conducted in a discrete fashion $t = 1, 2, \cdots, 2000$. The proximity graph Γ is formed by connecting each node with a given number $(k_N = \lfloor (\log N)^{1.2} \rfloor = 10)$ of nodes with the largest empirical cumulants (choosing $k_N = (\log N)^{1.2}$ is due to Theorem 1). Note that this approach does not require any apriori knowledge about the statistics of the random field.

Example 1 (Round clouds (Boolean model)). In this simulation, round clouds of random radii uniformly distributed on $[0, .2]$ are modeled as a Poisson field with intensity 30. Blue circles on Figure 1 depict a realization of the Boolean model. One can see that the proximity graph Γ shown on Figure 1 strongly resembles a nearest neighbor graph; there are very few edges connecting nodes far away, and almost all pairs of close nodes are connected.

In this example, on each corner of the unit square, a beacon with known position (illustrated in red) is shown. Though only four beacons are given, based on the proximity graph, we still can give a reasonably good estimation of the locations for most of the sensors. This result is presented in Section 5.1.

Example 2 (Big clouds). The big clouds in this simulation were modeled by half planes (a realization is shown in blue on Figure 2) bounded by lines with isotropic orientation. A visual inspection indicates a high similarity between the proximity graph and the nearest neighboring graph.

Example 3 (Random walkers). Consider $W = 10$ random walkers that are monitored by sensors of a sensing radius $r = .13$ with $\Delta = 2$. Yellow trajectories show part of the traces of the walkers.

One can see in Figure 6 that the quality of the proximity graph in this situation, even in the presence of a much slower convergence speed of this random field, is as good as that in the previous two examples, which is further illustrated in the scatter plot of cumulants in Figure 7.

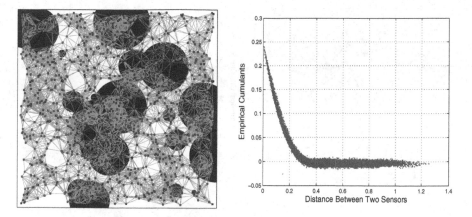

Fig. 2. Boolean model with round clouds **Fig. 3.** Cumulant-distance scatter plot for the round clouds model

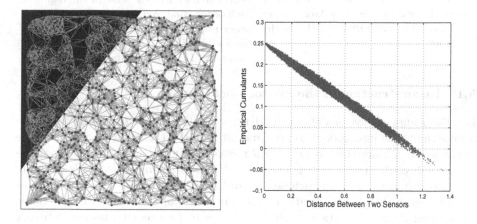

Fig. 4. Big clouds formed by isotropic **Fig. 5.** Cumulant-distance scatter plot for half-planes the big clouds model

A better feeling about the quality of recovering the distances by the inter node correlations can be gained from the scatter plots, which show the cumulant values versus the distances of pairs of sensors. The first two scatter plots, the round clouds model in Figure 3 and the large clouds model in Figure 5, show the results for *all pairs* of sensors. For the random walkers model, the scatter plot (Figure 7) only shows the cumulant-distance pairs with one of the nodes fixed. The heterogeneity of the occupation measure leads to significantly different ranges of the cumulants at different parts of the region. However, the cumulants still can be efficiently used for constructing the proximity graph, as the plot shows.

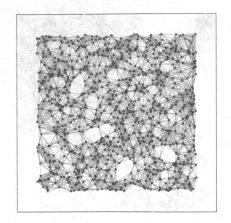

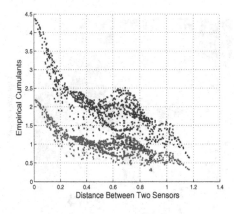

Fig. 6. Part of the traces of $W = 10$ random walkers (reflected at the boundary) are shown in yellow. The region in which the walkers move is larger than the region A to avoid irregularities at the boundary.

Fig. 7. Cumulant-distance scatter plot for all pairs of sensors with one fixed. Blue dots show the data for one node in the region of high occupation; red dots correspond to another node with low occupation. The plots are visibly similar, exhibiting a reliable estimation of the cumulants.

5.1   Reconstruction of the Sensor Locations

For the round clouds model, we used the cumulant-based proximity graph to approximate the internode distances and ultimately to reconstruct the positions of the sensors. Since the second step of reconstructing exact positions is not the main focus of this paper, we choose to do it in a rather naive way. We compute the hop distances between each sensor and the four beacons lying on the four corners using Dijkstra's algorithm. Assuming the hop distance is proportional to the real Euclidean distance (which is shown in Theorem 1), we can estimate the locations of all the sensors as being shown in Figure 8. One can see that boundary effects (caused by the inefficient algorithm we used here) are rather significant, yet in the interior of the area the positions are recovered quite well.

Here we did not use any sophisticated machinery for the second stage of the localization problem. A more holistic approach would be to generate a Gibbs measure, the ensemble of N nodes in A with a distribution consisting with the empirical measurements. Sampling from this distribution would give the most probable positions of the sensors in the region A.

6   Proof of Theorem 1

First, connecting each pair of sensors of distance less than $r(N) \triangleq \sqrt{(\log N)^c/\pi N}$, $c > 1$, one obtains a graph $G(N)$ with $h_{ij}^{(G)}$ being the hop distance between node i and j. We can prove that, for any $\epsilon > 0$,

$$\lim_{N \to \infty} \mathbb{P}\left[\left|h_{ij}^{(G)}r(N) - d_{ij}\right| < \epsilon, \forall Z_i, Z_j \in \mathcal{Z}\right] = 1. \tag{2}$$

The argument goes as follows. Choose $\rho(N) = \sqrt{\frac{c_1 \log N}{\pi N}}$, $c_1 > 2.5$. Connect sensor i and j by a sequence of circles of radius $\rho(N)$ in a way that the centers of the adjacent circles have a distance of $r(N) - 2\rho(N)$ with one of the distances possibly being less than $r(N) - 2\rho(N)$ (when N is large, $0 < r(N) - 2\rho(N) < \epsilon$). We call sensor i and j to be ρ-vicinity connected (denoted by $\{i \leftrightarrow j\}$) if there exists at least one sensor lying in each of the circles along the line (Z_i, Z_j), as is shown in Figure 9.

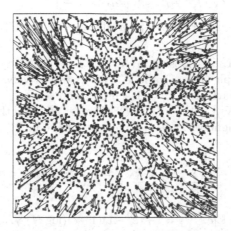

Fig. 8. Reconstruct the sensor positions in the second stage. The actual node positions (yellow squares) are connected to their estimated positions (green squares).

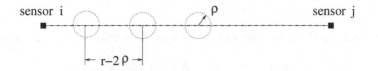

Fig. 9. ρ-vicinity connection

Therefore, for any i, j and large N, we have

$$\mathbb{P}\left[\{i \leftrightarrow j\}^C, |Z_i - Z_j| \geq r(N) - 2\rho(N)\right]$$
$$\leq \mathbb{E}[\text{number of empty circles along } (Z_i, Z_j), |Z_i - Z_j| \geq r(N) - 2\rho(N)]$$
$$\leq \frac{\sqrt{2}}{r(N) - 2\rho(N)}\mathbb{P}[\text{one given circle is empty}]$$
$$= \frac{\sqrt{2}}{r(N) - 2\rho(N)}(1 - \pi\rho(N)^2)^{N-2} \sim \sqrt{\frac{2\pi N}{(\log N)^c}}N^{-c_1} \leq N^{-(c_1 - \frac{1}{2})},$$

implying

$$\mathbb{P}\left[\{i \leftrightarrow j\}, \text{for all } |Z_i - Z_j| \geq r(N) - 2\rho(N)\right]$$

$$\geq 1 - \sum_{1 \leq i,j \leq N} \mathbb{P}\left[\{i \leftrightarrow j\}^C, |Z_i - Z_j| \geq r(N) - 2\rho(N)\right]$$

$$\gtrsim 1 - \binom{N}{2} N^{-c_1 + \frac{1}{2}} \to 1 \text{ as } N \to \infty,$$

which proves (2) by the following inequalities

$$1 \geq \lim_{N \to \infty} \mathbb{P}\left[\left|h_{ij}^{(N)} r(N) - d_{ij}\right| < \epsilon, \forall Z_i, Z_j \in \mathcal{Z}\right]$$

$$\geq \lim_{N \to \infty} \mathbb{P}\left[\{i \leftrightarrow j\}, \text{for all } |Z_i - Z_j| \geq r(N) - 2\rho(N)\right] = 1.$$

Next, for $0 < \epsilon < 1$, by choosing $r_1(N) = \sqrt{1 - \epsilon} \cdot r(N)$ and $r_2(N) = \sqrt{1 + \epsilon} \cdot r(N)$, we obtain graph $G^{-\epsilon}(N)$ and $G^{+\epsilon}(N)$, respectively. In graph $G^{-\epsilon}(N)$, define i.i.d. random variables $X_{ij} = \mathbf{1}$(sensor i and sensor j are connected), $j \neq i$ for sensor i with $\mathbb{E}[X_{ij}] = (1-\epsilon)(\log N)^c/N$ after ignoring the boundary effects. Let the number of neighbors in $G^{-\epsilon}(N)$ for sensor i to be Y_i, and then, for $\theta > 0$,

$$\mathbb{P}[Y_i < k_N] = \mathbb{P}\left[\sum_{j \neq i} X_{ij} < k_N\right] = \mathbb{P}\left[e^{\theta \sum_{j \neq i} X_{ij}} < e^{\theta(\log N)^c}\right],$$

which, by noting that $\mathbb{E}[e^{\theta X_{ij}}] = 1 + (e^\theta - 1)(1-\epsilon)(\log N)^c/N$ and using Chernoff bound, yields $\mathbb{P}[Y_i < k_N] < h e^{-\eta(\log N)^c}$ for some $h, \eta > 0$. Therefore, as $N \to \infty$,

$$\mathbb{P}[Y_i < k_N, 1 \leq i \leq N] \leq \sum_{i=1}^{N} \mathbb{P}[Y_i < k_N] \leq hN e^{-\eta(\log N)^c} \to 0. \tag{3}$$

From (3) and recalling the assumptions A and B in Subsection 2.2, we have,

$$\lim_{N \to \infty} \lim_{T \to \infty} \mathbb{P}[G^{-\epsilon}(N) \subset \Gamma(N)] = 1. \tag{4}$$

By the same argument, we can prove $\lim_{N \to \infty} \lim_{T \to \infty} \mathbb{P}[\Gamma(N) \subset G^\epsilon(N)] = 1$, which, combined with (2), (4) and passing $\delta \to 0$, completes the proof.

References

1. Albowicz, J., Chen, A., Zhang, L.: Recursive position estimation in sensor networks. In: Ninth International Conference on Network Protocols (November 2001)
2. Bulusu, N., Heidemann, J., Estrin, D.: GPS-less low cost outdoor localization for very small devices. IEEE Personal Communications Magazine 7(5), 28–34 (2000)
3. Stoyan, J.M.D., Kendall, W.S.: Stochastic geometry and its applications. Wiley, Chichester (1987)

4. Dohert, L., Pister, K., Ghaoui, L.: Convex position estimation in wireless sensor networks. INFOCOM'01 (April 2001)
5. Funiak, S., Guestrin, C., Paskin, M., Sukthankar, R.: Distributed localization of networked cameras. In: IPSN '06: Proceedings of the fifth international conference on Information processing in sensor networks, pp. 34–42. ACM Press, New York, USA (2006)
6. Girod, L., Bychkovskiy, V., Elson, J., Estrin, D.: Locating tiny sensors in time and space: A case study. In: Proceedings of the International Conference on Computer Design (ICCD2002), Freiburg, Germany (September 2002)
7. He, T., Huang, C., Blum, B., Stankovic, J., Abdelzaher, T.: Range-free localization schemes for large scale sensor networks. MobiCom'03, San Diego, CA, USA (August 2003)
8. Kallenberg, O.: Foundations of Modern Probability. Springer Series in Statistics. Probability and Its Applications (October 1997)
9. Langendoen, K., Reijers, N.: Distributed localization in wireless sensor networks: a quantitative comparison. Comput. Networks 43(4), 499–518 (2003)
10. Moses, R.L., Krishnamurthy, D., Patterson, R.: An auto-calibration method for unattended ground sensors. In: ICASSP, vol. 3, pp. 2941–2944 (May 2002)
11. Nagpal, R., Shrobe, H., Bachrach, J.: Organizing a global coordinate system from local information on an ad hoc sensor network. In: 2nd International Workshop on Information Processing in Sensor Networks (IPSN 03) (April 2003)
12. Niculescu, D., Nath, B.: Localized positioning in ad hoc networks. In: Sensor Network Protocols and Applications, Anchorage, Alaska (April 2003)
13. Patwari, N., A. O. H. III.: Using proximity and quantized RSS for sensor localization in wireless networks. WSNA'03.San Diego, CA,USA (September 2003)
14. Priyantha, N.B., Balakrishnan, H., Demaine, E., Teller, S.: Poster abstract: anchor-free distributed localization in sensor networks. In: SenSys '03: Proceedings of the 1st international conference on Embedded networked sensor systems, pp. 340–341. ACM Press, New York, USA (2003)
15. Priyantha, N.B., Miu, A.K.L., Balakrishnan, H., Teller, S.: The cricket compass for context-aware mobile applications. In: Proc. of the 6th ACM MOBICOM Conf. Rome, Italy (July 2001)
16. Savarese, C., Rabaey, J., Beutel, J.: Locationing in distributed ad-hoc wireless sensor networks. ICASSP (May 2001)
17. Savvides, A., Park, H., Srivastava, M.: The bits and flops of the N-hop multilateration primitive for node localization problems. WSNA'02, Atlanta, Georgia, USA (September 2002)
18. Shang, Y., Ruml, W., Zhang, Y., Fromherz, M.P.J.: Localization from mere connectivity. In: MobiHoc '03: Proceedings of the 4th ACM international symposium on Mobile ad hoc networking & computing, pp. 201–212. ACM Press, New York, USA (2003)
19. Stojanovic, M.: Acoustic (underwater) communications. In: Proakis, J.G. (ed.) Entry in Encyclopedia of Telecommunications, John Wiley & Sons, New York (2003)
20. Stoleru, R., He, T., Stankovic, J.A., Luebke, D.: A high-accuracy, low-cost localization system for wireless sensor networks. In: SenSys '05: Proceedings of the 3rd international conference on Embedded networked sensor systems, pp. 13–26. ACM Press, New York, USA (2005)

Mobile Anchor-Free Localization for Wireless Sensor Networks

Yurong Xu[1,2], Yi Ouyang[1,2], Zhengyi Le[1,2], James Ford[1,2],
and Fillia Makedon[1,2]

[1]DevLab, Computer Science Department, Dartmouth College
[2]Heracleia Lab, Univ. of Texas at Arlington
{yurong,ouyang,zyle,jford,makedon}@cs.dartmouth.edu

Abstract. In this paper, we consider how to localize individual nodes in
a wireless sensor network when some subset of the network nodes can be
in motion at any given time. For situations in which it is not practical or
cost-efficient to use GPS or anchor nodes, this paper proposes an Anchor-
Free Mobile Geographic Distributed Localization (MGDL) algorithm for
wireless sensor networks. Taking advantage of the accelerometers that
are present in standard motes, MGDL estimates the distance moved by
each node. If this distance is beyond a threshold, then this node will trigger a series of mobile localization procedures to recalculate and update
its location in the node itself. Such procedures will be stopped when the
node stops moving. Data collected using Tmote Invent nodes (Moteiv
Inc.) and simulations show that the proposed detection method can efficiently detect the movement, and that the localization is accurate and
the communication is efficient in different static and mobile contexts.

Keywords: Localization, Mobility, Wireless Sensor Networks.

1 Introduction

In recent years, Wireless Sensor Networks (WSNs) [18] have emerged as one of
the key enablers for a variety of applications such as environment monitoring,
vehicle tracking and mapping, and emergency response. One important problem
for such applications is how to locate a node's position. One example scenario is
that of a WSN deployed as part of the static infrastructure to detect fire as well
as to locate and guide fire fighters during fire emergencies by communicating
with mobile nodes they carry or wear.

Though many localization algorithms have been proposed for wireless ad hoc
networks or WSNs [4,5,6,7,15,16], they assume that the nodes inside of the networks are static. Little research has been presented on considering localization
in cases where the network cannot be assumed to be static. There are several
potential ways to provide localization for WSNs with mobile nodes:

(1) Let mobile nodes deploy global positioning system (GPS) to get their
locations. However, many applications require node mobility in environments
where GPS signals may not be available, our solution works with GPS, and can
even fill in completely where GPS is unreachable.

J. Aspnes et al. (Eds.): DCOSS 2007, LNCS 4549, pp. 96–109, 2007.

(2) Re-execute localization algorithms (such as [4,7,15]) periodically to compute the location of the mobile nodes. In order to locate potential fast moving nodes, localization algorithms will need to be restarted frequently. So, there will be a significant energy and communication cost for such localization scheme.

(3) Redesign localization algorithms that are particularly focused on the localization of mobile nodes. There are some existing mobile localization schemes, such as MCL [11], MCB [3] and ELA [17]. They rely on that there are a set of nodes (anchor nodes) which already knew their location. Our work focuses on finding a solution that does not require any anchor nodes in computation.

We use Anchor-Based (AB) to refer to the above methods that rely on some special nodes that already know their exact physical position. The alternative to AB localization is Anchor-Free (AF) localization which uses no specially designated reference nodes with known physical coordinates.

Because they assume that there are some anchor nodes that are aware of their exact physical location even when moving, MCL, MCB and ELA focus on ways to use these mobile anchor node to localize nearby nodes. There are several problems that may be faced under these localization algorithms: 1. The accuracy of AB schemes is related to the number of anchor nodes, so, in order to achieve high accuracy, AB schemes usually need large sets of anchor nodes. 2. The fact that AB schemes depend heavily on anchor nodes makes them vulnerable to the loss or malfunctioning of any of these anchor nodes. 3. Current AB solutions such as MCL, MCB and ELA need all nodes, or at least all anchor nodes, to broadcast their location periodically, even when there is no node movement.

This paper will talk about an AF localization called MGDL for WSNs with node mobility. We assume that the network is comprised mostly of low-mobility nodes which are embedded into the environment while some other nodes are carried by some mobile objects. Based on this assumption, MGDL generates a distance measurement for each node by combining local network connection information with hop numbers generated without the help of GPS devices or anchor nodes. Using the accelerometers deployed in motes, MGDL estimates the distance moved for each node after the whole network is started. If such distance is beyond a threshold, then a series of procedures will be triggered to recalculate/update the location for that node. This procedure will continue while movement continues to be detected.

In this paper, we make the following contributions: (i) We design a movement detection procedure for standard motes in WSNs to detect the movement of a node by using accelerometers. (ii) We propose an MGDL localization algorithm in detail, based on (i). simulations using the algorithm shows that the MGDL has high accuracy in localization in different placements of networks, as well as an efficient communication cost in different movement scenarios.

The remainder of the paper is organized as follows. Section 2 describes related work, Section 3 discusses the details of MGDL algorithm, and Section 4 reports simulation results. Finally, Section 5 gives our conclusions.

2   Related Work

The majority of prior research related to localization problems has focused on static sensor networks [4,5,6,7,15,16]. Recently, however, more attention has been paid to mobile environments.

S. Čapkun et al. proposed an anchor-free localization called SPA for mobile WSNs in [9], which localizes nodes in mobile sensor networks through triangulation of neighbor nodes. SPA first computes a relative coordinate system for each node, then converts the above relative coordinate systems in each node into a global coordinate system by calculating differences in terms of distance and direction between each node and a particular central node, or a dense group of nodes called a Location Reference Group (LRG). The problem for SPA is that if any nodes, especially those nodes inside an LRG, are moved, then a recalculation must be done in almost the whole network, which is costly and unnecessary.

In [11], Hu and Evans present a range-free anchor-based localization algorithm for mobile sensor networks based on the sequential Monte Carlo method [8]. The Monte Carlo method has been extensively used in robotics [5] where a robot estimates its localization based on its motion, perception and possibly a prelearned map of its environment. Hu and Evans extend the Monte Carlo method as used in robotics to support the localization of sensors in unmapped terrain. The authors assume a sensor has little control and knowledge over its movement, in contrast to a robot. A similar paper [3] shares the same idea.

By using an analogy with a system of springs and masses, the Elastic Localization Algorithm (ELA) [17], which is an anchor-based algorithm, tries to calculate locations using anchor nodes which already know their locations. A mobile version of ELA supports mobile localization by updating neighbors' locations in each node at fixed intervals. Such periodic updating in the whole network may lead to significant communication and computation costs.

Work in [12] proposes an anchor-free method which focuses on locating group movement—cases in which multiple nodes have a similar direction and velocity. By deploying a compass in each node to detect the direction of a node, each node computes the relative locations of its neighbors. The work in [12] pays particular attention to group movement, and does not consider independent movements by individual nodes, which is an important and common case in mobile networks.

3   Mobile Geographic Distributed Localization (MGDL)

3.1   Overview of MGDL Algorithm

We assume that a node in a WSN can be either in a mobile or a static state, we use "mobile" or "static" to distinguish whether a node is mobile or not. At the same time, we use "updated" or "non-updated" to distinguish whether a node is localized or not. Combining these labels gives four states, which are represented as S/N ("Static/Non-updated"), S/U ("Static/Updated"), M/N ("Mobile/Non-Updated") and M/U ("Mobile/Updated").

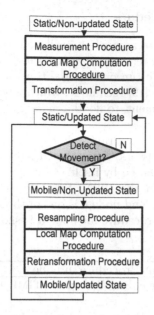

Fig. 1. Sensor States and MGDL Algorithm

Based on the assumption that each node in the networks stays in the "S/N" state initially, MGDL uses a technique similar to hop-counting as a measurement procedure (Section 3.2) to measure the distance from some bootstrap node to other nodes. After the running of the measurement procedure, each node will collect a set of hop coordinates from its neighbor nodes that are within one (or k) hop(s) distance of itself, and then it will run Dijkstra's algorithm to get the shortest path between each pair of nodes. After that, it will construct a local map (Section 3.3) using MDS (Multidimensional Scaling). A transformation procedure (Section 3.4) will merge the local map inside of each node into a global map. After all the above procedures are done, the state of the node will be set as "S/U".

In order to detect the movement of a node, we utilize an accelerometer, which is a standard component in many current motes (such as Moteiv's Invent [2]), to detect the movement of the node. If there is a node that is starting to move inside of the network, the accelerometer can detect its acceleration, and then our algorithm will compute the total distance moved based on the acceleration. If this distance is beyond a threshold, then the state of the node will be set as "M/N", and mobile localization procedures (Sections 3.3, 3.6, 3.7) will be triggered to resample hop-coordinates again, to recompute the local map and transformation matrix for that local map in order to merge this map into the global map. While movement continues to be detected, the above procedures will continue. Once movement is no longer detected, the above updating procedures will be stopped and the state of the node will be set back to "S/U". Fig. 1 shows the states and the whole MGDL algorithm for a node.

3.2   Measurement Procedure

In the measurement step, we assume that all nodes inside of the networks are static at this step (which is reasonable, since a measurement procedure is relatively fast, and need only be run once at the time of the network deployment). We use the hop-coordinates [19] technique, which is similar to hop-counting but has more accurate measurement, to flood a message to the network to finish the measurement. The basic idea is:

(i) In bootstrap node: A bootstrap node (x) creates a measurement message with ($i = x$) to flood the network. After that, the bootstrap node will drop any measurement message originated by itself.

(ii) In all other nodes in the WSN: Suppose that an arbitrary node a is calculating its hop distance, and node b is one of the neighbors of node a. Then the basic hop-coordinates procedure for node a is shown in Procedure 1.

Procedure 1. Measurement Procedure in node a

1: **for** message (hop_b) from any B $\in N_a$ and not TIMEOUT **do**
2:    **if** $hop_b < hop_a$ **then**
3:       $hop_a = hop_b + 1$
4:       forward(message(hop_a)) to MAC
5:    **else**
6:       drop(message(hop_b))
7:    **end if**
8: **end for**
9: **if** $|N_a| == 0$ **then**
10:    $offset_a = 0$
11: **else**
12:    $offset_a = \dfrac{\sum_{b \in N_a}(hop_b - (hop_a - 1)) + 1}{2(|N_a| + 1)}$
13: **end if**
14: **return** hop-coordinate $hop_a + offset_a$

Here, a is a node, hop_a is the minimum number of hops to reach node a counting from some bootstrap node (x), the combination of hop_a and $offset_a$ is the hop coordinate for node a, N_a is a set of nodes which can be reached by node a in one hop, and $|N_a|$ is the number of nodes in N_a.

The total cost for this step is as follows: a computational cost of $O(1)$, a communication cost of $O(|N_a|)$, a memory cost of $O(|N_a|)$ for each node.

3.3   Local Map Computation

In this step, each node will compute a local map for it's neighbors based on the hop-coordinate computed in the previous step. After the generation of hop-coordinates with Procedure 1, each node will send a request to its neighbor nodes that are within k hops to send back their hop coordinate from some bootstrap node (x).

After each node receives the hop coordinate from its neighbors, that node will compute shortest paths between all pairs of nodes k hops to that node, using Dijkstra's algorithm or other similar algorithms.

Then, we apply MDS to the $(|N_a|+1)\times(|N_a|+1)$ shortest path matrix (here $|N_a|$ is the number of nodes that can be reached by node A in k hops and retain the first two (or three) largest eigenvalues and eigenvectors to construct a 2-D (or 3-D) local map.

The total cost for this step is a computational cost of $O(|N_a|^3 n)$ and a memory cost of $O(|N_a|^2)$ per node, with no communication cost in this step.

3.4 Transformation Procedure

In this step, we will assemble the local maps that are computed and stored in each node into a global map through a transformation.

First, we will bootstrap from some node to compute the transformation matrix for that node and follow it by broadcasting a transformation message to its neighbors to let them start to compute the transformation matrix with their neighbors, too.

Then, after we find the set of neighboring nodes, we compute a transformation matrix for each node b in the neighbor node set N_a of node a. Suppose that there are two sets of neighbor nodes, N_a and N_b, that are within k hops of the nodes a and b, respectively. Their intersection is $I = N_a \cap N_b$, and we use matrix $I_a = \left\{ \ldots, (x_i, y_i)', \ldots \right\}$ (here (x_i, y_i) are the coordinates of node i) to represent the coordinates of nodes in the I generated by node a, and similarly $I_b = \left\{ \ldots, (\check{x}_i, \check{y}_i)', \ldots \right\}$ for node b. We can then compute a transformation matrix T such that it minimizes $\sqrt{\sum((x_i - \check{x}_i)^2 + (y_i - \check{y}_i)^2)}$, where $(x_i, y_i) \in I_a$, $(\check{x}_i, \check{y}_i) \in I_b$, and $(\check{x}_i, \check{y}_i) \in I_b \times T$. Procedure 2 gives the detail.

Procedure 2. Compute transformation matrix T in node a

Require: Input: matrix T from neighbor node
1: **if** this node is transformed **then**
2:    drop(matrix T)
3:    **return**
4: **end if**
5: **for** each node $b \in N_a$ **do**
6:    request N_b from node b
7:    $I = N_b \cap N_a$
8:    generate I_b and I_a from I
9:    compute transformation matrix T such that $\sqrt{\sum((x_i - \check{x}_i)^2 + (y_i - \check{y}_i)^2)}$ is minimized, here $(x_i, y_i) \in I_a$, $(\check{x}_i, \check{y}_i) \in I_b \times T$.
10:    send matrix T to node b
11: **end for**
12: set (this node is transformed)

If a node receives a transformation matrix T, then it will first check whether it is already transformed or not; if so, the node will drop such a message, and if not, it will apply Procedure 2. This will allow the node to compute the local map which it will then send to its $|N_a|$ neighbor nodes.

After we have flooded the network to the transformation step in the whole networks, we archive the global map for the whole network, which is stored in a distributed way in each node in the network.

Total cost for this step: computational cost of $O(|N_a|)$, memory cost of $O(|N_a|)$ for nodes, and communication cost of $O(|N_a|)$ for each node or $O(n)$ for whole network.

3.5   Mobile Measurement Techniques

Until now, we have only talked about how to do localization without considering mobility of nodes (after the previous procedures, current state of a node is S/U). In this section, we will solve the problem of how to decide whether a node has moved, then how to recompute the location of mobile nodes.

2D and 3D Accelerometer. In order to detect the movement of mobile nodes, we need some sensor which can detect and quantify node movement. In our algorithm, we make use of accelerometers installed in standard nodes to detect their movement. An accelerometer is a device that measures its own acceleration. We can use a 2D (X-Y) accelerometer to measure 2D acceleration, or a 3D (X-Y-Z) version to measure 3D acceleration. The component we used for 2D in this paper is a standard component in current commercial motes, such as the Moteiv Inventor mote [2]. In order to detect 3D movement of the node, one can install an inexpensive external 3D accelerometer, such as the MMA7260Q accelerometer [1] from Freescale.

Movement Detection. With the aid of a 2D accelerometer, we can roughly measure the acceleration vector $\boldsymbol{a}$ in a plane. Since the accelerometer can't detect the rotation of a node, it is of limited use as a direct way to measure position changes. Therefore, we use the integral of the absolute value of the $\boldsymbol{a}$ to compute an approximation of the moved distance $d = \int \int |\boldsymbol{a}| d^2 t$. If such distance is beyond a threshold, we then say this node has moved.

First we assume that at the beginning of time $t = 0$, every node inside of the network is still. Consider an arbitrary node: suppose its acceleration $\boldsymbol{a} = 0$, its velocity $v = 0$, the distance it has moved $d = 0$ for that node. Then, we will sample the accelerometer in that node periodically. Here, we assume that the interval time for sampling is dt, and the reading of acceleration from the accelerometer in that node is shown as $\boldsymbol{a}$, so current velocity for this node can be approximated as $v = \int |\boldsymbol{a}| dt$, and the distance moved from the beginning location (when $t = 0$), can be approximated as $d = \int \int |\boldsymbol{a}| d^2 t$. If d is beyond a threshold ϵ, we then say this node has moved, and we let $v = 0$ and $d = 0$ to restart the measurement. Thus, though the values of $\boldsymbol{a}$, v and d are not accurate, in comparison to their real values, they are sufficient to detect the movement of a node. The complete process is described in Procedure 3.

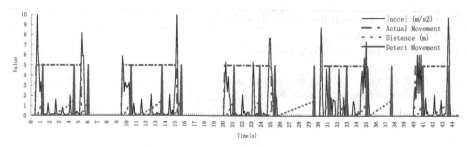

Fig. 2. First 40 Seconds of Experiment on Movement Detection with Accelerometer. Movement is detected within each movement block, although with a slight lag, and is only occasionally spuriously detected between movements. (Here $dt = 0.1$s, threshold $\epsilon = 1.5$m. If a node is moving then actual movement will be shown as value $=5$, otherwise actual movement will be shown as value $=0$.)

Procedure 3. Movement Detection Procedure

Require: this procedure will be invoked to read the accelerometer during the time period dt.

1: $v =$ previous $v + \int |a|dt$
2: $d =$ previous $d + \int v dt$
3: **if** $d >$ threshold ϵ **then**
4:     $v = 0$
5:     $d = 0$
6:     **return** "detect movement"
7: **end if**
8: **return** "un-detect movement"

If Procedure 3 detects that a node has moved, then MGDL in that node will recompute its location. Threshold ϵ will decide when a mobile node should recompute its location. The smaller the threshold is, the higher the frequency with which the node computes its location, leading to higher localization accuracy as well as more communication cost.

In order to evaluate how to detect the movement of a node with the above procedure, we set up an experimental environment in a long hallway (about 50 m). A node carried by one person works as a mobile node. The above movement detection procedure is running inside of that node with $dt = 0.1$s. At first, we will let the mobile node start to move from one end of the hall, and move for 5 seconds with walking speed, then stop for 5 seconds in the same hall, and so on until reaching the end of the hall, then turn back with the same movement. The first 40 seconds of this experiment is shown in Fig. 2.

From Fig. 2, we can see that this movement detection procedure can detect the movement of a node when that node is moving, albeit with an average delay of about 0.95 s.

Procedure 4. Resampling Procedure for Node a

1: **if** $|N_a| == 0$ **then**
2:     keep the previous $offset_a$ and hop_a
3: **else**
4:     **for** each node $b \in N_a$, for which the state of node b is "S/U" **do**
5:         request hop_b from node b
6:     **end for**
7:     $offset_a + hop_a = \frac{\sum_{b \in N_a}(hop_b + hop_b + offset_a)}{|N_a|}$
8: **end if**
9: **return** $hop_a + offset_a$

3.6   Resampling Procedure

If we detect that a node is moving with the movement detection as shown in Procedure 3, we will mark this node as "M/N" (Mobile but Non-updated). Then, we will resample the hop-coordinate for this node from its neighbors. In order to increase the accuracy of resampling, we only get hop-coordinates from nodes that are marked as "S/U", instead of from nodes marked "M/N" or "M/U".

The total cost for this step is as follows: computational cost of $O(1)$, communication cost of $O(|N_a|)$, memory cost of $O(|N_a|)$ for each node, and $O(n)$ for the whole network.

3.7   ReTransformation Procedure

After the resampling procedure and local map computation, we retransform the local map in this moved node into the global map. In this process we get a new transformation matrix for this local map, since the old transformation matrix is new out of date. Suppose this moved node is node a; first, we find a closest neighbor node with "S/U" state (here we assume it is node b); then compute a new transformation matrix T for node a. Suppose that there are two sets of neighbor nodes N_a and N_b for nodes a, b, respectively. Their intersection is $I = N_a \cap N_b$, and we use matrix $I_a = \left\{ \ldots, (x_i, y_i)', \ldots \right\}$ (here (x_i, y_i) are the coordinates of one node i) to represent the coordinates of nodes in the I generated by node a, and similarly $I_b = \left\{ \ldots, (\dot{x}_i, \dot{y}_i)', \ldots \right\}$ for node b. We can then compute a transformation matrix T such that it minimizes $\sqrt{\sum((x_i - \breve{x}_i)^2 + (y_i - \breve{y}_i)^2)}$, where $(x_i, y_i) \in I_a$, $(\dot{x}_i, \dot{y}_i) \in I_b$, and $(\breve{x}_i, \breve{y}_i) \in I_b \times T$. The procedure is given in Procedure 5.

Total cost for this step: computational cost of $O(|N_a|)$, memory cost of $O(|N_a|)$ for nodes, and communication cost of $O(|N_a|)$ for each node.

4   Simulation Result

4.1   Simulation Configuration

We implemented our localization algorithm as a routing agent and our bootstrap node program as a protocol agent in NS-2 version 2.29 [14] with 802.15.4

Procedure 5. Recalculate Transformation Matrix T for Node a

1: find node b from N_a, such that $|hop_a + offset_a - (hop_b + offset_b)|$ *is minimized.*
2: *request* N_b *from node* b
3: *request* T_b *from node* b
4: $I = N_b \cap N_a$
5: *generate* I_b *and* I_a *from* I
6: *compute transformation matrix* T *such that* $\sqrt{\sum((x_i - \tilde{x}_i)^2 + (y_i - \tilde{y}_i)^2)}$*is minimized, here* $(x_i, y_i) \in I_a \times T_a$, $(\tilde{x}_i, \tilde{y}_i) \in I_b \times T_b$.
7: *set this node to "M/U" state.*

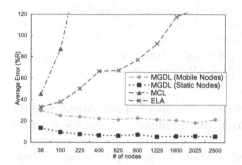
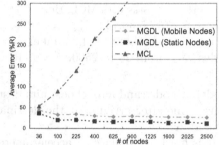

Fig. 3. Accuracy Comparison with MGDL, MCL and ELA. Here, ND=10, and V_{avg}= 1R/s, Number of Anchors = 4, Number of Nodes = 36,100,225,400, 625,900,1225,1600,2025,2500, ϵ = 0.1R.

Fig. 4. Overall Accuracy Comparison with MGDL and MCL. Here, ND=4.6, 6.5,10,17.5,38,144, V_{avg} = 1R/s, Number of Anchors = 4, Number of Nodes are from 36 to 2500, threshold ϵ = 0.1R.

MAC layer [20] and CMU wireless [10] extensions. The configuration used for NS-2 is RF range = 15 meters, propagation = TwoRayGround, antenna = OmniAntenna.

In our experiments, we used uniform placement—n nodes are placed on a grid with $\pm 0.5r$ randomized placement error. Here r is the width of a small square in the grid. We constructed a total of 60 placements with n =36, 100, 250, 400, 625, 900, 1600, 2250, and 2500, and with $r = 2, 4, 6, 8, 10$ and 12 meters, respectively. The reason we use uniform placement with $\pm 0.5r$ error is that usually such a placement produces both node holes and islands in one placement. To better simulate realistic mobile network situations, in each placement, we let most of the nodes inside of the network work as static nodes deployed in the environment, while about 10% of the nodes move inside the network under the following mobility model.

Mobility Model. Mobile nodes in the simulation move according to a model that is called as the "random waypoint" model [13]. It is one of the most commonly used mobility models for mobile ad hoc networks. In the random waypoint

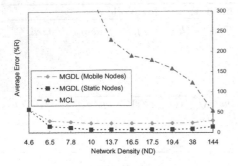 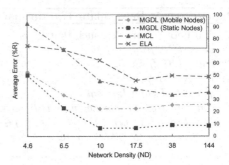

Fig. 5. Comparison of Accuracy Vs. Density of the Networks with MCL. Here, ND =4.6,6.5,10,17.5,38,144, V_{avg} = 1R/s, Number of Anchors = 4, number of nodes varis from 36 to 2500, ϵ = 0.1R

Fig. 6. Comparison of Accuracy Vs. Density of the networks with MCL. Here, number of nodes is 400 (40 anchor nodes inside them for ELA and MCL), V_{avg}=1R/s, threshold ϵ = 0.1R.

model, a node randomly chooses its destination, its speed of movement, and its pause time after arriving at the destination. In our simulation we use a pause time of zero.

It is hard to simulate an accelerometer in NS-2, so for simplification, we feed the moved distance of one node to the motion detection procedure in that node.

Each node controlled by the mobility model, begins the simulation by remaining stationary. It then selects a random destination in the network space and moves to that destination at a speed randomized from a uniform distribution between 0 and some maximum speed. Upon reaching the destination, the node selects another destination, again, and proceeds there as previously described, repeating this behavior for the duration of the simulation. Each simulation ran for 120 seconds of simulated time.

Simulation Parameters. We will control the following parameters in our simulations:

1. Average speed of nodes (V_{avg}): We represent the speed as the moving distance per time unit. A node's speed is chosen from a uniform distribution $[0, 2 * V_{avg}]$, so that the average velocity equals V_{avg}.

2. Node Density (ND): The average number of nodes in one hop transmission range. In our placements, we chose r = 2, 4, 6, 8, 10 and 12 meters, which corresponds to ND = 144, 38, 17.5, 10, 6.5, and 4.6, respectively.

3. Number of nodes: The total number of nodes inside a WSN.

4. Threshold (ϵ): A threshold is used to judge whether the current node is moving or not. We assume a fixed threshold ϵ = 0.1R for all simulations (except in varying of ϵ in Section 4.6).

4.2   Localization Accuracy

The key metric for evaluating a mobile localization technique is the accuracy of location when nodes are moving. Since MGDL is an AF localization, it does not use anchor nodes, while such anchor nodes will be needed inside of the MCL and ELA. In order to compare MGDL with MCL, and other localization algorithms, we assume that there are only 4 anchor nodes in both MCL and MGDL, in all placements, except as noted. Because at the same time there are some nodes moving and some nodes being still, we also compute the accuracy of localization for mobile nodes and still nodes, separately.

First we compare these three algorithms under different number of nodes. ELA only has data when ND = 10 available. Fig. 3 shows the comparison of localization accuracy of MGDL vs. MCL, ELA under different number of nodes = 36, 100, 225, 400, 625, 900, 1225, 1600, 2025, 2500, when ND = 10. Fig. 4 compares the localization accuracy of MGDL vs. MCL under different number of nodes with different ND = 144, 38, 17.5, 10, 6.5, 4.6. Both figures share same additional parameters as $V_{avg} = 1R/s$, and threshold $\epsilon = 0.1R$ for MGDL. From the figures we can see that MGDL, unlike MCL or ELA, has stable performance on localization accuracy even when the number of nodes is very large, while MCL and ELA usually only have good performance when the number of nodes is small. To achieve good performance with more nodes, it is necessary that they should increase the number of anchor nodes significantly.

4.3   Node Density

Since ELA only has data with number of nodes = 400 available for node density, Fig. 6 shows the comparison of localization accuracy of MGDL, MCL and ELA under number of nodes = 400. Even when there are about 10% anchor nodes inside of the network for ELA and MCL, MGDL can still achieve an improvement of about 20% of R on average in localization accuracy over MCL, and an improvement of about 28%R over ELA. Here we again note that the results from MCL and ELA are based on using about 10% nodes as anchor nodes, while our algorithm does't depend on anchor nodes when computing the global map for the networks.

Fig. 5 shows the impact of node density over all 60 placements on MGDL and MCL. MGDL is far beyond MCL in different node densities.

4.4   Node Speed

Fig. 7 compares the localization accuracy of MGDL vs. MCL under the varying of V_{avg} from 0 to 1R. Even when the number of anchor nodes for MCL is about 10% of the number of total nodes, we can see that MGDL shows low localization error when nodes are in low-mobility, while MCL is encoutering higher error. MGDL achieves an average of 22%R more location accuracy for the overall varying of V_{avg} from 0 to 1R.

4.5   Communication Overhead

One important consideration for WSNs is lowering communications overhead. Here, we measure communication overhead with the average number of messages

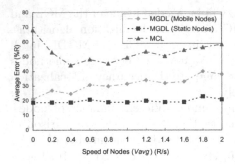 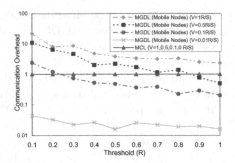

Fig. 7. Impact of Node Speed V_{avg} on MGDL and MCL. Here, # of anchor nodes for MCL is about 10% of # of total nodes.

Fig. 8. Communication Overhead with Different Threshold Value on MGDL and MCL under Different V_{avg}

transmitted by a node each second. From Fig. 8, we can see that the lower the speed of mobile nodes is inside of the network, the lower the communication overhead is in MGDL, while MCL keep the same communication overhead. This phenomenon shows that our algorithm can efficiently adjust the communication overhead to save more energy when localizing low-speed mobile nodes while keep reasonable accuracy for high-speed mobile nodes, as shown in Fig. 8. Also, at a given communication speed, we can decrease communication overhead by increasing the threshold ϵ, at the cost of a slower update frequency for mobile localization.

5 Summary

In this paper, we proposed an anchor-free localization for mobile WSNs. By making use of a standard device accelerometer, we proposed a set of movement detection algorithms, and by testing them on Moteiv's Invent motes, we verify that our approach is reasonable. Then based on such movement detection, we provided the whole AF localization algorithm called MGDL. Based on simulation in NS-2, we found that our algorithm has more accurate localization results than previous mobile localization algorithms. MGDL has flexible communication overhead for both high-mobility and low-mobility nodes, while MCL only has fixed communication overhead for both of them, which may impose overly high communication overhead in the latter case.

References

1. Freescale Inc., MMA7260Q 3D Accelerometer. URL http://www.freescale.com/
2. Moteiv Inc., Invent Motes. URL http://www.moteiv.com
3. Baggio, K.L.A.: Monte-carlo localization for mobile wireless sensor networks. In: 2nd Int. Conference on Mobile Ad-hoc and Sensor Networks (MSN 2006), Hong Kong, China (December 2006)

4. Ash, J., Potter, L.: Sensor network localization via received signal strength measurements with directional antennas. In: Proceedings of the 2004 Allerton Conference on Communication, Control, and Computing (2004)
5. Bahl, P., Padmanabhan, V.: RADAR: an in-building RF-based user location and tracking system. INFOCOM 2000. In: Nineteenth Annual Joint Conference of the IEEE Computer and Communications Societies. Proceedings. IEEE (2000)
6. Bischoff, R., Wattenhofer, R.: Analyzing connectivity-based multi-hop ad-hoc positioning. Pervasive Computing and Communications, 2004. PerCom 2004. In: Proceedings of the Second IEEE Annual Conference on, pp. 165–174 (2004)
7. Bruck, J., Gao, J., Jiang, A.: Localization and routing in sensor networks by local angle information. In: Proceedings of the 6th ACM international symposium on Mobile ad hoc networking and computing, pp. 181–192 (2005)
8. Burgard, W., Fox, D., Hennig, D., Schmidt, T.: Estimating the absolute position of a mobile robot using position probability grids. In: Proceedings of the Thirteenth National Conference on Artificial Intelligence, pp. 896–901 (1996)
9. Čapkun, S., Hamdi, M., Hubaux, J.: GPS-free Positioning in Mobile Ad Hoc Networks. Cluster Computing 5(2), 157–167 (2002)
10. Group, T.C.M.: Wireless and Mobility Extensions to ns-2. obtain from http://www.monarch.cs.cmu.edu/cmu-ns.html
11. Hu, L., Evans, D.: Localization for mobile sensor networks. In: Proceedings of the 10th annual international conference on Mobile computing and networking, pp. 45–57 (2004)
12. Akcan, H.B.A.D.H., Kriakov, V.: Gpsfree node localization in mobile wireless sensor networks. In: Proceedings of the 5th ACM international workshop on Data engineering for wireless and mobile access (MobiDE 2006) (2006)
13. Johnson, D., Maltz, D.: Dynamic source routing in ad hoc wireless networks. Mobile Computing 353, 153–181 (1996)
14. McCanne, S., Floyd, S.: ns-2 Network Simulator. Obtain via: http://www.isi.edu/nsnam/ns
15. Shang, Y., Ruml, W.: Improved MDS-based localization. INFOCOM 2004. In: Twenty-third AnnualJoint Conference of the IEEE Computer and Communications Societies, 4 (2004)
16. Shang, Y., Ruml, W., Zhang, Y., Fromherz, M.: Localization from mere connectivity. In: Proceedings of the 4th ACM international symposium on Mobile ad hoc networking & computing, pp. 201–212 (2003)
17. Vicaire, P., Stankovic, J.: Elastic Localization: Improvements on Distributed, Range Free Localization for Wireless Sensor Networks. Technical report, Tech. Rep. CS-2004-35, University of Virginia (2004)
18. Vieira, M., Coelho Jr, C., da Silva Jr, D., da Mata, J.: Survey on wireless sensor network devices. IEEE Emerging Technologies and Factory Automation, pp. 537–544 (2003)
19. Xu, Y., Ford, J., Makedon, F.S.: A Variation on Hop-counting for Geographic Routing. Embedded Networked Sensors, 2006. EmNetS-III. In: The third IEEE Workshop on (2006)
20. Zheng, J., et al.: 802.15.4 extension to NS-2. Obtain via: http://www-ee.ccny.cuny.edu/zheng/pub

Optimal Cluster Association in Two-Tiered Wireless Sensor Networks

WeiZhao Wang*, Wen-Zhan Song**, Xiang-Yang Li ***,
and Kousha Moaveni-Nejad***

weizhao@google.com
songwz@wsu.edu
xli@cs.iit.edu, moavkoo@iit.edu

Abstract. In this paper, we study the two-tiered wireless sensor network (WSN) architecture and propose the optimal cluster association algorithm for it to maximize the overall network lifetime. A two-tiered WSN is formed by number of small sensor nodes (SNs), powerful application nodes (ANs), and base-stations (BSs, or gateways). SNs capture, encode, and transmit relevant information to ANs, which then send the combined information to BSs. Assuming the locations of the SNs, ANs, and BSs are fixed, we consider how to associate the SNs to ANs such that the network lifetime is maximized while every node meets its bandwidth requirement. When the SNs are homogeneous (e.g., same bandwidth requirement), we give optimal algorithms to maximize the lifetime of the WSNs; when the SNs are heterogeneous, we give a 2-approximation algorithm that produces a network whose lifetime is within 1/2 of the optimum. We also present algorithms to dynamically update the cluster association when the network topology changes. Numerical results are given to demonstrate the efficiency and optimality of the proposed approaches. In simulation study, comparing network lifetime, our algorithm outperforms other heuristics almost twice.

1   Introduction

The deployment of tiny sensors to large scale wireless sensor networks raises massive challenges. Due to large scale, it is natural to adopt the two-tiered (even multiple tiered) architecture. A two-tiered WSN is formed by number of small sensor nodes (SNs), powerful application nodes (ANs), and base-stations (BSs, or gateways). SNs capture, encode, and transmit relevant information to ANs, which then send the combined information to BSs. In fact, some works have already addressed different issues regarding this hierarchical architecture, including minimizing the number of clusters [1,2], minimizing the total energy

---

* Google Inc.
** Washington State University, USA. The research of Wen-Zhan Song is supported in part by NASA ESTO 05-AIST05-0082.
*** Illinois Institute of Technology, USA. The research of Xiang-Yang Li was supported in part by NSF CCR-0311174.

J. Aspnes et al. (Eds.): DCOSS 2007, LNCS 4549, pp. 110–123, 2007.
© Springer-Verlag Berlin Heidelberg 2007

consumption [3] and maximizing the lifetime [4,5]. Following the work in [4], we consider how to associate the SNs to ANs such that the network lifetime is maximized while every node meets its bandwidth requirement. When the SNs are homogeneous (e.g., same bandwidth requirement), we give optimal algorithms to maximize the lifetime of the WSNs; when the SNs are heterogeneous, we give a 2-approximation algorithm that produces a network whose lifetime is within $1/2$ of the optimum. We also present algorithms to dynamically update the cluster association when the network topology changes. Numerical results are given to demonstrate the efficiency and optimality of the proposed approaches. In simulation study, comparing network lifetime, our algorithm outperforms other heuristics almost twice.

Problem Definition. A two-tiered wireless sensor network (WSN) consists of a set of small sensor nodes (SN), denoted as $S_M = \{s_1, s_2, \cdots, s_m\}$, a set of application nodes (AN), denoted as $V_N = \{v_1, v_2, \cdots, v_n\}$, and at least one base station (BS). The ANs and SNs form *clusters*, and in each cluster there are many SNs and one AN. For simplicity, we assume that the application node v_i is in cluster C_i and the set of small sensors in cluster C_i is $S_i \subseteq S_M$. A small sensor, once triggered by the internal timer or some external signals, starts to capture and encode the environmental phenomena (such as temperature, moisture, motion measure, etc) and broadcast the data directly to all ANs within its transmission range and to certain ANs via the relay of some other neighboring sensors. Here, if AN v_i can receive the data from the small sensor s_j, then we call v_i is a neighbor of s_j. Here sensor s_j may have to reach AN v_i via relay of other sensors. For notational simplicity, we use $N(v_i)$ to denote the neighboring small sensors of AN v_i. Remember that although several ANs can receive the data packets from the small sensor s_j, only the AN in the same cluster as s_j processes the information. Here, we assume that once formed, the cluster formation does not change over the time. We also let r_i be the data-rate of the small sensor s_i generates and $r(S) = \sum_{s_i \in S} r_i$ be the total data-rates produced by a set of small sensors S. Usually, the data-rate $r_i(t)$ is a function over the time t instead of a constant. However, if we average the rate over a period of time T, e.g., one day or one week, most often it is a constant. Thus, we can define the rate r_i as the the average rate over a period of time, i.e., $r_i = \frac{\int_{T_0}^{T+T_0} r_i(t)}{T}$.

It is reasonable to expect that the life time of an AN *decreases* when the number of small sensors in its cluster *increases*. Given an AN v_i, let $S_i \in S_M$ be the set of small sensors in its logical cluster. The power consumption of the AN v_i is a general function $p_i(r(S_i), N(v_i))$, where $r(S_i)$ is the total data-rate of the small sensors in S_i. Since $N(v_i)$ does not depend on the cluster formation and can be taken as a constant for a given application node v_i, we can simplify the power consumption function as $p_i(r(S_i))$. The only assumption in this paper is that function $p_i(x)$ should satisfy that $p_i(x) > p_i(x')$ when $x > x'$. Notice that, the above monotone increasing property is only assumed to be true for each AN. For two different ANs v_i and v_j, it is possible that $p_i(x) < p_j(x')$ when

$x > x'$. In this paper, we assume that $\mathcal{P}_i$ is the initial battery power level of the application node v_i and $p_i(r(S_i))$ is its *average* energy consumption rate when the set of small sensors S_i is in the cluster $\mathcal{C}_i$. The lifetime of an individual AN v_i is define as $l_i = \frac{\mathcal{P}_i}{p_i(r(S_i))}$. We adopt the following *network lifetime* definitions for theoretical analysis and simulations: (1) CRITICAL APPLICATION NODE LIFETIME (CANLT): The mission fails when any AN runs out of energy, i.e., the lifetime L_N is $L_N = \min_{i=1}^{N}\{l_i\}$. The first AN that run out of energy are denoted as the critical AN. (2) FULL COVERAGE LIFETIME (FCLT): A small sensor is called a *covered* sensor if it has at least one alive AN neighbor. The total sensing area of all *covered* sensors is called the covered area of the WSN here. The mission fails when the covered area of the WSN is smaller than the originally covered area. (FCLT).

Related Works. Numerous literatures have discussed efficient cluster formation for wireless ad hoc and sensor networks. Although almost all works assumed that there are some nodes acting as *clusterheads* who are in charge of gathering the information from other nodes and sending back to some base stations, the criteria of forming the clusters vary from case to case. One fundemental difference between the cluster formation problem studied in this paper and the traditional cluster formation problems is that *every* node could be a clusterhead in the traditional methods, while only the AN can be the clusterhead for the problems studied here.

In the Linked Cluster Algorithm (LCA) [1], a node becomes the clusterhead if it has the highest identity among all nodes within one hop of itself or among all nodes within one hop of one of its neighbors. This algorithm was improved by the LCA2 algorithm [6], which generates a smaller number of clusters. The LCA2 algorithm elects the node, with the lowest ID among all nodes which are not within 1-hop of any chosen clusterheads, as a new clusterhead. The algorithm proposed in [7], chooses the node with highest degree among its 1−hop neighbors as a clusterhead. In [8], the authors propose a distributed algorithm that is similar to the LCA2 algorithm. The Distributed Clustering Algorithm (DCA) uses weights associated with nodes to elect clusterheads [9]. It elects the node that has the highest weight among its 1-hop neighbors as the clusterhead. The DCA algorithm is suitable for networks in which nodes are static or moving at a very low speed.

Results reported in [4,5] are closest to this paper in spirit. In [4], Pan *et al.* studied the problem of maximizing lifetime of a two-tiered WSN with focus on the top-tier. By assuming the *prior known* fixed cluster formation, the authors mainly studied how to place the base-station in the network such that the lifetime of the WSN is maximized. The ANs are assumed to be homogenous in [4] and generalized to be heterogenous in [5]. The authors also discussed how to relay the packets via ANs to some fixed based stations. In this paper, we will focus on the lower-tier of the two-tiered WSN: how to form the cluster (associate small sensors to application nodes) so the network lifetime is maximized.

2   Homogeneous Small Sensors

In this section, we study the case when the small sensors are homogeneous, *i.e.*, all small sensors have the same data rate, say r. Thus $r(S) = r \cdot |S|$, where $|S|$ is the number of small sensors in the set S.

2.1   Homogeneous Application Nodes

In this subsection, we discuss how to maximize the lifetime of the WSN when all application nodes are homogeneous, i.e., their initial on-board energy are the same, say $\mathcal{P}$ and the energy consumption functions are the same, say $p(x)$. Remember that $L_N = \min_{i=1}^n \{l_i\} = \min_{i=1}^n \frac{\mathcal{P}}{p(r \cdot |S_i|))}$ and $p(x)$ is increasing. Thus maximizing the lifetime L_N of the WSN is equivalent to minimizing the maximum cluster size. For simplicity, we denote $x_{i,j} = 1$ if the sensor s_j belongs to cluster $\mathcal{C}_i$, and $x_{i,j} = 0$ otherwise. Let $N(v_i)$ be the set of sensors who are v_i's neighbors. We formalize the problem of maximizing L_N as the following Integer Programming.

$$\textbf{IP (1)}: \quad \min \max_{v_i \in V_N} \sum_{s_j \in S_M} x_{i,j}$$

Subject to constraint set (1)

$$\textbf{CS (1)}: \quad x_{i,j} = 0, \forall v_i, \forall s_j \notin N(v_i); \quad x_{i,j} \in \{0,1\}, \forall s_j, \forall v_i; \quad \sum_{v_i} x_{i,j} = 1, \forall s_j$$

Obviously, a feasible solution of the **IP** (1) problem is a *feasible* cluster formation. For simplicity, the set of small sensors in the cluster $\mathcal{C}_i$ is denoted as S_i in this paper, when no confusion is caused. For simplicity, let $\mathbf{x}^{min}$ be the solution to **IP** (1) and $T^{\min} = \min \max_{v_i \in V_N} \sum_{s_j \in S_M} \mathbf{x}_{i,j}^{min}$. Next we present two different approaches to solve the **IP** (1) exactly.

Efficient Centralized Approach. Note that $T^{\min}$ is a non-negative integer at most m, thus it could have m possible values. For an given integer k, if we can decide whether we can find a feasible solution $\mathbf{x}$ such that (1) it satisfies constraint **CS** (1); (2) $\min \max_{v_i \in V_N} \sum_{s_j \in S_M} \mathbf{x}_{i,j}^{min} = k$, then by performing a binary search on $T^{\min}$ we can find the exact value. Following, we use a Max-Flow approach to find a feasible solution for given integer k if it exists.

The idea is that we construct a flow network as shown in Figure 1 with s as the source and t as the sink. There is a directional link $\overrightarrow{sv_i}$, $1 \leq i \leq n$ with capacity k, a directional link between $\overrightarrow{v_i s_j}$ with capacity 1 if $s_j \in N(v_i)$ and a directional link $\overrightarrow{s_j t}$ with capacity 1. Usually, for a maximum flow problem, the flow on each directional link could be any real number. Fortunately, all capacities in the graph take on only integral values. Thus, the maximum flow f has the property that $|f|$ is integer-valued. Moreover, for all vertices u and v, the flow on edge uv is an integer. Therefore, each link $\overrightarrow{v_i s_j}$ is either 0 or 1. Remember

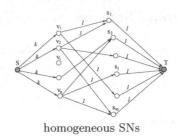

homogeneous SNs

Fig. 1. A flow network for two-tiered WSN

that the flow on link $\overrightarrow{v_i s_j}$ corresponds to $x_{i,j}$, which implies that $x_{i,j} \in \{0,1\}$ for every v_i and $s_j \in N(v_i)$. Thus, a flow in Figure 1 and a solution to **IP** (1) has an one to one mapping.

We can find the solution to **IP** (1) by solving $\log m$ max-flow problems for different values of T. Thus, the time complexity for Max-Flow approach is $m \cdot \log m \cdot (n + m)^3$, which is very expensive and impractical. Notice that the cluster formation problem with minimum cluster size becomes the Maximum Cardinality Matching problem in a bipartite graph [10]. In [10], Hopcroft and Karp presents the best known algorithm that achieves the time complexity $\sqrt{m} \cdot nm$. This reduces the time complexity from $O((n + m)^3)$ to $O(nm \cdot (n + m)^{1/2} \log(n + m))$ for a fix value T. Therefore, we can solve the **IP** (1) in time $O(n \cdot m^{3/2} \log^2(m))$.

Efficient Distributed Algorithm by Smoothing. Although the previous approach computes a clustering quickly in centralized manner, it may be too expensive to collect the necessary information. In this subsection, we propose a different approach that can be implemented efficiently in a distributed manner. The basic idea of this approach is to construct a virtual directed graph on ANs and iteratively move the sensors from those clusters who have the largest number of small sensors to smaller clusters. In the virtual directed graph, there is an edge $\overrightarrow{v_i v_k}$ from AN v_i to v_k if there is a sensor s_j that can be moved from the cluster of v_i to the cluster of v_k. The weight of the edge is the number of such small sensors that can be moved from the cluster of v_i to the cluster of v_k. Following algorithm presents the method constructing a virtual graph based on a feasible solution **x** to **CS** (1).

Algorithm 1. Constructing the virtual graph
___
1: Set V_N as the vertices for virtual graph VG.
2: **for** every pair of v_i and s_j such that $x_{i,j} = 1$ **do**
3:    **for** every v_k such that $s_j \in N(v_k)$ **do**
4:       **if** there is no directed edge $\overrightarrow{v_i v_k}$ from v_i to v_k **then**
5:          Add a directed edge $\overrightarrow{v_i v_k}$ from v_i to v_k. Set the weight of the edge to $c(v_i v_k) = 1$.
6:       **else**
7:          Update the weight as $c(v_i v_k) = c(v_i v_k) + 1$.

In the directed virtual graph $VG(\mathbf{x})$, if there is a path from v_i to v_j, then we say v_i *reaches* v_j. All vertices that v_i can reach forms a set $\mathcal{R}_i(\mathbf{x})$, called the *clique* centered at the AN v_i. Given a solution $\mathbf{x}$ of **CS** (1) and its corresponding virtual graph $VG(\mathbf{x})$, we have the following property about cliques (its proof is omitted due to space limit).

Lemma 1. *Given a feasible assignment $\mathbf{x}$ of small sensors to ANs and its corresponding virtual graph $VG(\mathbf{x})$, for any AN v_i and its clique $\mathcal{R}_i(\mathbf{x})$ in $VG(\mathbf{x})$, if $\mathbf{y}$ is also a feasible assignment of SNs to ANs, we have $\sum_{v_j \in \mathcal{R}_i(\mathbf{x})} |S_i(\mathbf{x})| \leq \sum_{v_j \in \mathcal{R}_i(\mathbf{y})} |S_i(\mathbf{y})|$*

The Algorithm relies on the relation between $\omega_i(\mathbf{x})$, $\omega_j(\mathbf{x})$ and T^{min} where v_i is the AN with the largest weight and v_j is the AN with the smallest weight in $\mathcal{R}_i(\mathbf{x})$.

Lemma 2. *Let v_i be the AN with the largest weight and v_j be the AN with the smallest weight in $\mathcal{R}_i(\mathbf{x})$ under any feasible assignment $\mathbf{x}$, then $|S_j(\mathbf{x})| \leq T^{min} \leq |S_i(\mathbf{x})|$.*

Proof. The proof is omitted here due to space limit. Please refer to full version of the paper for more details.

Given a virtual graph constructed by Algorithm 1 based on a feasible assignment of SNs to ANs, our approach to find a better solution is to iteratively apply a process called SMOOTH to reduce the maximum weight of the application nodes if possible. Here, the weight of an application node v_i under a feasible assignment $\mathbf{x}$ is the number of small sensors assigned to the cluster $\mathcal{C}_i$, denoted as $\omega_i(\mathbf{x})$.

Algorithm 2. Smooth Algorithm

1: Construct virtual graph $VG(\mathbf{x})$ based on the $\mathbf{x}$ using Algorithm 1.
2: **repeat**
3:     Find any AN with the largest weight, say v_i.
4:     Find any AN with the smallest weight in $\mathcal{R}_i(\mathbf{x})$, say v_j.
5:     Apply procedure SMOOTH$(v_i, v_j, VG(\mathbf{x}), \mathbf{x})$.
6: **until** $\omega_i(\mathbf{x}) \leq \omega_j(\mathbf{x}) + 1$

Algorithm 3. SMOOTH$(v_i, v_j, VG(\mathbf{x}), \mathbf{x})$

1: Let $v_{i_0} v_{i_1} \cdots v_{i_k}$ be the path connecting v_i and v_j with the minimum number of hop. Here, $v_{i_0} = v_i$ and $v_{i_k} = v_j$.
2: **for** $t = 0$ to $k - 1$ **do**
3:     Assume that $x_{t,l} = 1$ for some SN s_ℓ with $s_\ell \in N(v_t)$ and $s_\ell \in N(v_{t+1})$. Set $x_{t,l} = 0$ and $x_{t+1,l} = 1$, i.e., move s_ℓ from cluster $\mathcal{C}_t$ to cluster $\mathcal{C}_{t+1}$.
4:     **for** every v_a such that $s_\ell \in N(v_a)$ **do**
5:         Update $c(\overrightarrow{v_{i_t} v_a}) = c(\overrightarrow{v_{i_t} v_a}) - 1$. Remove directed link $\overrightarrow{v_{i_t} v_a}$ if $c(\overrightarrow{v_{i_t} v_a}) = 0$.
6:         Update $c(\overrightarrow{v_{i_{t+1}} v_a}) = c(\overrightarrow{v_{i_{t+1}} v_a}) + 1$. Add a directed link $\overrightarrow{v_{i_{t+1}} v_a}$ if $c(\overrightarrow{v_{i_{t+1}} v_a}) = 1$.
7: Set $\omega_j(\mathbf{x}) = \omega_j(\mathbf{x}) + 1$ and $\omega_i(\mathbf{x}) = \omega_i(\mathbf{x}) - 1$.

Theorem 1. *Algorithm 2 terminates after at most m iterations, with an solution to **IP** (1).*

Proof. From Lemma 2, we have $\omega_j(\mathbf{x}) \le T^{\min} \le \omega_i(\mathbf{x})$. If Algorithm 2 does not stop at this iteration, we have $\omega_i(\mathbf{x}) > \omega_j(\mathbf{x}) + 1$, which implies that $T^{min} > \omega_j(\mathbf{x}) + 1$. For a feasible solution $\mathbf{x}$, we define $\delta_i(\mathbf{x}) = |\mathcal{S}_i(\mathbf{x})| - |\mathcal{S}_i(\mathbf{x}^{\min})|$ if $\omega_i(\mathbf{x}) > T^{min}$ and 0 otherwise. Let $\Delta(\mathbf{x}) = \sum_{v_i \in V_N} \delta_i(\mathbf{x})$, it is not difficult to observe that $\Delta(\mathbf{x})$ will be decreased by 1 for each iteration. Thus, Algorithm 2 terminates after at most m iterations.

Remember when Algorithm 2 terminates, we have $\omega_i(\mathbf{x}) \le \omega_j(\mathbf{x}) + 1$. Combining with the relation $\omega_j(\mathbf{x}) \le T^{\min} \le \omega_i(\mathbf{x})$, we have $\omega_j(\mathbf{x}) \le T^{\min} \le \omega_i(\mathbf{x}) \le \omega_j(\mathbf{x}) + 1$. This implies that $T^{\min} = \omega_i(\mathbf{x}) - 1$ or $T^{\min} = \omega_i(\mathbf{x})$. First, we consider the case when $T^{\min} = \omega_i(\mathbf{x}) - 1$. In this case, we have $\omega_j(\mathbf{x}) \ge T^{\min}$ for every $v_j \in \mathcal{R}_i(\mathbf{x})$ which implies $\sum_{v_j \in \mathcal{R}_i(\mathbf{x})} |\mathcal{S}_i(\mathbf{x})| \ge T^{\min} \cdot |\mathcal{R}_i(\mathbf{x}) - 1| + T^{\min} + 1$. For the solution $\mathbf{x}^{min}$ of **IP** (1), every AN's weight is not greater than T^{min}. Thus, $\sum_{v_j \in \mathcal{R}_i(\mathbf{x})} |\mathcal{S}_i(\mathbf{x}^{min})| \le T^{\min} \cdot |\mathcal{R}_i(\mathbf{x})|$. From Lemma 1, we have $\sum_{v_j \in \mathcal{R}_i(\mathbf{x})} |\mathcal{S}_i(\mathbf{x}^{min})| \ge \sum_{v_j \in \mathcal{R}_i(\mathbf{x})} |\mathcal{S}_i(\mathbf{x})|$. This implies that $T^{\min} \cdot |\mathcal{R}_i(\mathbf{x})| \ge T^{\min} \cdot |\mathcal{R}_i(\mathbf{x}) - 1| + T^{\min} + 1$, which is a contradiction. Thus, $T^{\min} = \omega_i(\mathbf{x})$. Remember that $\mathbf{x}$ is a solution to the **CS** (1). Therefore, $\mathbf{x}$ is a solution to **IP** (1). This finishes our proof.

Now we analyze the time complexity of Algorithm 2. In procedure SMOOTH $(v_i, v_j, VG(\mathbf{x}), \mathbf{x})$, there are at most n nodes on the path between v_i, v_j and up to n iterations in the "FOR" loop between line 4-7. Thus, the time complexity of SMOOTH$(v_i, v_j, VG(\mathbf{x}), \mathbf{x})$ is $O(n^2)$. From Theorem 1, it takes at most $O(m \cdot n^2)$ for Algorithm 2 to terminate. Constructing the virtual graph based on a feasible solution $\mathbf{x}$ could take time $O(m \cdot n^2)$. Thus, the total time complexity of smoothing algorithm is also $O(m \cdot n^2)$. If $n = o(\sqrt{m})$, then Algorithm 2 outperforms the best known max-flow algorithm by $\log^2 m$; when n is a constant the time complexity becomes $O(m)$ which is optimal.

Efficient Distributed Implementation. So far we have illustrated the basic idea of the Smoothing algorithm, which clearly can be implement in a distributed manner. In the remainder of the section, we will describe how this method can be implemented efficiently. Given an AN v_i, we say v_i is adjacent to AN v_j if there is a small sensor s_k in the cluster $\mathcal{C}_i \bigcap N(v_j)$. If v_i and v_j are not adjacent, then we define the distance between v_i and v_j as the smallest number of hops between them if we consider the adjacent graph of the ANs. For an AN v_i that is adjacent v_j, let ℓ be the largest non-negative integer such that $\frac{p_j(r \cdot \sum_{s_k \in S_M} x_{j,k} + \ell \cdot r)}{\mathcal{P}_j} < \omega_i(\mathbf{x})$. We define the *difference* of v_i and v_j as $\text{dif}_{i,j}(\mathbf{x}) = \ell$. Based on the notation of difference, we have following localized algorithm.

Regarding the distributed Algorithm 4, we have the following theorem.

Theorem 2. *Algorithm 4 converges in at most $m \cdot n$ rounds and total message complexity is $O(n^2 \cdot m)$ if the ANs are homogeneous.*

Algorithm 4. Distributed Smoothing algorithm for AN v_i

1: When v_i receive an UPDATE-LEAVE or UPDATE-JOIN message from an adjacent AN v_j, it updates $\gamma_{i,j}$ if necessary.
2: Let v_j be one of v_i's adjacent AN with the maximum difference. Here, we break the tie arbitrarily.
3: **if** $\text{dif}_{i,j}(\mathbf{x}) \geq 1$ **then**
4:     Send a REQUEST message to AN v_j.
5: When v_j receives all REQUEST messages the ANs that adjacent to it, it sends out an ACK message to the AN that has the maximum weight and REJECT messages to all other ANs.
6: **if** v_i receives an ACK message from v_j **then**
7:     Choose one SN, say s_k, in $\mathcal{C}_i \cap N(v_j)$. Set $x_{i,k} = 0$ and send SUCC message with the ID k to v_j.
8:     Upate $\gamma_{i,j} = \gamma_{i,j} - 1$ and send the UPDATE-LEAVE message with ID k to all adjacent ANs.
9: When v_j receives the SUCC message from v_i with ID k, it first sets $x_{j,k} = 0$ and $\gamma(j,i) = \gamma(j,i) + 1$. After that it also sends UPDATE-JOIN message with ID k to all adjacent ANs.
    Remark: Afterward, we also say that the small sensor s_k is *migrating* from cluster $\mathcal{C}_i$ to $\mathcal{C}_j$.

Proof. Given an assignment $\mathbf{x}$, we denote $\kappa_i(\mathbf{x})$ as the number of small sensors in ith largest cluster. Let $\Gamma_i(\mathbf{x}) = \sum_{j=1}^{i} \kappa_j(\mathbf{x})$, and $\mathbf{x}^k$ be the assignment of sensors in round k. Considering $\Gamma^k = \sum_{i=1}^{n} \Gamma_i(\mathbf{x}^k)$. If there is a small sensor joining $\mathcal{C}_{i^k}$ and leaving $\mathcal{C}_{j^k}$ in round k, then $|S_{i^k}| > |S_{j^k}| + 1$ and $i^k < i^k$. Notice that after the small sensor migrating from cluster $\mathcal{C}_{i^k}$ to $\mathcal{C}_{j^k}$, $\Gamma_\ell(\mathbf{x}^k)$ decreases by 1 if $j < \ell \leq i$ and does not change otherwise. Thus, Γ^k decreases by 1 for every small sensor migrating. It is not difficult to observe that if there is no small sensor migrating in round k, then Algorithm 4 terminates. Since $\Gamma^1 < n \cdot m$, Algorithm 4 terminates in at most $n \cdot m$ rounds.

In every round, every AN sends only one REQUEST message and receives at most one REJECT message. Thus, there is at most $O(n)$ REQUEST and REJECT messages. It is also not difficult to observe that every AN sends at most one ACK messages. Thus, there are at most $O(n^2 \cdot m)$ REQUEST, ACK and REJECT messages in total. On the other hand, there is exact one UPDATE-LEAVE and UPDATE-JOIN message for every small sensor migrating. Thus, there are at most $O(n \cdot m)$ UPDATE-LEAVE and UPDATE-JOIN messages. Therefore, the overall message complexity is $O(n^2 \cdot m)$.

Notice that the message complexity analysis is very pessimistic. In simulations, it is much smaller than the worst case analysis. Observe that when Algorithm 4 terminates, it not necessarily gives an optimal solution. However, Algorithm 4 gives the best solution among all localized algorithms in which every AN can only know the information of its adjacent ANs. Furthermore, if we define the diameter of the network as the largest distance of the ANs, we have the following theorem (its proof is omitted due to space limit).

Theorem 3. *When Algorithm 4 terminates, it gives an assignment with maximum cluster size at most $T \leq T^{\min} + D$ where D is the diameter of the network.*

2.2 Heterogeneous Application Nodes

In subsection 2.1, we discuss how to form the clusters when both the small sensors and application nodes are homogeneous. However, in practice, such node homogeneity cannot always be guaranteed. For example, the initial onboard energy of ANs built by different vendors may not be proportional to the bit-rate at which they generate, or the application nodes could be redeployed (e.g., new ANs join the system long after old ANs have been activated). Furthermore, two different application nodes may consume different energy to receive, process and send the information to the base station even given the same set of small sensors. Thus, it is more practical to assume the application nodes are heterogenous. In this paper, we consider the heterogeneity in two ways: the initial on board energy $\mathcal{P}$ and energy consumption function $p(x)$ where x is the sum of the rate of the small sensors in the cluster.

In this subsection, we redefine *weight* of a AN v_i for assignment $\mathbf{x}$ as $\omega_i(\mathbf{x}) = \frac{p_i(r \cdot \sum_{s_j \in S_M} x_{i,j})}{\mathcal{P}_i}$, where $\mathcal{P}_i$ is the initial onboard energy and $p_i(x)$ is energy consumption function. Here, the lifetime of the network is defined as $L = \max \min_{v_i \in V_N} \frac{\mathcal{P}_i}{p_i(r \cdot \sum_{s_j \in S_M} x_{i,j})} = \min \max_{v_i \in V_N} \omega_i(x)$

Thus maximizing the lifetime is equivalent to minimizing the maximum weight over all ANs. Similar to the approach for the homogenous application node case, we formalize the problem as an Integer Programming as follows.

$$\text{IP (2)}: \quad \min \max_{v_i \in V_N} \frac{p_i(r \cdot \sum_{s_j \in S_M} x_{i,j})}{\mathcal{P}_i}$$

Subject to constraint set (2):

$$\text{CS (2)}: \quad x_{i,j} = 0, \forall v_i, \forall s_j \notin N(v_i); \quad x_{i,j} \in \{0,1\}, \forall s_j, \forall v_i; \quad \sum_{v_i} x_{i,j} = 1, \forall s_j$$

Algorithm 5. Smoothing algorithm for heterogenous ANs

1: Find a feasible solution $\mathbf{x}$, *e.g.*, randomly assign every SN to a neighboring AN.
2: Construct a virtual graph $VG(\mathbf{x})$ based on $\mathbf{x}$ by applying Algorithm 1.
3: **repeat**
4:     Choose any one of AN with the largest weight randomly, say v_i.
5:     Define $\omega_k^+(\mathbf{x}) = \frac{p_k(r \cdot \sum_{s_j \in S_M} x_{k,j} + r)}{\mathcal{P}_k}$.
6:     Find any AN v_j with the smallest $\omega_j^+(\mathbf{x})$ in $\mathcal{R}_i(\mathbf{x})$. If there are more than one such ANs, choose one randomly.
7:     Apply SMOOTH_HETE$(v_i, v_j, VG(\mathbf{x}), \mathbf{x})$ if $\omega_i(\mathbf{x}) > \omega_j^+(\mathbf{x})$
8: **until** $\omega_i(\mathbf{x}) \leq \omega_j^+(\mathbf{x})$

Algorithm 6. SMOOTH_HET$(v_i, v_j, VG(\mathbf{x}), \mathbf{x})$

1: Let $v_{i_0}(v_i)v_{i_1} \cdots v_{i_k}(v_j)$ be the path connecting v_i and v_j with the minimum number of hop. Here, $v_{i_0} = v_i$ and $v_{i_k} = v_j$.
2: **for** $t = 0$ to $k - 1$ **do**
3:     Assume $x_{t,l} = 1$ and $s_\ell \in N(v_{t+1})$. Set $x_{t,l} = 0$ and $x_{t+1,l} = 1$.
4:     **for** every v_a such that $s_\ell \in N(v_a)$ **do**
5:         Update $c(v_{i_t}v_a) = c(v_{i_t}v_a) - 1$. Remove directed link $v_{i_t}(v_a)$ if $c(v_{i_t}v_a) = 0$.
6:     **for** every v_b such that $s_\ell \in N(v_b)$ **do**
7:         $c(v_{i_{t+1}}v_b) = c(v_{i_{t+1}}v_b) + 1$. Add a directed link $v_{i_{t+1}}v_b$ if $c(v_{i_{t+1}}v_b) = 1$.
8: Update $\omega_j(\mathbf{x}) = \omega_j^+(\mathbf{x})$ and $\omega_i(\mathbf{x}) = \dfrac{p_i(r \cdot \sum_{s_j \in S_M} x_{i,j} - r)}{\mathcal{P}_i}$.

Smoothing Algorithm. In this subsection we shows that our smoothing Algorithm 2 also applies to the heterogenous case with only minor modification.

Lemma 3. *Let v_i be the AN with the largest weight and v_j be the AN with the lowest weight that is reachable by v_i in a feasible assignment $\mathbf{x}$. Then $\omega_j(\mathbf{x}) \leq T^{min} \leq \omega_i(\mathbf{x})$.*

Theorem 4. *Algorithm 5 outputs a solution of $\mathbf{IP}$ (2) and terminates after m iterations.*

The proof of this theorem is omitted here due to space limit. Surprisingly, the time complexity of Algorithm 5 is also $O(m \cdot n^2)$, which is exactly the same as in the homogenous case. This reduces the time complexity by an order of $\sqrt{m} \log^2 m$ and more importantly, Algorithm 4 also works for the heterogenous case with only modification of the definition of difference. However, we only have the following conjecture for the convergence and message complexity of localized smoothing algorithm. It is an open and interesting problem to either prove or disprove the following conjecture.

Conjecture 1. Algorithm 4 terminates after at most $n \cdot m$ rounds and the total message complexity $O(n^2 \cdot m)$ when the ANs are heterogenous.

3   Heterogeneous Small Sensors

Usually in WSNs, several different kinds of sensors cooperate together to fulfill some certain goals. Some sensors may generate data at a higher rate than others do, *e.g.*, the visual sensors have a bit-rate that is much higher than the bit-rate generated by a temperature sensor. Even in scenarios when all small sensors are of same type, sometimes sensors located at different locations may need to sample the data at a different time interval. Thus, it is more reasonable to assume that in a WSN different type of sensors produce different bit-rates.

By assuming that every small sensor has its own data rate r_i, we formalize the problem of maximizing the lifetime as an Integer Programming as follows:

$$\mathbf{IP}\ (3):\quad \min_{v_i \in V_N} \max \frac{p_i(\sum_{s_j \in S_M} r_j \cdot x_{i,j})}{\mathcal{P}_i}$$

Subject to constraint set (3)

CS (3) : $x_{i,j} = 0, \forall v_i, \forall s_j \notin N(v_i);$ $x_{i,j} \in \{0, 1\}, \forall s_j, \forall v_i;$ $\sum_{v_i} x_{i,j} = 1, \forall s_j$

Unlike the case for homogenous SNs in which we can find the solution that maximizes the lifetime exactly, Theorem 5 shows that it is NP-Hard to find the solution to **IP** (3).

Theorem 5. *We can not find the solution of **IP** (3) in polynomial time if $P \neq NP$.*

Proof. We consider the special case when application nodes are homogeneous. In this case, since $p_i(x) = p(x)$ is increasing, it is equivalent to minimizing the maximum $\sum_{s_j \in S_M} r_j \cdot x_{i,j}$ subject to constraints set (1). If every AN v_i satisfies that $N(v_i) = S_M - v_i$, then the problem becomes the traditional job scheduling problem [11,12], which is known to be NP-Hard. This finishes our proof.

Since solving **IP** (3) is NP-hard, we will present an algorithm approximating the optimal solution by borrowing some ideas from job scheduling [13,14]. Again we transform **IP** (3) into Integer Programming as follows.

$$\textbf{IP } (4): \quad \min T$$

Subject to constraints set (4)

$$\textbf{CS } (4): \quad x_{i,j} = 0, \forall v_i, \forall s_j \notin N(v_i); x_{i,j} \in \{0, 1\}, \forall s_j, \forall v_i;$$
$$\sum_{v_i} x_{i,j} = 1, \forall s_j; \sum_{s_j \in S_M} r_j \cdot x_{i,j} \le k_i, \forall v_i$$

Here $k_i = p_i^{-1}(\mathcal{P}_i \cdot T)$. Let $\mathbf{x}^{min}$ be the solution to **IP** (4) and T^{min} be the $\min T$ under solution $\mathbf{x}^{min}$. It is easy to observe that $\mathbf{x}_{ij}^{min}$ satisfies the following constraint.

$$x_{i,j} = 0 \qquad \forall v_i, \forall s_j \quad r_j > k_i \tag{1}$$

If we relax the constraint $x_{i,j} \in \{0, 1\}$, we obtain a Linear Programming (4). Let $x^\star$ be the solution to **LP** (4) plus constraint 1 and $T^\star$ be the value of $\min T$ under solution $x^\star$. Then $T^\star \le T^{min}$. By binary search on $T^\star$ we can find the solution $x^\star$ to **LP** (4) plus constraint 1 in polynomial time. Furthermore, we can find a solution $x^\star$ that has some special properties. For a small sensor s_j, if there exists an AN v_i such that $0 < x_{i,j} < 1$, we call s_j is fractionally assigned to cluster $\mathcal{C}_i$. We construct a graph with vertex $V_N \bigcup S_M$ and add an edge $s_j v_i$ if and only if $0 < x_{i,j} < 1$. Obviously, it is a bipartite graph and it is generally known [14,15] that we can transform the solution $x^\star$ to another solution $x^\star$ such that its corresponding bipartite graph is composed of forests with(or without) a line. Remember that every node in S_M connects to at least two nodes in A_N, thus there is a matching such that every node in S_M can connect to a distinct node in

A_N. The final solution is to assign s_j to cluster with head v_i if one of the following two conditions holds: (1) $x_{ij}^* = 1$ (2)s_j is connected with v_i in the matching.

In this section, to make sure that we can guarantee the performance of the above job-scheduling based approach, we add one more requirement for the power consumption function p_i. We assume that the marginal cost of $p_i(x)$ is not increasing, i.e., for $x_1 \geq x_2$, $p_i(x_1 + \delta) - p_i(x_1) \leq p_i(x_2 + \delta) - p_i(x_2)$. This assumption is almost universally satisfied. If this assumption is not satisfied, we can construct examples to show that the above approach (based on job scheduling) cannot provide any theoretical performance guarantees, although its practical performance may still be good.

Theorem 6. *Our job scheduling based method produces a cluster formation such that the lifetime of the WSN is at least $\frac{1}{2}$ of the maximum lifetime of the WSN.*

4   Performance Studies

We mainly study the case with heterogeneous application nodes and homogeneous sensor nodes. We randomly placed 2000 sensor nodes in a $800 \times 800 feet$ square region, the transmission range of each sensor node is set to $50 feet$ and the sensing range is set to $10 feet$. Then we put a different number of application nodes, from 150 to 300 (with incremental 25) and measured the network lifetime. In addition, the initial battery power of each sensor node is a random value between 100 units and 200 units. A SN node is called an *alive sensor* if it has power remaining and has at least one alive application node in neighborhood. We compare our Algorithm 2 with other heuristics listed below: (1) *[-Nearest]* Each sensor node is assigned to the nearest AN. (2) *[-Arbitrary]* Each sensor node is *randomly* assigned to one of the neighboring application nodes. (3) *[-Smart-Arbitrary]*: The probability of a SN s_j assigned to a neighboring AN v_i is the ratio of the remaining power of v_i over the total remaining power of all neighboring ANs of this SN s_j. (4) *[-All]* Here, each sensor node is assigned to *all* the application nodes that are inside the sensor node's transmission range. This is clearly the worst method. Thus we will not compare with this method in most simulations.

Lifetime. We compare the lifetime of four different methods under two different definitions of lifetimes: CANLT, FCLT. Figure 2 (a), (b) show the lifetime

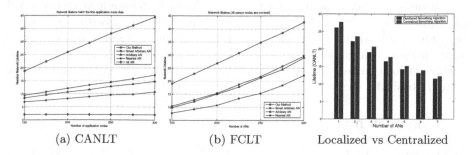

(a) CANLT   (b) FCLT   Localized vs Centralized

Fig. 2. Comparison of lifetime for different methods

of different assignment methods under lifetime definition CANLT, FCLT respectively. We generate 100 random WSNs and all results are the average over the performance of these 100 WSNs.

As can be seen, the network lifetime increases almost linearly with the number of application nodes available initially for all methods, except the simplest ALL approach that does not perform any logic cluster at all. A striking observation is that, as we expected, our smoothing based method outperforms all other tree methods under all four definitions of lifetimes regardless of the density of the application nodes. In all simulations, we found that our method generally outperforms the other methods by almost 100%. In other words, the network lifetime is almost *doubled* when our method is used to form the cluster.

We also compare the performance of the Centralized Smoothing Algorithm 2 (CSA) and Localized Smoothing Algorithm 4 (LSA). We fixed the number of the ANs to 50 and varies the number of SNs from 200 to 500. Figure 3 (c) shows difference of the lifetime (CANLT) between CSA and LSA, and it is not difficult to observe that the lifetime of LSA and CSA only differs about 5% to 8%. This corroborates our theoretical analysis and we will only compare the lifetime of CSA with other four methods afterwards.

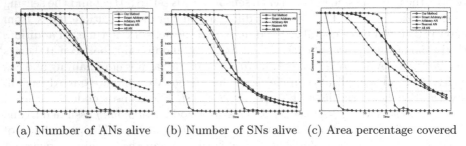

(a) Number of ANs alive   (b) Number of SNs alive   (c) Area percentage covered

Fig. 3. Comparison for different methods

Load Balancing. As mentioned in Section 2.2, for heterogeneous application nodes case, application nodes have different initial battery powers, and the objective of the Algorithm 2 is to assign less sensor nodes to application nodes that have lower remaining battery power and more sensor nodes to application nodes that have higher battery power. To see how good the load balancing of our algorithm is, we run simulation for the networks with 150 application nodes till all application nodes die. As can be seen in Figure 3, our algorithm achieves a very good load balancing meaning that all application nodes consume energy at a rate proportional to their initial battery power and then they all die together. The result for number of alive sensor nodes and also the percentage of coverage area are basically the same as shown in Figure 3 (b) and (c).

5   Conclusion

In this paper, we studied how to organize the WSN to form logic clusters to maximize the lifetime of the networks. We also showed that it is NP-hard to find the optimum cluster formation. Our theoretical results are corroborated by extensive simulation studies. Our simulations show that our algorithms actually perform very well.

References

1. Baker, D.J., Ephremides, A.: The architectural organization of a mobile radio network via a distributed algorithm. IEEE Transactions on Communications 29(11), 1694–1701 (1981)
2. Parekh, A.K.: Selecting routers in ad-hoc wireless networks. In:Proceedings of ITS (1994)
3. Bandyopadhyay, S., Coyle, E.: An energy efficient hierarchical clustering algorithm for wireless sensor networks. In: Proceedings of the 22nd Annual Joint Conference of the IEEE Computer and Communications Societies (Infocom) (2003)
4. Cai, L., Shi, Y., Pan, J., Hou, Y.T., Shen, S.X.: Topology control for wireless sensor networks. In: Proceedings of the 9th Annual International Conference on Mobile Computing and Networking, pp. 286–299. ACM Press, New York (2003)
5. Cai, L., Shi, Y., Pan, J., Hou, Y.T., Shen, S.X.: Optimal base-station locations in two-tiered wireless sensor networks. In: IEEE TRANSACTIONS ON MOBILE COMPUTING (To appear)
6. Wieselthier, J., Ephremides, A., Baker, D.J.: A design concept for reliable mobile radio networks with frequency hopping signaling. Proceedings of IEEE 75, 56–73 (1987)
7. Parekh, A.K.: Selecting routers in ad-hoc wireless networks. In: Proceeding ITS (1994)
8. Lin, C.R., Gerla, M.: Adaptive clustering for mobile wireless networks. IEEE Journal on Selected Areas in Communications 15, 1265–1275 (1997)
9. Basagni, S.: Distributed clustering for ad hoc networks. In: Proceedings of the 1999 International Symposium on Parallel Architectures, Algorithms and Networks (ISPAN '99), p. 310. IEEE Computer Society, Washington (1999)
10. Hopcroft, J.E., Karp, R.M.: $n^{5/2}$ algorithm for maximum matchings in bipartite graphs. SIAM Journal on Computing 2, 225–231 (1973)
11. Graham, R.: Bounds for multiprocessing timing anomalies. In: SIAM Journal on Applied Mathematics 17 (1969)
12. Hochbaum, D.S., Shmoys, D.B.: Using dual approximation algorithms for scheduling problems theoretical and practical results. J. ACM 34(1), 144–162 (1987)
13. Tardos, E., Lenstra, J.K., Shmoys, D.B.: Approximation algorithms for scheduling unrelated parallel machines. Math. Program 46(3), 259–271 (1990)
14. Jansen, K., Porkolab, L.: Improved approximation schemes for scheduling unrelated parallel machines. In: Proceedings of the Thirty-first Annual ACM Symposium on Theory of Computing, 1999th edn., pp. 408–417. ACM Press, New York (1999)
15. Shmoys, D.B., Tardos, E.: An approximation algorithm for the generalized assignment problem. Math. Program 62(3), 461–474 (1993)

Distributed Facility Location Algorithms for Flexible Configuration of Wireless Sensor Networks

Christian Frank and Kay Römer

Department of Computer Science
ETH Zurich, Switzerland
{chfrank,roemer}@inf.ethz.ch

Abstract. Many self-configuration problems that occur in sensor networks, such as clustering or operator placement for in-network data aggregation, can be modeled as facility location problems. Unfortunately, existing distributed facility location algorithms are hardly applicable to multi-hop sensor networks. Based on an existing centralized algorithm, we therefore devise equivalent distributed versions which, to our knowledge, represent the first distributed approximations of the facility location problem that can be practically implemented in multi-hop sensor networks with local communication. Through simulation studies, we demonstrate that, for typical instances derived from sensor-network configuration problems, the algorithms terminate in only few communication rounds, the run-time does not increase with the network size, and, finally, that our implementation requires only local communication confined to small network neighborhoods. In addition, we propose simple extensions to our algorithms to support dynamic networks with varying link qualities and node additions and deletions. Using link quality traces collected from a real sensor network deployment, we demonstrate the effectiveness of our algorithms in realistic multi-hop sensor networks.

1 Introduction

An important problem in wireless sensor networks [1] is *self-configuration* [2], where network nodes take on different functions to achieve a given application goal. One example is *clustering* [3], where some nodes are elected as *cluster leaders*, serving as communication hubs for nearby nodes. A similar problem is aggregator placement [4], where some nodes are elected as *aggregators* that collect and aggregate sensor data from nearby sensor nodes. Recently, tiered sensor networks [5] have been proposed, consisting of resource-poor sensor nodes in the first tier and powerful hub nodes in the second tier. In these networks, every sensor node is assigned to and controlled by a hub node. Note that in all of the above examples, self-configuration consists in electing some nodes as *servers* while the remaining *client* nodes are assigned to a server.

While many proposals exist for finding such network configurations, they often do not pay attention to optimizing the *overall cost* of these configurations, which consists of two components: on the one hand, the costs of operating the servers (e.g., representing the servers' increased communication load as these forward traffic for many clients), and, on the other hand, the costs of communication between clients and their server. In wireless networks, the latter cost can be dependent on the physical distance between a

J. Aspnes et al. (Eds.): DCOSS 2007, LNCS 4549, pp. 124–141, 2007.

client and its server (as a longer wireless link requires higher transmit power and thus increased energy consumption), on the number of hops in a multi-hop network graph, or on interference and network congestion. In all cases, lowering communication costs by means of additional hub nodes may prove beneficial.

Our goal is the provision of a *generic* and *practical* mechanism for finding cost-optimized solutions to the above self-configuration problems. Our approach is based on the observation that the above optimization problem can be modeled as an (uncapacitated) *facility location problem*. There, we are given a set F of *facilities*, a set C of *clients* (also known as cities or customers), a cost f_i for opening a facility $i \in F$ and connection costs c_{ij} for connecting client j to facility i. The objective is to open a subset of facilities in F and connect each client to an open facility such that the sum of connection and opening costs is minimized.

Although the facility location problem has been studied extensively in the past, no *practical* solutions exist that would be suitable for multi-hop sensor networks. While distributed algorithms for facility location exist, they are either not generally applicable [6], require a certain (albeit small) amount of global knowledge [7], require impractical communication models [7,8], or (based on the provided approximation factor [8]) might not improve over existing configuration heuristics for sensor networks.

We therefore contribute a local facility location algorithm that lends itself well for implementation in multi-hop sensor networks and provides an approximation factor of 1.61 for metric instances. By means of an experimental study, we show that the algorithm terminates after few communication rounds for typical problem instances derived from sensor network configuration problems.

While the above view adopts a static graph model of sensor networks, practical sensor networks are rather dynamic: nodes may fail and the quality of wireless links fluctuates over time. To make our algorithm applicable to such realistic settings, we propose a set of rules to repair a sensor network configuration in case of node failures, additions, and link quality changes. Also, we study the optimality of our algorithm using link quality traces collected from a real sensor network deployment.

2 Preliminaries

We model the multi-hop network subject to configuration as a graph $G = (V, E)$. In our application of the facility location problem, a network node takes on the role of a client and that of a potential facility at the same time, that is, $F = C = V$. In some cases, only a subset nodes have the necessary capabilities (e.g., remaining energy, available sensors, communication bandwidth, or processing power) to execute a service. In such cases, the nodes eligible as facilities can be selected beforehand based on their capabilities [9], which results in $F \subseteq C = V$. When clients are connected to facilities, we will use $\sigma(j)$ to refer to the facility that connects a client j.

Based on the problem at hand, one may choose a particular setting of opening costs f_i and connection costs c_{ij}. In most settings, for a network link $(i, j) \in E$, the respective communication cost c_{ij} will be set to some link metric that can be determined locally at the nodes, e.g., based on dissipated energy or latency. Some approximation algorithms require that the costs c_{ij} constitute a *metric* instance. A *metric* instance requires that, for any three nodes i, j, k, the direct path is shorter than a detour ($c_{ij} \leq c_{ik} + c_{kj}$).

However, if connection costs c_{ij} should represent the transmit power used for sending, these are often proportional to the square of the geographic distance between i and j, which results in non-metric instances, for example:

If the input to a facility location algorithm is non-metric, the problem is particularly hard to solve (see Section 3 below). However, one may obtain a *metric* instance by ignoring non-metric links and setting c_{ij} to the cost of a shortest path between two nodes i and j. In multi-hop networks, the required shortest-paths computation can be achieved using a local flood around the current node.

When addressing settings in which facilities and clients can be an arbitrary number of network hops apart, we will always compute c_{ij} via shortest-paths. We refer to this metric problem setting as *multi-hop*. Alternatively, we will consider a second (constrained) version of the problem, in which we require that every client is connected to a facility which is its direct network neighbor. We denote this constrained problem definition as *one-hop*. One-hop instances are inherently non-metric, as missing links $(i, j) \notin E$, modeled by $c_{ij} = \infty$, violate the metric property.

3   Related Work

An ample amount of literature exists on (centralized) approximation algorithms for the NP-hard facility location problem [10]. Such centralized algorithms are not applicable as these would require a prohibitive communication overhead associated with collecting the whole network topology at a single point (e.g., at the network basestation).

For *non-metric* instances of the facility location problem, even approximations are hard to come by: As the set cover problem can be reduced to (non-metric) facility location, the best achievable approximation ratio (even with a centralized algorithm) is logarithmic[1] in the number of nodes [11]. A classic and simple algorithm [12] already comes close to this lower bound. Distributed approximations are rare: [7] solve non-metric facility location even in a constant number of communication rounds. However, the algorithm requires that a coefficient ρ, which is computed from a global view of the problem instance, is distributed to all nodes before algorithm execution – which prevents it from being used "as-is" in practice. Moreover, the algorithm requires global communication among all relevant clients and facilities and therefore can only efficiently be used in the *one-hop* setting where such communication can be implemented efficiently by wireless broadcast. Finally, the best approximation factor it can obtain, which is independent of the problem instance, is on the order of $O(\log(m+n) \log(mn))$ where m and n denote the number of facilities and clients, respectively.

For *metric* instances of the facility location problem, much better approximation factors $\in O(1)$ can be achieved. While it has been shown [13] that a polynomial-time algorithm cannot obtain an approximation ratio better than 1.463, a centralized algorithm [14] already provides a solution that is at most a factor of 1.52 away from the optimum. For the metric case, to our knowledge only one distributed algorithm has

---

[1] This holds unless every problem in NP can be solved in $O(n^{O(\log \log n)})$ time.

been mentioned [8] which solves only a constrained version of the problem in which facilities and clients may be at most 3 hops away. It provides a $3 + \epsilon$ approximation factor derived from a parallelized execution of a respective centralized algorithm [15] and is formulated in terms of a synchronous message passing model. The same paper [8] includes additional versions, which restrict the facility location problem in one way or another. Only recently, a highly-constrained version of the facility location problem has been addressed in a distributed manner [6]. Finally, a distributed algorithm based on hill-climbing [16] addresses a version of the problem in which exactly k facilities are opened. However, the worst-case time complexity and the obtained approximation factor are not discussed explicitly.

In this paper, we develop a distributed version of a centralized algorithm [17] which provides an 1.61 approximation factor with metric instances. Compared to related work, our work improves on the approximation factor achievable in a distributed manner. Moreover, we provide the adaptations required to execute this algorithm in multi-hop networks for which, to our knowledge, no efficient algorithm with guaranteed worst-case approximation factor exists. Finally, compared to [7,8], our algorithms do not require a synchronous message passing model. Instead, they perform synchronization among network neighbors implicitly as nodes wait for incoming messages.

In the remainder of the paper, we briefly summarize the centralized approximation algorithms [17] our work is based on in Section 4. We then describe their distributed re-formulation in two steps. The first variant, in Section 5, still requires global communication, namely that all clients communicate with all relevant facilities in each step, and is therefore only applicable to the *one-hop* setting, where this can be efficiently implemented as a wireless broadcast. In the second step, we use this algorithm as a subroutine in the algorithms of Section 6, which distribute messages only to a local neighborhood around the sending node and may therefore be used in multi-hop networks. Finally, we provide experimental results in Section 7 and an outlook to future work in Section 8.

4   Centralized Algorithms

Jain et al. [17] devised two centralized approximation algorithms for the facility location problem. Both use the notion of a *star* (i, B) consisting of a facility i and an arbitrary choice of clients $B \subseteq C$ (in clustering terminology, a star corresponds to a cluster leader and a set of associated slave nodes). The first is shown in Algorithm 1. In its core step (line1.3), the algorithm selects the star (i, B) with best (lowest) cost efficiency. The cost efficiency of a star is defined as

$$c(i, B) = \left(f_i + \sum c_{ij} \right) / |B| \qquad (1)$$

and represents the average cost per client which this star adds to the total cost.

Therefore, in each step, the algorithm selects the most cost-efficient star (i, B), opens the respective facility i, connects all clients $j \in B$ to i (sets $\sigma(j) = i$), and from this point on disregards all (now connected) clients in B. The algorithm terminates once all clients are connected.

Algorithm 1. Centralized 1.861-approximation algorithm [17]

1.1 set $U = C$
1.2 **while** $U \neq \emptyset$ **do**
1.3   find most cost-efficient star (i, B) with $B \subseteq U$
1.4   open facility i (if not already open)
1.5   set $\sigma(j) = i$ for all $j \in B$
1.6   set $U = U \setminus B$
1.7   set $f_i = 0$

Note that in spite of there being exponentially many sets $B \subseteq U$, the most efficient star can be found in polynomial time: For each facility i, clients j can be sorted by ascending connection cost to i. Any most cost-efficient star spanning some $k = |B|$ clients will consist of the first k clients with lowest connection costs – all other subsets of k clients can be disregarded as these cannot be more efficient. Hence, at most $|C|$ different sets must be considered.

When a facility i is opened, its opening cost f_i is set to zero. This allows facility i to be chosen again to connect additional clients in later iterations, based on a cost-efficiency that disregards i's opening costs f_i – as the facility i has already been opened before in order to serve other clients. For metric instances, Algorithm 1 provides a 1.861 approximation factor. Note that line 1.7 constitutes the only difference to a classic algorithm [12], whose approximation factor for metric instances is much worse. An even better approximation factor of 1.61 can be obtained when changing the above algorithm to additionally take into account the benefit of opening a facility i for clients that are already connected to some other facility. This involves two changes.

First, this requires that a revised cost-efficiency definition is used in line 1.3. We let $B(i)$ denote the set of clients j which are already connected to some facility $\sigma(j)$ and would benefit if i would be opened as their connection cost to i would be lower than their current connection cost $c_{\sigma(j)j}$, i.e.,

$$B(i) = \left\{ j \in C \text{ with } \sigma(j) \neq none \text{ and } c_{ij} < c_{\sigma(j)j} \right\}. \tag{2}$$

The cost efficiency of a star (i, B) can now be restated as

$$c(i, B) = \left(f_i + \sum_{j \in B} c_{ij} - \sum_{j \in B(i)} (c_{\sigma(j)j} - c_{ij}) \right) / |B|. \tag{3}$$

A second analogous change is made to line 1.5. In addition to the clients which are part of the most-efficient star (i, B), all already-connected clients $B(i)$ which benefit from switching are connected to i. For this, line 1.5 becomes

$$\text{set } \sigma(j) = i \text{ for all } j \in B \cup B(i).$$

The authors prove [17] that this change improves the approximation factor to 1.61 for metric instances. In the following, we will present a distributed version of this 1.61-algorithm. In the discussed distributed adaptations, we will always use the revised cost-efficiency definition of Eq. (3).

5   One-Hop Approximation

Consider the distributed algorithms given in Algorithm 2 (for facilities) and 3 (for clients). We will show below that they perform the exact same steps as the centralized Algorithm 1. While these algorithms require that each client communicates with each facility and vice versa, the algorithms can be also applied "locally" such that each node communicates only with its network neighbors. This way, they can be used to compute a solution to the *one-hop* version of the facility location problem, for example, to compute an energy-efficient clustering that takes costs of individual links into account. Unfortunately, this constrained problem version results in a non-metric instance (see Section 2) and thus the approximation guarantee of 1.61 cannot be preserved. However, in the next section, we will use these algorithms as a subroutine to obtain an algorithm that maintains the approximation factor of 1.61 for multi-hop sensor networks. Moreover, we will show that it computes good solutions, nevertheless, in our experimental results of Section 7.

We assume that after an initial neighbor discovery phase, each client j knows the set of neighboring facilities, which it stores in the local variable F_j, and the connection costs c_{ij} to facilities $i \in F_j$. Vice versa, each facility i knows the set of neighboring clients C_i and c_{ij} of all $i \in C_i$. In the following we will simply write C and F, as the respective indices i and j can be deduced from the context.

Algorithm 2. Distributed formulation of Algorithm 1 for Facility i

---

2.1   set $U = C$
2.2   **repeat**
2.3   find most cost-efficient star (i, B) with $B \subseteq U$
2.4   **send** $c(i, B)$ to all $j \in U$
2.5   **receive** "connect-requests" from set $B^* \subseteq U$
2.6   **if** $B^* = B$ **then**
2.7   open facility i (if not already open)
2.8   **send** "open" to all $j \in F$
2.9   set $U = U \setminus B$
2.10   set $f_i = 0$
2.11   **receive** $\sigma(j) \neq none$ from set C_a
2.12   set $U = U \setminus C_a$
2.13 **until** $U = \emptyset$

As in Algorithm 1, this time each facility i maintains a set U of unconnected clients which is initially equal to C (line 2.1). Facilities start a round by finding the most cost-efficient star (i, B) with respect to U and sending the respective cost efficiency $c(i, B)$ to all clients in B (lines 2.3-2.4). In turn, the clients can expect to receive cost-efficiency numbers $c(i, B)$ from all facilities $i \in F$ (line 3.2). In order to connect the most cost-efficient star among the many existing ones, clients reply to the facility i^* that has sent the lowest $c(i^*, B)$ with a "connect request" (line 3.4). In turn, facilities collect a set of clients B^* which have sent these "connect requests" (line 2.5). Intuitively, a facility should only be opened if $B = B^*$, that is, if it has connect requests from all clients B in its most efficient star (line 2.6). This is necessary, as it could happen that some clients

---

Algorithm 3. Distributed formulation of Algorithm 1 for a Client j

---

3.1 **repeat**
3.2   **receive** $c(i, B)$ from all $i \in F$
3.3 $i^* = \mathrm{argmin}_{i \in F}\, c(i, B)$ // *use node ids to break ties among equal $c(i, B)$*
3.4   **send** "connect-request" to i^*
3.5   **if** *received "open" from i^** **then**
3.6   set $\sigma(j) = i^*$
3.7   **send** $\sigma(j)$ to all $i \in F$
3.8 **until** *connected*
3.9 **on** "open" from i with $c_{ij} < c_{\sigma(j)j}$
3.10   set $\sigma(j) = i$
3.11   **send** $\sigma(j)$ to all $i \in F$

---

in B have decided to connect to a different facility than i as this facility spans a more cost efficient star. So, if all clients in B are ready to connect, facility i opens, notifies all clients in B about this, removes the connected clients B from U, and sets its opening costs to 0 (lines 2.7-2.10) as in the centralized algorithm.

If a client j receives such an "open" message from the same facility i^* which it had previously selected as the most cost efficient, it can connect to i^* (lines 3.5-3.6). Further, in line 3.7, client j notifies all facilities that it is now connected to i^*, which update their sets of unconnected clients U in lines 2.11-2.12.

Once connected, clients simply switch the facility they are connected to in case a closer facility becomes available (lines 3.9-3.10). This feature enables the 1.61 approximation factor. Note that whenever a client changes its facility $\sigma(j)$, it informs all facilities about this (lines 3.7 and 3.11). All these $\sigma(j)$ messages include the associated connection costs $c_{\sigma(j)j}$ and will be received in line 2.11 of the facility algorithm. By the next iteration, facilities will have received $\sigma(j)$ and $c_{\sigma(j)j}$ from all relevant clients, and will therefore be able to correctly compute the most cost-efficient star (line 2.3) according to Eq.(3).

Discussion. In the following, we argue that the distributed and the centralized versions are equivalent. For this, we denote one execution of the inner loops at Algorithms 3 and 4 as a round. Note that the distributed version opens some stars out-of-order, that is, earlier than the centralized version. The following lemma states that these stars are disjoint from any star that might follow and has lower cost-efficiency.

Lemma 1. *Let U^k be the set of uncovered clients prior to the beginning of round k. If a client j is part of a star (i, B) opened by the distributed algorithm in round k, then there is no star (i', B') considering $B' \subseteq U^k$ with $j \in B'$ and $c(i', B') < c(i, B)$.*

Proof. Assume the contrary, namely that a star (i', B') exists with $c(i', B') < c(i, B)$ and say j is a client in $B' \cap B$. Note that $B' \subseteq U^k$, and therefore i' will choose some star (i', B'') with cost-efficiency $c(i', B'') \leq c(i', B')$ in line. However, as (i, B) is opened in round k, client j has sent its connect request to i and not to i', which implies $c(i', B') \geq c(i, B)$ and contradicts the assumption.

Given the above, we can show that the stars opened by the distributed algorithm can be re-ordered to correspond to the execution of the centralized algorithm.

Theorem 1. *The distributed and centralized versions are equivalent.*

Proof. We sequentialize the distributed algorithm as follows: In the sequentialized version we open only one star (the globally most cost-efficient star) per round. Further, we postpone opening a star (i, B) which has been opened in parallel by the distributed algorithm to a later round prior to which all stars (i', B') with $c(i', B') < c(i, B)$ have been processed. Let (i', B') denote one such star. Because of Lemma 1, $B' \cap B = \emptyset$, and therefore opening (i', B') ahead of time does not remove any client in B from U and therefore does not interfere with opening (i, B). Similarly, postponing any (i, B) will not allow that a more cost-efficient star including elements of B is formed earlier – again by Lemma 1. Postponing (i, B) can further influence (raise) the cost-efficiency of the stars (i', B') as it changes the set $B(i)$ for these facilities and thus may change the order in which these are processed. However, as by Lemma 1 all these stars are mutually disjunct, the order in which they are opened does not affect total costs. Finally, all stars opened in parallel are disjunct and re-ordering them does not change algorithm execution.

Therefore, the sequentialized version opens the same stars as the distributed algorithm. Moreover, as the sequentialized version opens the most cost-efficient star in every round, it implements the execution of the centralized algorithm.

Nevertheless, the worst-case number of rounds required by Algorithms 2 and 3 remains linear in the number of nodes, because there can be a unique point of activity around the globally most cost-efficient facility i^* in each round: Consider for instance a chain of m facilities located on a line, where each pair of facilities is interconnected by at least one client, and assume that facilities in the chain have monotonously decreasing cost efficiencies. Each client situated between two facilities will send a "connect-request" to only one of them (the more cost efficient), thus the second cannot open. In this example, only the facility at the end of the chain can be opened in one round. Similarly, once at least one facility is open, it could happen that in each round only one client connects to this facility. The worst-case runtime is therefore $O(n)$, in which n is the number of network nodes.

The linear number of rounds required in the worst-case would constitute a very high overhead in large-scale sensor networks. However, a worst-case configuration on a larger scale is highly improbable (as we will show in Section 7), and the approximation factor inherited from the centralized version is intriguing, particularly because the algorithm performs even much better than 1.61 on average instances. We will evaluate the *average* number of rounds required for typical instances in sensor networks and the optimality gap when the algorithm is executed with such instances in Section 7.

As we mentioned, however, the above algorithm only retains its approximation factor with metric instances, and as any metric instance is essentially a complete graph, it requires global communication between all clients and facilities. This is only efficient in few settings, for example when all nodes hear each other over the wireless broadcast medium. In the next section we use the algorithms of this section as subroutines in an adapted "local" version that functions properly in multi-hop networks.

6 Multi-hop Approximation

The described algorithm can be changed to work in multi-hop settings using only a slight adaptation. As it turns out, if connection costs represent shortest paths between network nodes, the communication performed by the algorithms can be restricted to small network neighborhoods. Specifically, if one is interested in determining whether a facility i has a cost-efficiency of less than a certain threshold s, it is sufficient to consider only clients j that are reachable by i over a path with costs of at most s, i.e., clients j with $c_{ij} \leq s$. To see this, consider the definition of a facility's cost-efficiency and assume that some star's cost efficiency $c(i, B) \leq s$. One can always obtain an even smaller cost-efficiency once one removes the clients $j \in B'$ which have $c_{ij} > s$, that is, $c(i, B \setminus B') < c(i, B)$. Similarly, given a facility i, the clients with $c_{\sigma(j)j} > c_{ij}$ will not occur in the set $B(i)$ of Eq. (3). Therefore, it is sufficient that clients j which are newly connected to $\sigma(j)$ distribute $\sigma(j)$ only to facilities i with cost $c_{ij} < c_{\sigma(j)j}$.

In an outer loop added around Algorithms 2 and 3, we therefore exponentially increase the communication scope s, that is, the maximum distance over which messages are forwarded. Specifically, given a certain scope s, a message is only flooded within a localized neighborhood $N_s(i)$ around the sending node i, where $N_s(i) := \{j \in V \text{ with } c_{ij} \leq s\}$. Note that if the direct link (i, j) is not present in the network graph, c_{ij} representing the shortest path from j to i can be determined on the fly while flooding a message within $N_s(j)$. Nodes simply stop forwarding a message if it has covered a distance of larger than s or if it has already been received over a shorter path.

The updated versions are given in Algorithm 4 (clients) and Algorithm 5 (facilities). In the following, we will respectively use C_s and F_s to refer to client and facility nodes within scope s of the current node.

Algorithm 4. Multi-Hop Adaptation of Algorithm 3 for a Client j

---

4.1 set $s = 1$, set $\sigma(j) =$ none
4.2 **repeat**
4.3   set $s = s \times a$
4.4   **send** "start(s)" to all $i \in F_s$
4.5   **if** no "begin(s)" received **then continue**
4.6   **repeat**
4.7   **receive** $c(i, B)$ from all facilities F_s
4.8   set $F_a = \{i \in F_s \text{ with } c(i, B) \leq s\}$
4.9   **if** $F_a \neq \emptyset$ **then**
4.10 $i^* = \text{argmin}_{i \in F_a}\, c(i, B)$ // use node ids to break ties
4.11   **send** "connect-request" to i^*
4.12   **if** received "open(s)" from i^* **then**
4.13   set $\sigma(j) = i^*$
4.14   **send** $\sigma(j)$ to all $i \in F_s$
4.15   **until** connected or $F_a = \emptyset$
4.16 **until** connected
4.17 **on** "open(s^*)" from i with $c_{ij} < c_{\sigma(j)j}$
4.18   set $\sigma(j) = i$
4.19   **send** $\sigma(j)$ to all $i \in F_{s^*}$

Algorithm 5. Multi-Hop Adaptation of Algorithm 2 for Facility i

---

5.1   set $s = 1$
5.2   **repeat**
5.3   set $s = s \times a$
5.4   **if** *"start(s)" received* **then send** "begin(s)" to all $j \in C_s$ **else continue**
5.5   **query** $\sigma(j)$ from all $j \in C_s$
5.6   set $U_s = \{j \in C_s$ with $\sigma(j) = none\}$
5.7   **repeat**
5.8   find most cost-efficient star (i, B) with $B \subseteq U_s$
5.9   **send** $c(i, B)$ to all $j \in U_s$
5.10   **if** $c(i, B) \leq s$ **then**
5.11   **receive** "connect-requests" from set $B^* \subseteq U_s$.
5.12   **if** $B^* = B$ **then**
5.13   open facility i (if not already open)
5.14   **send** "open(s)" to all $j \in C$
5.15   set $U_s = U_s \setminus B$, set $f_i = 0$
5.16   **receive** $\sigma(j) \neq none$ from some clients $B' \subseteq U_s$
5.17   set $U_s = U_s \setminus B'$
5.18   **until** $U_s = \emptyset$ or $c(i, B) > s$
5.19   **until** $s > s_{\max}$

In the outer loop, the considered scope s is raised exponentially (lines 4.3 and 5.3). To initialize an outer round, clients, which have not yet been connected, send a "start" message containing their current scope s to all facilities in scope (line 4.4). In turn, facilities wait for at least one such "start" message for a certain time (line 4.5) upon which they reply "begin(s)". The waiting period must be long enough to allow relevant clients to send the respective start messages and finish earlier rounds. If no "start" messages were received, facilities simply advance to the next outer round (line 5.4) to wait for "start" messages from a larger scope. Clients, analogously, wait and then skip the current round if no neighboring facility has sent "begin".

A start message sent by a client j thus triggers execution of one outer round at all the facilities in scope F_s. Facilities then query all clients in scope for their status $\sigma(j)$ in line 5.5 and compute the set of yet unconnected clients U_s. This query-reply cycle allows the facility to wait for all relevant clients to catch up to the current scope s. Clients reply to this query once they have reached scope s – note that we have omitted the respective code in the client algorithm. Similarly clients can wait for facilities lagging behind in line 4.7 where they expect to receive a message from all facilities in scope.

After this initialization, facilities execute Algorithm 2 in an inner loop (lines 5.7-5.18) and clients react accordingly (lines 4.6-4.15) implementing Algorithm 3. Compared to Algorithms 2 and 3 the termination conditions of the inner loops must be changed to allow clients and facilities to proceed to a larger scope in a properly synchronized manner. As with the 1-hop version, clients terminate their inner loop once they are connected (line 4.15) and facilities once no active clients remain in scope (line 5.18). In addition, within an inner-loop with scope s, the algorithm should only consider stars (i, B) with cost-efficiency $c(i, B) < s$. Therefore, facilities only proceed with the current inner loop as long as they are efficient enough for this scope (lines 5.10 and 5.18) while in turn clients only proceed with their inner loop as long as there is a facility in scope that is efficient enough to connect them (lines 4.8,4.9 and 4.15).

Finally, once a client has been connected (4.17-4.19), it acts analogously to Algorithm 3: It simply changes its facility if this is beneficial and notifies all relevant facilities about it. Here the client can synchronize to the scope s^* of the sending facility as it is included in the received "open" message to ensure that all relevant facilities are informed. Note that the messages sent in line 4.19 are also received by facilities still performing their inner loop in line 5.16.

Discussion. The algorithms presented in this section enhance Algorithms 3 and 4 by making them "local", meaning that they do not need to communicate with all relevant facilities but only to the ones within a confined neighborhood. This allows to perform shortest-paths computations in these confined neighborhoods which, in turn, give rise to metric instances and preserve the approximation factor of Algorithm 1.

An additional outer loop provides for both, an adequate expansion of the involved communication scope and for sufficient synchronization of the nodes in scope without depending on a synchronized communication model. Because clients and facilities may repeatedly have to wait in lines 4.5 and 5.4, respectively, the worst-case runtime becomes $O(n \log_a s_{\max})$ where $s_{\max}$ denotes the cost efficiency of the least efficient star which occurs in the network and n denotes the total number of participating nodes. However, the maximum number of rounds involving actual communication is smaller. If no unconnected clients or eligible facilities are present, the involved nodes do not communicate in their inner loop at all. Instead, they simply skip the inner loop. In turn, in rounds involving communication, a client or facility can be a single point of activity only once during algorithm execution. Therefore, the number of required communication rounds is still in $O(n)$.

Dynamic Re-configuration. In real-world deployments of sensor networks, link qualities change over time and nodes may fail. To accommodate for *major* changes in the network topology, the algorithms are re-executed at regular intervals. As such re-starts involve relatively high overhead, these are performed only infrequently (e.g., once a day). In between such re-starts, a client j combines periodic re-evaluations of link costs c_{ij} (within a local scope of size $c_{\sigma(j)j}$) with a liveness check on the facility $\sigma(j)$. In both cases, if $\sigma(j)$ has failed or a closer open facility has been found, client j re-connects to the closest open facility. In Section 7, we will show that such adaptations suffice to maintain a close-to-optimal configuration over longer periods of time.

7   Experimental Results

In the following, we show results from two distinct sets of experiments. The first, detailed in Section 7.1, is based on simulations which test the scalability of the proposed algorithms. The second, detailed in Section 7.2 tests the applicability of the proposed algorithms to operational networks with dynamic links.

7.1 Scalability

In the experiments based on simulations, we uniformly deployed a variable number of nodes (x-axis) onto a 300m by 300m area. The network graph has an edge $(i, j) \in E$ if the nodes i and j are less than $30m$ apart (this number stems from a model that is based on the characteristics of the CC1000 transceiver used on BTnodes [18] and Berkeley Motes). Assuming that nodes can control their transmit power, for $(i, j) \in E$, we set connection costs $c_{ij} \sim g(i, j)^2$ where $g(i, j)$ denotes the distance in meters between i and j and normalize them such that $c_{ij} \in [0, 1]$.

Scenarios. To test our algorithms with a range of applications, we examined three different parameterizations of the facility location problem, of which qualitative results are shown in Figure 1. In the first, we set opening costs $f_i = 1$ and additionally require that clients and facilities must be neighbors. We show a solution obtained by the one-hop Algorithms 2 and 3 on such an instance in Figure 1(a).

Further, we tested the multi-hop Algorithms 4 and 5 in two different settings. In the first, we set $f_i = 5$ to denote that a high effort is required to operate a cluster leader, of which an example result is shown in Figure 1(b). In the second scenario, shown in Figure 1(c), we assumed that cluster leaders must send much data to the network basestation and therefore their operation costs increase with their network distance to the sink (yielding smaller stars close to the sink and larger ones further away).

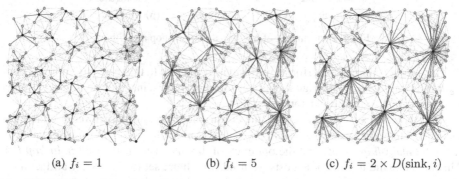

(a) $f_i = 1$ (b) $f_i = 5$ (c) $f_i = 2 \times D(\text{sink}, i)$

Fig. 1. Effects of varying opening costs ($D(sink, i)$ denotes the shortest-path distance to the sink, which is located in the upper left corner of the simulated area)

One-Hop Clusters. In the *one-hop* setting (Figure 1(a)), we evaluated the costs of configurations produced by different algorithms while varying the number of nodes in the simulation area (that is, the node density). The results are given in Figure 2(a) which shows the costs obtained with the following five methods.

One-hop denotes the simple one-hop algorithms of Section 5. Respectively, *one-hop IP* refers to the optimal configuration of the constrained case which requires clients to connect to facilities which are direct network neighbors. Further, *multi-hop* denotes the multi-hop algorithm described in Section 6, which has a 1.61 approximation guarantee. Here, clients may connect to facilities which are an arbitrary number of hops away. Respectively, *multi-hop-IP* computes the optimal solution to the facility location problem, in which facilities and clients may be multiple hops apart and the instance is made metric by a centralized shortest-paths computation. Finally, *MDS-IP* denotes the optimal solution to the minimum dominating set problem, in which dominator nodes represent open facilities and slave nodes are clients that connect to the closest dominator node. The costs are computed using the original (non-metric) instance.

The costs of a minimum dominating set (*MDS-IP*) which suffer from expensive long links mark one end of the optimization spectrum. Here we argued that facility location can provide a more energy efficient configuration. On the other hand, the optimal facility-location based configuration (*multihop-IP*) marks the other end as it represents a lower bound for the employed approximation algorithms.

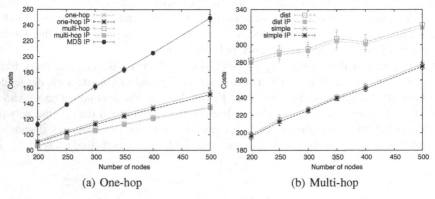

(a) One-hop   (b) Multi-hop

Fig. 2. Performance of one-hop and multi-hop algorithms

The *one-hop* algorithm performs well and is even close to the respective optimal configuration *one-hop IP*, although it operates on a non-metric instance and thus without a guaranteed approximation factor.

Note that in this particular setting, the constrained versions, which require facilities and clients to be direct neighbors (*one-hop* and the optimal *one-hop IP*), are not far away from the *multi-hop* results and the optimum of the unconstrained case (*multi-hop IP*). This is due to the low opening costs we used, which are set to $f_i = 1$ for all facilities. With larger opening costs, multi-hop solutions would benefit more from larger stars.

Multi-Hop Clusters. In the experiments shown in Figure 2(b), we additionally evaluate the quality of the solutions obtained by the multi-hop algorithm with the two different opening cost settings shown in Figures 1(b) and 1(c). In the first (denoted as *simple*) we set opening costs to a constant $f_i = 5$ which corresponds to configurations as shown in Figure 1(b). In the second, denoted as *dist*, we apply the heuristic shown in Figure 1(c), where the opening costs correspond to twice the costs of the shortest path to the sink. In both cases, the results of the distributed implementation are very close to the achievable optimum computed by CPLEX on the same instance.

Runtime and Overhead. In the experiments shown in Figure 2(b), the scope s started out with 0.2 and a was set to 2, thus doubling the scope in each outer round. Note, however, that these two parameters do not influence the quality of the obtained solution. Rather, they determine the trade-off achieved between the runtime of the algorithms and the scope within which messages are sent. On the one hand, the smaller a is set, the more one may be sure that scopes are not increased too far (in vain). On the other hand, the required number of outer rounds until termination increases with lower a-values.

Figure 3 demonstrates this trade-off as observed in the simulation run corresponding to Figure 2(b). In Figure 3(a) we show the average scope with which messages were sent during algorithm execution, given different settings of a (the scope s always starts at 0.2). The lower we set a, the better the results as the scope is increased by smaller amounts. Note that in general, the effort involved in the execution of our algorithm is proportional to the "locality" implied by the problem instance: On the one hand, if opening costs are high (here $f_i = 5$), a facility will generally connect clients in a larger neighborhood (as seen in Figure 1(b)). On the other hand, the experienced scopes are even much lower with small opening costs (e.g., for $f_i = 1$, not shown).

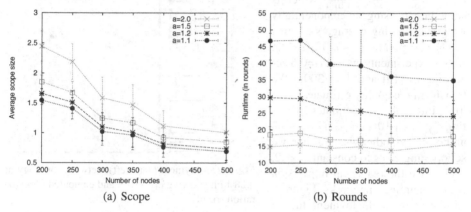

(a) Scope   (b) Rounds

Fig. 3. Average scope size vs. total runtime (in rounds). In Figure 3(b) the error bars denote the maximum and the minimum that occurred.

In contrast, in Figure 3(b), we show the runtime in rounds (one round corresponds to one execution of the inner loop) of the multi-hop algorithm on the same instances. Note that, while previously the error bars indicated confidence intervals of 95%, we use them in Figure 3(b) to mark the maximum and minimum values that occurred in 10 random instances (as we are particularly interested in the maximum value). The results show that – while in theory the worst case runtime can be large – in typical instances based on multi-hop networks the runtime is sufficiently small and does not even grow with the number of nodes. Moreover, based on the trade-off between runtime and scope size, the runtime improves with higher a values. Finally, the scope size decreases with increasing network density. This is due to the fact that, given certain opening costs, the algorithms will connect stars of around the same size (namely, facilities are opened once enough clients are connected to pay for opening them). Therefore, smaller stars are opened in denser networks and the cumulated communication overhead stays the same.

7.2 Network Dynamics

One open question is whether such, albeit close-to-optimal solutions, can provide a benefit for real-world deployments in which the network topology changes over time. To obtain realistic link qualities, we extended a testbed of 13 TMote Sky modules that gather temperature, humidity, and light measurements from our office premises to record network topology information as well. Next to its sensor measurements, every 5 seconds, a node reports the set of nodes from which an application-layer message has been received since the last update.

Such topology information received from each node i allows to compute a (packet-level) link quality estimate $e_{ij}(t)$ for each network link directed from j to i [19]. The estimate e_{ij} (t) is based on the packet success rate $r_{ij} = \frac{\text{packets received in T}}{\text{packets expected in T}}$ which is smoothened using an exponentially weighted moving average such that $e_{ij}(t) = \alpha r_{ij}(t) + (1-\alpha)e_{ij}(t-1)$. In our experiments, we set α=0.6 according to [19] and T to 300 s. We transform the quality estimates $e_{ij} \in$ $[0,1]$ into link cost estimates by setting $c_{ij} = 1 + 10(1 - e_{ij})$ if $e_{ij} > 0.5$ and $c_{ij} = \infty$, otherwise. Further, we set opening costs to constant $f_i = 2$.

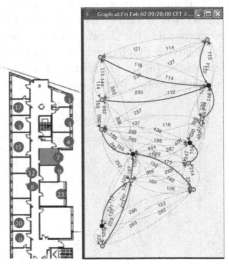

Fig. 4. Deployment plan (left); network topology at 9:28 a.m. showing $c_{ij} \times 100$ and computed configuration (right)

To give the reader an impression of the examined networks, Figure 4 shows our mote deployment, the resulting network topology, and a configuration computed by the multi-hop algorithms.

Given the link costs $\{c_{ij}(t_0)\}$ observed at a certain time of the experiment t_0, we let the presented multi-hop algorithms compute a configuration (a set of open facilities and assigned clients), whose costs $C(t_0, t)$ vary with t as link qualities change over time. Once a configuration has been computed, only small dynamic adaptations (detailed in Section 6) are performed.

In Figure 5(a), we show the ratio between $C(t_0, t)$ and the costs of an optimal configuration C_{opt} computed by CPLEX – for configurations computed at three arbitrarily chosen instants of time t_0. Observe how at $t = t_0$, e.g. at 7:46 or at 11:42, the respective optimality gap is close to 1. As expected, however, this is not always the case. For example the configuration obtained at t_0=9:28 is not optimal even at this time.

In Figure 5(a), one can observe how the time t_0 at which the initial configuration is computed influences the respective outcome of $C(t_0, t)$. To obtain more general results, t_0 is randomly drawn from the total 24 hour interval corresponding to available topology data and used to compute the respective curve $C(t_0, t)$ in 20 repeated simulation runs. The ratio of the average $C(t_0, t)$ to the costs of the optimal configuration is shown Figure 5(b). In addition, Figure 5(b) shows the costs C_{MDS} of a minimum dominating set computed by CPLEX for each instant of experiment time. The latter costs can be

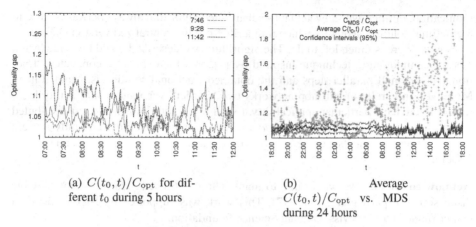

(a) $C(t_0, t)/C_{\text{opt}}$ for different t_0 during 5 hours

(b)   Average $C(t_0, t)/C_{\text{opt}}$ vs. MDS during 24 hours

Fig. 5. Solutions' optimality over time

used as an assessment of whether a much faster MDS approximation, which can be re-executed frequently, could out-perform a facility location algorithm executed more rarely. As said earlier, however, MDS-based configurations require slaves to use expensive links (with poor link quality estimates) to communicate with their cluster leader. Such "bad" links are often the most volatile and cause the costs of an MDS-based configuration to diverge significantly from an optimal configuration. While this is not always the case (Figure 5(b) has portions in which MDS is close-to-optimal), one can observe that facility-location based configurations, which focus on high-quality links, are robust with respect to varying link qualities. The observed gap to an optimal configuration remains small – in the observed 24 hours it stayed below 10% at all times.

8   Conclusion and Outlook

In this paper, we motivated the use of facility location algorithms to address configuration tasks in multi-hop networks as they can flexibly implement many sensor-network configuration problems, such as an energy-efficient clustering, a clustering in which cluster leaders can connect nodes through multiple hops, or a configuration in which cluster leaders are chosen based on their distance to the sink. We claim that many more such applications of the problem can be found.

Further, we have shown that algorithms which are very good in theory (with an approximation factor of 1.61 whilst the theoretically best polynomial algorithm cannot be better than 1.463) can be feasibly transformed for distributed execution. The transformations we described resulted in (to our knowledge) the first facility location algorithm which can be efficiently executed in multi-hop networks.

In the experimental evaluation, we were able to show that although our algorithm exhibits a linear worst-case runtime, in typical sensor-network instances it terminates in only few communication rounds. Moreover, by analyzing the scopes within which messages were forwarded during algorithm execution, we showed that the devised algorithm, although equivalent to its centralized ancestor, requires only very local communication. Further, we showed that the distributed algorithm always performs close to the optimal solution, a quality which it inherits from the centralized version [17].

Finally, there is much left to do. The algorithms we described could be made faster, possibly employing a technique inspired by [7], in which stars are connected "fractionally" in small parallel steps and the obtained fractional solution is rounded later. Moreover, in wireless multi-hop networks two "harder" versions of the facility location problem have particular applicability, for which, to our knowledge, no distributed algorithms exist at all: The *capacitated* version, in which a facility can only serve a limited number of clients and the *robust* version, in which every client is connected by k facilities.

Acknowledgments. We would like to thank Thomas Moscibroda for many valuable discussions, in particular on [7,8,17]. This work was supported by NCCR-MICS, a center funded by the Swiss National Science Foundation.

References

1. Karl, H., Willig, A.: Protocols and Architectures for Wireless Sensor Networks. Wiley, Chichester (2005)
2. Cerpa, A., Estrin, D.: ASCENT: Adaptive Self-Configuring Sensor Networks Topologies. In: Proceedings of the 21st Annual Joint Conference of the IEEE Computer and Communications Societies (INFOCOM'02), New York, NY, USA (June 2002)
3. Basagni, S., Mastrogiovanni, M., Petrioli, C.: A performance comparison of protocols for clustering and backbone formation in large scale ad hoc networks. In: Proceedings of the 1st IEEE International Conference on Mobile Ad-hoc and Sensor Systems (MASS'04) (2004)
4. Madden, S., Franklin, M.J., Hellerstein, J.M., Hong, W.: TAG: a Tiny AGgregation Service for Ad-Hoc Sensor Networks. In: Proceedings of the 5th Symposium on Operating Systems Design and Implementation (OSDI'02), Boston, MA, USA (December 2002)
5. Gnawali, O., Greenstein, B., Jang, K.Y., Joki, A., Paek, J., Vieira, M., Estrin, D., Govindan, R., Kohler, E.: The TENET architecture for tiered sensor networks. In: Proceedings of the 4th International Conference on Embedded Networked Sensor Systems (SENSYS'06), Boulder, CO, USA (November 2006)
6. Gehweiler, J., Lammersen, C., Sohler, C.: A distributed O(1)-approximation algorithm for the uniform facility location problem. In: Proceedings of the 18th Annual ACM Symposium on Parallel Algorithms and Architectures (SPAA'06), Cambridge, MA, USA (2006)
7. Moscibroda, T., Wattenhofer, R.: Facility location: Distributed approximation. In: Proceedings of the 24th ACM Symposium on Principles of Distributed Computing (PODC'05), pp. 108–117 (2005)
8. Chudak, F., Erlebach, T., Panconesi, A., Sozio, M.: Primal-dual distributed algorithms for covering and facility location problems. Unpublished Manuscript (2005)
9. Frank, C., Römer, K.: Algorithms for generic role assignment in wireless sensor networks. In: Proceedings of the 3rd International Conference on Embedded Networked Sensor Systems (SENSYS'05), San Diego, CA, USA (November 2005)
10. Vygen, J.: Approximation algorithms for facility location problems. Technical Report 05950-OR, Research Institute for Discrete Mathematics, University of Bonn (2005)
11. Feige, U.: A threshold of ln n for approximating set cover. Journal of the ACM 45(4) (1998)
12. Hochbaum, D.S.: Heuristics for the fixed cost median problem. Mathematical Programming 22(1), 148–162 (1982)

13. Guha, S., Khuller, S.: Greedy strikes back: Improved facility location algorithms. Journal of Algorithms 31, 228–248 (1999)
14. Mahdian, M., Ye, Y., Zhang, J.: Improved approximation algorithms for metric facility location problems. In: Jansen, K., Leonardi, S., Vazirani, V.V. (eds.) APPROX 2002. LNCS, vol. 2462, Springer, Heidelberg (2002)
15. Jain, K., Vazirani, V.V.: Primal-dual approximation algorithms for metric facility location and k-median problems. In: Proceedings of the 40th Annual IEEE Symposium on Foundations of Computer Science (FOCS'99), pp. 2–13 (October 1999)
16. Krivitski, D., Schuster, A., Wolff, R.: A local facility location algorithm for sensor networks. In: Prasanna, V.K., Iyengar, S., Spirakis, P.G., Welsh, M. (eds.) DCOSS 2005. LNCS, vol. 3560, Springer, Heidelberg (2005)
17. Jain, K., Mahdian, M., Markakis, E., Saberi, A., Vazirani, V.V.: Greedy facility location algorithms analyzed using dual fitting with factor-revealing LP. Journal of the ACM 50, 795–824 (2003)
18. BTnodes (2006), www.btnode.ethz.ch
19. Woo, A., Tong, T., Culler, D.: Taming the underlying challenges of reliable multihop routing in sensor networks. In: Proceedings of the 1st International Conference on Embedded Networked Sensor Systems (SENSYS'03), Los Angeles, CA, USA (November 2003)

SNTS: Sensor Network Troubleshooting Suite

Mohammad Maifi Hasan Khan, Liqian Luo, Chengdu Huang,
and Tarek Abdelzaher

University of Illinois at Urbana Chamapign, Department of Computer Science, USA
mmkhan2@uiuc.edu, lluo2@uiuc.edu, chuang30@uiuc.edu, zaher@cs.uiuc.edu

Abstract. Sensor network troubleshooting is a notoriously difficult task, further exacerbated by resource constraints, unreliable components, unpredictable natural phenomena, and experimental programming paradigms. This paper presents *SNTS* (Sensor Network Troubleshooting Suite), a tool that performs *automated failure diagnosis* in sensor networks. SNTS can be used to monitor network conditions using simple visualization techniques as well as to troubleshoot deployed distributed sensor systems using data mining approaches. It is composed of (i) a data collection front-end that records events internal to the network and (ii) a data processing back-end for subsequent analysis. We use data mining techniques to automate failure diagnosis on the back-end. The assumption is that the occurrence of execution conditions that cause failures (e.g., traversal of an execution path that contains a "bug" or occurrence of a sequence of events that a protocol was not designed to handle) will have a measurable correlation (by causality) with the resulting failure itself. Hence, by mining for network conditions that correlate with failure states the root causes of failure are revealed with high probability. To evaluate the effectiveness of the tool, we have used it to troubleshoot a tracking system called EnviroTrack [4], which, although performs well most of the time, occasionally fails to track targets correctly. Results show that SNTS can identify the major causes of the problem and give developers useful hints on improving the performance of the tracking system.

Keywords: Sensor network, Data mining, Distributed troubleshooting.

1   Introduction

In this paper we present a distributed troubleshooting tool for automated sensor network failure-diagnosis. The goal is to reduce the effort needed for sensor network software development. Paradigms for sensor network programming have received a lot of attention, including node-based [14], group-based [16,31], event-based [8,9,19], database-centric [24,23], state-centric [20], and virtual-machine-based approaches [18]. In contrast, efforts to facilitate program *troubleshooting* have been rather sparse. This imbalance is not congruent with the development needs. Most sensor network programmers who developed large applications would agree that sensor network troubleshooting is a notoriously difficult task and in many instances, most of the programmer's time is spent not on writing new code, but on making current code operate correctly.

J. Aspnes et al. (Eds.): DCOSS 2007, LNCS 4549, pp. 142–157, 2007.
© Springer-Verlag Berlin Heidelberg 2007

The presented tool is aimed at addressing the void in sensor network development support brought about by the lack of appropriate network troubleshooting tools. With few exceptions [28], current troubleshooting support mostly revolves around laboratory testbeds [33,13], simulation systems [17,32] or emulation-based tools [1,26]. These may help locate failures by testing or stepping through instruction execution, but they do not suggest where a programmer should look in order to find a failure cause. Hence, locating the "bugs" remains an expensive trial-and-error process. In a distributed system, root causes of failure can be subtle. They may arise because of the complex and unexpected ways different individually sound components interact, as opposed to because of some coding error within one component. This makes it hard to identify problems in distributed protocols.

To complement the above techniques, this paper describes a tool that performs *automated failure diagnosis* in sensor networks. It is composed of (i) a data collection front-end that records events internal to the network and (ii) a data processing back-end for subsequent analysis. We use machine learning techniques to automate failure diagnosis on the back-end. *Failure*, in the context of this paper, refers to any deviation from what the developer deems to be correct behavior of the system. For example, it could refer to inability to synchronize clocks, inability to elect a unique leader, incorrect data aggregation, occurrence of routing cycles, or unusually low link utilization in the presence of backed-up demand.

Machine learning techniques allow the diagnostic tool to automatically extract conditions (e.g., sequences of events or ranges of measured system state parameters) that correlate highly with the occurrence of failures and bring those conditions to the attention of the developer. The assumption is that the occurrence of execution conditions that cause failures (e.g., traversal of an execution path that contains a "bug" or occurrence of a sequence of events that a protocol was not designed to handle) will have a measurable correlation (by causality) with the resulting failure itself. Hence, the conditions identified by the implemented tool will include root causes of failure with high probability. Of the previous debugging tools, Sympathy [28] is closest in spirit to our automated failure analysis. However, it is restricted to reasoning about crashed nodes or disconnected links in the event of reduced network throughput. The presented tool can investigate and address a broad array of failure types, offering diagnostic capabilities to the programmer of distributed applications and protocols.

Our approach is supported by an encouraging preliminary evaluation that leverages advances in data mining and machine learning techniques that have been applied to failure diagnosis in large systems [6,5,7]. As a proof-of-concept, our diagnostic tool was applied to a tracking system. The tracking system operated well most of the time, but would occasionally fail to track targets correctly. The tool uncovered two conditions under which failure occurred with high probability. Analysis revealed that these conditions corresponded to two corner cases not considered in the design of the tracking protocol. Understanding and quantifying practical cases in which a distributed protocol fails is thus one of the fundamental contributions of the work described in this paper.

While formal methods can also be applied to verify protocol correctness, they have two limitations. First, in complex concurrent systems, these methods often make simplifying assumptions (e.g., regarding the sensors or the model of the physical environment). Hence, properties that have been verified in theory may not hold in practice. Second, often failures occur because of conditions that have not been anticipated in the design and protocol specification phase. Hence, while a protocol can be verified to conform to its specifications, it may still fail in practice because the specifications are incomplete. Our tool takes the more pragmatic approach of analyzing problems that do occur and identifying their causes.

Equally important is to understand which potential problems are not the cause of a current failure. For example, the design of a distributed protocol might use a simplified model of communication or a simplified failure hypothesis. It is important to understand, which of the simplifications, if any, are causing failures and which are not. Since the sensor nodes are very resource constrained, simplifications that do not cause problems in practice are very welcome.

The remainder of the paper is organized as follows. Section 2 reviews related work. Section 3 describes the scope of the fault diagnosis tool. Section 4 presents the system architecture and design details of SNTS. A case study of using SNTS to troubleshoot EnviroTrack is discussed in Section 5. Section 6 presents a discussion on the generalizability of SNTS and possible future extensions. Section 7 concludes the paper.

2   Related Work

Debugging sensor networks is a promising research area with very few existing tools in use at the present time. Current tools and middleware that aid the debugging and evaluation of sensor network applications can generally be divided into four categories: simulators, emulators, test-beds and services [22].

Simulators are popular tools in debugging and evaluation of sensor network applications since they don't usually require the deployment of sensor hardware. NS-2 [3], GloMoSim [34], TOSSIM [17] and S^2DB [32] are good examples. NS-2 is a discrete event simulator supporting various networking protocols over wired and wireless networks. GloMoSim focuses more on mobile, wireless networks. It allows comparison of multiple protocols at a given layer. TOSSIM is a simulator especially designed for TinyOS applications, which provides scalable simulations of sensor network software. S^2DB has extensive functionality for both software and hardware debugging. Current simulators, however, do not adequately capture the real behavior of sensor networks. This is due to the difficulty in modeling practical imperfections such as radio irregularity as well as due to the lack of good models of environmental inputs. Hence, when used in debugging, they might not uncover certain failures caused by unexpected considerations in the real world. More importantly, failures that are revealed by simulation remain unexplained by the tool. It is up to the programmer to infer the causes of simulated failures.

Another category of debugging and performance evaluation tools in sensor networks is emulators that mimic sensor devices either in software or hardware. AVR JTAG ICE [1], a real time in-circuit emulator, is a good representative of hardware emulators. It uses the JTAG interface to enable a user to do real-time emulation of the microcontroller of sensor devices. A drawback of such in-circuit emulators is that they have to be physically connected to emulated devices, which causes logistical difficulties in conducting experiments especially for large-scale applications covering a wide field. Atemu [26] is a software emulator for AVR-processor-based systems that emulates AVR processors as well as other peripheral devices on the MICA2 platform. Like TOSSIM, Atemu also simulates wireless communication. Such software emulators do not introduce the logistical difficulties exhibited in hardware emulators, but they are usually less realistic in reproducing network behavior.

The final stages of debugging and performance tuning typically use actual testbeds to evaluate sensor network applications. For example, Motelab [33] is a public testbed using MICA2 platforms, which allows users to upload executables and receive execution results via the Internet. Kansei [25] is another testbed. It employs XSM, MICA2, and Stargate platforms. EmStar [13] is a combination of emulators and testbeds for Linux-based sensor network applications, which runs applications using either a modeled radio channel or the channel of real nodes. EmTOS [13] extends EmStar to run TinyOS applications by compiling them into EmStar binaries. These testbeds ease evaluation a lot without requiring full-scale deployment. However, they merely uncover failures without attempting to diagnose the root cause. One exception is Sympathy [28], which attempts to identify node or link failures that result in decreased throughput at the base-station.

We categorize all other software facilitating field tests of sensor network applications as services. EnviroLog belongs to this category. Monitoring tools such as Message Center [30] aid field tests by capturing messages in the air, filtering and displaying them to users. Closest to EnviroLog is TOSHILT [15], a middleware for hardware-in-the-loop testing. TOSHILT defines emulated stimuli to replace the real environmental events, so that applications can be evaluated repeatedly before the final deployment.

The tool suite described in this paper is unique in that it automates the process of reasoning about possible *causes* of observed failures. Hence, unlike many other tools that help detect, record, reproduce, or step through failure scenarios, ours will be one that *diagnoses* the failure (using techniques borrowed from machine learning).

The idea of diagnosing software failures using various statistical and machine learning approaches is not new. However, it has not yet been applied to sensor networks. For example, algorithms for detecting performance bottlenecks in distributed systems are presented in [5]. Their authors rely on statistical convolution techniques to identify message (load) pathways through opaque components. Automated diagnosis of performance anomalies in server farms is presented in [6,7].

Unlike many existing debugging tools for sensor network, ours is geared to provide insight into the *root causes* of failure (as opposed to uncovering the failure location or giving access to system state around failure time). Hence, the approach offers much help in debugging when there is a fault in the design of the protocol. It saves the developer the process of guessing the cause by trial and error, which can be very time consuming and frustrating.

3   Scope of the Fault Diagnosis Tool

Our tool is motivated by the difficulties in sensor network troubleshooting. Unlike other computing systems, sensor networks interact with their physical environment in a distributed fashion, intimately combining computation, communication and sensing. Debugging tools for traditional distributed systems are, therefore, largely inadequate for sensor networks. It is these deficiencies that we try to address in our work.

There are several reasons why sensor network code may fail. The first category is what we call "single node errors" (i.e., code errors that manifest themselves on one node). Examples might be infinite loops, dereferencing invalid pointers, or running out of memory. Current debugging tools are, for the most part, well-equipped to help find such errors. One can make the argument that since such errors can manifest themselves on one node in isolation, they are easier to test for and therefore will more likely be eliminated at an earlier stage of code development. These errors are therefore not the focus of our tool.

Our tool is geared for uncovering errors that occur in distributed component interaction. These are typically errors in the design or execution of distributed protocols. For example, the design of the protocol might not have considered a particular corner-case or sequence of distributed events that then leads to incorrect behavior. Since this sequence of events might occur only occasionally, the error is generally not repeatable. Distributed interaction errors are harder to debug. They require, possibly, a large-scale system prototype and more testing to induce and observe a manifestation of the error.

A particularly hard-to-debug category of distributed component interaction errors are those that are environmentally induced. Distributed sensor network protocols often make implicit assumptions on factors such as node connectivity, communication error probability, effective sensing range, communication range, or sensory signatures of environmental events. These factors are typically dependent on conditions in the physical environment. Environmental execution conditions are hard to reproduce in the lab. Hence, often manifestations of environmentally-induced errors do not occur until the network is installed in the field.

In short, sensor networks often exhibit errors that occur (i) because of unexpected interactions between the system and its physical environment, (ii) because of unexpected interactions between subsystems (e.g., the tracking software and the kinetic properties of the hardware enclosure), or because of violations in implicit assumptions (e.g., assumptions regarding sensor performance) made in the design or specifications of distributed protocols. It is generally very hard to uncover such errors. It is also hard to formally verify the protocol or discover failure by simulation. For example, in a previous actual deployment, the cause of a failed magnetic tracking subsystem was eventually attributed to windy weather conditions; the wind was causing device antennas to move which recorded a magnetic reading on the local magnetometer and generated a deluge of false alarms. Formal methods and simulations would have failed to uncover this error since it is not immediately obvious to a designer that wind has something to do with magnetic target detection and hence needs to be reflected in the specifications or introduced in simulations. Tools are needed to help focus the operator's attention on what might be causing the system malfunction when it occurs at run-time. Such as tool is described below.

4   System Architecture

The basic idea used in this paper for automated diagnosis of failures in distributed sensor network protocols is very simple. At a high-level, the tool would monitor many internal system metrics (which collectively define *system state*) and divide these observations into two sets, one for "good" behavior and one for "bad" behavior, as classified by an operator or by some specified Boolean condition on measurements. The tool would then automatically analyze each class to determine the salient features that are different between the two classes. Chances are, these conditions will be indicative of what has gone wrong in the "bad" cases. They will provide useful leads on where to focus debugging effort. Therefore, SNTS consists of two major components: a data collection component to collect the system metrics and a data analysis component to carry out the automated analysis.

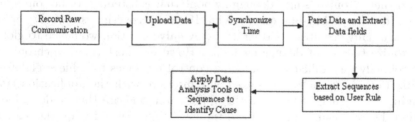

Fig. 1. System architecture of SNTS

4.1   Data Collection Component

The front end of the tool is a data collection component to collect all the communication traffic in the network, which is accomplished by deploying extra debugging nodes in real field along with the application nodes, as depicted in Figure 2. The debugging nodes passively listen to the communication channel, and record every message into their local nonvolatile storage devices, say, flash. Every message is timestamped using the local clocks. Once the data collection is done, we upload the data to a PC. The data collection component is completely application independent. So once installed, the same set of debugging nodes can be used repeatedly for different applications. Below, we describe the major challenges in designing the component and our solutions to them.

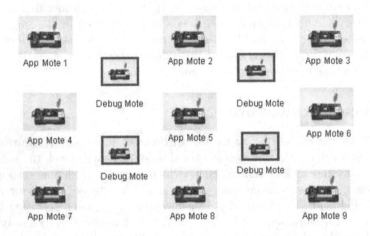

Fig. 2. Deployed debugging nodes

Time Synchronization. As we are analyzing time series data, a big challenge is how to serialize the collected messages that are timestamped by unsynchronized local clocks. Inaccurate serialization may lead to erroneous analysis of the data. If there is only a single debugging node that is listening to the communication traffic, the problem becomes trivial. However, covering the whole deployed network usually requires multiple nodes. A naive solution would be to globally synchronize the set of debugging nodes. However, most time synchronization protocols rely on explicitly exchanging control messages to achieve global synchronization. Such control messages may interfere with the application traffic, changing the normal behavior of the application, and are thus undesirable. To address this problem, we first record the local times of all the debugging nodes and the local time of the base node (connected to a PC, not deployed in the field) before deployment. After data collection, we record again the local times of all the debugging nodes and the base node. Based on these collected local times we are able to calibrate all the timestamps according to the local clock of the base node, thus serializing all the messages correctly.

Eliminating Duplicate Records. In the case of multiple debugging nodes, it is possible that more than one node record the same message. Such duplicate records should be discarded after data collection to improve the accuracy of data analysis. Therefore, after data collection, we sort all messages based on their calibrated timestamps. For messages with extremely close timestamps (i.e., the difference is less than a heuristically determined value ε), we discard all but one if they contain the same content.

Placement of Debugging Nodes. There can be many ways we can place the debugging nodes in the field to cover the whole network. The simplest is to use one debugging node for each application node. However, this will require as many debugging nodes as there are application nodes, which makes it expensive to deploy. Rather, we deploy one node at the center of a cell (assuming application nodes are deployed in a grid structure the cell has four corner nodes) and thus one debug node can listen to many (e.g., four) application nodes.

4.2   Data Analysis Component

This component is the data processing back-end which runs on a PC. It performs preprocessing, parsing and analysis of the data to generate the rules and to identify potential causes of the problem in the network. We describe its inputs and design in the following sections.

Inputs. Consider a sensor network application developer who is trying to troubleshoot their system. The format of the communication packets exchanged between application nodes are known to the developer and hence can be made available to our tool as a header file. We also assume that developer "sees a problem with the system". In other words, the developer expects a certain behavior of his/her protocol that might be getting violated some of the time. The developer can specify what behavior to expect at a high level as a function of metrics available to the diagnosis tool. For example, in a leader-election algorithm the developer might specify that only one leader ID should be measured in the system at a given time. In a real-time system, the developer might specify that the difference between message origination timestamp and its delivery timestamp should be less than some given bound. These timestamps and IDs should be parameters that the diagnosis system can access. For example, they might be parameters in normal application message headers or payload, or parameters explicitly recorded by the collection front-end, such as timestamps. This enables our tool to automatically recognize good and bad behavior. The message formats and definition of correct behavior are the only two application-specific pieces of information that our tool needs in order to analyze the data.

Preprocessing. First, After uploading the data from the debugging nodes, the data-processing back-end first discards all the redundant communication headers and then based on a header file supplied by the developer parses the raw data to extract meaningful information. This header file describes the data formats of the raw communication packets.

Second, after extracting data, the data analysis component uses user specified rules to partition the data into separate piles. It separates the sequences of events which conform to the rules specified by the developer (i.e., good behavior sequences) from the sequences of events that do not and label as good or bad.

Finally, the tool will generate the metrics that are going to be used by the machine learning algorithms to correlate with good or bad behavior. Metrics generated for each sequence (good and bad) for diagnosing the EnviroTrack application in our case study are shown in Table 1. These are the metrics that we will feed to the machine learning algorithm. Currently the user has an interface to suggest suitable metrics. In the absence of user hints, all measured metrics can be considered.

Data Analysis. At this point, we are ready to analyze the data and may apply various machine learning algorithms to identify potential causes of the problem. Among the many choices of machine learning algorithms, we choose the PART [12] algorithm. PART infers rules by repeatedly creating partial decision trees. PART does not perform global optimization. Rather it adopts the divide and conquer strategy. It builds a rule, removes the instances it covers and continues with the remaining instances. For details of the algorithm, interested readers are referred to [12]. We choose it because it is is faster than C4.5 [27] and RIPPER [11] and because accuracy is similar to C4.5 and higher than RIPPER.

As any machine learning algorithm, PART can generate trivial and redundant rules (conditions that correlate with success or failure) as well as important ones. Even obvious conditions are important as they indicate (in the case of success conditions) which parts of the protocol are working.

We also provide trivial network performance metrics like throughput, list of active nodes in the network, signal strength, etc. We have used the Weka software [2], a freeware machine learning tool, to analyze the data.

4.3   Interface

The data collection component can be installed on MicaZ motes as a regular application. As soon as these debugging motes are on, they are going to send the local time to the base station. After that, they can be deployed in the field for data collection. Once the data is collected, these motes can upload data to PC using any TinyOS listening utility program such as Listen tool.

As mentioned above, the user needs to provide two pieces of information. First, the user has to provide a text file describing the packet formats. Ultimately, this file is used to infer the byte offset of a particular field in the raw packet and the length of the field. Second, user specifies boolean rules that can be used to extract good and bad sequences from the time series data. In the following, a case study is described based on a tracking application.

5   Case Study: EnviroTrack

This section describes an illustrative application example of using the presented tool. We use SNTS to troubleshoot EnviroTrack [4] which is a distributed target tracking application.

5.1 Failure of the Tracking Protocol

EnviroTrack [21] implements a distributed tracking protocol whereby a detected target is assigned a new ID. The target should be referred to consistently by this ID while it remains in the field. This is the cornerstone of EnviroTrack's unique target to object mapping. It is key to the programming paradigm and to application semantics supported by the middleware in which EnviroTrack operates. If a second (spurious) ID was generated in the network to refer to the same target, the system will report two detected objects while in reality there is only one. The problem was that the tracking protocol did not ensure unique mapping in some deployments, which made it a good candidate for our case-study.

To appreciate the help offered by automated failure diagnosis, a word is due on the internals of the target tracking protocol. Briefly, to achieve unique mapping in EnviroTrack, whenever a set of nodes detect a target, they form an implicit group and elect a leader who assigns the target an object ID and shares it with group members. As the target moves, membership of the group changes (being restricted only to those nodes that sense the target). When a current leader is no longer able to sense the target, it sends a leader hand-off message, which starts a new leader election among current group members. Election occurs by setting a random timeout such that the first member to time out wins. The new leader continues to use the same ID for the target, ensuring continuity of the unique target-to-object mapping.

Obviously, the correctness of the above protocol hinges on several simplifying assumptions. For example, there is an assumption on message reliability (we do send multiple copies of each critical message but if all copies are lost, there is no way to agree on the target ID). There is also an assumption on the relation between sensing and communication ranges. Namely, leader election assumes that all group members (i.e., nodes that can sense a target) can directly hear each other. If two nodes that sense the same target are outside each other's communication range, they may generate different IDs for that target. Our earlier debugging efforts were directed at these potential causes with no success.

Interestingly, our tool revealed that the above probable causes were not the real reasons behind the observed tracking failures. Most failures were in fact attributed to causes not anticipated in protocol design. The tool identified that a surprising 80% of all failures to maintain unique identity were attributed to two corner cases. The first is when a group contained only the leader and no members. The hand-off failed in this case, because it did not consider the case of a singleton group. Also, when the target was small enough that it was not picked up by any of the sensors for a fraction of the time, the protocol failed. Indeed, the designer had assumed a 100% sensory coverage. This pragmatic evidence was very valuable to the programmer. It identified what specific problem with the protocol needed to be fixed.

Table 1. Set of Metrics

Generated Metrics
No Of Messages transmitted between Leader Handoff
Average Time between Each Message between Leader Handoff
Time difference between the Last message of Last leader and First Message of Current Leader
No of LTM types Message
No of MTL types Message
No of LTB types Message
Geometric Distance between Last Leader and Current Leader
Type of Sequence(Good or Bad)

5.2   Failure Diagnosis Scenario

The details of the above case study are now presented. In general, the tool collects state from the network. In the implemented prototype, we restricted ourselves to collecting externally measurable state. We deployed EnviroTrack (on sixteen nodes), and deployed additional debugging motes to collect messages communicated between EnviroTrack nodes. The tool would then analyze the types, sequences, and timing of these messages. We moved a single target in the network and collected the communication data. The experiment was repeated multiple times. We uploaded the data to a PC and used a data-mining tool (Weka [2]) to analyze it.

With the help of a header file, supplied by the EnviroTrack developer (that describes the format of message headers), it was possible to automatically distinguish different types of messages and different header fields in each message-type transmitted. There are three different types of messages in EnviroTrack, which we call MTL (Member to Leader), LTM (Leader to Member) and LTB (Leader to Base). Data mining tools could then be used to determine sequences of events or conditions on event types that were predictive of (and hence correlated with) bad behavior. Currently we only logged the message-type field, the object ID, the leader ID, the sender field, and the receiver field of message headers. The sender and receiver identity was used to infer geographic locations of the nodes involved. We also time-stamped all message events.

State was collected at given intervals. The developer supplied a simple Boolean function to classify sequences into good and bad. The debugging tool itself did not need to understand this function. It merely applied it to the message sequence in each interval and recorded the result. In our case, since we knew that there was only one target in the experiment, the rule stated that if the object ID (reported in leader messages) changed within the interval, then an error was reported and the interval was marked "bad". Table 1 shows the metrics collected for each interval.

Our algorithm extracted bad sequences of events and good sequences of events across all tests. We subsequently applied several different rule learning algorithms to determine a set of models that correctly classify maximum number of instances based on the above metrics. A *model*, in this context, refers to a set of Boolean conditions on collected metrics. Table 2 shows the output generated by PART [12], when applied on the metrics for EnviroTrack application. Boolean conditions suitable for classification are then extracted.

Table 2. Rules generated by PART Algorithm

Rules Generated by PART Algorithm	Accuracy
$(MTL{>}0 \ AND \ Time\_diff \le 0.58sec) \Rightarrow OK$	83.94%
$(MTL = 0) \Rightarrow Fail$	
$(Distance \ge 1.41) \Rightarrow Fail$	
$(Time\_diff \ge 0.83sec) \Rightarrow Fail$	

Table 3. Leader Handoff Performance Before and After Parameter Tuning

Target Type	Failure Before Parameter Tuning	Failure After Parameter Tuning
Light	37.64%	21.40%
Sound	50.60%	44.70%

5.3   Interpretation of the Rules

If we analyze the generated rules, we can see that the absence of member-to-leader (MTL) messages is the first predictor of failure. When this rule was presented to the EnviroTrack developer, the developer realized that the protocol failed because of a lack of members in the group (which would explain the absence of member-to-leader messages). Indeed, the case of a singleton group had not been considered in the design of the protocol. Other failure conditions include the case where the geometric distance between successive leaders is more than a threshold $(Distance \ge 1.41)$. This means that the target is moving too quickly for the leader hand-off to occur correctly. It is therefore picked up by nodes outside the group of the previous leader, who then create a new ID. This was expected. A more interesting case is $(Time\_diff \ge 0.83sec)$. By protocol design, leaders must periodically send messages at a rate higher than the above interval, as long as they sense a target. Such a large time difference between successive leader messages is a strong indication that the target must have disappeared in the middle of the field (as would be the case if sensory coverage was less than 100%, creating blind spots between nodes), or moved too slowly, eventually causing the group to disband. Indeed, this case was not considered in the design. It is to be noted that the minimum node distance or the time difference between two leader handoffs to fail the protocol vary from experiment to experiment. This is due to the nature of the experiment and each time the target moving behavior is somewhat different. So each time we collect data, we may get a different value for threshold but rules are similar in nature and obvious to interpret consistently.

5.4 Effect of Parameter Tuning

After analyzing the rules generated by the machine learning algorithm, we tuned the protocol parameters. Since absence of multiple members in a group is the primary reason for failure, we tuned the protocol parameters to increase leader broadcasts, and increased member timeout. We then tested the tracking application using two types of sensing modalities - light and sound. As we can see from Table 3 , the number of failed Leader handoffs for Light targets decreases significantly after tuning the parameters. For Sound targets, the improvement is not that much because acoustic range is too large violating the required tracking assumption that communication range exceeds sensing range. It helps to realize that the current protocol is not good enough to track a certain type of targets. In the future, we plan to implement an auto-tuner that will tell the developer what the parameter value should be to detect a target of a particular type and speed, and whether a particular failure can be fixed by parameter tuning or not.

6   Discussion

SNTS is designed keeping in mind that it should be usable for troubleshooting many different applications. Data collection and the machine learning part are totally independent of the application. The only two pieces of information we need at the data processing stage are a header file describing packet format and rules to extract good or bad sequences. Thus, the tool is quite general. We plan to extend the tool so that it can use clustering algorithms to identify deviant behavior without the help of the boolean rules.

The tool is very easy to use and only needs the header format file and the boolean rules to adapt to any applications. The user can also specify which of the fields are important enough to apply data mining. As a general rule, the more information the user provides, the more accurate diagnosis can be performed.

We envision building a complete debugging suite that can provide support at different levels of development. The aforementioned diagnosis tool will be complemented by one for distributed data recording and replay (the *Logger*), as well as a tool for system auto-tuning (the *Auto-tuner*). In many cases, the main reason for the poor performance or failure of a distributed protocol lies in poor settings of protocol parameters. The objective of the auto-tuner tool would be to determine which parameter settings are culprit and auto-tune them to current operational conditions.

7   Conclusion

The implemented tool shows great promise for both troubleshooting purposes and validation purposes. From the case study, it is clear that SNTS can help a developer better understand his/her own protocol and tune the protocol parameters for better performance. The generated rules are very useful for understanding how the implemented protocol behaves in real life, what the limitations are and to provide useful insights about tuning the protocol parameters for better performance. The developer does not need to modify the original code and the tool does not have any impact on the performance of the original application. SNTS provides an effective way to troubleshoot sensor network applications in real life.

Acknowledgement

The work reported in this paper was funded in part by NSF grants CNS 06-13665, DNS 05-54759 and CNS 06-26342.

References

1. Atmel corporation. mature avr jtagice.
   http://www.atmel.com/dyn/products/tools-card.asp?toolid=2737
2. http://www.cs.waikato.ac.nz/ml/weka/
3. ns-2. The Network Simulator. http://www.isi.edu/nsnam/ns/
4. Abdelzahera, T., Blum, B., Cao, Q., Evans, D., George, J., George, S., He, T., Luo, L., Son, S., Stoleru, R., Stankovic, J., Wood, A.: Envirotrack: Towards an environmental computing paradigm for distributed sensor networks. In: Proceedings of the 24th International Conference on Distributed Computing Systems, Japan (March 2004)
5. Aguilera, M.K., Mogul, J.C., Wiener, J.L., Reynolds, P., Muthitacharoen, A.: Performance debugging for distributed systems of black boxes. In: Proc. 19th ACM SOSP (2003)
6. Bodk, P., Fox, A., Jordan, M.I., Patterson, D., Benerjee, A., Jagannathan, R., Su, T., Tenginakai, S., Turner, B., Ingalls, J.: Advanced tools for operators at amazon.com. In: The First Annual Workshop on Autonomic Computing, Dublin, Ireland (2006)
7. Bodk, P., Friedman, G., Biewald, L., Levine, H., Candea, G., Fox, A., Jordan, M.I., Patterson, D., Patel, K., Tolle, G., Hui, J.: Combining visualization and statistical analysis to improve operator confidence and efficiency for failure detection and localization. In: Proc. 2nd International Conference on Autonomic Computing(ICAC05) (2005)
8. Boulis, A., Han, C.-C., Srivastava, M.B.: Design and implementation of a framework for efficient and programmable sensor networks. In: In MobiSys '03: Proceedings of the 1st international conference on Mobile systems, applications and services, pp. 187–200. ACM Press, New York, USA (2003)
9. Cheong, E., Liebman, J., Liu, J., Zhao, F.: Tinygals: a programming model for eventdriven embedded systems. In: SAC '03: Proceedings of the 2003 ACM symposium on Applied computing, pp. 698–704. ACM Press, New York, USA (2003)
10. Cohen, I., Goldszmidt, M., Kelly, T., Symons, J., Chase, J.S.: Correlating instrumentation data to system states: A building block for automated diagnosis and control. In: Sixth Symposium on Operating Systems Design and Implemntation, San Francisco,CA (December 2004)

11. Cohen, W.W.: Fast effective rule induction. In: Proceedings of the 12th International Conference on Machine Learning, pp. 115–123. Morgan Kaufmann, San Francisco (1995)
12. Frank, E., Witten, I.H.: Generating accurate rule sets without global optimization. In: Shavlik, J. (ed.) Machine Learning: Proceedings of the Fifteenth International Conference, Morgan Kaufmann Publishers, San Francisco (1998)
13. Girod, L., Stathopoulos, T., Ramanathan, N., Elson, J., Estrin, D., Osterweil, E., Schoellhammer, T.: A system for simulation, emulation, and deployment of heterogeneous sensor networks. In: Proceedings of the Second ACMConference on Embedded Networked Sensor Systems (SenSys'04) (November 2004)
14. Gummadi, R., Gnawali, O., Govindan, R.: Macro-programming wireless sensor networks using kairos. In: Prasanna, V.K., Iyengar, S., Spirakis, P.G., Welsh, M. (eds.) DCOSS 2005. LNCS, vol. 3560, pp. 126–140. Springer, Heidelberg (2005)
15. Jia, D., Krogh, B.H., Wong, C.: Toshilt:middleware for hardware-in-the-loop testing of wirelesssensor networks.
    http://www.ece.cmu.edu/webk/sensor-networks/toshilt/toshilt.html
16. Whitehouse, K., Sharp, C., Brewer, E., Culler, D.: Hood: a neighborhood abstraction for sensor networks. In: In MobiSys '04: Proceedings of the 2nd international conference on Mobile systems, applications, and services, pp. 99–110. ACM Press, New York, USA (2004)
17. Levis, P., Lee, N., Welsh, M., Culler, D.: Tossim: Accurate and scalable simulation of entire tinyos applications. In: First International Conference on Embedded Networked Sensor Systems (SenSys'03) (November 2003)
18. Levis, P., Culler, D.: Mat: A tiny virtual machine for sensor networks. In: ASPLOS-X: Proceedings of the 10th International Conference on Architectural Support for Programming Languages and Operating Systems (2002)
19. Li, S., Lin, Y., Son, S.H., Stankovic, J., Wei, Y.: Event detection services using data service middleware in distributed sensor networks. Telecommunication Systems, Special Issue on Information Processing in Sensor Networks (2004)
20. Liu, J., Chu, M., Liu, J., Reich, J., Zhao, F.: State-centric programming for sensor-actuator network systems. Pervasive Computing, IEEE (2003)
21. Luo, L., Abdelzaher, T., He, T., Stankovic, J.A.: Envirosuite: An environmentally immersive programming framework for sensor networks. ACM Transactions on Embedded Computing Systems (to appear, 2006)
22. Luo, L., He, T., Zhou, G., Gu, L., Abdelzaher, T., Stankovic, J.A.: Achieving repeatability of asynchronous events in wireless sensor networks with envirolog. Infocom (April 2006)
23. Madden, S.R., Franklin, M.J., Hellerstein, J.M., Hong, W.: Tinydb: An acquisitional query processing system for sensor networks. ACM Transactions on Database Systems (2005)
24. Madden, S., Franklin, M.J., Hellerstein, J.M., Hong, W.: Tag: a tiny aggregation service for ad-hoc sensor networks. SIGOPS Oper. Syst. Rev (2002)
25. Ohio State University. Kansei: Sensor Testbed for At-Scale Experiments (Febuary 2005)
26. Polley, J., Blazakis, D., McGee, J., Rusk, D., Baras, J.S.: Atemu: A fine-grained sensor network simulator. In: First International Conference on Sensor and Ad Hoc Communications and Networks (October 2004)
27. Quinlan, J.R.: C4.5: Programs for Machine Learning. Morgan Kaufmann, San Mateo, CA (1993)

28. Ramanathan, N., Chang, K., Kapur, R., Girod, L., Kohler, E., Estrin, D.: Sympathy for the sensor network debugger. SenSys'05. UCLA Center for Embedded Network Sensing, San Diego, California, USA (2005)
29. Silverstein, C., Brin, S., Motwani, R., Ullman, J.: Scalable techniques for mining causal structures. In: Proceedings of the 24th VLDB Conference, New Yorkm, USA (1998)
30. Vanderbilt University. Message Center
    http://www.isis.vanderbilt.edu/projects/nest/msgctr.html
31. Welsh, M., Mainland, G.: Programming sensor networks using abstract regions. NSDI '04: Proceedings of the First USENIX/ACMSymposium on Networked Systems Design and Implementation (March 2004)
32. Wen, Y., Wolski, R.: s^2db: A novel simulation-based debugger for sensor network applications. UCSB, 2006-01 (2006)
33. Werner-Allen, G., Swieskowski, P., Welsh, M.: Motelab: A wireless sensor network testbed. In: Proceedings of the Fourth International Conference on Information Processing in Sensor Networks (IPSN05), Special Track on Platform Tools and Design Methods for Network Embedded Sensors (SPOTS) (April 2005)
34. Zeng, X., Bagrodia, R., Gerla, M.: Glomosim: A library for theparallel simulation of large-scale wireless networks. In: Proceedings of the 12th Workshop on Parallel and Distributed Simulation (PADS98) (May 1998)

Design and Implementation of a Flexible Location Directory Service for Tiered Sensor Networks

Sangeeta Bhattacharya, Chien-Liang Fok, Chenyang Lu,
and Gruia-Catalin Roman

Department of Computer Science and Engineering
Washington University in St. Louis

Abstract. Many emergent distributed sensing applications need to keep track of mobile entities across multiple sensor networks connected via an IP network. To simplify the realization of such applications, we present MLDS, a Multi-resolution Location Directory Service for tiered sensor networks. MLDS provides a rich set of spatial query services ranging from simple queries about entity location, to complex nearest neighbor queries. Furthermore, MLDS supports multiple query granularities which allow an application to achieve the desired tradeoff between query accuracy and communication cost. We implemented MLDS on Agimone, a unified middleware for sensor and IP networks. We then deployed and evaluated the service on a tiered testbed consisting of tmote nodes and base stations. Our experimental results show that, when compared to a centralized approach, MLDS achieves significant savings in communication cost while still providing a high degree of accuracy, both within a single sensor network and across multiple sensor networks.

1 Introduction

Many emerging distributed sensing applications require the capability of keeping track of a large number of mobile entities over a wide area that is covered by tiered sensor networks. Let's consider the specific example of co-ordinating doctors over multiple make-shift clinics, set up after a natural calamity. Such clinics are often short of doctors and so the doctors may move between the various clinics, depending on the need of the clinics. In such a scenario, there is often a need to keep track of the doctors, as they move within and between clinics, so that it is possible to find a particular doctor or the nearest available doctor. Existing infrastructure (e.g. phone lines and cell phone towers) is often destroyed or overloaded in such scenarios, requiring the deployment of sensor networks connected via ad hoc IP networks to achieve the objective. As another example, consider the tracking of tools that are shared between various workshops spread across a manufacturing facility. The tools are usually moved around within one or more workshops by the workers. Hence, it is very difficult to locate a particular tool when it is needed. In such a situation it would be helpful to keep track of the

J. Aspnes et al. (Eds.): DCOSS 2007, LNCS 4549, pp. 158–173, 2007.

location of the tools as they are moved within and across workshops. This would allow a worker to easily find the nearest available tool that he needs. Sensor networks help realize such applications by providing the capability to sense and identify the mobile entities. However, to fully realize such applications, it is essential to provide a location directory service that can efficiently maintain the location information of mobile entities as they move *across multiple sensor networks* as well as support a broad range of *spatial queries* concerning the mobile entities. Our goal is to realize exactly such a service. The primary contribution of our work is the design, implementation, and empirical evaluation of MLDS, the first Multi-resolution Location Directory Service for *tiered sensor networks*. The key contributions of our work include (1) Design of MLDS, which efficiently maintains location information of mobile entities across multiple sensor and IP networks *and* supports a rich set of multi-granular spatial queries; (2) Implementation of MLDS on tiered sensor networks composed of resource constrained sensor networks and IP networks and (3) Empirical evaluation of MLDS on a tiered testbed of 45 tmote nodes. Our empirical results show that MLDS can maintain a high degree of accuracy at low communication cost, both within a single sensor network and across multiple sensor networks.

2   Services

MLDS can support multiple sensor networks connected by IP networks. Each sensor network, consisting of stationary location-aware sensor nodes and a base station, is assumed to have a unique name that maps to the base station's IP address. We assume that the sensor networks track mobile entities in the physical environment using existing tracking algorithms [1,2,3] or RFID technology. Furthermore, in our implementation of MLDS, we assume that mobile entities are represented by mobile agents in the sensor network. A mobile agent is a software process that can migrate across nodes while maintaining its state. Mobile agents present a convenient way of representing mobile entities (e.g. cars, people and wild fire) in the sensor network [4]. For instance, in the make-shift clinic example described above, mobile agents may be created to shadow the doctors. Users can then query the locations of doctors by querying the locations of the corresponding mobile agents, through MLDS. Note that even though MLDS is implemented to work with mobile agents, it can be easily extended to work with other programming models for mobile entity tracking such as EnviroSuite [1] and others based on message passing [2,3].

MLDS supports four types of flexible spatial queries that include (i) finding the location of a particular agent, (ii) finding the location of all agents, (iii) finding the number of agents and (iv) finding the agent that is closest to a particular location. To meet the needs of diverse applications, all of these queries support different scopes and granularities that can be specified by the application. MLDS supports two query scopes, (i) *local scope* i.e. within a single sensor network and (ii) *global scope* i.e. across all sensor networks. It supports three query granularities, *fine*, *coarse* and *network*. The query result of a fine query is based on the exact locations of the mobile agents while the query result of a coarse query is based on the approximate locations of the mobile agents. The query result of a network query, on the other hand, is based only on the knowledge of the sensor networks that the agents are in. MLDS supports queries issued from both within a sensor network and from outside a sensor network (e.g. by an agent or user on the IP network). The scope and granularity of a query are set via parameters S and G, respectively. Queries can also be limited to a "class" of mobile agents through a parameter C. The API of the four spatial queries are as below:

1. **GetLocation**(id, S, G) returns the location of an agent with ID id.
2. **GetNum**(C, S) returns the number of class C agents.
3. **GetAll**(C, S, G) returns the location of all class C agents.
4. **GetNearest**(C, L, S, G) returns the location of the class C agent that is closest to the location L.

3   Design

MLDS is designed for common sensor network tracking applications like vehicle and personnel tracking for security, emergency care etc. Due to the high mobility of agents in these systems, the location information update rate is expected to be much higher than the query rate in these systems. Hence, MLDS is specifically tailored for systems in which the location information update rate is greater than the query rate. To optimize the operation of such systems, MLDS adopts a hierarchical architecture with multi-resolution information storage. As a result (1) it can support multi-granular spatial queries which enables applications to achieve the desired tradeoff between location information accuracy and communication cost, (2) location information update is not always propagated to the upper tiers of the hierarchy, which significantly reduces communication cost and (3) queries are answered at the closest tier of the hierarchy that meets the query scope and granularity requirements, thus reducing both communication cost and query latency. Note that while MLDS' hierarchical directory structure bears some resemblance to the Domain Name System (DNS) in the Internet and cellular networks, its novelty lies in the fact that it is specifically designed and implemented for tiered sensor networks consisting of resource constrained sensor platforms. In particular, our goal was to minimize communication cost without considerable loss in data accuracy. Moreover, MLDS provides a rich set of multi-granular spatial queries, which is not supported by the above systems.

3.1   Architecture

MLDS has a four tiered hierarchical architecture. The topmost tier of the hierarchy is a central registry that stores information about the different sensor networks. The base stations of the different sensor networks, that are connected by IP networks, form the second tier of the hierarchy. The other two tiers of the hierarchy lie within the sensor networks and are formed by a clustering algorithm that groups the sensor nodes into non-overlapping 1-hop clusters. The clusterheads of these clusters form the third tier of the hierarchy while the cluster members form the fourth tier. Note that the system consists of heterogeneous nodes, with nodes at higher tiers having more resources than nodes at lower tiers. For example, the clusterheads are resource constrained sensor nodes; the base stations are more powerful computers such as PCs or stargates; while the registry, is stored at a server or server cluster.

MLDS stores location information at different resolutions, at different tiers of the hierarchy. Clusterheads store the exact location of the agents in their cluster while base stations store only the IDs of the clusters that the agents in their network belong to. The registry on the other hand stores the IDs of the networks (denoted by the network base station IP address) that all agents in the system belong to. A base station also maintains the location of the clusterhead and the minimum bounding rectangle (MBR) of each cluster in its network. While the registry also stores the MBR of all the connected sensor networks. The network and cluster MBRs are needed to answer nearest neighbor queries, as explained later in Section 3.3.

3.2 Location Information Maintenance

Since MLDS maintains less accurate information at higher tiers of the hierarchy, location information is not always propagated to the upper tiers, which significantly reduces communication cost. In the following we describe how MLDS maintains agent location information at different tiers of the hierarchy.

A node hosting an agent periodically sends location update messages to its clusterhead, at an interval ΔT. Note, periodic messages are required to maintain the directory in the face of node/agent failures. The location update messages contain the agent ID, class and location, which is set to the location of the host node. When a clusterhead receives a location update message, it first updates it's directory with the agent information. If the agent has just entered its cluster, it then sends a message to the base station containing the agent ID and class, and it's own ID, instead of the agent location. The base station in turn updates its directory on receiving this information and also updates the registry if an entry for the agent did not exist in its directory, previously.

Agent location information at a clusterhead expires after a period $2\Delta T$. Thus, if a clusterhead does not receive location update messages from an agent for a period $2\Delta T$, it assumes that the agent has left its cluster and hence deletes the agent from its directory. A clusterhead may therefore have stale information for a maximum period of $2\Delta T$. This design trades off accuracy for lower communication cost and was preferred over other options that provide higher accuracy but at a higher communication cost.

3.3 Query Processing

MLDS answers a query at the closest tier of the hierarchy that meets the query scope and granularity requirements. For queries issued from within(outside) a sensor network, the closest tier would be the lowest(highest) tier of the hierarchy that meets the query scope and granularity requirements. This approach reduces both communication cost and query latency. All queries issued by an agent from within a sensor network are first sent to the clusterhead of the cluster that the agent is in. If the query type is GetLocation or GetNearest, the clusterhead checks if it can answer the query. If it can, it sends the query reply to the querying agent, otherwise it forwards the query to the base station. On the other hand, if the query type is GetAll or GetNum the query is directly forwarded to the base station. The base station processes the query and sends the reply to the clusterhead that sent the query, which in turn forwards the reply to the querying agent. Queries issued by an external agent or user on the IP network are sent to the relevant base stations that process the queries and route the result back to the querying agent/user.

We now explain how MLDS processes a query when the query is issued by an agent within a sensor network. Since a base station processes in-network-queries the same way that it processes out-of-network-queries, the later process can be derived from the description of the former, and hence is not explicitly described. Moreover, due to space limitations, we only describe the GetNearest and GetLocation query types in detail. The GetNum query is the simplest of all queries and just involves querying the base station, while the GetAll query is a simple extension of the GetLocation query. In the following discussion, we assume that the ID of the querying agent is q. We also assume that for any agent with ID i, C_i denotes the clusterhead of the cluster that agent i is in and B_i denotes the base station of the network that agent i is in.

GetLocation. When clusterhead C_q receives a GetLocation(id, S, G) query from agent q, it checks if agent id is in its cluster. If the agent is in its cluster, it sends a query reply to agent q. If agent id is not in agent q's cluster, then C_q forwards the query to the base station B_q. On receiving this query, B_q checks if agent id is in its network. If the agent is in the network, B_q sends a reply containing either the location of the clusterhead C_{id}, if the query is coarse or the exact location of the agent, which it obtains from C_{id}, if the query is fine. In the case that agent id is not in the local network (i.e. $B_{id} \neq B_q$), B_q finds out B_{id} from the registry, and forwards the query to B_{id}. B_{id} processes the query as explained above and sends the result to B_q. B_q sends the query result to C_q, which forwards it to agent q.

GetNearest. When C_q receives a GetNearest(C, L, S, G) query, it checks if there are class C agents in its cluster. If there are such agents, C_q finds the agent that is geographically closest to location L and sends a reply to agent q. If there are no class C agents in the cluster, C_q forwards the query to B_q.

Let's first see how B_q handles local queries. If the query is coarse, B_q just returns the location of the clusterhead, whose cluster contains class C agents and whose location is geographically closest to location L. However, if the query is fine, B_q finds the answer by using the branch and bound technique [5]. The intuition behind this technique is to query only those clusters that contain class C agents whose locations could be closest to location L. These clusters are found by first obtaining a set of clusters that contain class C agents and then looking at the minimum and maximum distances of the MBRs of the clusters in this set, from L. Clusters whose minimum distances are greater than the maximum distance of the cluster that has the least minimum distance, are discarded. B_q queries the clusterheads of the remaining clusters and waits for a certain time period to hear from them. When B_q hears from all the clusterheads (before the end of the time period) or at the end of the time period, B_q computes the agent that is closest to location L based on the information obtained in the query replies and sends the reply to C_q. Note that although the MBR of a cluster does not accurately represent the cluster boundary, it is preferred over other complex methods like the convex hull due to its low computational complexity.

B_q handles global queries similarly, by first looking up the registry to find the networks that contain class C agents and then applying the above branch and bound technique at the network level. Note that by design, this query returns the approximate geographically closest agent. This design achieves lower communication cost by trading off accuracy.

3.4   Cost Benefits over a Centralized Directory Approach

In this section we discuss key benefits of MLDS when compared to a centralized directory (CD) approach. In CD, location information is stored in a centralized directory maintained at the base stations of the sensor networks. Hence, all location information and queries in a sensor network are sent to the base station in CD. MLDS has the following key properties when compared to CD:

- MLDS has significantly lower location update cost, when compared to CD, especially when agents move locally (within a cluster) most of the time, which is common in many application scenarios.
- MLDS achieves a lower total communication cost compared to CD when the location update rate is higher than the query rate. Moreover, for a given update and query rate, the savings in communication cost increase with increasing locality of movement and also with increasing network size. Thus, MLDS is more scalable in comparison to CD.
- Coarse and network query cost in MLDS is close to the query cost in CD. The cost of fine queries is low, when answered locally (by the clusterhead), but high otherwise. Thus, applications that can tolerate coarse query results benefit the most from using MLDS.

A more in-depth theoretical comparison of MLDS and CD can be found in [6].

4   Implementation

We have implemented and integrated MLDS with Agimone, a unified middleware that integrates sensor and IP networks. In this section, we first give an overview of Agimone and then describe the implementation details of MLDS.

Agimone [7] combines two mobile agent middlewares called Agilla [4] and Limone [8]. Agilla is optimized for resource-constrained sensor networks and

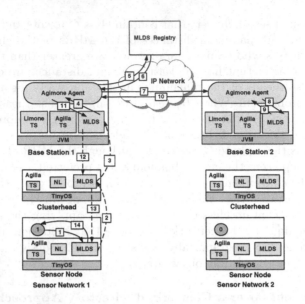

Fig. 1. Interaction between MLDS and Agimone modules when the **GetLocation(0, "global", "coarse")** query is issued by agent 1.(TS: Tuple Space, NL: Neighbor List).

is implemented in nesC on the TinyOS platform. Limone is designed for more powerful nodes (e.g. PDAs, stargates and laptops) connected by IP networks and is implemented in Java on standard Java Virtual Machines (JVMs). In Agimone, creation and deployment of mobile agents within a sensor network is done using Agilla, while migration of mobile agents across sensor networks via an IP network, is done using Limone. Agilla provides primitives for an agent to move and clone itself from sensor node to sensor node while carrying its code and state, effectively reprogramming the network. To facilitate inter-agent coordination within a sensor network, Agilla maintains a local tuple space and neighbor list on each sensor node. Multiple agents can communicate and coordinate through local or remote access to tuple spaces. In Agimone, the base stations communicate through Limone tuple spaces maintained at the base stations. Specific Limone agents called AgimoneAgents that reside at the base stations provide an interface between Agilla and Limone and enable the migration of Agilla agents across an IP network. Agimone maintains a central registry for the registration and discovery of sensor networks over the IP network.

MLDS is integrated with the Agimone modules that run on the sensor nodes and base stations. It is implemented in nesC on the sensor nodes and in Java on the base station. MLDS also extends the Limone registry to serve as the registry for its location directories. Figure 1 shows the interaction between the MLDS and Agimone modules at different tiers of the hierarchy when the **GetLocation(0, "global", "coarse")** query is issued by an agent with ID 1. Agent 1 is in sensor network 1 while agent 0 is in sensor network 2, in the figure. Note that the agents are Agilla agents. Steps 1-3 in the figure show the query message being propagated up the hierarchy to the base station. Once it reaches the MLDS module at the base station, control is transfered to the AgimoneAgent (step 4), since agent 0 is not found in sensor network 1. The AgimoneAgent then queries the registry to find out which network agent 0 is in (steps 5-6). Once it finds that out, it sends the query to the AgimoneAgent at the base station in sensor network 2 (step 7). The AgimoneAgent in base station 2 queries the local MLDS module to obtain the result of the query (steps 8-9) and sends the result back to the AgimoneAgent in base station 1 (step 10). The AgimoneAgent in base station 1 then sends the query reply to the local MLDS module (step 11). After that, the query reply is forwarded down the hierarchy to agent 1 (steps 12-14).

MLDS adapts Agimone's sensor-network-discovery and neighborhood-maintenance mechanisms, to build and maintain its hierarchical structure. The upper two tiers of the hierarchy are formed via the sensor-network-discovery process, in which the base stations register themselves with the registry. The lower two tiers of the hierarchy that lie within individual sensor networks are formed via a simple clustering algorithm. MLDS uses Agimone's neighborhood-maintenance process to achieve clustering at minimum communication cost. Agimone maintains neighborhood information at each node through a periodic beaconing process. Each node periodically broadcasts *beacon messages* containing its ID and hop count to the base station. The hop count information is used for routing messages to the base station. Details of the clustering algorithm are left out due to space limitations but can be found in [6].

5   Experimental Results

We evaluated MLDS through two sets of experiments. The first set of experiments compares MLDS' performance to the centralized approach (CD) within a single sensor network. Recall that in CD, location information in a sensor network is stored only at the base station. All location information and queries are thus sent to the base station in CD. The second set of experiments evaluate MLDS' ability to keep track of mobile agents across sensor networks. In both experiments, tmotes were arranged in a grid, with the gateway node at one corner of the grid. The gateway node is the tmote that acts as a gateway between the sensor network and the PC which serves as the base station. Multi-hop communication between the nodes was achieved by setting a filter at the nodes, that accepted packets only from neighboring nodes on the grid. In order to collect trace data, all nodes in a sensor network were connected to a PC via USB ports.

We use the following four metrics to evaluate query performance, in our experiments. (1) **Success Ratio**: the ratio of the number of queries that returned the accurate result and the total number of queries issued. Network query results are considered accurate if they contain the correct network name; coarse query results are considered accurate if they contain the correct cluster information and fine query results are considered accurate if they contain the correct agent location. (2) **Average Error**: the average error among all queries for which a query result is received, in term of hops. Fine query error is computed as the

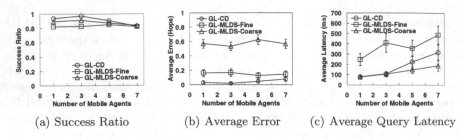

(a) Success Ratio   (b) Average Error   (c) Average Query Latency

Fig. 2. Performance of local GetLocation queries

number of hops between the location returned in the query result and the actual location of the agent. Coarse and network query error is computed as the number of hops the agent is from the clusterhead and from the base station, respectively. (3) **Communication Cost** includes *Location Update Cost* and *Query Cost*. Location Update Cost is the total number of location information messages sent per experiment while Query Cost is the total number of query messages and query result messages sent per experiment. (4) **Average Query Latency**: the average query latency among all queries for which a query result is received. Query latency is the time interval between the issuance of a query and the arrival of the query result, at the querying node. We present 90% confidence intervals for both average error and query latency.

5.1   Single Sensor Network

This set of experiments was carried out on a testbed of 24 tmote nodes, arranged in a 6×4 grid, with a PC as the base station. In each of these experiments, we deployed one stationary agent two hops from the gateway, and n ($1 \leq n \leq 7$) mobile agents. The mobile agents were programmed to follow a random movement pattern over the sensor network at a speed of 1 hop every 5s. Queries were issued by the stationary agent at the rate of 0.2 queries/s. 200 queries were issued in each experiment. Note that by varying the number of mobile agents from 1 to 7 in the experiments, we vary the total location update rate from 0.2 updates/s to 1.4 updates/s and hence evaluate the performance of MLDS under varying network loads. We evaluate only the performance of the **GetLocation** and **GetNearest** queries in these experiments. Since the GetNum query is the same in both MLDS and CD by design and the GetAll query is just an extension of the GetLocation query, we do not evaluate them. We evaluate the performance of the **GetLocation** and **GetNearest** queries, at both fine and coarse granularities, in MLDS. However, only fine queries are evaluated in the centralized approach since it does not support coarse queries. We refer to the GetLocation query in the centralized approach as GL-CD, and the GetLocation fine and coarse queries in MLDS as GL-MLDS-Fine and GL-MLDS-Coarse, respectively. Similarly, the GetNearest queries are referred to as GN-CD, GN-MLDS-Fine and GN-MLDS-Coarse.

GetLocation Query Results. Figures 2 and 3 show the results obtained for the GetLocation query. From Figure 2(a) we see that the success ratio of

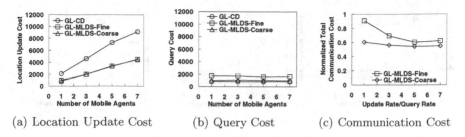

(a) Location Update Cost   (b) Query Cost   (c) Communication Cost

Fig. 3. Communication Cost of local GetLocation queries

GL-CD is higher than that of GL-MLDS-Fine, when there are fewer mobile agents in the network. GL-MLDS-Fine has a lower success ratio partly because a clusterhead retains outdated location information of an agent that has left its cluster, for a maximum time period $2\Delta T$. Interestingly, as the number of mobile agents increases, the success ratio of GL-CD decreases and approaches that of GL-MLDS-Fine. This is because, as the number of mobile agents increases, the number of location information messages also increases. Since all these messages are sent to the base station in CD, there is an increased number of collisions and message loss in the network, which lowers the success ratio of GL-CD. In contrast, the success ratio of GL-MLDS-Fine remains almost constant with the increase in the number of mobile agents, due to its hierarchical architecture. The success ratio of GL-MLDS-Coarse is higher than that of GL-MLDS-Fine and only slightly lower than that of GL-CD. However, since GL-MLDS-Coarse returns an approximate location, it's average error is higher than that of GL-MLDS-Fine, as shown in Figure 2(b). The query reply error of GL-MLDS-Coarse is mostly 1 hop, since MLDS constructs 1-hop clusters. Figure 2(c) displays the query latencies. As expected, GL-MLDS-Fine has the longest query latency since most queries and query results of this type take a longer path. The query latencies of GL-CD and GL-MLDS-Coarse are nearly the same when there are few mobile agents in the system. However, the query latency of GL-CD becomes higher than that of GL-MLDS-Coarse when the number of agents increases, as a result of increased network load.

Figures 3(a), 3(b) and 3(c) show the location update cost, query cost and total communication cost incurred by the GetLocation queries, respectively. From Figure 3(a) we see that MLDS achieves about 55% savings in location update cost when compared to CD. This is due to MLDS' hierarchical architecture, by virtue of which, a large number of location information messages are only sent to the clusterheads and are not forwarded to the base station. Figure 3(b) shows that the query cost of fine queries is higher in MLDS than in CD. This is because GL-MLDS-Fine queries that are routed to the base station, get further routed to a clusterhead. Likewise, the query results of these queries take a longer route to reach the querying agent. Comparatively, the query cost of GL-MLDS-Coarse is much lower and is close to that of GL-CD, since these queries are routed only up to the base station. Overall, MLDS achieves significantly lower

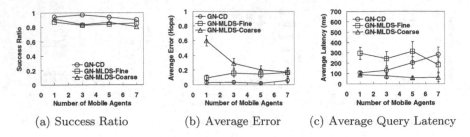

(a) Success Ratio   (b) Average Error   (c) Average Query Latency

Fig. 4. Performance of local GetNearest queries

total communication cost than CD as shown in Figure 3(c). The figure shows the total communication cost of MLDS normalized by the total communication cost of CD for varying ratios of total update rate and query rate. Note that the update rate increases due to the increase in the number of mobile agents in the system. From the figure, we see that the total communication cost of MLDS is lower than that of CD even when the update rate is the same as the query rate and decreases further as the update rate becomes higher than the query rate.

GetNearest Query Results. Figure 4 shows the performance of the GetNearest queries. From Figure 4(a) we see that the success ratios of GN-MLDS-Fine and GN-MLDS-Coarse are almost the same, and remain above 80%, irrespective of the number of mobile agents in the network. The average errors of GN-CD and GN-MLDS-Fine reflect the same trend as their success ratios, as shown in Figure 4(b). What is interesting is the trend in the average error of GN-MLDS-Coarse. The average error of GN-MLDS-Coarse is higher than that of GN-MLDS-Fine when there are fewer mobile agents in the system. However, it decreases as the number of mobile agents in the system increases. This trend is due to the fact that as the number of mobile agents increases, the probability of a mobile agent being in the same cluster as the querying agent also increases and so more number of queries are answered directly by the clusterhead of the cluster that the querying agent is in. Thus, with the increase in the agent density, a higher percentage of the coarse query replies contain exact agent locations, which in-turn reduces the error.

The query latency of GN-MLDS-Fine is higher than that of GN-CD when there are few mobile agents in the network, as shown in Figure 4(c). However, as the number of mobile agents increases, the query latency of GN-CD increases considerably whereas the query latency of GN-MLDS-Fine decreases. The query latency of GN-MLDS-Fine becomes less that of GN-CD when there are 7 mobile agents in the network. The increase in the query latency of GN-CD with the increase in the number of mobile agents is a result of increased network load. The reason for the decrease in query latency of GN-MLDS-Fine, with the increase in mobile agents in the network, is the increase in the percentage of queries that get answered locally, by the clusterhead. This same reason also causes the decrease in the query latency of GN-MLDS-Coarse as the number of mobile agents in the network increases. Thus, the GN query benefits significantly from local responses, made possible by MLDS' hierarchical architecture. The benefit is not only decreasing query latency but also decreasing query cost (not shown here), with increasing agent density.

In summary, MLDS consistently achieves success ratios above 80% in all our experiments, at significantly lower total communication cost than the centralized approach. In particular, coarse queries supported by MLDS achieved the lowest communication cost and query latency, while introducing an average error of less than 1 hop. Thus, applications that can tolerate a small amount of location error gain the most from using MLDS. Furthermore, MLDS' hierarchical architecture enables efficient execution (low cost and latency) of GetNearest queries, especially when the density of mobile agents is high.

5.2 Multiple Sensor Networks

We now evaluate MLDS' performance across multiple sensor networks. In these experiments, mobile agents move between three sensor networks via an IP network running over 100Mbps Ethernet. The IP network is private with a single Linksys WRT54G router and an 8 port switch. These experiments were carried out on a testbed of 45 tmote nodes, equally divided into three sensor networks arranged in a 5×3 grid. Each sensor network has a PC connected via USB to one of its corner motes that serves as a base station. These base station PCs are connected to each other via the IP network. A fourth PC on the IP network serves as the registry. Evaluating MLDS' performance requires comparing its results with the ground truth. The ground truth is obtained by connecting every mote except those directly attached to a base station to the registry PC via USB. The motes are programmed to send trace messages identifying key events like agent movement and query activities over their USB port. The registry PC monitors these connections for incoming trace data and saves them into a file. In addition, it also accepts trace messages over the IP network, which the base stations use to record trace messages generated by the motes they are attached to. The registry PC serves as a central aggregation point for the trace data. Each trace event is time stamped and saved for off-line analysis.

Like the single sensor network experiments, we evaluate only the performance of the GetLocation (GL) and the GetNearest (GN) queries. Both the network and coarse granularity versions of the queries are evaluated. In each of these experiments, the workload is varied by varying the number of mobile agents in the system from 1 to 21 in increments of 3. The mobile agents move 10 hops randomly in a sensor network before randomly migrating to another sensor network and repeating. Initially, the mobile agents are distributed evenly across the three sensor networks. The GetLocation and GetNearest queries are issued at a rate of 1 query every 5s by an external agent running on the registry PC. Note that this differs from the single network experiments where the querier was located within the sensor network. By placing the querier on the registry, the query messages only travel down the hierarchy. Scenarios where the query

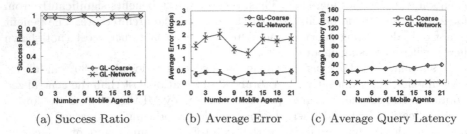

(a) Success Ratio   (b) Average Error   (c) Average Query Latency

Fig. 5. Performance of global GetLocation queries

messages travel up the hierarchy were already evaluated in the single-network experiments. Each experiment is repeated 100 times.

GetLocation Query Results. For these experiments, a querier located on the registry periodically issues a GL query for a particular mobile agent (termed the *target agent*) within the sensor networks. The success ratio of the GL query is shown in Figure 5(a). Both the coarse (GL-Coarse) and network (GL-Network) granularity versions of GL achieve nearly perfect success. GL-Coarse has a slightly lower success ratio because it attempts to return a more accurate location of the agent. However, it has approximately 3 times lower error, as shown in Figure 5(b). Notice that GL-Network has a higher average error variance and that GL-Coarse has an average error variance of less than one. This is because an agent may be multiple hops away from the network base station, but can be at most one hop away from its cluster head. GL-Coarse queries have significantly longer latency as shown in Figure 5(c). The latency of GL-Network queries is negligible since the querier is located on the registry and can query the registry locally to determine which network the target agent is in. For GL-Coarse queries, the agent must first lookup which network the agent is in, then query that network's base station to determine which cluster the agent is in. As the number of mobile agents increases, the latency also increases due to increased network congestion.

GetNearest Query Results. The GN experiments are the same as the GL experiments except the target agent is a stationary agent that resides two hops away from the base station on one of the sensor networks. The querier on the registry periodically searches for the mobile agent closest to the target agent. The results are shown in Figure 6. As the number of mobile agents increase, the success ratio of GN-Coarse decreases due to network congestion preventing updates from propagating up the hierarchy, as shown in Figure 6(a). On the other hand, GN-Network queries almost always succeed since it involves at most 1 call to the registry. The average error of GN-Coarse queries remains roughly less than 1 hop regardless of the number of agents as shown in Figure 6(b). This is expected since the cluster members are at most 1 hop away from the cluster head. The average error of GN-Network queries is also close to 1, but is dependent on the size of the network. As the number of mobile agents increase, the probability

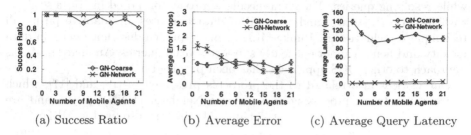

(a) Success Ratio   (b) Average Error   (c) Average Query Latency

Fig. 6. Performance of global GetNearest queries

of finding an agent in the same network as the target increases, decreasing the latency of GN-Coarse queries, as shown in Figure 6(c). The latency of GN-Network queries is negligible because in our experiments, an agent is always present in the target agent's network and hence the queries are always answered locally by the base station.

6 Related Work

MLDS is related to data-centric storage (DCS) systems like GHT [9], DIFS [10], DIMENSIONS [11] and DIM [12]. GHT hashes data by name to nodes in the network and provides no index for accessing the data. Hence it is unsuitable for storing and accessing location information. DIFS leverages on GHT and maintains a hierarchical index of histograms to support multi-range queries. DIM, on the other hand, uses a locality-preserving hash function that maps a multi-attribute event to a geographic zone. It divides the network into zones and maintains a zone tree to resolve multi-dimensional range queries. However, the index of neither DIF nor DIM can efficiently support spatial queries. DIMENSIONS hashes sensor data to nodes in the network and maintains a multi-resolution hierarchical index that enables it to efficiently answer queries by drilling down to the appropriate nodes. However, DIMENSIONS was not designed for storing location information and hence does not support spatial queries such as **GetNearest**. Thus, the key differences between the above systems and MLDS are (1) the above systems store sensor data while MLDS is specifically tailored for storing location information of mobile entities, (2) MLDS supports a broad range of flexible spatial queries which cannot be supported efficiently by the above systems, and (3) MLDS builds a distributed directory over multiple sensor networks connected by an IP network, while the above systems systems are designed for a single sensor network. TSAR [13] is another in-network storage architecture, which stores sensor data at a lower tier consisting of sensor nodes and stores only meta data at a higher tier consisting of a network of proxies. Unlike the above approaches, TSAR maintains a distributed index at the proxies. TSAR differs from MLDS in that it is not tailored for storing location information nor does it support spatial queries. Moreover, unlike MLDS, TSAR does not have an in-network tier that enables the system to take advantage of data locality while resolving queries. The comb-needle approach proposed in [14] also deals with in-network storage and retrieval of data and uses an adaptive push-pull technique to achieve this. Unlike MLDS, this approach does not maintain data locality and hence cannot efficiently support spatial queries. An analysis of this approach to other DCS approaches has been provided in [15].

Our work is also related to the protocols presented in [16] and [17], which address in-network processing of K-Nearest Neighbor (KNN) queries and are based on the branch-and-bound technique [5] which is also used in MLDS. Another related service is EASE [18], which keeps track of mobile entities within a single sensor network through in-network storage and supports multi-precision queries that fetch the location of a specified mobile entity. MLDS differs from the above protocols in the following three important ways: (1) MLDS presents an architecture for storing location information in sensor networks that enables efficient computation of nearest-neighbor as well as other multi-resolution spatial queries, (2) MLDS is the first location directory service that can keep track of mobile entities across multiple sensor networks, and (3) we implemented and integrated MLDS with a mobile agent middleware and present experimental results on a physical testbed. In contrast, the above protocols are only evaluated through simulations.

7   Conclusion

We have developed MLDS, a Multi-resolution Location Directory Service for tiered sensor networks comprising multiple sensor networks connected via IP networks. MLDS has several salient features: (1) it is the first system that maintains location information of mobile entities across sensor and IP networks, (2) it supports a range of multi-granular spatial queries that can span multiple sensor networks and (3) it has low communication cost. We integrated MLDS with Agimone, a mobile agent middleware for sensor and IP networks, and evaluated its performance on a testbed of tmote nodes. The empirical results obtained show that MLDS successfully keeps track of mobile agents across single and multiple sensor networks at significantly lower communication cost than a centralized approach. Most importantly, MLDS enables applications to achieve the desired tradeoff between accuracy and communication cost, which is particularly useful for resource constrained sensor networks. Currently, MLDS is optimized for systems that have a higher location update rate. As future work, we plan to extend MLDS such that it dynamically adapts to the query and the update rate and hence performs well under all conditions. We wish to achieve this by using a push-pull strategy that dynamically adjusts the location and granularity of location information based on the query and update load.

Acknowledgments. This work is funded by the NSF under the ITR grant CCR-0325529 and the NOSS grant CNS-0520220.

References

1. Luo, L., Abdelzaher, T., He, T., Stankovic, J.A.: Envirosuite: An environmentally immersive programming framework for sensor networks. TECS (2006)
2. Liu, J., Reich, J., Zhao, F.: Collaborative in-network processing for target tracking. Journal of Applied Signal Processing (2003)
3. Pattem, S., Poduri, S., Krishnamachari, B.: Energy-quality tradeoffs for target tracking in wireless sensor networks. In: Zhao, F., Guibas, L.J. (eds.) IPSN 2003. LNCS, vol. 2634, Springer, Heidelberg (2003)
4. Fok, C.L., Roman, G.C., Lu, C.: Rapid development and flexible deployment of adaptive wireless sensor network applications (In: ICDCS'05)
5. Roussopoulos, N., Kelly, S., Vincent, F.: Nearest neighbor queries (In: SIGMOD'95)
6. Bhattacharya, S., Fok, C.L., Lu, C., Roman, G.C.: Mlds: A flexible location directory service for tiered sensor networks. Technical Report WUCSE-2007-1, Dept. of Computer Science and Engineering, Washington University in St. Louis (2007)
7. Hackmann, G., Fok, C.L., Roman, G.C., Lu, C.: Agimone: Middleware support for seamless integration of sensor and ip networks (In: DCOSS'06)
8. Fok, C.L., Roman, G.C., Hackmann, G.: A lightweight coordination middleware for mobile computing (In: Coordination'04)
9. Ratnasamy, S., Karp, B., Yin, L., Yu, F.: GHT: A geographic hash table for data-centric storage (In: WSNA'02)
10. Greenstein, B., Estrin, D., Govindan, R., Ratnasamy, S., Shenker, S.: DIFS: A distributed index for features in sensor networks (In: SNPA'03)
11. Ganesan, D., Estrin, D., Heidemann, J.: DIMENSIONS: Why do we need a new data handling architecture for sensor networks? (In: HotNets-I'02)
12. Li, X., Kim, Y.J., Govindan, R., Hong, W.: Multi-dimensional range queries in sensor networks (In: SenSys'03)
13. Desnoyers, P., Ganesan, D., Shenoy, P.: Tsar: A two tier storage architecture using interval skip graphs (In: SenSys'05)
14. Liu, X., Huang, Q., Zhang, Y.: Combs, needles, haystacks: Balancing push and pull for discovery in large-scale sensor networks (In: SenSys'04)
15. Kapadia, S., Krishnamachari, B.: Comparative analysis of push-pull query strategies for wireless sensor networks. In: Gibbons, P.B., Abdelzaher, T., Aspnes, J., Rao, R. (eds.) DCOSS 2006. LNCS, vol. 4026, Springer, Heidelberg (2006)
16. Demirbas, M., Ferhatosmanoglu, H.: Peer-to-peer spatial queries in sensor networks (In: P2P'03)
17. Winter, J., Xu, Y., Lee, W.C.: Energy efficient processing of k nearest neighbor queries in location-aware sensor networks (In: Mobiquitous'05)
18. Xu, J., Tang, X., Lee, W.C.: EASE: An energy-efficient in-network storage scheme for object tracking in sensor networks (In: SECON'05)

A Semantics-Based Middleware for Utilizing Heterogeneous Sensor Networks

Eric Bouillet, Mark Feblowitz, Zhen Liu, Anand Ranganathan, Anton Riabov, and Fan Ye

IBM T.J. Watson Research Center, Hawthorne, NY 10532, USA
{ericbou, mfeb, zhenl, arangana, riabov, fanye}@us.ibm.com

Abstract. With the proliferation of various kinds of sensor networks, we will see large amounts of heterogeneous data. They have different characteristics such as data content, formats, modality and quality. Existing research has largely focused on issues related to individual sensor networks; how to make use of diverse data beyond the individual network level is largely unaddressed. In this paper, we propose a semantics-based approach for this problem and describe a system that constructs applications that utilize many sources of data simultaneously. We propose models to formally describe the semantics of data sources, and processing modules that perform various kinds of operations on data. Based on such formal semantics, our system composes data sources and processing modules together in response to users' queries. The semantics provides a common ground such that data sources and processing modules from various parties can be shared and reused among applications. We describe our system architecture, illustrate application deployment, and share our experiences in the semantic approach.

1 Introduction

Increasingly ubiquitous sensors and sensor networks bring us data sources of various content, formats, modality and quality. They provide vast amount of information about events and phenomena in the physical world. Current sensor network research has mostly focused on issues pertaining to individual sensor networks. As a result, many applications are tied closely to one or a few sensor networks. Large numbers of diverse data sources, however, present the opportunities for new kinds of applications that utilize many data source simultaneously, thus achieving functions not possible by using any single sensor network.

Consider the following motivating scenario. Local, state and federal transportation departments deploy cameras, motion / magnetic detectors and temperature sensors along highways and roads. Individually, each data source (i.e., a sensor or sensor network) provides traffic information with limited geographic coverage, sensing modality and data quality. None of them can individually satisfy the diverse needs: drivers want real time driving instructions; the Highway Patrol wants videos of accident scenes or plates of speeding vehicles; the Department of Transportation wants long-term traffic statistics for road expansion

J. Aspnes et al. (Eds.): DCOSS 2007, LNCS 4549, pp. 174–188, 2007.

plans. These applications require the use of heterogeneous data sources in an integrated manner.

In this paper, we propose a semantics-based approach to this problem and describe a system that constructs such applications on the fly. These applications ingest data from many distributed, heterogeneous sensors and sensor networks. They use interconnected software modules (called *Processing Elements* (PEs)), which take data of certain content and format, and perform various operations, from elementary filtration to complex analysis. Finally they produce the highly summarized end results needed by users.

Our approach is based on the use of *ontologies*, a formal method for describing the terms and relations relevant to a certain domain of interest. In our system, we use ontologies described in OWL [1], a standard representation language in the Semantic Web. Descriptions of data sources, PEs and users' queries use the terms and relations defined in the OWL ontologies. Data sources are described by the semantics of typical data objects they produce; PEs are described by the semantics of data objects they consume and produce; queries express the semantics of end results users desire. An AI planning algorithm, enhanced to utilize semantic descriptions, automatically composes appropriate data sources and PEs together as applications that answer users' queries.

Automatic construction of applications based on formal semantics has many advantages. With diverse and heterogeneous forms and content of sensed data, and large numbers of PEs, the composition of such applications becomes a grand challenge. It is infeasible for a human user to sort out, from thousands of sensor data sources and PEs, which ones are appropriate for his needs, and the correct and efficient ways to interconnect them. It is through the formal semantic descriptions of sources and PEs that the planning algorithm is able to compose them in legitimate and efficient ways so that they collectively produce meaningful end results.

The formal semantics also enables the reuse of sources and PEs among applications. We envision many parties will provide data sources or PEs of various kinds. By tapping on the growing reservoir of data sources and PEs, more and more powerful applications can be built. Without a common ground for the description of sources and PEs, one party's sources and PEs cannot be reused by others.

We make several contributions in this paper. We propose a semantic model for the formal description of data sources and processing elements so that they can be reused. We also define the conditions for legitimate connection among sources and PEs, and devise an efficient semantic planning algorithm to the automatic construction of applications. We also build a system that implements these ideas and proves the feasibility of our approach.

The rest of the paper is organized as follows. We give an overview of the system in Section 2. In section 3 we present the semantic model used to describe data sources and PEs. Section 4 explains how to use the semantic descriptions to construct applications automatically. Section 5 demonstrates an example application deployment. We compare with related work in Section 6 and conclude in Section 7.

2   System Overview

We have built our middleware of the semantics-based approach and deployed it in
System S [2], which is a distributed stream processing system (Figure 1). Users
submit *inquiries* that describe the formal semantics of desired end results. A
planner constructs applications on the fly based on formal semantic descriptions
of PEs and data sources. The sources are managed by a Data Source Manager
component. Finally, applications are deployed by Job Manager (JMN) and executed on a Stream Processing Core (SPC), running in a cluster of machines.

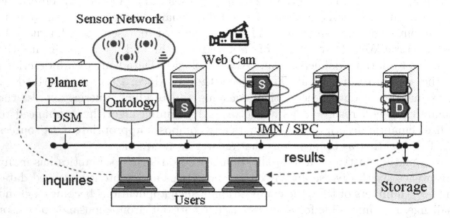

Fig. 1. The architecture of System S. Users submit inquiries to specify the end results
they need. A planner constructs applications from data sources and PEs, whose semantics are formally described and stored in the ontology. A Data Source Manager (DSM)
manages connections to data sources. Finally applications are deployed through Job
Manager (JMN) and run on the underlying Stream Processing Core.

Our system interacts with sensors and sensor networks on the data level and
lower level issues such as in-network processing are transparent to our system.
As long as they can be accessed through some well-defined interface (e.g., a
gateway or base station), they are treated as data sources providing data whose
semantics can be formally described using ontologies. PEs perform different kinds
of processing on data. Source PEs (denoted by "S" in Figure 1), a special type
of PEs, talk the protocols of these data sources. They can access data sources
and package raw data into internal *Stream Data Objects (SDOs)*, a lightweight
data container format that is used within the system and universally understood
by all PEs. Finally Sink PEs (denoted by "D" in Figure 1) collect and deliver
results to end-users.

Our system uses a small set of commonly agreed upon ontologies. They can
be defined through collaborative efforts, and can reuse existing ontologies from
different domains, such as the NCI cancer ontology [3]; the GALEN medical
ontology [4], geographical metadata, dependable systems[1] , etc. The problem of

---

[1] http://protege.cim3.net/cgi-bin/wiki.pl?ProtegeOntologiesLibrary

exactly how to define ontologies falls in the area of ontology engineering, which is not our focus in this paper. We focus on how to use ontologies for describing components and for the automatic construction of applications.

To automatically construct applications, we need to address several critical issues: 1) How to formally describe the semantics of data sources and PEs; 2) How to decide the legitimate connection between PEs, i.e., which PEs' output streams match the types and semantics of required input streams of other PEs; 3) how to compose PEs and form a processing graph that can produce the desired final results.

Before we explain how we address these issues in the following sections, we use an exemplary inquiry to illustrate how it is processed in our system. Consider an inquiry that requests traffic congestion reports for a particular road intersection. The final flow graph depicting which PEs are used and how they interconnect to produce the results is illustrated in Figure 2.

The system needs to know which data sources provide relevant data. A sound sensor and a video camera around that intersection provide audio and video raw data from which congestion levels can be extracted. The system discovers relevant sources through their semantic descriptions. Similarly, the system needs to know which PEs can process such data. It can identify an Audio Pattern Analysis PE from its semantics as

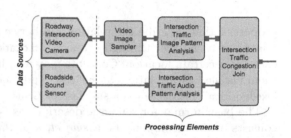

Fig. 2. A Stream Processing Graph example for an inquiry requesting traffic congestion report at an intersection

taking roadside sound and producing traffic pattern. For video, two PEs (Video Image Sampler and Image Pattern Analysis) have to be connected to produce traffic pattern. The system needs their semantics to know they can be legitimately connected and produce meaningful results. Finally, a join PE is used to combine the results from the two chains of analysis.

3   Semantic Model of Data Sources and PEs

We use OWL ontologies [1] as the basis for the formal representation of the semantics of data sources and PEs. OWL ontologies describe *concepts* (or *classes*), *properties* and *individuals* (or *instances*) relevant to a domain of interest.

A concept is the abstraction for a kind of entities sharing common characteristics. In the Traffic services example (Figure 3), Sensor, Location and MultimediaData are concepts. Individuals are specific entities that belong to certain concepts. Traffic Camera 10036-1 is an individual belonging to concept Sensor, and BwayAt42nd (the intersection of Broadway and 42nd Street) is an individual belonging to concept Location. Concepts are associated with each other through

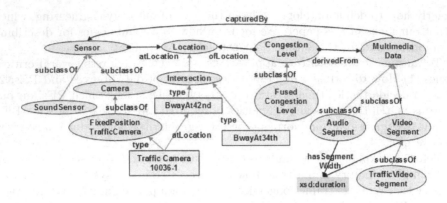

Fig. 3. Example ontology in Traffic Services domain defining concepts like Sensor and Location, properties like capturedBy and atLocation, and individuals like BwayAt42nd

properties that describe the relationship between them. Sensor and Location are related to each other through property atLocation, meaning that a sensor is located at a certain location. Concepts may also have hierarchical relation among them via subclassOf relation. SoundSensor and Camera are subclasses of Sensor, and FixedPositionTrafficCamera is a subclass of Camera.

The basic format for semantic descriptions is an *RDF triple*. An RDF triple consists of three components: a *subject*, a *predicate* and an *object*. The subject and object can be concepts or individuals, and the predicate is the property that associates them. RDF triples can describe OWL *axioms* and OWL *facts*, semantic information about concepts and individuals, respectively. For example, (Camera subClassOf Sensor) is an OWL axiom; and (TrafficCamera10036-1 atLocation BwayAt42nd) is an OWL fact.

A set of RDF triples can be represented as a graph, called *RDF graph*. The nodes in the graph are subjects and objects in the triples, and the edges are predicates (properties) between them. One example is shown in Figure 3.

3.1   Descriptions of Data Sources

In our model, a data source produces a continuous stream of SDOs, each of which contains several data elements. A data source's semantics lists what data elements are contained in an SDO, and an RDF graph that describes the characteristics (or semantics) of the data elements.

As an example (Figure 4), consider the semantic description of the data source VideoCameraBway-42nd, a video camera located around Broadway and the 42nd Street. It produces a stream Bwy-42ndVideoStream. A typical SDO in this stream contains two data elements: _VideoSegment_1 and _TimeInterval_1. The characteristics of these two data elements are described in the form of an RDF graph. The characteristics include the concepts they belong to and the values of various properties. For example, the data element _VideoSegment_1 belongs to the concept VideoSegment and is captured by a camera Traffic Camera 10036-1, which

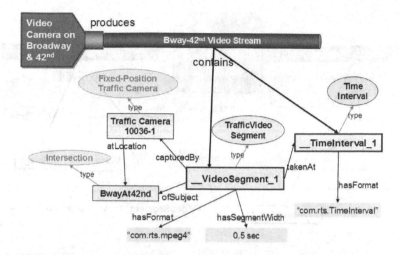

Fig. 4. Semantic description of a camera data source describing the stream it produces, the data elements contained in the stream (__VideoSegment_1 and __TimeInterval_1) and the properties of these data elements described as an RDF graph

is a Fixed Position Traffic Camera and located at BwayAt42nd, the intersection of Broadway and the 42nd Street.

Observe that a typical SDO is described as containing two typical data elements: __VideoSegment_1 and __TimeInterval_1. We refer to these typical data elements as *exemplars*, since they provide an example description of the elements in the stream. Actual SDOs in the stream may replace these exemplars by actual values. For example, they may have specific VideoSegment instances that share the semantics of the exemplar, e.g. they have the same values of the ofSubject and capturedBy properties. Exemplars are represented as individuals that belong to the special concept called "Exemplar". Syntactically, we denote exemplars with a preceding double underscore (i.e. "__").

The RDF graph that describes the semantics of the stream is based on domain ontologies, such as the one in Fig 3. The domain ontologies provide the common "language" for the interoperability of sources from different parties. When they describe their sources using the same ontology, we are sure that the same term has exactly the same meaning, even if they come from different parties.

To summarize, the semantic description of a data source (or its stream) is a pair, (D, G), where

- D is the set of data elements contained in the typical SDO in the stream. These data elements are represented as exemplar individuals.
- G is an RDF Graph that describes the semantics of the data elements on the typical SDO as a set of OWL facts.

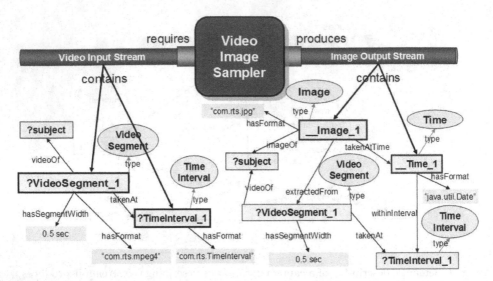

Fig. 5. Semantic description of VideoImageSampler PE that takes as input a stream containing a video segment and a time interval and produces a stream containing an image and a time

3.2   Descriptions of PEs

A Processing Element takes some number of input streams and produces some number of output streams. We treat PEs as "black boxes" and describe PEs through the semantics of their input streams and output streams. Each input and output is described by a *stream pattern*, which specifies the kinds of streams that can be fed to the PE as legitimate input, and the kinds of streams the PE produces as output. The stream patterns are defined using variables, which are represented with a preceding question mark ("?").

As an example, the VideoImageSampler PE in Figure 5 is defined as requiring a stream whose SDO contains two data elements: ?VideoSegment_1 and ?TimeInterval_1. In addition, it specifies several constraints on the semantics of these elements. For example, ?VideoSegment_1 should be of type VideoSegment and ?VideoSegment_1 should be taken at ?TimeInterval_1, a certain time interval. When an input stream that satisfies these constraints is connected to this PE, the PE produces a single output stream that contains two exemplar data elements: _Image_1 and _Time_1. These exemplars are associated with a semantic description that describes their characteristics and also relate them back to the input data elements. E.g., (_Image_1 extractedFrom ?VideoSegment_1) means that _Image_1 element in output stream is extracted from the video segment element ?VideoSegment_1 in input stream. They also share the same subject ?subject.

The input stream pattern describes the set of constraints that must be satisfied by any stream that can be legitimately connected as input to the PE. It may be regarded as a semantic query that must be satisfied by the description of an input stream. The output stream pattern describes new streams produced by a

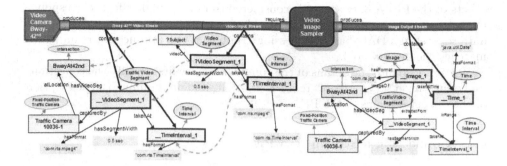

Fig. 6. Semantically matched data source and PE. The dashed lines show the variable substitutions that allow the input graph pattern of the PE to be embedded in the output graph of the data source, after DLP reasoning.

PE as the result of connecting compatible input streams to the PE. We now give a formal definition of stream patterns.

A *triple pattern* is an RDF triple where the subject or the object is a variable. An example is (?videoSegment_1 takenAt ?timeInterval_1). A *graph pattern* is a set of triple patterns. A *stream pattern* has two elements: the set of data elements and their semantics. It is a pair of the form (VS, GP) such that

- VS is a set of variables and exemplars representing the data elements in a typical SDO in the stream.
- GP is a graph pattern describing the semantics of the data elements.

A PE is described as a 2-tuple of the form $(\langle ISR \rangle, \langle OSD \rangle)$ where

- $\langle ISR \rangle$ is a set of input stream requirements, where each input stream requirement is represented as a stream-pattern.
- $\langle OSD \rangle$ is a set of output stream descriptions, where each output stream description is represented as a stream-pattern.

There is a subtle difference between our use of variables and exemplars. Variables (like ?VideoSegment_1) represent the elements in an input stream pattern, and they may be carried forward to the output stream pattern. Exemplars (like _Image_1)represent those elements that are newly created by the PE in the output stream pattern.

We have developed a language for representing the semantic descriptions of PEs and data sources called SGCDL (Semantic Graph-based Component Description Language). The language is based on OWL and allows importing various domain ontologies and using terms defined in these ontologies for describing PEs and sources. Due to space limit, we do not go into details in this paper.

4   Semantic Composition of Applications

Our model of PEs and data sources makes them amenable to composition using AI planning. The input and output stream-patterns act as preconditions and

effects of the PE. A data source is considered to have a single effect, corresponding to the stream it produces. The problem now becomes composing PEs and
data sources in a DAG-based processing graph that can produce a stream which
matches the inquiry goal.

Before we go into the details of the composition, we first specify the conditions
for connecting a set of streams to a PE as input, and determining the description
of the output streams of the PE.

4.1   Connecting a Stream to a PE

A stream can be connected to a PE if the stream-pattern describing the PE's input requirements can be *matched* to the description of the stream. The matching
process involves finding a substitution of all the variables in the stream pattern
such that it can be *inferred* from the stream. Definitions in the domain ontology
can be used in the inference process.

Let us first define the notion of a substitution of variables. A substitution function $(\theta : V \to RDF_T)$ is defined from the set of variables (V) to the set of RDF
terms (RDF_T). For example, mappings defined in a possible definition of θ in
the example PE include : $\theta(?videoSegment\_1) = \_VideoSegment\_1$, $\theta(?timeInterval\_1)$
$= \_TimeInterval\_1$ and $\theta(?subject) = BwayAt42nd$.

The result of replacing a variable, v, is represented by $\theta(v)$. The result of
replacing all the variables in a graph pattern, GP, is written as $\theta(GP)$.

Consider a stream-pattern, $SP(VS, GP)$ described by a set of variables, VS
and a graph pattern, GP. Also, consider a stream, $S(D, G)$, with a set of data
elements, D, and a semantic graph, G, describing the elements. We define SP
to be *matched* by S, based on an ontology, O, if and only if there exists a
substitution, θ, defined on all the variables in GP, such that following conditions
hold:

- $S \supseteq \theta(VS)$, i.e. the stream contains at least those elements that the pattern
  says it must contain.
- $G \cup O \models \theta(GP)$ where O is the common ontology and $\theta(GP)$ is a graph
  obtained by substituting all variables in GP. $\models$ is an entailment relation
  defined between RDF graphs, i.e. it defines a logical reasoning framework
  by which it is possible to determine whether the set of facts in the left hand
  side (i.e. $G \cup O$) can be used to infer the set of facts in the right hand side
  (i.e. $\theta(GP)$).

One way of looking at the above definition is that the stream should be more
specific than the stream pattern. The stream should have at least as much data
and as much semantic information as described in the pattern.

The use of reasoning allows matching streams to stream patterns even if they
use different terms or graph structures. The exact logical reasoning that can be
performed using OWL is some subset of description logics (DL), e.g. based on
RDFS, OWL-Lite, OWL-DLP, OWL-DL [5], etc. At the minimum, the reasoning
can involve determining a graph-embedding relationship, i.e. can the substituted

graph pattern, $\theta(GP)$, be embedded in the graph G. More sophisticated reasoning mechanisms allow inferring additional facts from G, which can then be used to determining the same graph-embedding relationship. Our matching system uses DLP reasoning [6], which allows making inferences based on subclass and subproperty relationships, symmetric, transitive and inverse property definitions, domain and range definitions, value restrictions, etc.

An example of this process is shown in Fig 6. Here, the stream produced by VideoCameraBway-42nd Data Source is matched to the stream-pattern describing the input requirements of the VideoImageSampler PE. The variables ?VideoSegment_1 and ?TimeInterval_1 are mapped to the exemplars _VideoSegment_1 and _TimeInterval_1 respectively. The dashed arrows show the variable substitutions. In order to make the match, the system must perform DL reasoning based on subclass and inverse property relationships defined in the domain ontology. For example, the triple _VideoSegment_1 videoOf BwayAt42nd is inferred, since videoOf is declared to be an inverse property of hasVideoSeg in the ontology. Also, the triple _VideoSegment_1 type VideoSegment is inferred, since TrafficVideoSegment is declared to be a subclass of VideoSegment. Once the inferences are done, it is clear to see that the substituted graph pattern can be embedded into the graph describing the stream produced by the camera; hence a match is obtained.

4.2   Automatic Composition of Applications

The goal of a composition process is to produce a processing graph that generates streams that satisfy some high-level information need. This high-level information need, or inquiry, is represented as a semantic stream pattern that describes the kind of data elements and their semantics that some user or application is interested in. This stream pattern becomes a goal for our planner. The stream pattern is represented in a syntax that is similar to SPARQL [7], a semantic query language for RDF. An example goal for real-time traffic congestion levels at the Broadway-42'nd St intersection is

PRODUCE ?congestionLevel, ?time
WHERE (?congestionLevel type CongestionLevel) , (?time type Time),
    (?congestionLevel ofLocation BwayAt42nd) , (?congestionLevel atTime ?time)

Note that this stream pattern does not contain any reference to the actual data elements, or their formats, in which the congestion level and time may be represented. This means that users can frame information needs without knowing the exact formats (e.g. Java classes) for different messages. They can describe the information need in terms of the semantics without having to know the syntax of the final messages.

In previous sections, we defined the conditions under which two PEs can be connected to each other, based on the matching between the output streams and input stream patterns. At a high level, the planner works by checking if a set of streams can be given as input to a PE. If so, it generates new streams corresponding to the outputs of the PE. It performs this process recursively and keeps generating new streams until it produces one that matches the goal pattern, or until no new unique streams can be produced, or the plan size exceeds a pre-specified maximum size.

We have developed a planner that employs a two-phase approach to generate plans. In the first phase, which occurs offline, it does pre-reasoning on the output descriptions of different PEs to generate additional facts about the streams produced by these PEs. The exact flavor of reasoning performed is OWL-DLP (Description Logic Programs), which is known to be decidable, complete and to take polynomial time.

The original and the inferred facts about components are translated into a language called SPPL (Stream Processing Planning Language) [8]. SPPL is a variant of PDDL (Planning Domain Definition Language) and is specialized for describing stream-based planning tasks. It models the state of the world as a set of streams and different predicates are interpreted only in the context of a stream. The SPPL descriptions of different components are persisted and re-used for multiple queries.

The second phase is triggered whenever an information request is submitted to the system. During this phase, the planner translates the query into an SPPL planning goal. It then calls SPPL solver [8] to produce a plan consisting of actions that correspond to components. The plan is constructed by recursively connecting components to one another based on their descriptions until a goal stream is produced. In our implementation, this plan is then deployed in the System S stream processing system [2]. The main reason for the two-phase planning process is to achieve scalability in the presence of possibly time consuming reasoning.

5   Experiments

In this section we measure the automatic composition capability with an application domain we call Realtime Traffic Services (RTS). Applications in RTS provide vehicle routing services based on the analysis of real-time data obtained from sensors. The target deployment testbed used in our experiments consists of four 4-way 3GHz Intel Xeon(TM) machines and five 2-way 2.4GHz AMD Opteron (TM) 250 machines, running the Linux Suse 9.3 operating system and interconnected with 1Gbs network cards via a Cisco Catalyst 6509 switch. The planner was running on a Pentium M 2Ghz laptop with 2GB of RAM.

Figure 7 shows a screenshot of a tool that visualizes processing graphs deployed on the System S infrastructure. PEs are grouped by host, and represented with a distinct color for each inquiry. Semioval PEs represent the source (left semioval) and the sink PEs (right semioval) which interface the RTS stream processing application with external data sources and consumers. The figure depicts the processing graph for a route-update inquiry (i.e. an inquiry for the best route from the current location to the final destination). As described in previous sections, this processing graph is automatically composed by planner to satisfy a set of results prescribed by the inquiry. The automatic placement of the PEs to their respective hosts is coordinated by a separate resource management

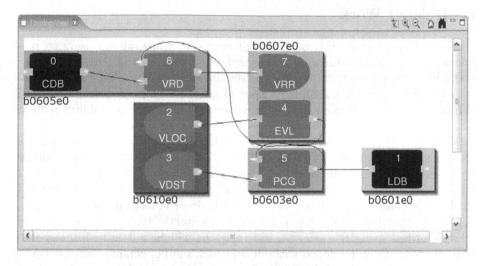

Fig. 7. The flow graph for Route Update Service that takes as input the vehicle positions and destinations from source PEs 2 and 3 respectively, and outputs a best route (PE 7), and generates a list of locations to monitor(PE 1)

component not described in this paper. The PEs of the resulting graph are the analytic modules that receive the streaming data from the current vehicle location (source PE 2) and destination (source PE 3), extract the GPS coordinates of the vehicle location (PE 4), generate the K best potential travel corridors using information from a map database(PE 5), receive updates on traffic conditions in the relevant locations (PE 0), and decide on routes based on vehicle size (PE 6). The two main results of this inquiry are route updates for the vehicles (sink PE 7) and updates to a list of currently relevant locations (cached in PE 1) for which updates about traffic and other conditions are required.

The RTS scenario contained descriptions of 19 PEs and 6 sources, described using 246 OWL classes, 156 OWL properties and 385 OWL individuals. In this ontology we specified on average 13 triple patterns to describe a PE input, and 23 triple patterns to describe a PE output, and each PE description used an average of 7 variables and 2 exemplars. The offline phase of the planner took 42.51 seconds. The planning times (in seconds) for several inquiries are presented in the table below, together with graph size (number of PEs and sensors):

Inquiry	Time	Graph Size
Get Best Route Updates from current locn (Fig. 7)	2.51	8
Get Traffic Conditions Update near current location	2.32	12
Get Weather Conditions Update near current location	2.15	12
Get All Conditions Update near current location	2.5	18
Get Conditions and Best Route from current location	2.33	19

The results show that the response time of the planner is consistently around 2.5 seconds, which is acceptable for an end-user submitting the inquiry.

6   Related Work

One closely related work is Semantic Streams [9], which allows users to pose queries based on the semantics of sensor data. It uses a Prolog-based language and defines logic rules that describe the semantics, such as the type and location of sensor data sources, and input/output of inference units (equivalent to PEs). While that model allows a highly expressive query language, scalability remains an open question, since Prolog logic programs can be undecidable. Our model uses OWL, based on description logics, which is known to be decidable. The computational advantage of OWL combined with the use of scalable planning algorithms allows our middleware to construct processing graphs quickly even when there are a large number of PEs and data sources. The choice of OWL as the representation medium also allows utilizing the large number of existing ontologies that have been developed in the Semantic Web. Data sources and PEs described using these terms and relations will be easier to inter-operate than those that are developed without any such shared, common knowledge.

Programming models for individual sensor networks have received much attention. TinyDB [10] proposes a SQL based query interface to extract data from sensor networks. Welsh et al. [11] describe "abstract regions" to support the programming of a collection of related sensors. Although a non-exhaustive list, the above mostly deal with programming within individual sensor networks. They are not designed for applications that utilize data from many sensor networks and apply complex processing on the data.

Other efforts have focused on how to utilize data from large numbers of heterogeneous sensors and sensor networks. Hourglass [12] proposes an Internet based infrastructure to inter-connect geographically diverse sensor networks, where applications collect and process data from them. Medusa [13] propose architecture and systems for distributed data processing applications that take data from many geographically distributed sensor networks. They do not use a semantic approach, nor do they address the problem of automatic construction of applications from declarative, semantic queries.

Various stream query languages and stream processing architectures have been proposed. Aurora [14] lets human users create a network of stream operators to process incoming data. TelegraphCQ [15] proposes a declarative language for continuous queries that uses relational operations and expressive windowing constructs. These systems process structured data using relational and sliding window operators. In contrast, our system supports complex processing of unstructured data, such as speech-to-text and image recognition. It is highly extensible and supports processing beyond relational and sliding window operators.

Prior work on composition by planning in Grid and Web Services includes [16,17]. The major difference of our work from web service composition is the emphasis on the semantics of the data. Semantic web service models like OWL-S only associate concepts in an ontology with inputs and outputs, while our model associates more expressive RDF graphs with variables.

7 Conclusion

In this paper we describe a semantics-based approach to automatically constructs applications that utilize data from heterogeneous sensors and sensor networks. We use a semantic model to formally describe desired end results, data sources and PEs. Given an inquiry, a planner can automatically compose relevant PEs and data sources to form applications. We have developed a prototype and have demonstrated its ability to flexibly construct applications in different domains and manage semantically rich and diverse sensors and sensor networks.

The semantic model helps in describing the diverse formats and meanings of sensor data sources, and the nature of possibly complex processing needed by applications. It also separates processing functions from the query model. New functions can be added by enriching the ontology with semantic descriptions of PEs; neither the query model nor the planner are affected. As our system continues to be applied in a wider range of application scenarios, we will continue to evaluate and improve this promising approach.

References

1. McGuinness, D., van Harmelen, F.: Owl web ontology language overview. In: W3C Recommendation (2004)
2. Jain, N., Amini, L., Andrade, H., King, R., Park, Y., Selo, P., Venkatramani, C.: Design, implementation, and evaluation of the linear road benchmark on the stream processing core. In: SIGMOD'06 (June 2006)
3. National Cancer Institute Center for Bioinformatics: NCI thesaurus. http://www.mindswap.org/2003/CancerOntology/
4. Rector, A.L., Horrocks, I.R.: Experience building a large, re-usable medical ontology using a description logic with transitivity and concept inclusions. In: AAAI (1997)
5. Baader, F., Calvanese, D., McGuinness, D.L., Nardi, D., Patel-Schneider, P.F.: The Description Logic Handbook: Theory, Implementation, and Applications. In: Baader, F., Calvanese, D., McGuinness, D.L., Nardi, D., Patel-Schneider, P.F. (eds.) The Description Logic Handbook, Cambridge University Press, Cambridge (2003)
6. Grosof, B., Horrocks, I., Volz, R., Decker, S.: Description logic programs: combining logic programs with description logic. In: WWW'03, pp. 48–57 (2003)
7. Prud'hommeaux, E., Seaborne, A.: SPARQL Query Language for RDF. In: W3C Working Draft (2006)
8. Riabov, A., Liu, Z.: Planning for stream processing systems. In: AAAI'05 (July 2005)
9. Whitehouse, K., Zhao, F., Liu, J.: Semantic streams: A framework for composable semantic interpretation of sensor data. In: Römer, K., Karl, H., Mattern, F. (eds.) EWSN 2006. LNCS, vol. 3868, pp. 5–20. Springer, Heidelberg (2006)
10. Madden, S., Franklin, M., Hellerstein, J., Hong, W.: TinyDB: An acquisitional query processing system for sensor networks. TODS'05 (2005)
11. Welsh, M., Mainland, G.: Programming Sensor Networks Using Abstract Regions. In: NSDI'04 (March 2004)

12. Shneidman, J., Pietzuch, P., Ledlie, J., Roussopoulos, M., Seltzer, M., Welsh, M.:
    Hourglass: An Infrastructure for Connecting Sensor Networks and Applications.
    Technical Report TR-21-04, Harvard EECS Dept (2004)
13. Zdonik, S., Stonebraker, M., Cherniack, M., Cetintemel, U., Balazinska, M., Balakr-
    ishnan, H.: The Aurora and Medusa projects. Bulletin of the Technical Committe
    on Data Engineering, IEEE Computer Society (March 2003)
14. Abadi, D., Carney, D., Cetintemel, U., Cherniack, M., Convey, C., Lee, S., Stone-
    braker, M., Tatbul, N., Zdonik, S.: Aurora: a new model and architecture for data
    stream management. The VLDB Journal (2003)
15. Chandrasekaran, S., Cooper, O., Deshpande, A., Franklin, M.J., Hellerstein, J.M.,
    Hong, W., Krishnamurthy, S., Madden, S.R., Raman, V., Reiss, F., Shah, M.A.:
    TelegraphCQ: Continuous dataflow processing for an uncertain world. In: CIDR'03
    (January 2003)
16. Gil, Y., Deelman, E., Blythe, J., Kesselman, C., Tangmurarunkit, H.: Artificial
    intelligence and grids: Workflow planning and beyond. IEEE Intelligent Systems
    (January 2004)
17. Pistore, M., Traverso, P.: Bertoli, P.: Automated composition of web services by
    planning in asynchronous domains. In: ICAPS'05 (2005)

A Compilation Framework
for Macroprogramming Networked Sensors*

Animesh Pathak[1], Luca Mottola[2], Amol Bakshi[1],
Viktor K. Prasanna[1], and Gian Pietro Picco[3]

[1] Ming Hsieh Department of EE-Systems, University of Southern California, USA
{animesh, amol, prasanna}@usc.edu
[2] Dipartimento di Elettronica ed Informazione, Politecnico di Milano, Italy
mottola@elet.polimi.it
[3] Department of Information and Communication Technology, University of Trento, Italy
picco@dit.unitn.it

Abstract. Macroprogramming—the technique of specifying the behavior of the *system*, as opposed to the *constituent nodes*—provides application developers with high level abstractions that alleviate the programming burden in developing wireless sensor network (WSN) applications. However, as the semantic gap between macroprogramming abstractions and node-level code is considerably wider than in traditional programming, converting the high level specification to running code is a daunting process, and a major hurdle to the acceptance of macroprogramming.

In this paper, we propose a general compilation framework for a data-driven macroprogramming language that allows for plugging in different modules implementing various stages of compilation. We also demonstrate an actual instantiation of our framework by showing an end-to-end solution for compiling macroprograms. Our compiler provides the final code to be deployed on real nodes as well as an estimate of the costs the running system will incur, e.g., in terms of messages exchanged. We compared the auto-generated code against a hand-coded version for the same application behavior to verify the outcome of our compiler.

1 Introduction

Macroprogramming refers to a set of programming techniques whose objective is to increase application developers' productivity and allow non-expert programmers to write distributed, sense-and-respond applications easily. Abstractions are provided to specify the high-level collaborative behavior at the *system level*. Most of the low-level details concerning state maintenance or message passing are intentionally hidden from the programmer. As a result of this, macroprogramming is emerging as a viable technique for developing complex embedded applications, as demonstrated by the several efforts [2, 11, 20] currently underway in this field.

---

* This work is partially supported by the European Union under the IST-004536 RUNES project and by the National Science Foundation, USA, under grant number CCF-0430061.

J. Aspnes et al. (Eds.): DCOSS 2007, LNCS 4549, pp. 189–204, 2007.

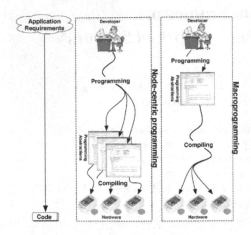

Fig. 1. Comparing node-centric and macro- programming

As illustrated in Fig. 1, the ease of design provided by macroprogramming comes at a cost when compared to traditional node-centric programming. In the former approach, application developers reason at a high level of abstraction, while the process of converting the high level representation to that of the individual nodes is delegated to a *compiler*. The higher the level of abstraction, the more work needs to be done by the compiler. This makes the process of generating the final running code more difficult than in the node-level compilers currently seen in WSNs.

In the context of macroprogramming for WSNs, we define compilation as the *semantics-preserving transformation of a high level application specification into a distributed software system collaboratively hosted by the individual nodes*. In [22], we summarized the challenges faced by the designers of compilation frameworks for macroprogramming languages. As illustrated in Sect. 3, the process of semantics-preserving transformation itself involves addressing challenges of correct and efficient conversion of representation. In addition, developers should be given the ability to express performance goals for the deployed system (e.g., in terms of expected network lifetime or latency) that the compiler should consider in optimizing the configuration of individual nodes and the allocation of different functionality to them.

In this paper, we present the design, implementation and evaluation of a compilation framework to support macroprogramming. Specifically, we focus on a data-driven macroprogramming model called the *Abstract Task Graph* (ATaG) [2], whose salient features are described in Sect. 2. We make two *contributions* in this paper:

- We propose a general framework for compilation used for data-driven macroprogramming languages like ATaG. An overview of the compilation process is given in Sect. 3. Our framework breaks down the process of converting the high-level specification to node-level functionality into a set of independent procedures—such as optimizing the placement of functionality on the real nodes, or predicting communication costs. These different stages are connected through well-defined interfaces, that allow for plugging in different modules implementing the various steps of compilation. Our compilation framework is described in Sect. 4.
- We demonstrate the flexibility and generality of our framework by describing an end-to-end solution for compiling ATaG macroprograms. Our proof-of-concept compiler, obtained by instantiating the different modules in our framework, provides the code to be deployed on each node, as well as an estimate of the message passing costs of the same. Moreover, the resulting code can be deployed on real world nodes as well as in a simulation environment. As described in Sect. 5, the

functionality of our compiler is assessed by inspecting and comparing the auto-generated code against a manually developed version of the same.

Compilation of macroprograms is still in its formative stages, and there is great variety in both the current work and future directions in the community. A discussion of related work is presented in Sect. 6. Section 7 concludes this paper.

2 ATaG: Abstract Task Graph

Macroprogramming of WSNs is an active area of research, with several programming paradigms currently being investigated [2, 11, 20]. In this work, we focus on ATaG (Abstract Task Graph) [2], a data-driven macroprogramming framework. ATaG includes an extensible, high-level *programming model* to specify the application behavior, and a corresponding node-level *run-time support*, called DART [1]. The compilation of ATaG programs consists of mapping the high-level ATaG abstractions to the functionality provided by DART. We now provide some background on these topics, as they represent the inputs and outputs of the transformation process, respectively.

2.1 Programming Model

ATaG provides a data driven programming model and a mixed *imperative-declarative* program specification. A *data driven* model provides natural abstractions for specifying reactive behaviors, while *declarative specifications* are used to express the placement of processing locations and the patterns of interactions.

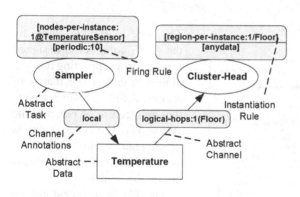

Fig. 2. ATaG program for data-gathering

The concept of *abstract data items* and *abstract tasks* are integral to specifying applications in ATaG. The former represents the information generated and communicated in the system, while the latter is a logical entity encapsulating the processing of one or more data items. The processing within a task is expressed using an imperative language. The flow of information between tasks is defined by *abstract channels*, which connect a task to a data item when the task *produces* that item, or vice versa when the task *consumes* it. Not that in an ATaG program, a data item can have more than one *consumers*, but only one *producer*.

Figure 2 illustrates an example ATaG program specifying a data gathering application [5] for building environment monitoring. Sensors within a cluster take periodic temperature readings, which are then collected by the corresponding cluster-head. The former aspect is encoded in the *Sampler* task, while the latter is represented by *Cluster-Head*. The *Temperature* data item is connected to both tasks using abstract channels.

Tasks are annotated with *firing* and *instantiation rules*. The former specify when the processing in a task must be triggered. In our example, the *Sampler* is triggered every 10 seconds according to the **periodic** rule. Differently, the **any-data** rule requires *Cluster-Head* to run when a data item is ready to be consumed on *any* of its incoming channels. The instantiation rules govern the placement of tasks on real nodes, whose characteristics (e.g., sensing device attached) are encoded using node attributes. The **nodes-per-instance:q@Device** rule requires the task to be instantiated once every q nodes equipped with a specific *device*. According to **@TemperatureSensor**, the *Sampler* task in our example will be instantiated on every node equipped with a temperature device. Differently, the programmer requires a single *Cluster-Head* to be instantiated on every floor in the building. The **partition-per-instance:1/Floor** construct is used for this purpose. Its semantics is to derive a system partitioning based on the values of the node attribute provided (**Floor**). In this case, the programmer requires only *one* task to be instantiated in each partition.

Abstract channels are annotated to express the *interest* of a task in a data item. In our example, the *Sampler* task generates data items of type *Temperature* kept **local** to the node where they have been generated. The *Cluster-Head* collects data not only from its own partition (floor), but also from adjacent ones. The **logical-hops:1(Floor)** annotation specifies a number of hops counted in terms of how many system partitions can be crossed, independent of the physical connectivity. Since *Temperature* data items are to be used within *one* partition (floor) from where they generated, they will be delivered to cluster-heads running on the same floor as the task that produced them, as well as adjacent floors.

2.2 Runtime System

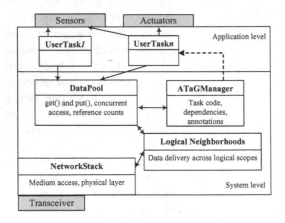

Fig. 3. DART: Data-driven ATaG run-time system

The node-level code output by the ATaG compiler is designed to run atop a supporting runtime hiding the underlying, platform-specific details. Figure 3 depicts the architecture of our runtime system [1]. The functionality is divided into a set of modules to facilitate customization to various deployments.

The *ATaGManager* stores the declarative portion of the user-specified ATaG program that is relevant to the particular node. This information includes task annotations such as firing rule and I/O dependencies, and the annotations of input and output channels associated with the data items that are produced or consumed by tasks on the node. The *DataPool* is responsible for managing all instances of abstract data items produced or consumed at the node. The

LogicalNeighborhoods [18, 17] module handles data delivery by implementing a dedicated routing scheme. In particular, the inputs to this module include the data items and the *scope specifications* those are addressed to. A scope identifies, in a logical manner, the nodes an item is addressed to by referring to the relevant node attributes. For instance, a scope may specify all the nodes running the *Cluster-Head* tasks deployed on first *Floor* as intended recipients. Finally, the *NetworkStack* is in charge of communication with other nodes in the network, and manages the physical layer protocols. Note that by itself, ATaG does not deal with fault tolerance. However, the runtime system and compiler developers are free to provide the user with an implementation that takes desired fault-tolerance requirements and support them by techniques such as task migration.

3   Compilation of Data-Driven Macroprograms: Overview

In this section, we provide an overview of the compilation process using the application given in Fig. 2 as example. Formally, an abstract task graph $A(AT, AD, AC)$ consists of a set AT of abstract tasks and a set AD of abstract data items. The set of abstract channels AC can be divided into two subsets – the set of *output channels* $AOC \subseteq AT \times AD$ and a set of *input channels* $AIC \subseteq AD \times AT$. In our example, the *Sampler* is AT_1 and *Cluster-Head* is AT_2, while *Temperature* is AD_1. AOC is $\{AT_1 \rightarrow AD_1\}$ and AIC is $\{AD_1 \rightarrow AT_2\}$. The compiler generates a set of node-level programs based on AT and the description N of the target system.

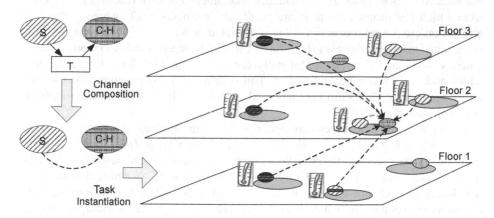

Fig. 4. An example illustrating the compilation process of our sample program

Composition of Channels. Owing to ATaG's purely data-driven programming model, the developer only specifies relations between tasks and the data items they are producing (via AOC) and consuming (via AIC). While this provides a clean model to the developer, traditional task allocation techniques work on task graphs with *direct dependency* links between tasks. To address the problem of generating such task graphs, we convert each *path $AT_i \rightarrow AD_k \rightarrow AT_j$* to an *edge $AT_i \rightarrow AT_j$*.

Since the channels in ATaG have logical scopes associated with them, composing two channels into one poses its own set of challenges. The basic process of composing channels results in the (composed abstract channel) CAC_{ijk} being annotated with the union of *three* constraints. The first is that the node should have task AT_j assigned to it. The second(third) constraint is obtained by combining the instantiation rule of $AT_i(AT_j)$ with the annotation on the abstract channel connecting it to AD_k. For instance, in our example, after composition, AC_{121} is {*(Cluster-Head is instantiated)* && *(Floor = Floor of Sampler or ±1)*}. Depending on the complexity of scopes used in the channels, the resultant constraint can be further simplified by set operations to get a more compact constraint for the composed channel.

This task graph with composed channels is then instantiated on the given target network. Figure 4 illustrates an example of a target network. The nodes are on three different floors, and those marked with a thermometer have temperature sensors attached to them.

ITaG: Instantiated Task Graph. The intermediate representation used for applying task-allocation techniques is called the *instantiated task graph* (ITaG). It is a representation of the target system, with the tasks assigned to each node and communicating with each other. It consists of multiple copies of each abstract task specified in the ATaG program, each assigned to a particular node. The (directed) edges of the ITaG connect each task to the tasks that depend on it, i.e., the tasks that a) consume the data item produced by it, and b) belong to the logical scope specified by the constraints in the connecting composed channel. Formally, the ITaG $I(IT, IC)$ consists of a set IT of instantiated tasks and a set IC of instantiated channels. For each task AT_i in the ATaG from which I is instantiated, there are $f(AT_i, N)$ elements in IT, where f maps the abstract task to the number of times it is instantiated in N. $IC \subseteq IT \times IT$ connects the instantiated version of the tasks. The ITaG I can also be represented as a graph $G(V, E)$, where $V = IT$ and $E = IC$. Additionally, each IT_j in the ITaG has a *label* indicating which node in N it is to be deployed on. This overlay of communicating tasks over the target deployment allows us to use modified versions of classical techniques meant for analysing task graphs.

In our example, since there are seven nodes with attached temperature sensors, $f(AT_1, N) = 7$. Similarly, $f(AT_2, N) = 3$, since the *Cluster-Head* task is to be instantiated once on each of the three floors. The figure shows one allocation of the tasks in IT, with arrows representing the instantiated channels in IC (we have showed channels leading to only one instance of AT_2 for clarity). Note that the although the ITaG notation captures the information stored in the abstract task graph (including the instantiation rules of the tasks and the scopes of the connecting channels) it does not capture the *firing rules* associated with each task. The compiler's task involves incorporating the firing rule information while making decisions about allocating the tasks on the nodes.

In summary, the compiler is responsible for generating an efficient task placement, ensuring that the composed channels are consistent with the semantics specified by the application developer in the abstract channels, and configuring the runtime system modules. An added complexity in the compilation process is brought by the large space of *optimizations* possible in the process to meet the user-specified performance goals (e.g. energy efficiency). Note that although tasks are assigned fixed locations at the end of the compilation process, *task migration* can happen later if the the underlying system supports it. Even in such situations, a *good* initial task placement by a compiler using global knowledge can go a long way in creating efficient systems. In the following section, we describe how the components of the compilation framework work to produce the outputs from the inputs, using the ITaG notation internally.

4   Compilation Framework

ATaG is designed to enable the addition of domain-specific constructs, and customize the abstractions offered depending on the application requirements. This requires a flexible and extensible approach to the compilation problem. Ideally, the system designer should be given the ability to add new language constructs by implementing the required mappings without modifying any of the pre-existing compilation mechanisms. For instance, creating a new instantiation rule should not require modifications to the algorithms used to map tasks to nodes using an existing rule.

To address this issue, we first identified the different steps involved in the compilation of ATaG programs by factoring out orthogonal concerns and mechanisms. Next,

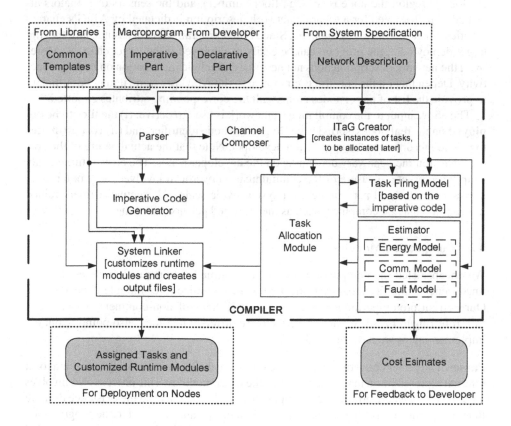

Fig. 5. The ATaG compilation framework

considering the decomposition obtained, we designed a modular compilation framework, upon which we based the construction of our ATaG compiler. In this section, we first illustrate the input and output of our framework (illustrated in Fig. 5), and then proceed to the description of the different modules implementing the compilation itself.

4.1 Compilation Input and Output

The information provided at the beginning of a compilation effort are:

ATaG declarative specification: consisting of the abstract task graph itself, i.e., the set of abstract tasks and abstract data items, connected by abstract channels.

ATaG imperative code: namely, the description of the actions taken when each task is fired, expressed in an imperative language.

DART run-time templates: including both the node-level code later customized by the compiler, and generic supporting mechanisms, e.g., for routing messages.

Network description: containing information on the target deployment scenario, e.g., number, location and attributes of nodes. The attributes may contain information about the logical region the node is in (e.g., floor number), and the sensors or actuators attached to it. The need for a separate network description is dictated by ATaG's characteristic of being *deployment agnostic*. Since ATaG programs do not assume any specific target deployment, the program can be easily re-deployed if the target changes. Moreover, the network description does not necessarily include information on node connectivity. Depending on the constructs employed in the ATaG program, it may be sufficient to provide the list of target nodes along with the corresponding attributes exported.

The above input to the compilation framework is used to derive: (i) the **files to be deployed** on the real nodes, sorted according to the node identifier, and (ii) **cost estimates** to provide feedback to the application developer. Note that the actual nature of the cost estimates returned can vary depending on the developer needs. The costs returned may simply represent a measure of the communication overhead involved, e.g., in terms of messages exchanged per minute on a system-wide scale. Alternatively, finer-grained information may be computed, such as the expected per-node lifetime.

4.2 Compilation Modules

We encapsulated the compilation stages we identified in separated modules, and defined generic interfaces between them so as to minimize inter-module dependencies. Our current prototype implementation has 2677 lines of non-commented Java code. Still referring to Fig. 5, we now describe these different modules, also pointing out the implementations we have realized so far.

Parser. The parser converts text files containing the declarative part of the program to an internal representation that is then used by the other modules. This process also involves a *syntax check* where errors such as duplicate task/data names and the existence of more than one producer task for one data item are identified and reported to the programmer.

In our current implementation, the declarative part of the ATaG program is specified using XML. This will allow an easy integration of tools for the automated generation of XML specifications from graphical representations. Our parser module is a simple XML parser that performs the aforementioned checks, assigns unique IDs to tasks and data items, and populates an internal data structure with the information.

Imperative Code Generator. Based on the parser output, the imperative code generator creates a set of files containing the basic declaration of the variables associated with each task and data items. The imperative part of the code provided by the programmer can then be plugged into these templates.

In our prototype implementation, the imperative part of an ATaG program is expressed using Java. As such, our current code generator creates Java files with unique numerical constants for each abstract task and data item corresponding to their id. Then, it creates a separate class for each abstract task with basic functionality filled in (e.g., a thread instance with a loop for periodic tasks).

Channel Composer. Looking at the declarative part of the ATaG program returned by the parser, this module performs the *composition* of channels to and from each data item to form edges of the ITaG, as described in Sect. 3.

Depending on the actual channel annotations supported, our prototype implementation may perform a range of operations, from a simple concatenation to complex operations that also consider the instantiation rules of the producer/consumer tasks.

ITaG Creator. Based on the network description and the output of the channel translator, the ITaG creator first computes the number of distinct *target regions* for each task, i.e., the set of candidate nodes for hosting a given task. For instance, tasks instantiated with **nodes-per-instance:x** as instantiation rule have the entire system as target region. For tasks assigned by **partition-per-instance:x/PLabel**, each set of nodes with the same value for **PLabel** is a target region. The ITaG creator then instantiates as many copies of the task as the product of the number of target regions and the number of instances per target region required in the ATaG program. Note that, at this stage, tasks are instantiated but *not yet* assigned to nodes. That is done by the task allocator module, discussed next.

Our implementation of this module performs the above operations using the network description read from a text file containing basic information on the nodes, e.g., their identifier, and set of attributes describing their characteristics, such as sensing devices installed.

Task Allocation Module. As such, the allocation module constitutes the core of the compilation process, since its job is to output a mapping from the set of instantiated tasks to the set of nodes. Note the task instantiation rules can be characterized as either *fixed* location (e.g., **nodes-per-instance:1**) or *variable* location (e.g., **nodes-per-instance:3**), depending on whether there is a unique way of instantiating the copies of a task given the network description. In this respect, an extremely large problem space exists depending on the annotations used, metrics to be optimized, and properties of the network. To perform its job, the allocation module relies on two further modules—the estimator and the task firing model–described next.

In our implementation, this module performs task allocation in two passes. In the first pass, it assigns all the tasks with *fixed* locations. In the second pass, it assigns *vari-*

able location tasks. For the latter, we currently employ a simple randomized assignment policy, with each node in the target region having an equal probability of hosting the instances of the task. However, due to the generality of our framework, more sophisticated mechanisms can be plugged in to achieve performance goals specified by the application designer. This is among our immediate research goals.

Estimator. Taking as inputs the network description and the task placement returned by the allocation module, the estimator computes the cost metric returned at the end of the compilation process. Our framework gives great flexibility in instantiating this module, as its interface is designed to be generic w.r.t. the nature of information required. This allows application developers to explore the trade-off between the *quality* of the estimate obtained, and the *time* required to obtain it. For instance, during the early design stages it is usually helpful to have a quick estimate of the communication costs, so that many alternative solutions can be explored. In this case, a simple but fast *estimation algorithm* can be employed that does not account for message losses. Conversely, when the application developer is to fine-tune the application, an actual *simulation* of the deployed application can be run within the estimator.

In our prototype system, we implemented both ends of the spectrum. Specifically, we realized a naive estimator returning communication costs as if all the tasks produced data when fired and the underlying routing mechanisms were able to identify the optimal message routes. On the other hand, we also implemented a wrapper around SWANS/Jist [3]: a simulator able to run unmodified Java code on top of a simulated network. This plug-and-play capability highlights the power of our framework.

Task Firing Model. It would appear that if we know the exact paths taken by the data items, we can precisely estimate the cost of running a given task allocation. However, not all instantiated tasks produce data when they fire. For instance, although a *Temperature Sampler* task may produce a *Temperature* data item whenever it fires, an *Alarm* task may or may not produce an alarm depending on whether or not the temperature of the region is high enough. The task firing model's function is to assign probabilities to the firing of various tasks in the program. Although this module is not mandatory for a working compiler, various approaches can be used to obtain the needed information - ranging from the developer providing profiling data obtained from previous runs of the system, to static code analysis techniques [4, 6].

System Linker. At the end of the whole process, the linker module combines the information generated by the various paths of the compilation into the actual code to be deployed on the real nodes. More specifically, it configures the *ATaGManager* and *DataPool* modules in the node-level run-time depending on the task and data items handled at each node, and merges the imperative code provided by the application developer with the templates generated by the imperative code generator.

In our implementation, the output of this module is a set of Java packages for each node. Note that these files are not binaries. They still need to be *compiled* in the classical sense, but that can be done by any node-level compiler designed for the target platform.

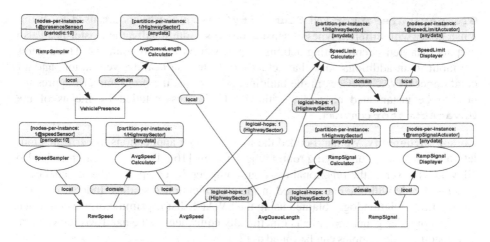

Fig. 6. An ATaG program for highway traffic management

5   Demonstration

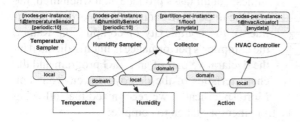

Fig. 7. An ATaG program for building environment management

To demonstrate the effectiveness of our prototype compiler, we consider two non-trivial applications, and report on the *functionality* of the code generated, as well as the *performance* of the compilation process.

The first application, illustrated in Fig. 6, describes a highway traffic management system. In this case, two different sub-goals must be achieved - regulating the speed of vehicles on the highway by controlling speed limit displays, and controlling the access to the highway by means of red/green signals on the ramps. The highway is divided into sectors, and sensors are deployed on the highway lanes and ramps to sense the speed and presence of vehicles, respectively. The sensed data goes through a multi-stage process where it is first aggregated w.r.t. a single sector to derive an average measure (*AvgSpeedCalculator* and *AvgQueueLengthCalculator* tasks), and then delivered to tasks deciding the actions taken in adjacent highway sectors (*SpeedLimitCalculator* and *RampSignalCalculator* tasks). Note the latter is expressed using the **logical-hops** construct relative to the **HighwaySector** attribute. Finally, data items describing the actions to perform are delivered to dedicated tasks instantiated on nodes equipped with the corresponding device, i.e., speed limit displays for the *SpeedLimitDisplayer*, and ramp signals for the *RampSignalDisplayer*.

The second application, depicted in Fig. 7, targets a *building environment management* system. Essentially, the processing is similar to the cluster-based data aggregation of Fig. 2, but now gathering data from two different types of sensors. The

@TemperatureSensor and **@HumiditySensor** constructs are used to distinguish nodes with different types of sensing devices. Additionally, the cluster-head also outputs data items representing actions to perform on the environment. These items are input to an additional task that actually operates the heating, ventilation, and air conditioner (HVAC) devices in the building. As for this, the programmer requires the task to be instantiated on nodes with HVAC devices installed by means of the **@hvacActuator** construct.

Code Functionality. We hand-coded the logic for both applications to perform simulation studies on the underlying routing mechanisms [16]. The hand-written code also allowed us to verify the functionality of our compiler, by comparing the automatically generated code with the one we used in the aforementioned studies. Indeed, by comparing the simulation logs obtained using the SWANS/Jist [3] simulator, we confirmed that the compiler-generated code is functionally equivalent to the hand-written version. The specific code samples can be found at [21].

	Building	Traffic
Abstract Tasks	4	8
nodes-per-instance:x@PLabel	3	4
partition-per-instance:x/PLabel	1	4
Abstract Data Items	3	6
Abstract Channels	6	14
local	3	6
domain	3	4
logical-hops:1(PLabel)	0	4

Fig. 8. Sample applications

Settings for Performance Studies. We look at the *time* and *memory* taken to compile the above ATaG programs. Since our task firing model assumes that all tasks produce data when fired, the specific imperative code of the tasks does not influence the complexity of compilation. Rather, the compiler's performance is mainly dictated by the declarative part of an ATaG program and the characteristics of the deployment environment. More specifically, we recognized the following factors are pivotal in determining the time/memory taken to compile:

1. the number of abstract *tasks*, *data items*, and *channels*,
2. the nature of *instantiation rules* and *channel interests*, and
3. the *number of nodes* specified in the network description.

The complexity of the compilation task comes from different sources. The effort in composing channels is dependent on the actual channel annotations used, as well as the number of channels themselves. The ITaG creation stage becomes more complex as the complexity of the network grows. Note that this includes the number of logical regions the network can be divided into, as well as the variation in the attributes of the nodes. The size of the problem addressed by the task allocation module depends both on the network size as well as the constraints used in the program. For instance, placing a task whose instantiation rule is in the form **partition-per-instance:x/PLabel** requires more processing than placing a task with **nodes-per-instance:1**. All this in turn affects the performance of the system linker as it customizes the run-time on each node. Figure 8 reports the values of these factors seen in our sample applications.

In our tests, the compilation framework has been instantiated with the prototype implementations we described in Sect. 4 for each module. In particular, we have chosen to employ the naive estimator and an *always-firing* task firing model. For each test we performed, we repeated the compilation process 500 times to account for fluctuations due

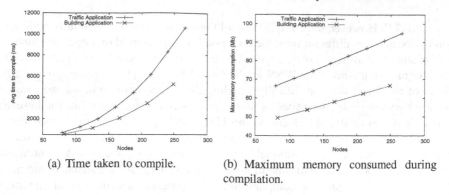

(a) Time taken to compile.　　　　　(b) Maximum memory consumed during compilation.

Fig. 9. Compiler performance

to concurrent processes. The experiments were on a Pentium IV HT 3.2 Ghz running Gentoo Linux 2.16.15, using the DJProf [7] profiler.

Performance Results. Figure 9 illustrates the performance of our compiler as a function of the number of target nodes. As expected, the time taken to compile an ATaG program grows quadratically as the number of nodes increases. This is due to the naive estimator we used, that computes the all-to-all shortest path with an algorithm whose time complexity is quadratic w.r.t. the number of vertices. However, fairly large instances can be compiled in reasonable time. For instance, slightly more than ten seconds are needed to compile the traffic application for a target system with > 250 nodes.

In addition, the memory consumed during the compilation process exhibits a linear increase with respect to the number of nodes in the deployed system. The source of this behavior is in the data structures we employed in the ITaG creator and allocation modules, that allocate a fixed amount of data for each target node. The memory consumed is always well within the limits of standard desktop PCs (< 100 MB).

In exchange for the above costs in term of memory and time, the framework buys the developer *ease-of-use* in implementing the application using ATaG macroprograms. To reassert this fact, we note that looking at the number of Java classes compiled to deploy our traffic application on a single node, it turns out only 15 out of a total of 51 classes are the direct result of the developer's effort. The others are the implementation of the DART run-time system. Furthermore, considering the actual number of lines of non-commented code, only about 12% of the imperative code is hand-written by developers, whereas the rest is either part of the run-time support, or automatically generated.

6   Discussion

Initial programmming of WSNs was done by the nesC [8] language and the tinyOS operating system [12], and helped a wide research community build and test applications and system components for networked sensing [9, 13, 14, 23]. Over time, tools such as SNACK [10] were developed to support the programmers of such systems, and sensor nodes supporting more traditional programming languages such as Java have also emerged [24]. However, the compilers of all these languages are essentially node-level compilers, not very different from the common C compiler used on larger machines.

Various macroprogramming approaches have been proposed recently to alleviate the programming burden for WSN application developers [11, 20]. Since we are not aware of published work specifically detailing their compilation process, we compare our work with the issues we expect would be addressed by similar systems for these languages based on existing literature. *Kairos* [11] is an imperative, control driven macroprogramming language where the application designer can write a single program in a Python-like language with additional keywords to express parallelism. A 'centralized' program describes the activities at all nodes in the system and is translated into node-level binaries by a dedicated compiler. Since the program is written in an imperative form, and whether the action will be performed at a particular node or not is decided by conditions mentioned in the macroprogram itself, the issues faced by the compiler are very different from ours. For example, there is no channel composition to be done and no specific tasks to be allocated.

Regiment [20] is a functional programming language, with support for region-based functions like filtering, aggregation and function-mapping. The Regiment primitives operate on a model of the sensor network as a set of continuous data streams. In [19], the authors introduced the TML intermediate language to represent the actions being performed at individual nodes. The authors state that Regiment programs can be seen as *data flow graphs*, with primitives such as **afold** combining functions and data on actual nodes to produce data. Although the functional programming approach of Regiment is very different from the data-driven approach of ATaG, the above similarity (ATaG tasks combine data produced at other nodes to produce more data) might lead to some re-use of our ideas in the compilation of Regiment macroprograms.

EnviroSuite [15] is an object-based programming system that introduces the *environmentally immersive paradigm*. Its abstractions revolve directly around elements of the environment as opposed to sensor network constructs, such as regions, neighborhoods, or sensor groups. Object instances float across the network following (geographically) the elements they represent. The EnviroSuite Compiler (EIPLC) is essentially a translator that takes EnviroSuite code as input and outputs desired environmental monitoring applications in nesC, which then can be compiled by a standard nesC compiler and uploaded to the motes.

This paper does not claim to completely solve the problem of compilation of macroprograms for WSN applications. Our main focus is to present a clear set of subtasks involved in the process, and the interrelationships of the modules implementing them. We believe that this will contribute towards the achievement of two goals. By clearly identifying the modules, we can help researchers in the community attack the particular subtasks involved in compilation. Clearly, more efficient techniques are required to provide the functionalities of the *Estimator*, *Task Firing Model*, and the *Task Allocation* module. Another issue that remains to be addressed is the possibility of *timing conflicts* among the tasks that are instantiated on a node, which is part of our future work. Further, by presenting a proof of concept implementation of the compiler, domain experts can begin to use the ATaG macroprogramming framework and provide us feedback on the language, the compiler as well as the runtime system. Although our current implementation runs on a simulator, the nature of the SWANS/Jist system is such that the same code can be run on actual nodes. We indend to present a demo of our approach on SunSPOT [24] nodes in the near future.

7 Concluding Remarks

In this paper, we presented a general compilation framework for a data-driven macro-programming language for sensor networks. We demonstrated the feasibility of our approach by developing a compiler that can convert macroprograms written in ATaG into a running sensor system. Our experiments indicate that the time taken to compile the macroprogram depends closely on the complexity of both the macroprogram and that of the target sensor system.

Our compilation framework currently assumes a *static* network structure, which greatly limits the class of applications that we can address using this approach. Even in those applications, issues such as faults cannot be addressed by the current approach. Our immediate future work will involve exploring *on-line* task migration algorithms that can continually work for optimizing the task allocation, in addition to efficient algorithms for ascertaining *good* initial task placements.

References

1. Bakshi, A., Pathak, A., Prasanna, V.K.: System-level support for macroprogramming of networked sensing applications. In: Int. Conf. on Pervasive Systems and Computing (PSC) (2005)
2. Bakshi, A., Prasanna, V.K., Reich, J., Larner, D.: The abstract task graph: A methodology for architecture-independent programming of networked sensor systems. In: Workshop on End-to-end Sense-and-respond Systems (EESR) (June 2005)
3. Barr, R., Haas, Z.J., van Renesse, R.: Jist: an efficient approach to simulation using virtual machines. Softw. Pract. Exper. 35(6) (2005)
4. Bernat, G., Burns, A., Wellings, A.: Portable worst-case execution time analysis using java byte code. In: Proc. of the 12nd Euromicro Conf. on Real-Time Systems (2000)
5. Choi, W., Shah, P., Das, S.: A framework for energy-saving data gathering using two-phase clustering in wireless sensor networks. In: Proc. of the 1st Int. Conf. on Mobile and Ubiquitous Systems: Networking and Services (MOBIQUITOUS) (2004)
6. Corbett, J.C., Dwyer, M.B., Hatcliff, J., Laubach, S., Pasareanu, C.S.: Robby, and H. Zheng. Bandera: extracting finite-state models from java source code. In: Proc. of the 22nd Int. Conf. on Software Engineering (ICSE) (2000)
7. DJProf Java Profiler, http://www.mcs.vuw.ac.nz/~djp/djprof/
8. Gay, D., Levis, P., von Behren, R., Welsh, M., Brewer, E., Culler, D.: The nesC language: A holistic approach to networked embedded systems. In: Proceedings of Programming Language Design and Implementation (PLDI) (2003)
9. Habitat Monitoring on the Great Duck Island, www.greatisland.net
10. Greenstein, B., Kohler, E., Estrin, D.: A sensor network application construction kit (SNACK). In: 2nd ACM Conference on Embedded Networked Sensor Systems (2004)
11. Gummadi, R., Gnawali, O., Govindan, R.: Macro-programming wireless sensor networks using Kairos. In: Prasanna, V.K., Iyengar, S., Spirakis, P.G., Welsh, M. (eds.) DCOSS 2005. LNCS, vol. 3560, pp. 126–140. Springer, Heidelberg (2005)

12. Hill, J., Szewczyk, R., Woo, A., Hollar, S., Culler, D., Pister, K.: System architecture directions for networked sensors. SIGOPS Oper. Syst. Rev. 34(5), 93–104 (2000)
13. Karp, B., Kung, H.T.: GPSR: Greedy perimeter stateless routing for wireless networks. In: Proc. ACM/IEEE MobiCom (August 2000)
14. Krishnamachari, B.: Networking Wireless Sensors. Cambridge University Press, Cambridge (2006)
15. Luo, L., Abdelzaher, T.F., He, T., Stankovic, J.A.: Envirosuite: An environmentally immersive programming framework for sensor networks. Trans. on Embedded Computing Sys. 5(3), 543–576 (2006)
16. Mottola, L., Pathak, A., Bakshi, A., Prasanna, V.K., Picco, G.: Enabling Scoping in Sensor Network Macroprogramming. Technical report. Submitted for publication, (2006) Available at http://indus.usc.edu/atag
17. Mottola, L., Picco, G.P.: Logical Neighborhoods: A programming abstraction for wireless sensor networks. In: Gibbons, P.B., Abdelzaher, T., Aspnes, J., Rao, R. (eds.) DCOSS 2006. LNCS, vol. 4026, pp. 150–168. Springer, Heidelberg (2006)
18. Mottola, L., Picco, G.P.: Programming wireless sensor networks with logical neighborhoods. In: Proc. of the 1st Int. Conf. on Integrated Internet Ad hoc and Sensor Networks (InterSense) (2006)
19. Newton, R., Arvind, Welsh, M.: Building up to macroprogramming: An intermediate language for sensor networks. In: Proc. of the 4th Int. Conf. on Information Processing in Sensor Networks (IPSN) (2005)
20. Newton, R., Welsh, M.: Region streams: Functional macroprogramming for sensor networks. In: Proc of the 1st Int. Workshop on Data Management for Sensor Networks (DMSN) (2004)
21. Pathak, A., Mottola, L., Bakshi, A., Prasanna, V.K., Picco, G.P.: Compiling macroprograms using the ATaG compilation framework Technical report, University of Southern California (2007), http://indus.usc.edu/atag
22. Pathak, A., Prasanna, V.K.: Issues in Designing a Compilation Framework for Macroprogrammed Networked Sensor Systems. In: Proc. of the the 1st Int. Conf. on Integrated Internet Ad hoc and Sensor Networks (InterSense) (2006)
23. Rahimi, M., Hansen, M., Kaiser, W., Sukhatme, G., Estrin, D.: Adaptive sampling for environmental field estimation using robotic sensors. In: IEEE/RSJ International Conference on Intelligent Robots and Systems (IROS) (August 2005)
24. Small Programmable Object Technology (Sun SPOT), http://www.sunspotworld.com

Passive Inspection of Sensor Networks*

Matthias Ringwald[1], Kay Römer[1], and Andrea Vitaletti[2]

[1] Institute for Pervasive Computing, ETH Zurich, Switzerland
[2] Department of Informatics, University of Rome "La Sapienza", Italy
{mringwal,roemer}@inf.ethz.ch, andrea.vitaletti@dis.uniroma1.it

Abstract. Deployment of sensor networks in real-world settings is a labor-intensive and cumbersome task: environmental influences often trigger problems that are difficult to track down due to limited visibility of the network state. In this paper we present a framework for passive inspection (i.e., no instrumentation of sensor nodes required) of deployed sensor networks and show how this framework can be used to inspect data gathering applications. The basic approach is to temporarily install a distributed network sniffer alongside the inspected sensor network, with overheard messages being analyzed by a data stream processor and network state being displayed in a graphical user interface. Our tool can be flexibly applied to different sensor network operating systems and protocol stacks, and can deal well with incomplete information.

1 Introduction

Deployment of sensor networks in real-world settings is typically a labor-intensive and cumbersome task [1, 2, 3, 4, 5, 6, 7, 8, 9]. While simulation and lab testbeds are helpful tools to test an application prior to deployment, they fail to provide realistic environmental models (e.g., regarding radio signal propagation, sensor stimuli, chemical/mechanical strain on sensor nodes). Hence, environmental effects often trigger bugs or degrade performance in a way that could not be observed during pre-deployment testing. To track down such problems, a developer needs to inspect the state of network and nodes. While this is easily possible during simulation and experiments on lab testbeds (wired backchannel from every node), access to network and node states is very constrained after deployment.

Current practice to inspect a deployed sensor network requires *active* instrumentation of sensor nodes with monitoring software and monitoring traffic is sent in-band with the sensor network traffic to the sink (e.g., [6, 10, 11]). Unfortunately, this approach has several limitations. Firstly, problems in the sensor network (e.g., partitions, message loss) also affect the monitoring mechanism, thus reducing the desired benefit. Secondly, scarce sensor network resources (energy, cpu cycles, memory, network bandwidth) are used for inspection. In Sympathy [6], for example, up to 30% of the network bandwidth is used for monitoring traffic. Thirdly, the monitoring infrastructure is tightly interwoven with the application. Hence, adding/removing instrumentation may change

---

* The work presented in this paper was partially supported by the by the Swiss National Science Foundation under grant number 5005-67322 (NCCR-MICS).

J. Aspnes et al. (Eds.): DCOSS 2007, LNCS 4549, pp. 205–222, 2007.

the application behavior in subtle ways, causing probe effects. Also, it is non-trivial to adopt the instrumentation mechanism to different applications. For example, [6, 10] assume a certain tree routing protocol being used by the application and reuse that protocol for delivering monitoring traffic.

In contrast to the above, we propose a *passive* approach for sensor network inspection by overhearing and analyzing sensor network traffic to infer the existence and location of typical problems encountered during deployment. To overhear network traffic, a so-called *deployment support network* (DSN) [12] is used: a wireless network that is temporarily installed alongside the actual sensor network during the deployment process. The DSN may be removed as soon as initial problems have been fixed and the sensor network is operational. Each DSN node provides two different radio frontends. The first radio is used to overhear the traffic of the sensor network, while the second radio is used to form a robust and high-bandwidth network among the DSN nodes to reliably collect overheard packets. A data stream framework performs online analysis of the resulting packet stream to infer and report problems soon after their occurrence.

This approach removes the above limitations of active inspection: no instrumentation of sensor nodes is required, sensor network resources are not used. The inspection mechanism is completely separated from the application, can thus be more easily adopted to different applications, and can be added and removed without altering sensor network behavior. Online analysis (as opposed to long periods of data collection followed by offline analysis) contributes to a more effective deployment process, as it allows an engineer to go out and study affected nodes while a problem is still present. Also, problems can be fixed in an incremental fashion as they occur, thus reducing the chance for complex aftereffects. Besides these advantages, we need to address a number of challenges:

Incomplete information. The DSN may fail to overhear some packets and messages might not contain all information that is needed to infer a problem. To support robust problem detection nonetheless, we provide appropriate loss-tolerant data stream operators.

Flexibility. There is no established protocol stack for sensor networks – a large variety of radio configurations, MAC, routing, and application layer protocols are in use. To support this open protocol space, we provide a packet capturer that works with a large variety of MAC protocols and radio configurations, as well as a flexible packet decoder.

Reliability The DSN should provide reliable wireless communication. We use Bluetooth for this purpose, which has been designed as a cable replacement, employing frequency hopping and other techniques to minimize loss.

In the first part of this paper we present a concrete instance of the above approach called SNIF (Sensor Network Inspection Framework) which is – as the name suggests – intended as a widely applicable framework for passive inspection. The second part of the paper contains an extensive case study of how SNIF can be applied to so-called data gathering applications. In particular, our case study can detect similar problems as approaches for active inspection in [6, 10].

2   SNIF

SNIF is a general framework for passive inspection of multi-hop sensor networks to detect problems related to individual nodes (e.g., reboot, death), wireless links, paths (e.g., routing failures, loops), or global problems (e.g., partitions). SNIF consists of a deployment support network (DSN) that acts as a distributed network sniffer. Each of the DSN nodes implements the receiver part of the sensor network protocol stack, namely a receive-only *physical layer* and *media access*. All overheard packets are routed to the SNIF sink, which executes a *data stream processor* to analyze packet streams for problems. The results of this analysis are displayed by a *user front-end*. Below we give an overview of these components. More details can be found in a technical report [13].

2.1   Deployment Support Network (DSN)

To overhear the traffic of multi-hop networks, multiple radios are needed, forming a distributed network sniffer. We use a so-called deployment support network for this purpose, a wireless network of DSN nodes, each of which provides two radios. The first radio (DSN radio) is used to form a wireless network among the deployment support nodes, while the second radio (WSN radio) is used to overhear the traffic of the sensor network. Both radios should be free of interference (e.g., operate in different frequency bands). Also, the DSN radio should support the formation of a robust network with negligible message loss and high bandwidth. Since the data stream processor needs to examine temporal relationships between packets overheard by different DSN nodes, internal time synchronization of DSN nodes is necessary. The DSN is installed alongside the actual sensor network and may be removed as soon as deployment is finished and the sensor network works as expected. Thus, the lifetime of the DSN is typically much shorter than the lifetime of the sensor network and energy efficiency is not that much of an issue.

Our current implementation of a DSN is based on the BTnode Rev. 3 [14], which provides two radio front-ends: a Zeevo ZV 4002 Bluetooth 1.2 radio which is used as the DSN radio, and a Chipcon CC 1000 (e.g., also used on MICA2) which is used as the WSN radio. Using a scatternet formation algorithm, the DSN nodes form a robust Bluetooth scatternet (see [12] for details). A laptop computer with Bluetooth acts as the SNIF sink that connects to a nearby DSN node. This DSN node thereupon acts as the DSN sink and forms the root of an overlay tree spanning the whole DSN. The SNIF sink can send data to DSN nodes down the tree while DSN nodes send overheard packets up the tree to the sink. Time synchronization exploits the fact that Bluetooth uses a TDMA MAC protocol and thus performs clock synchronization internally, providing an interface to read the Bluetooth clock and its offset to the clocks of network neighbors. We use this interface to compute the clock offset of each DSN node to the DSN sink. Bluetooth provides an accuracy of 1.25 milliseconds per hop.

One might argue that the deployment of the DSN may be as difficult and error-prone as deploying the sensor network itself. However, as the lifetime of the DSN is short (in the order of days), energy and resource constraints are not a primary issue here. This enables us to use more reliable networking technologies such as Bluetooth. In fact, Bluetooth has been designed as a cable replacement and employs techniques such as frequency hopping and forward error correction to provide highly reliable data transmission.

2.2 Physical Layer and Medium Access

DSN nodes need a receive-only implementation of the physical (PHY) and MAC layers in order to overhear sensor network traffic. Due to the lack of a standard protocol stack, many variants of PHY and MAC are in use in sensor networks. Hence, we need a flexible implementation that can be easily configured for the sensor network under inspection.

Our generic PHY implementation supports configurable carrier frequency, baud rate, and checksumming details. We assume that the sensor network uses a single frequency for communication (which is the case with current implementations) such that a single-channel radio is sufficient to overhear WSN traffic.

Regarding MAC, we exploit the fact that – regardless of the specific MAC protocol used – a radio packet always has to be preceded by a preamble and a start-of-packet (SOP) delimiter to synchronize sender and receiver. In our generic MAC implementation, every DSN node has its WSN radio turned to receive mode all the time, looking for a preamble followed by the SOP delimiter in the received stream of bits. Once an SOP has been found, payload data and a CRC follow. This way, DSN nodes can receive packets independent of the actual MAC layer used.

Fig. 1 shows an excerpt of a sample configuration file for inspecting a TinyOS 1.x application running on MICA2 motes. The first five lines set the carrier frequency of the WSN radio to 868.000 Mhz and a data rate of 19200 bits/second, and instruct the packet sniffer to check for a start-of-packet sequence of 0xcc33. The 16 bit CRC-CCITT polynomial $x^{16} + x^{12} + x^5 + 1$ (0x1021) is used as checksum algorithm.

```
1   // PHY+MAC parameters                          13  const int LADV = 2; // LinkAdvertisement packet
2   cc.freq = 868000000; cc.baud =19200;           14  default.packet = "TOS_Msg"; // default packet type
3   cc.sop  = 0xcc33; cc.crc  = 0x1021;            15  struct TOS_Msg {
4   // encoding: endianness + alignment             16      uint16_t addr;
5   enc.endianness = " little "; enc.alignment = 1; 17      uint8_t type , group, length;
6   // type definitions and constants               18      int8_t data[length]; // variable payload size
7   typedef uint16_t  mote_id_t;                    19      uint16_t crc;
8   typedef uint8_t   quality_t;                    20  };
9   struct  link_quality_t {                        21  struct LinkAdv : TOS_Msg.data (type == LADV) {
10      mote_id_t id;                               22      mote_id_t id;
11      quality_t quality;                          23      struct link_quality_t links []; // var. size
12  };                                              24  };
```

Fig. 1. A SNIF configuration file

2.3 Packet Decoder

Again, since no standard protocols exist for sensor networks, we need a flexible mechanism to decode overheard packets. Since most programming environments for sensor nodes are based on the C programming language or a dialect of it (e.g., nesC for TinyOS), it is common to specify message contents as (nested) C structs in the source code of the sensor network application. Our packet decoder uses an annotated version of such C structs as a description of the packet contents. This way, the user can copy and paste packet descriptions from the source code.

The configuration of the packet decoder consists of some global parameters (such as byte order and alignment), type definitions, and one or more C structs. One of these structs is indicated as the default packet layout. Note that such a struct can contain nested other structs, effectively implementing a discriminated union.

Consider Fig. 1 for an example, which describes link advertisement packets used by the Multihop routing service implemented in ESS [15]. Line 14 defines the struct TOS_Msg as the default packet layout. The LinkAdv PDU used by ESS, is encapsulated in the field TOS_Msg.data, but only if the TOS_Msg.type is equal to LADV. Arrays of variable size are supported, where the size is either contained in the packet (e.g., for TOS_Msg.data), or inferred from the packet size (e.g., for LinkAdv.links).

At startup of SNIF, the configuration file is parsed and the default packet type is investigated. If the default packet type is of fixed size, the packet size is computed. Otherwise, size and position of the packet length indicator (e.g., TOS_Msg.length in the example) is computed. This information, along with the parameters for the physical layer are then broadcast to all DSN nodes, allowing them to correctly receive WSN traffic. All overheard WSN packets are then annotated with reception time and routed to the SNIF sink.

2.4 Data Stream Processor

The DSN outputs a stream of overheard packets that needs to be analyzed to detect problems in the WSN. To enable an efficient deployment process, this analysis should be performed *online*, allowing an engineer to go out and study and fix affected nodes while the problem is still present.

Given these preconditions, we decided for a data stream processor to perform online analysis of packet streams. Here, a data stream is an unbounded sequence of records. A data stream processor provides three basic abstractions: *sources* that produce data streams, *sinks* that consume data streams, and *operators* that modify data streams. Sinks and operators can *subscribe* to sources and operators, such that a data stream output by the subscriber acts as input for the subscriber. That is, sources, operators, and sinks form a directed *operator graph* with data streams flowing from sources through operators towards sinks. Mainly motivated by practical considerations (Java as implementation language, stability, open-source availability) we chose the PIPES data stream processor [16] for use with SNIF.

In SNIF, we model the DSN as a data stream source. An operator graph (being executed on the SNIF sink) processes this data stream to detect indicators for problems, and sink nodes act as an interface to the user. A data stream record in SNIF is a list of attribute-value pairs with two special attributes holding record type and time stamp. The DSN produces records of type Packet with attributes holding the contents of an overheard packet. The syntax of the latter attribute names follows C syntax for accessing a field of a structure (e.g., TOS_Msg.addr to access the source address of a packet in Fig. 1).

The data stream processor provides a number of basic operators to manipulate data streams, such as *Mapper* to rename record attributes, *Union* to merge multiple data streams into one where records are sorted by increasing time stamps, or *Filter* to drop records that do not match a given predicate. *TimeWindowAggregator* groups records according to a given attribute, removes duplicates, and computes aggregates over a time window. *ArrayIterator* provides access to array elements by creating N copies of each input record holding an array, where in the i-th output copy the array is replaced with element i of the array with size N.

Besides these generic operators, SNIF provides several data stream sources. The output of *DSNSource* consists of the packets overheard by the DSN, with records being sorted by increasing time stamp and duplicate packets (resulting from two or more DSN nodes overhearing the same sensor node) being removed. *EmSource* provides a similar interface to the EmStar [17] sensor network simulator, but is otherwise identical to *DSNSource*.

A typical application of SNIF is to infer the current state of inspected sensor nodes (e.g., node dead, node has no neighbors, etc.). To infer the state of a node, typically multiple data streams must be considered (e.g., a stream of periodic beacon packets to decide if a node is dead, a stream of neighborhood announcement packets to decide if a node has any neighbors). To this end, SNIF provides an operator *StateDetector* which groups records by type and node and stores the last record in each group. Whenever a group changes, an evaluation based on a configurable decision tree is invoked to decide on the node state. We will refer to the above operators in Sect. 3.4.

2.5   User Interface

To display problems in the sensor network that have been detected by the data stream processor, SNIF provides a configurable user interface, which allows to display a real-time view of the network topology graph, where nodes and links can be annotated with application-specific information (e.g., state of a node, packet loss of a link) using a simple API. Also, logging and later replay of execution traces is supported. Fig. 2 shows an instance of this user interface for a typical data gathering application as discussed in the next section.

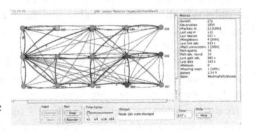

Fig. 2. An instance of SNIF's user interface

3   Case Study: Data Gathering Applications

Almost all existing non-trivial deployments are data gathering applications (e.g., [7, 9, 18]), where nodes send raw sensor readings at regular intervals along a spanning tree across multiple hops to a sink. In this case study we will therefore consider how SNIF can be applied to this application class. We first characterize the application in more detail and define the problems we want to detect. We then describe application-specific data stream operators to detect these problems and how they are used to form an operator graph. Finally, we evaluate the resulting inspection tool.

3.1 Application Model

Two prominent implementations of data gathering applications are the Extensible Sensing System (ESS) [15] using beacon-based multi-hop routing for data collection, and Surge using MintRoute [19] for data collection. Both implement a similar multi-hop tree routing scheme as described below. We will use ESS as an example throughout the paper, but our approach can be readily applied to other, similar implementations.

In ESS, all nodes broadcast *beacon messages* at regular intervals. To discover neighbors, nodes overhear these messages and estimate the quality of incoming links from neighbors based on message loss. Nodes then broadcast *link advertisement messages* at regular intervals, containing a list of neighbors and link quality estimates. Overhearing these messages, nodes compute the bidirectional link quality to decide on a good set of neighbors. To construct a spanning tree of the network with the sink at the root, nodes broadcast *path advertisement messages*, containing the quality of their current path to the sink. Nodes overhearing these messages can then select the neighbor with the best path as their parent and broadcast an according path advertisement message. All this is executed continuously to adapt neighbors and paths to changing network conditions. Finally, *data messages* are sent from nodes to the sink along the edges of the spanning tree across multiple hops.

In ESS, beacons are sent every 10 seconds, path advertisements and link advertisements every 80 seconds, data message are generated every 30 seconds. All messages except data messages are broadcast messages and contain per-hop source address. Data messages contain the address of the originator of the sensor data and the per-hop destination address, but not the per-hop source address. In addition, beacon messages and data messages contain a sequence number.

3.2 Problems and Indicators

In [13] we studied existing deployments to identify common problems and passive indicators that allow to infer the existence of a problem from overheard network traffic. Below we summarize the problems that are considered in our case study and give passive indicators for their detection. Note the similarity to problems that can be detected by tools for active inspection [6, 10].

Node death (fail stop). An affected node will not send any messages.

Node reboot. After reboot the sequence number contained in beacon messages will be reset.

Isolated node. The node is not listed as a neighbor in any link advertisement messages send by other nodes.

Node has no parent. The node fails to send path advertisement messages.

No path from node to sink. Data messages sent by the node are not forwarded to the sink.

Node's path to sink loops. A data message originating from the node is sent twice to the same destination by different senders. Note that this is a special case of "no path from node to sink".

Node partitioned from sink. A node on the path from the node to the sink died and there is no alternate path available. Note that this is a special case of "no path from node to sink".

Although the above indicators are straightforward from a conceptual point of view, incomplete information makes their implementation less obvious as discussed in the following section.

3.3 Application-Specific Operators

This section presents application-specific operators that assist in detecting the problems described in Sect. 3.2. The primary challenge here is to deal with incomplete information due to i) the DSN failing to overhear packets, and due to ii) information that would be needed to detect a problem not being explicitly included in messages.

```
on receive beacon(src , seq , t ):
  if ( exists n[src ]) {
    if ( seq < n[src ]. seq) {
      if ( n[src ]. seq < maxSeq − C)
        emit reboot(src , t );
      else if ( t − n[src ]. t <
           (seq − n[src ]. seq) % maxSeq * n[src]. ival )
        emit reboot(src , t );
    }
    n[src ]. ival ← min (n[src]. ival ,
           (t − n[src ]. t )/( seq − n[src ]. seq ));
  } else
    n[src ]. ival ← ∞;
  n[src ]. seq ← seq;
  n[src ]. t ← t;
```

Fig. 3. SeqReset operator

```
on receive data(dst , seq , orig , t ):
  if ( exists p[seq| orig ]) {
    if (p[seq| orig ]. dst = dst)
      emit retransmission (dst , seq , orig , t );
    src ← p[seq|orig]. dst ;
    p[seq| orig ]. dst ← dst;
  } else {
    src ← orig;
    p[seq| orig ]. dst ← dst;
  }
  emit data( src , dst , seq , orig , t );
```

Fig. 4. PacketTracer operator

SeqReset. This operator detects node reboots exploiting the fact that the sequence number contained in beacon messages will be reset after reboot. The main challenge here is to tell apart a wrap-around of the sequence number from reboot in case of lost beacon messages. The algorithm in Fig. 3 maintains a data structure n that holds for each node i the last sequence number $n[i].seq$, last time stamp $n[i].t$, and minimum interval $n[i].ival$ between successive beacons. Whenever a beacon with source address src, sequence number seq, and time stamp t is received, the algorithm checks if seq is smaller than the last sequence number $n[src].seq$ seen for this node. If the last sequence number is far apart from maximum sequence number $maxSeq$ (parameter C must be selected such that loss of C consecutive beacon messages is highly unlikely), then src has rebooted. Otherwise, we apply an additional check to distinguish reboots from wrap-arounds with lost messages. In case of a wrap-around, the time between the last and current beacon messages $t - n[src].t$ must be greater than or equal to the minimum beacon interval $n[src].ival$ times the number of beacon messages that were lost plus one $(seq - n[src].seq)$ % $maxSeq$.

PacketTracer. To reconstruct the multi-hop path of a message through the network, we need to know source and destinations addresses of each message. Unfortunately, data messages do not contain per-hop source addresses (as message receipt is not acknowledged). Also, messages not overheard by the DSN result in "gaps" in the multi-hop path. PacketTracer infers a source address for each packet, making sure that there are no gaps in the multi-hop path. The algorithm in Fig. 4 exploits the fact that each multi-hop message contains the address of the originator *orig*, a sequence number *seq*, and per-hop destination address *dst*. The operator maintains a data structure *p* that contains the last destination address *p[seq|orig].dst* for each multi-hop message uniquely identified by the concatenation *seq|orig* of sequence number and originator address. If an entry for packet *seq|orig* doesn't exist yet, then the sender is set to the originator of the packet, otherwise the sender is set to the destination of the previous packet. If no messages are lost, then this approach obtains correct sender addresses. Otherwise, packets may span multiple hops, resulting in a multi-hop path without gaps. The following operators rely on this property. PacketTracer uses a timeout-based garbage collector to reclaim memory for past multi-hop packets (not shown).

```
on receive data ( src , dst , t ):
  if ( dst ∈ n[src].desc ) {
    emit routingloop ( src , dst , t );
    remove dst from n[ src ]. desc ;
  }
  desc ← (src, t ) ∪ n[src].desc ;
  foreach ( dn, dt ) ∈ desc {
    if ( dst = sink ) {
      if ( dn ∉ n[sink].desc )
        emit goodpath ( dn, t );
      else if ( dt > n[sink ]. desc[dn])
        emit goodpath ( dn, t );
    }
    n[ dst ]. desc ← n[dst].desc ∪ (dn, max (n[dst ]. desc[dn ], nt ));
  }
}
```

```
on receive data ( src , dst ):
  n[ dst ]. nb ← n[dst].nb ∪ src;
  reset timeout ( dst , src );

on timeout ( dst , src ):
  remove src from n[ dst ]. nb;

on receive nodestate ( src , state ):
  if ( state = "dead" ) n[src ]. nb ← ∅;

periodically :
  DFS (n, sink );
  foreach unvisited node nn
    emit partitioned ( nn);
```

Fig. 5. PathAnalyzer operator   **Fig. 6.** TopologyAnalyzer

PathAnalyzer. This operator checks if a node has a good path to the sink and also detects routing loops. Here, a good path between a node and the sink exists if a sequence of packets $p_1, ..., p_n$ with increasing time stamps has been observed, such that the source address of p_1 equals the address of the node, the destination address of p_n is the sink, and the destination address of p_i equals the source address of p_{i+1}. The algorithm in Fig. 5 maintains a set *n[i].desc* of routing tree descendants for each node *i*, where each descendent is a pair *(j, tj)* of a node *j* and time stamp *tj*, meaning that *j* had a good path to *i* at time *tj* according to the above definition. When a data message with source address *src* (obtained by PacketTracer), destination address *dst*, and time stamp *t* is received, we first check if *dst* is among *src*'s descendants, which indicates a routing loop. Then we add *src* and all of *src*'s descendants to *dst*'s descendants, updating the time stamps accordingly. Whenever a new descendant is added to the sink or the time stamp of an existing descendent of the sink is incremented, this indicates a good path from this descendent to the sink.

TopologyAnalyzer. This operator detects network partitions between a node and the sink caused by dead nodes in cases where PathAnalyzer does not find a good path to sink for this node. The algorithm in Fig. 6 maintains an approximate set of downstream neighbors $n[i].nb$ for each node i. When a data packet with source address src and destination address dst is received, src is added to dst's neighbors and a (user-defined) timeout is activated to remove this neighbor unless another packet with same src and dst is received before the timeout expires. TopologyAnalyzer is also subscribed to a data stream of records holding node states (see Sect. 3.4 for details). Whenever such a node state record is received indicating death of node src, the neighbor set of src is emptied. Periodically, TopologyAnalyzer performs a depth-first search on the graph given by $n[].nb$ starting at the sink and marking all visited nodes. All nodes that have not been visited are reported as partitioned.

3.4 Operator Graph

Our inspection tool will compute the state of each node, which is either "node ok" or one of the problems described in Sect. 3.2. In this section we outline the data stream operator graph that computes these states. Eventually, this graph will generate a record describing a node's current state whenever the state of the node changes.

The node state is derived using the binary decision tree depicted in Fig 7 which is similar to the one used by Sympathy [6]. The leaves of this tree represent possible states of a node. The decision tree is implemented using the StateDetector operator described in Sect. 2.4. Each decision in the tree requires an operator graph that extracts the required information from the stream of observed packets. Below we describe how each of these decisions is implemented with an operator graph. Note that the individual operator graphs described below partially overlap. These common subgraphs are instantiated only once.

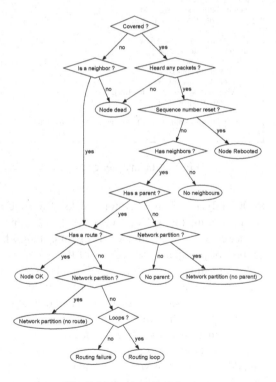

Fig. 7. Node state decision tree

Covered? This test examines whether a sensor node can be observed with sufficient quality by the DSN by examining the percentage of beacon messages that have been received from this node. To implement this test, DSNSource is filtered for beacon messages. The stream of beacon messages is then fed to a TimeWindowAggregator to compute the fraction of beacon messages that have been received. The test succeeds for a node if the fraction for this node is above a given threshold.

Heard any packets? This test succeeds if any packet from a sensor node could be overheard. Since data messages do not contain the per-hop source address, DSNSource is filtered for data packets and PacketTracer is applied to reconstruct the source address. Also, DSNSource is filtered for the remaining packet types (beacon, link and path advertisements) that do already contain the per-hop source address. The resulting data streams are merged with the Union operator to obtain a stream of all packets containing source addresses. This stream is then fed to a TimeWindowAggregator to count the number of packets per node using the count aggregation function.

Sequence number reset? This test succeeds is the node rebooted. To implement this test, DSNSource is filtered for beacon packets and SeqReset is applied to the resulting data stream.

Is a neighbor? This test checks whether a sensor node is listed as a neighbor of any other node in the network. DSNSource is filtered for link advertisement packets. Since each link advertisement contains an array of neighbors, the ArrayIterator operator is used to create one record for each node being listed as a neighbor. Using TimeWindowAggregator with the count aggregation function we obtain the number of times a node is listed as a neighbor.

Has any neighbors? This test examines whether a node has any neighbors. DSNSource is filtered for link advertisement packets containing at least one neighbor. Using TimeWindowAggregator, the number of such advertisements per node is computed. The test succeeds for a node if at least one non-empty link advertisement was heard from this node.

Has a parent? This test examines whether a node has a parent in the tree. DSNSource is filtered for path advertisement packets. Using TimeWindowAggregator, the number of such advertisements per node is computed. The test succeeds for a node if at least one path advertisement was heard from this node.

Has a route? This test checks whether a node recently had a routing path to the sink. DSNSource is filtered for data messages. PacketTracer is applied to reconstruct the source address. PathAnalyzer is applied and its output filtered for good route reports. Using TimeWindowAggregator, the number of good route reports per node is counted.

Loops? This test checks whether the path from a node to the sink recently had any loops. DSNSource is filtered for data messages. PacketTracer is applied to reconstruct the source address. PathAnalyzer is applied and its output filtered for routing loop reports. Using TimeWindowAggregator, the number of routing loop reports per node is counted. The test succeeds for a node if a routing loop was reported at least twice for this node to accomodate infrequent route changes.

Network partition? This test checks if a bad path from a node to the sink was caused by a network partition. DSNSource is filtered for data messages. PacketTracer is applied to reconstruct the source address. TopologyAnalyzer is applied to detect partitions. TopologyAnalyzer is also subscribed to the output of StateDetector in order to obtain *node death* events. The test succeeds for a node if the last record received from TopologyAnalyzer says that this node is partitioned.

In the above operator graphs, the time windows for TimeWindowAggregator are set to W times the interval of the packets they consider. For example, the time window in *Has a parent?* is set to $W \times 80$ seconds, since path advertisement messages are considered which are sent every 80 seconds. That is, W is a global parameter and we will study its performance impact in Sect. 3.5.

The structure of the decision tree is motivated by the desire to find and report the *root cause* of a failure. For example, a dead node (root cause) also has a routing problem (consecutive fault). Here, we want node death to be reported, but not the routing problem. Hence, in the decision tree the checks to detect node death are located above the checks to detect a routing problem.

In addition to the above operator graph, we introduce several data stream sinks (not shown) to display relevant information in the graphical user interface as shown in Fig. 2. For example, node color indicates state (green: ok, gray: not covered by DSN, yellow: warning, red: severe problem), detailed node state can displayed by selecting nodes. Thin arcs indicate what a node believes are its neighbors, thick arcs indicate the paths of multi-hop data messages.

3.5 Evaluation

To evaluate our case study, we used the same experimental setup as described in [6], where the Extensible Sensing System (ESS) [15] is executed in the EmStar emulator [17]. The reason for choosing EmStar instead of the real DSN as a data source for evaluation is the ease of injecting failures in a reproducible way with EmStar.

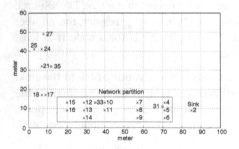 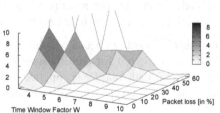

Fig. 8. Experiment setup: WSN (2-27) and DSN (31-35)

Fig. 9. Number of false reports as a function of packet loss and time window factor W

As depicted in Fig. 8, we consider a network of 21 nodes forming a multi-hop topology with a diameter of 7 hops. Node 2 acts as the sink. We added three DSN nodes (nodes 31, 33, and 35 marked with squares in Fig. 8). The link dump files of the DSN nodes generated by EmStar were used as input to the inspection tool. Since some sensor nodes could be overheard by more than one DSN node, the DSN received 1.3 ± 0.5 copies of each sensor network message during the experiments, while 4% of the beacon messages were lost (i.e., not overheard by any DSN node).

Accuracy and Latency. We study the accuracy (number and type of false error reports) and latency (time between failure injection and report) of our inspection tool. These metrics mainly depend on two parameters: the size of time windows used in the operator graph (i.e., the value of the time window factor W) and the amount of packet loss (i.e., fraction of sensor network messages that were not overheard by DSN nodes).

As most decisions regarding node state are based on packets received during a fixed time window, increasing W should improve accuracy (as operators then have more packets to base their decision on) and increase latency linearly (as more packets need to be collected before a decision is made). Increasing packet loss should degrade accuracy (as operators with fixed time windows then have less packets to base their decision on) and decrease latency (e.g., since node death is reported when no packets are received from a node during a time window, loss of the last packets sent by a node before death will decrease latency).

In general, the latency to detect a problem is determined by the path of decisions leading to this problem in the binary decision tree depicted in Fig. 7. For example, the decision *Network partition?* leading to state *Network partition (no parent)* can only be made when the previous decision *Has a parent?* has been made with a result of *no*. That is, the latency for detecting a given problem is a function of the maximum latency of the decisions in the decision tree on the path from the root to the leaf denoting this problem. In turn, the latency of a decision is determined by the size of time window(s) of the associated operator graph.

In order to assess the impact of W and packet loss on accuracy and latency, we ran a set of experiments injecting three types of faults into the network: node failure, network partition, and no data. The duration of each experiment was 30 minutes with faults being injected randomly between 10 and 15 minutes after experiment begin. In addition to the (small) packet loss of the DSN, we introduced additional packet loss by uniformly dropping a given fraction of the overheard packets. We report averages and standard deviation over multiple runs.

To guide the selection of W for a given amount of packet loss, we ran a first experiment without injecting any faults, varying both W and packet loss, counting the number of (false) error reports for each parameter choice. The averaged results over 10 runs are depicted in Fig. 9. The flat area of the graph shows feasible values for W given a certain packet loss. For a packet loss of 30% (a common value in single-hop sensor networks [7]), no errors were reported for $W \geq 7$, motivating our choice of $W = 8$ to study the impact of message loss in more detail as depicted in Fig. 10 top. Similarly, we chose a packet loss of 30% for a more detailed study of the impact of W as depicted in Fig. 10 bottom.

In the first experiment, we performed 40 runs and injected a single node failure per run, such that all nodes but the sink failed twice. All node crashes were correctly detected and no false errors were reported. The latency of the reports is mainly determined by the size of the time window used to implement the *Heard any packets?* test which is $W \times 10$s. For $W = 8$ and a beacon period of 10 seconds, we expect the latency to be between 70 and 80 seconds, which is confirmed by the experiments. Increasing packet loss does not have a significant impact on latency. The number of false positives is negligible until 30% of packet loss and raises significantly with more than 50% as depicted

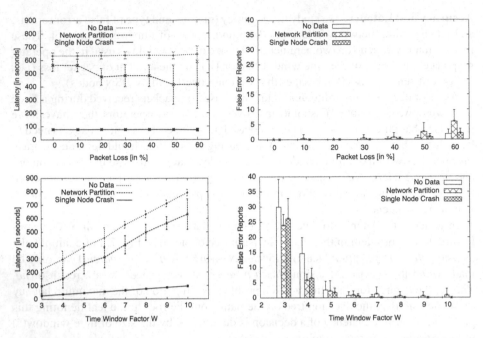

Fig. 10. Reporting latency and number of false reports as a function of packet loss for W=8 (top) and as a function of time window factor W for 30% packet loss (bottom)

in Fig. 10 top. We analyzed the generated error reports and observed that for up to 70% of packet loss, we only observed *no neighbor* and *no parent* reports. These reports are caused by missing link and path advertisements, respectively, which are rarely sent (every 80s). For higher packet loss, we found *node dead* reports for working nodes. We never observed any false negatives. When varying W, we find (as expected) a linear increase of latency and an improvement of accuracy as depicted in Fig. 10 bottom.

In the second experiment we made nodes 4-16 fail at random times to partition nodes 17-27 from the remainder of the network. We would expect a *network partition* error for nodes 17-27. We report the latency until the first node was classified as partitioned. As explained above, the latency of partition detection is bounded by the latencies of preceding decisions in the decision tree, namely *Has a parent?* and *Has a route?*, which both use a time window of $W \times 80$ seconds. As *Has a route?* basically tracks multi-hop data packets which are sent often (every 30s by all nodes), it reacts shortly before 640 seconds. The *Has a parent?* fails, if no path announcements were observed during the time window. As explained above, increasing packet loss results in reduced detection latency.

In the third experiment, we injected faults into the Multihop routing component of single nodes such that an affected node stops sending data messages, while still broadcasting beacons and advertisements. We would expect a *no route* error for the affected node and all other nodes whose paths contain the former. We report the time until the affected node is marked with *no route*. In this experiment, the latency is determined by the window size of the *Has a route?* test which is set to $W \times 80$ seconds. As most nodes in the network forward packets for other nodes and data packets are sent every

30 seconds, the DSN should observe data packets until the fault is injected and the average latency should be close to the window size. The average of 633 ± 24 seconds for $W = 8$ and no packet loss confirms this. Again, in Fig. 10 bottom, the accuracy improves and latency increases linearly with W as expected.

SNIF Performance. We also studied the performance overhead of SNIF itself. During one 30 minute experiment run without any fault injections, the DSN collected 261 kB of data, resulting in an average data rate of 1.2 kbps including duplicate packets. Note that this equals about 0.3% of the effective Bluetooth 1.2 bandwidth of 400 kbps. SNIF was executing on a 2 GHz PC using Java 1.5. The total cpu time for processing the above amount of data was about 13 seconds, which equals about 0.7% of the experiment duration of 30 minutes.

A Bug in ESS. In the course of our experiments, we encountered a bug in ESS Multihop. At one point we decided to upgrade to a new version of EmStar that fixed a bug with collision handling. After the upgrade, we suddenly observed a large number of *no parent* error reports without injecting any faults. As SNIF was still receiving close to 100% of all beacon packets and link advertisements, we concluded that this problems was caused solely by the path advertisement component. By examining the source code of Multihop, we learned that nodes react to receipt of a path advertisement message by updating their parent selection and broadcasting their updated path advertisement immediately without any delay. Here, the original path advertisement broadcast results in an implicit synchronization of all receivers, such that the secondary path advertisements collide with high probability without being retransmitted. By adding a random jitter, we were able to fix this problem.

4   Related Work

Complementary to SNIF is work on active debugging of sensor networks, notably Sympathy [6] and Memento [10]. Both systems require active instrumentation of sensor nodes and introduce monitoring protocols in-band with the actual sensor network traffic. Also, both tools support a fixed set of problems, while SNIF provides an extensible framework. Tools for sensor network management such as NUCLEUS [11] provide read/write access to various parameters of a sensor node that may be helpful to detect problems. However, this approach also requires active instrumentation of the sensor network.

Also complementary to SNIF is work on simulators (e.g., SENS [20]), emulators (e.g., TOSSIM [21]), and testbeds (e.g., MoteLab [22]) as they support development and test of sensor networks *before* deployment in the field. In particular, testbeds typically provide a wired backchannel from each node, such that sensor nodes can be instrumented to send status information to an observer. EmStar [17] integrates simulation, emulation, and testbed concepts into a common framework where some nodes physically exist in a testbed or in the real world, while the majority of nodes is being emulated or simulated. Physical nodes need instrumentation and a wired backchannel. In [23], a deployment support network is used to provide a wireless backchannel to deployed sensor nodes. However, sensor nodes need to be physically wired to DSN nodes (requiring as many DSN nodes as there are sensor nodes) and sensor node software must be instrumented.

Passive observation by means of packet sniffing has also been applied to wireless (and wired) LANs [24]. However, sensor networks differ substantially from wireless LANs. While typical wireless LANs are single-hop networks that can be observed with one or few sniffers, sensor networks are typically multi-hop networks. Also, many of the problems encountered during deployment of sensor networks are not present in WLANs. Very recently, two systems for passive analysis of WLANs have been proposed that use an approach similar to ours, namely WIT [25] and JIGSAW [26]. WIT follows an offline approach, merging redundant traces of network traffic collected by distributed sniffers. Using a detailed model of the 802.11 MAC, WIT then infers which packets have actually been received by the respective destination nodes and derives different network performance metrics. JIGSAW uses a similar approach to collect and merge traces, but then focuses on online inference of link-layer and transport-layer connections and their characteristics, also using a detailed model of the 802.11 MAC. In contrast, our approach is largely independent of the actual MAC used. Also, we focus on detecting a different set of problems as discussed in Sect. 3.2.

In the more general context of management and debugging of distributed systems, a large body of related work exists. Due to space constraints, we limit our discussion to very closely rated work. One such class of closely related work is performance debugging of distributed systems (e.g., [27,28]) where message traces are used to reconstruct causality paths and their latencies. While in principle applicable to sensor networks, these approaches are narrowly focused on a very specific problem and analysis is performed offline. In contrast, we provide a framework for online traffic analysis. A number of data stream management systems have been specifically developed for network traffic analysis (e.g., [29], [30]). However, we found it difficult if not impossible to express stateful SNIF operators using the SQL variants of these systems.

5   Conclusions

We presented a framework for passive inspection of deployed sensor networks, consisting of a distributed network sniffer, data stream processor, and user interface. The key advantage of this framework is that sensor networks need not be instrumented for inspection. The framework has been specifically designed to support different protocol stacks and operating systems. We showed how this framework can be applied to data gathering applications, demonstrating the our approach can detect typical problems encountered during deployment timely and accurately even in case of incomplete information. Using this tool, we found a bug in the ESS application. SNIF has been fully implemented and demonstrated at EWSN 2007 [31].

References

1. Buonadonna, P., Gay, D., Hellerstein, J.M., Hong, W., Madden, S.: Task: Sensor network in a box. In: EWSN 2005 (2005)
2. Greenstein, B., Kohler, E., Estrin, D.: A sensor network application construction kit (snack). In: SenSys 2004 (2004)
3. Mainwaring, A., Culler, D., Polastre, J., Szewczyk, R., Anderson, J.: Wireless sensor networks for habitat monitoring. In: WSNA 2002 (2002)

4. Padhy, P., Martinez, K., Riddoch, A., Ong, H.L.R., Hart, J.K.: Glacial environment monitoring using sensor networks. In: REALWSN 2005 (2005)
5. Polastre, J., Szewczyk, R., Mainwaring, A., Culler, D., Anderson, J.: Analysis of wireless sensor networks for habitat monitoring. In: Raghavendra, C.S., Sivalingam, K.M., Znati, T. (eds.) Wireless Sensor Networks, Kluwer Academic Publishers, Dordrecht (2004)
6. Ramanathan, N., Chang, K., Kapur, R., Girod, L., Kohler, E., Estrin, D.: Sympathy for the sensor network debugger. In: SenSys 2005 (2005)
7. Szewcyk, R., Mainwaring, A., Polastre, J., Anderson, J., Culler, D.: An analysis of a large scale habitat monitoring application. In: SenSys 2004 (2004)
8. Tateson, J., Roadknight, C., Gonzalez, A., Fitz, S., Boyd, N., Vincent, C., Marshall, I.: Real world issues in deploying a wireless sensor network for oceanography. In: REALWSN 2005 (2005)
9. Tolle, G., Polastre, J., Szewczyk, R., Culler, D., Turner, N., Tu, K., Burgess, S., Dawson, T., Buonadonna, P., Gay, D., Hong, W.: A macroscope in the redwoods. In: SenSys 2005 (2005)
10. Rost, S., Balakrishnan, H.: Memento: A Health Monitoring System for Wireless Sensor Networks. In: SECON 2006 (2006)
11. Tolle, G., Culler, D.: Design of an application-cooperative management system for wireless sensor networks. In: EWSN 2005 (2005)
12. Beutel, J., Dyer, M., Meier, L., Thiele, L.: Scalable topology control for deployment-sensor networks. In: IPSN 2005 (2005)
13. Ringwald, M., Römer, K., Vialetti, A.: Snif: Sensor network inspection framework. Technical Report 535, ETH Zurich, Zurich, Switzerland (2006)
14. BTnodes: A distributed environment for prototyping ad hoc networks, http://www.btnode.ethz.ch
15. Guy, R., Greenstein, B., Hicks, J., Kapur, R., Ramanathan, N., Schoellhammer, T., Stathopoulos, T., Weeks, K., Chang, K., Girod, L., Estrin, D.: Experiences with the extensible sensing system ess. Technical Report 61, CENS 2006 (2006)
16. Cammert, M., Heinz, C., Krämer, J., Markowetz, A., Seeger, B.: Pipes: A multi-threaded publish-subscribe architecture for continuous queries over streaming data sources. Technical report, University of Marburg, Germany (2003)
17. Girod, L., Elson, J., Cerpa, A., Stathapopoulos, T., Ramananthan, N., Estrin, D.: EmStar: A software environment for developing and deploying wireless sensor networks. In: USENIX 2004 (2004)
18. Langendoen, K., Baggio, A., Visser, O.: Murphy loves potatoes: Experiences from a pilot sensor network deployment in precision agriculture. In: WPDRTS 2006 (2006)
19. Woo, A., Tong, T., Culler, D.: Taming the underlying challenges of reliable multihop routing in sensor networks. In: SenSys 2003 (2003)
20. Sundresh, S., Kim, W., Agha, G.: SENS: A Sensor, Environment and Network Simulator. In: Annual Simulation Symposium (2004)
21. Levis, P., Lee, N., Welsh, M., Culler, D.: TOSSIM: Accurate and Scalable Simulation of Entire TinyOS Applications. In: SenSys 2003 (2003)
22. Werner-Allen, G., Swieskowski, P., Welsh, M.: Motelab: a wireless sensor network testbed. In: IPSN 2005 (2005)
23. Dyer, M., Beutel, J., Kalt, T., Oehen, P., Thiele, L., Martin, K., Blum, P.: Deployment support network - a toolkit for the development of wsns. In: EWSN 2007 (2007)
24. Henderson, T., Kotz, D.: Measuring wireless LANs. In: Shorey, R., Ananda, A.L., Chan, M.C., Ooi, W.T. (eds.) Mobile, Wireless, and Sensor Networks, Wiley, Chichester (2006)
25. Mahajan, R., Rodrig, M., Wetherall, D., Zahorjan, J.: Analyzing the mac-level behavior of wireless networks. In: SIGCOMM 2006 (2006)
26. Cheng, Y.C., Bellardo, J., Benkö, P., Snoeren, A.C., Voelker, G.M., Savage, S.: Jigsaw: Solving the puzzle of enterprise 802.11 analysis. In: SIGCOMM 2006 (2006)

27. Aguilera, M.K., Mogul, J.C., Wiener, J.L., Reynolds, P., Muthitacharoen, A.: Performance debugging for distributed systems of black boxes. In: SOSP 2003 (2003)
28. Barham, P.T., Donnelly, A., Isaacs, R., Mortier, R.: Using magpie for request extraction and workload modelling. In: ODSI 2004 (2004)
29. Cranor, C., Johnson, T., Spatcheck, O., Shkapenyuk, V.: Gigascope: A Stream Database for Network Applications. In: SIGMOD 2003 (2003)
30. Sullivan, M., Heybey, A.: Tribeca: A System for Managing Large Databases of Network Traffic. In: USENIX 1998 (1998)
31. Ringwald, M., Cortesi, M., Römer, K., Vialetti, A.: Demo abstract: Passive inspection of deployed sensor networks with snif. In: EWSN 2007 (2007)

Separating the Wheat from the Chaff: Practical Anomaly Detection Schemes in Ecological Applications of Distributed Sensor Networks

Luís M.A. Bettencourt[1], Aric A. Hagberg[1], and Levi B. Larkey[2]

[1] Mathematical Modeling and Analysis, Theoretical Division
Los Alamos National Laboratory, Los Alamos, NM 87545
[2] Modeling, Algorithms, and Informatics, Computer and Computational Sciences Division
Los Alamos National Laboratory, Los Alamos, NM 87545

Abstract. We develop a practical, distributed algorithm to detect events, identify measurement errors, and infer missing readings in ecological applications of wireless sensor networks. To address issues of non-stationarity in environmental data streams, each sensor-processor learns statistical distributions of differences between its readings and those of its neighbors, as well as between its current and previous measurements. Scalar physical quantities such as air temperature, soil moisture, and light flux naturally display a large degree of spatiotemporal coherence, which gives a spectrum of fluctuations between adjacent or consecutive measurements with small variances. This feature permits stable estimation over a small state space. The resulting probability distributions of differences, estimated online in real time, are then used in statistical significance tests to identify rare events. Utilizing the spatio-temporal distributed nature of the measurements across the network, these events are classified as single mode failures - usually corresponding to measurement errors at a single sensor - or common mode events. The event structure also allows the network to automatically attribute potential measurement errors to specific sensors and to correct them in real time via a combination of current measurements at neighboring nodes and the statistics of differences between them. Compared to methods that use Bayesian classification of raw data streams at each sensor, this algorithm is more storage-efficient, learns faster, and is more robust in the face of non-stationary phenomena. Field results from a wireless sensor network (Sensor Web) deployed at Sevilleta National Wildlife Refuge are presented.

1 Introduction

Wireless sensor networks consist of multiple sensor-processor nodes that communicate with each other using radio frequencies. Sensor nodes, at present and in the envisioned future, are simple devices that operate within limitations in local memory storage and processing. These constraints, although by no means fundamental, are often the result of the practical considerations of producing devices that are inexpensive, small, and autonomous. In addition, sensor operations, and their communication in particular, are also limited by battery capacity or by the ability to harvest power, e.g. through solar panels.

J. Aspnes et al. (Eds.): DCOSS 2007, LNCS 4549, pp. 223–239, 2007.
© Springer-Verlag Berlin Heidelberg 2007

Networks of distributed sensors are a promising technology because they can sense environments—natural and human made—over an unprecedented range of spatial and temporal scales [1, 2]. The large number of nodes required to cover large areas, over long times, places practical constraints on their individual cost. The drive for low-cost sensors and the need for unattended operation, frequently in harsh environments, requires simple and robust devices. Even the most robust devices, however, are subject to operational faults. Under these circumstances it is crucial that isolated errors in individual components do not jeopardize the operation of the whole network. Thus, an important issue for this emerging technology is data quality assurance and robustness of operation under point failures [3, 4, 5].

A general approach for robustness to point failures is to create partial functional redundancy among nodes in a sensor network. In some distributed sensor applications this emerges naturally because neighboring nodes measure local environments that are temporally and/or spatially correlated [6, 7]. Then, measurements at adjacent sensors, and at the same sensor over time, although potentially stochastic and non-stationary, display significant amounts of mutual information. Hence data quality can be assured through state co-inference between multiple, partially redundant and correlated readings from neighboring nodes, or from the same node at consecutive times [8, 9].

This paper presents a practical, distributed algorithm for detecting measurement anomalies - corresponding to both point failures and common mode events - and for estimating erroneous or missing data in ecological applications of wireless sensor networks. The algorithm has been designed for ecological sensing at the Sevilleta Long Term Ecological Research (LTER) site by a Sensor Web developed at NASA JPL [10, 2, 11]. Because it is designed to work under current technological constraints on memory and processing, the algorithm is intentionally simple and easy to implement. Processing can be performed locally on each node and requires only communication between proximal sensors. Such local, distributed algorithms are desirable for wireless sensor networks, where minimizing the amount of wireless communication is a necessary operational constraint [12].

The remainder of the paper is organized as follows. First, we describe related work on ecological applications of distributed sensor networks, and associated requirements for autonomous operation with emphasis on sensor measurement error detection and correction. We review related approaches in other contexts that use the distributed nature of the network for practical state co-inference, learning, and quality assurance and the performance and implementation requirements of direct Bayesian classifiers. Next, we describe the characteristics of the method, which performs automatic inference and prediction at a given sensor based on the distributions of differences of its measurements in time and in space relative to its neighbors. Finally we give several illustrations of the method's application to real data streams from a Sensor Web deployed at the Sevilleta LTER site, summarize our results, and discuss the outlook for future work.

2   Related Work

Ecological and habitat monitoring are natural applications for wireless sensor networks since the data often must be collected from remote areas that have little or no communication infrastructure and from sensing systems that are often distributed over large geographic areas. Among other advances, wireless sensor networks permit better sensor placement, unhindered by wires, and may use on-board computational power to processing running statistics, perform hypothesis testing and even operate the experiments themselves [8, 13].

Present deployments are still far from fulfilling this promise, but have been invaluable in providing experience and highlighting the difficulties that arise from measuring data streams in the physical world [14, 15, 16]. Most of these problems arise from sensors and networks operating unattended in harsh, real-world conditions, with inadequate error identification and correction capabilities, and without sufficient algorithms to automatically quantify and actively reduce uncertainty [8, 13].

Several algorithms have recently been proposed that utilize statistical models to selectively acquire and summarize data in distributed sensor networks [17, 18]. Because of common climatic drivers, environmental signals at neighboring sensors are usually spatially and temporally correlated. Some methods explicitly explore the correlated nature of raw signals to reconstruct missing or erroneous readings [19]. Environmental data streams pose the additional challenge that signals are non-stationary, driven by diurnal and seasonal cycles, and by climatic events that never quite repeat. These features are typical of other sensing problems measuring physical and/or social environments. Here we propose an approach based on difference techniques, similar to those found in image [20] and signal processing [21], to factor out common drivers and capture the statistics of correlations between neighboring sensors. We show that this approach, complemented with the use of statistical tests to detect anomalous measurements, naturally leads to the identification of events with different structure, that can correspond to point sensor failures, or common mode events. The common mode anomalies may be erroneous or result from real spatio-temporally coherent events. In this way, missing or erroneous measurements at a sensor can also be automatically inferred via the joint consideration of neighboring readings and learned difference probability distributions.

Because of these general properties of environmental data streams, the straightforward application of standard statistical learning methods to environmental data streams must be performed with care. For example, Bayesian classifier methods [22] are a powerful way to perform sequential estimation, and are therefore a natural formalism for devising learning algorithms in distributed sensor networks. However, the direct implementation of such methods tends to run into the *practical* limitations of these simple devices. A recent proposal for *context-aware sensors* based on Bayesian classifiers uses statistical correlations between sensor readings to detect outliers and approximate missing readings [23]. We briefly review this method in the next section in order to provide context to the conceptual differences of our approach.

3   Bayesian Classifier Method

Assume that sensor measurements take values in the interval $[l, u]$, and let $R = \{r_1, ..., r_m\}$ be a disjoint cover of this interval. Each subinterval in R is considered a discrete class, with average precision $(u - l)/m$. Each node has its own classifier, consisting of the state of that node's previous reading, h, and of the measurements from two (indistinguishable) nearby sensors, denoted as $n \in \{(r_i, r_j) \in R \times R, i \leq j\}$.

By Bayes' theorem, the conditional probability of a reading r_i, given the previous value h at that sensor and readings n from two nearby neighbors, is

$$P(r_i|h,n) = \frac{P(h,n|r_i)P(r_i)}{P(h,n)}. \tag{1}$$

In addition, to reduce the state space for inference, it is assumed in [23] that the neighbor's spatial measurements and the temporal information contained in the previous reading are conditionally independent,[1] given the reading of the sensor at the present time, yielding the "Naive Bayes" classifier

$$P(r_i|h,n) = \frac{P(h|r_i)P(n|r_i)P(r_i)}{P(h)P(n)}. \tag{2}$$

The output of the classifier is inferred using the method of maximum *a posteriori* (MAP) estimation [24], and is given by

$$\arg\max_{r_i \in R} P(r_i|h,n) = \arg\max_{r_i \in R} \frac{P(h|r_i)P(n|r_i)P(r_i)}{P(h)P(n)} = \arg\max_{r_i \in R} P(h|r_i)P(n|r_i)P(r_i), \tag{3}$$

where the denominator can be omitted from the optimization because it does not depend on r_i.

This method is exhaustive and powerful in classifying all possible states of the system and learning their likelihood, but runs into practical implementation problems. To see this, consider that each node must learn the parameters of its classifier online. To learn $P(r_i)$, a node keeps a count of the number of times r_i occurs for each of m possible values. To learn $P(h|r_i)$, a node also keeps a count of the number of times h and r_i occur together for each of m^2 possible combinations. Similarly, to estimate $P(n|r_i)$, a node must keep a tally of the number of instances n and r_i occur together, for each of $(m^3 + m^2)/2$ possible states. Finally, to compute probabilities for outlier detection, a node learns $P(n)$ online by keeping a count of the number of times n occurs for each of $(m^2 + m)/2$ values. $P(h)$ is given by $P(r_i)$ where $r_i = h$ and a node must also keep a count of the total number of instances observed. Thus the total number of states stored is $m^3/2 + 2m^2 + 3m/2 + 1$. This expression was obtained by considering the measurements of a node relative to *two* neighbors. For $k > 2$ neighbors, the corresponding expression scales with leading exponent $k + 1$.

The size of the state space required for inference is important for two reasons. First, nodes typically have limited storage capacity, which in turn limits precision. Consider the example of covering a range of 100 degrees with 1 degree precision. Then a classifier would have to store 520,151 counts, or roughly 2 megabytes. Secondly, the amount of learning data required to populate the state space is prohibitive in many cases. In the same example at least 5 million learning instances would be necessary for estimation (taken here to be roughly an order of magnitude greater than the size of the state space).

---

[1] We note that these assumptions do not apply to ecological environmental data under most circumstances.

To put this into perspective, consider that a node taking a reading every five minutes (e.g., [2]) would require about 47 years to populate its state space.

The issue of learning is even more critical in cases involving non-stationary phenomena because the learning rate cannot be slower than the rate at which parameters evolve. For example, in the case of outdoor air temperature, conditions change throughout the day as the sun rises, moves across the sky (e.g., placing sensors in and out of shadows), and sets. In addition, conditions also change with season and from year to year, such that combinations of data that occur frequently during a hot summer appear rarely during a cold winter, and will differ to the next summer. Thus an important discriminating criterion for any data quality assurance method is that it must operate on a timescale commensurate with that of any non-stationary phenomena being measured. For ecological sensing this time scale is typically less than a few hours.

4   A Method Based on the Statistics of Differences Between Sensor Measurements

We now propose a method for performing automatic event detection and data quality assurance, in which each node learns statistical distributions of *differences* between its readings and those of its neighbor's, and also between its own measurements at different times. Such distributions, together with current measurements are then used to identify anomalous measurements and to infer missing values. The inference of statistical distributions for measurement differences helps bypass issues of non-stationarity in environmental data streams, and leads, in general, to smaller ranges of statistical variables and better sampling for smaller datasets.

The crucial assumption required for the method to work is that the observed phenomena are spatiotemporally coherent, so that the measurements at neighboring sensors, and at the same sensor over time, display a large amount of mutual information. This is true of ecological applications, where typical node-to-node spacings are in the range of 100-200 meters or less. Moreover, environmental variables such as air temperature, relative humidity, light flux, soil temperature, and soil moisture display a substantial amount of temporal correlation as a result of common climatic drivers. It is assumed below that measurements at different sensors are performed at time intervals which are much smaller than the temporal correlation time of acquired signals, which we measured to be of order 1 hour. This is a characteristic of Sensor Web measurements, which are synchronous across the entire network and measurements can be taken every few minutes. An additional final assumption of the method is that the probability density of the differences has a peak near the mean and tails that taper as differences deviate away from it (e.g., see Fig. 1). That is, the method assumes that the probability of observing a difference decreases with the distance between that difference and the mean of all observed differences. This is not a strong assumption and could easily be relaxed in more complex circumstances if judged necessary.

Under these circumstances spatial and temporal measurement differences display a (much more) stationary distribution when compared to individual sensor readings. This permits more stable estimation of the statistics of differences over a much smaller state space. The estimation of differences between sensors placed at different micro-environments, or between those and experimental controls can also capture quantities of direct ecological interest [13], e.g. by comparing control plots to treatments.

To set the context and notation for the method presented below consider then a node with k neighbors. Let ϕ be the node's reading, ϕ_0 be its previous measurement, and $\phi_i, i = 1, ..., k$, be the readings of its neighbors. At each new measurement the node computes the difference between its current reading and its previous measurement and between its reading and each of its neighbor's $d_i = \phi - \phi_i, i = 0, ..., k$. Given knowledge of the distribution of differences each new observation can be tested for errors. The probability of observing a difference d as or more extreme than d_i is its p-value, p_i

$$p_i = \min [P(d \leq d_i), P(d \geq d_i)], \qquad (4)$$

where the probability P may refer to temporal differences $i = 0$ or differences with neighbor $i > 0$. For example, consider the distribution shown in Fig. 1, in which 88 percent of differences fall between -2 and 3, with 7 percent of differences less than or equal to -2, and 5 percent greater than or equal to 3. If $d_i = -2$, then $P(d \leq -2) = 0.07$ and $P(d \geq -2) = 0.93$. Thus $p_i = \min[0.07, 0.93] = 0.07$. Similarly, if $d_i = 3$, then $P(d \leq 3) = 0.95$ and $P(d \geq 3) = 0.05$. Thus $p_i = \min[0.95, 0.05] = 0.05$. The value of p_i in each instance is compared to a chosen significance level α. The measurement is flagged as anomalous if $p_i < \alpha$. We discuss how the combination of such p-tests between a sensor and all its neighbors identifies types of events below.

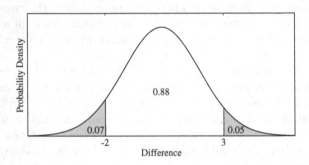

Fig. 1. An example of a probability density distribution illustrating the likelihood of observing an extreme difference. In this example, 88% of differences are between -2 and 3, with 7% of differences less than or equal to -2, and 5% greater than or equal to 3.

4.1   Statistical Inference

Each probability distribution $P(d)$ is learned from observed differences. There are several standard ways to implement this estimation, depending on the degree of prior knowledge. If the distributions are known to be well described by particular class of functions, then learning consists of estimating corresponding parameters. Filters, which specify sequential rules for parameter estimation, can then usually be constructed and optimized in order to minimize memory storage. If no parametric representation is adequate standard methods to construct non parametric distributions, in terms of frequency histograms, are employed.

Parametric estimation. If the distributions of differences are well fit by known distributions, estimation can be cast in terms of computation of distribution parameters from data. From the point of view of minimizing storage, estimation should be performed sequentially, so that only distribution parameters and current measurements are kept in memory at each single time. This can be achieved via the construction of filters to update estimators for distribution parameters [25].

Because distributions of differences of environmental variables are usually characterized by a small variance it is suggestive that, for sufficient number of observations, their shape may be well described by Gaussians. For a normal distribution $P(d)$ is defined by its mean and variance, which may be computed via standard maximal likelihood (unbiased) standard estimators, from t measurements as

$$\mu_{i,t} = \frac{1}{t} \sum_{k=1}^{t} d_{i,k}, \qquad \sigma_t^2 = \frac{1}{t-1} \sum_{k=1}^{t} (d_{i,k} - \mu_{i,t})^2, \tag{5}$$

which can be written using sequential updates as

$$\mu_{i,t} = \frac{(t-1)\mu_{i,t-1} + d_{i,t}}{t} \equiv \mu_{i,t-1} + K_t (d_{i,t} - \mu_{i,t-1}), \tag{6}$$

$$\sigma_{i,t}^2 = \frac{1}{t-1} \left[(t-2)\sigma_{i,t-1}^2 + \frac{t}{t-1} (d_{i,t} - \mu_{i,t})^2 \right]$$

$$\equiv \frac{1}{1-K_t} \left[(1 - 2K_t)\sigma_{i,t-1}^2 + \frac{K_t}{1-K_t} (d_{i,t} - \mu_{i,t})^2 \right], \tag{7}$$

where t indexes times when differences are observed (for simplicity, assumed here to be synchronous across the network), and $\mu_{i,0} = \sigma_{i,0}^2 = \sigma_{i,1}^2 = 0$. K_t is usually referred to as the gain factor in the context of filters. In the familiar case of t observations which are equally weighted the maximum likelihood estimator implies that $K_t = 1/t$.

Because our observations are correlated, we use the functional freedom introduced by K_t to optimize inference of missing or erroneous values. (Similar procedures can be applied to parameters of other distributions.) By varying the gain factor K_t we obtain the best estimator for the distribution parameters under the joint constraints of a limited number of samples and non-stationary data. The limit as $K_t \to 0$ corresponds to no update of the distribution resulting from the current reading. Even if perfect *a priori* knowledge of the parameters is given at some time, this eventually fails because of the non-stationarity of environmental data streams. As a consequence, the error between actual and predicted data must increase, eventually, as $K_t \to 0$. On the other extreme, when $K \to 1$, only the current measurement is used in predicting the distribution. This fails because of the standard estimation problem that a small sample of realizations generates imprecise parameter determinations. This reasoning indicates that there is an intermediate value for K_t that minimizes the error between actual and inferred measurements. We illustrate these features in the next section with data from the Sensor Web deployed at the Sevilleta LTER site.

Estimation of non-parametric distributions. When the distributions are not known to belong to a particular class, non-parametric estimation is still straightforward, although resulting in larger memory requirements [26].

Here we perform the estimation of the probability density for differences as a simple frequency histogram by dividing the interval of possible differences, $[l_d, u_d]$, into m subintervals, dictated in most cases by the corresponding sensor resolution. In this sense discretization of measurements is unavoidable in practice and the non-parametric estimation introduces no further approximation. We should nevertheless keep in mind that binning of data to construct frequency histograms is usually acceptable only when the underlying distribution $P(d)$ is approximately constant over the bin size [26]. As discussed below (see Fig. 3) the sensor precision may suffice to satisfy this criterion Figs. 3 (b)-(d), or have single bins with considerable excess of observations [Fig. 3 (a)].

The average precision, $(u_d - l_d)/m$ achieved in the estimation of differences, is generally much higher than that of the Bayesian classifier, $(u - l)/m$, because $u_d - l_d$ is typically much less than $u - l$. For example, while temperature readings may range from 0 to 100 degrees, differences between temperature readings at neighboring sensors may only vary between -5 and 5 degrees. Thus using 100 subintervals yields an average precision of 0.1 degrees for this method versus 1 degree for the Bayesian classifier.

Sample size and memory requirements. The advantage of using a parametric estimation, whenever it is applicable, is that a node is not required to store previously observed differences; only the current estimates for the distribution parameters and the number of utilized instances are required. For a normal distribution this is μ_i and σ_i^2 for differences in time and differences in space relative to each neighbor, and also t. Thus the total storage required in this case is $2(k+1)$ floating point numbers and an integer, roughly 24 bytes for a node with two neighbors. In addition, the mean and variance can be approximated from as little as 10 observed differences. Other distributions which may be relevant in sensing problems such as Laplace, Poisson, or negative binomial, require similar or smaller estimation effort and memory storage.

To approximate $P(d)$ without parametric assumptions, as a frequency histogram, a node keeps a count of the number of times observed differences fall in each subinterval. The probability $P(d \leq d_i)$ is the sum of counts for subintervals overlapping $(-\infty, d_i]$, normalized by the sum of all counts. Therefore, in the non-parametric case, a node needs to store $m(k+1)$ integers or roughly $4m(k+1)$ bytes. For example, to cover a range of differences spanning 10 degrees with one degree precision, a node with 2 neighbors would have to store 30 states or roughly 120 bytes, whereas the Bayesian classifier would have to store roughly 2 megabytes. In addition, the amount of learning data required to populate the counts is much smaller than for the Bayesian classifier. For example, to cover a range of differences spanning 10 degrees with 1 degree precision would require about 100 observations (roughly an order of magnitude greater than the size of the state space), versus about 5 million learning instances for the Bayesian classifier. In terms of learning time for a node taking a reading every five minutes, this method would require about 9 hours, versus 47 years for the Bayesian classifier. In some cases, a number of measurements commensurate with the size of the state space may suffice, resulting in learning times an order of magnitude below these numbers; however, the ratio between the learning times for each method would be the same.

4.2 Statistical Anomalies: Error and Event Detection

The estimated distributions of differences enable the acceptance or rejection of new measurements based on their likelihood. We adopt a simple p-value test, as described above, to determine if a new measurement difference is significant. If the new difference fails the significance test it is flagged as anomalous. Table 4.2 illustrates how different event types are encoded in the structure of these tests between a reference node l and the ensemble of its neighbors. We consider three characteristic situations.

First, for a standard measurement all observed differences at all nodes are significant. We refer to this situation as a global significance consensus because all tests agree and are significant. In this situation readings should be accepted and used to update statistics. Next, if there is a single point failure at sensor l then it will observe a global failure consensus, indicating an anomaly in time, relative to its earlier reading, and to each of its neighbors. In this situation sensor l identifies its measurement as anomalous, and may discard it. Furthermore, and assuming no other point failures for simplicity, each of the neighbors of l observes that each of its observed differences is significant, except for that to sensor l. This allows them to identify an error at l and produce their estimate of l's correct reading. We return to this point below. Finally, if there is a common mode event across the network, an anomaly may be detected for temporal differences but a spatial significance consensus will still be observed. Each sensor observers this same structure of p-value tests. This type of event may indicate a common mode failure or a real event, such as rain. Such discrimination may be identifiable through the consideration of correlations across different types of sensors (air temperature, relative humidity, soil moisture) but lies beyond the scope of this work. Ambiguous events may also take

Table 1. Determination of event types from combined p-value tests

Event type	Pod l	Neighboring Pods
Standard measurement	$p_0 > \alpha,\ p_{i \neq 0} > \alpha$	$p_0 > 0,\ p_{j \neq 0} > \alpha$
Point failure	$p_0 < \alpha,\ p_{i \neq 0} < \alpha$	$p_{j=l} < \alpha,\ p_{j \neq l} > \alpha$
Common event	$p_0 < \alpha,\ p_{i \neq 0} > \alpha$	$p_0 < \alpha,\ p_{j \neq 0} > \alpha$

place, where a fraction of all differences may fail significance tests, but not be easily classifiable as a single point failure or common mode event.

It may be desirable to combine various combinations of p-value tests in time and in space to each sensor's neighbor into a single significance test, that e.g. identifies consensus. The combination of multiple p-value tests into a single significance test has a long history in statistics going back to the work of Tippett and Fisher in the early 1930s [27]. Fisher's method is still probably the most widely used procedure. It assumes that the p_i are independent and uniformly distributed and so the combination

$$-2 \sum_{i=1}^{k} \ln(p_i), \tag{8}$$

is distributed as a χ^2_{2k} distribution with $2k$ degrees of freedom. The significance of the joint p-value tests is then computed as the probability of obtaining a value as or more extreme than that of expression (8) for a χ^2_{2k} distribution. Because this method of combining likelihood tests involves the geometric average of the p_i it is biased towards lower values of p_i and is not a good identifier of global or spatial consensus which, as indicated in Table 4.2, are the salient features of our expected events [27].

Several combinations of the set p_i which avoid these biases and are good identifiers of consensus have been proposed to address this issue. Among these, the z-transform test and the sum of p-values are the most widely used [28]. The z-transform test averages normal variables z each corresponding to a p_i and evaluates the significance level of this combination for a Gaussian distribution. Although the z-transform method is feasible, a much simpler method is the consideration of the sum

$$\bar{p} = \frac{1}{k} \sum_{i=0}^{k} p_i, \tag{9}$$

which can be compared to a desired significance level, typically of order α. This is the procedure we adopt below, guided essentially by simplicity. We emphasize, however, that many subtleties arise when taking into account the possible dependence of the several tests, which conditions the distribution of the variable combining the p_i, and consequently the nature of its significance test and choice of significance level as a function of those for individual tests. We intend to study these issues in future work with expanded datasets.

As a final remark, we note that if it is practical to perform the temporal and spatial (relative to neighboring sensors) significance tests independently, then a simple hierarchical structure for event classification becomes apparent. A temporal anomaly $p_0 < \alpha$ indicates an event. The event can be a point failure at the present sensor if there is also a spatial failure consensus, or a common mode if there is a spatial significance consensus. If no spatial consensus of either type is present the event is ambiguous and may be flagged for further study and possible creation of a new event class.

4.3   Inference of Missing Readings

As mentioned above the structure of temporal and spatial anomalies in the statistics of differences between a node and its neighbors allow a sensor to identify an error in its own measurement (global failure consensus) and its neighbors to identify the offending sensor and supply it with their estimation of its probable correct reading.

The most natural estimator of a sensor's missing or incorrect reading by neighbor i is simply

$$\hat{\phi}(i) = \phi_i + d_i, \tag{10}$$

where d_i is drawn from the distribution of differences between the two nodes. Averaging over d_i and over all neighbors leads to

$$\hat{\phi}_{av} = \frac{1}{k} \sum_{i=0}^{k} (\phi_i + \mu_i), \tag{11}$$

where $\hat{\phi}$ is the reading estimate and μ_i is the mean difference relative to the ith neighbor, or if $i = 0$, ϕ_0 is the previous reading and μ_0 is the mean difference between the current and previous measurements. A weighted average based on a measure of mutual information (e.g. smaller variance) between the nodes could also be adopted, but we use the simplest scheme here. In the case where the distribution class is known, μ_i is a stored value. If instead the distribution class is not known, the mean difference can be approximated by the usual maximum likelihood estimator

$$\mu_i = \frac{1}{m} \sum_{j=1}^{m} c_j m_j, \tag{12}$$

where c_j is the count for the jth subinterval and m_j is the midpoint of the jth subinterval.

5   Application to Ecological Data from Sevilleta LTER Site

In this section, we test the method using ecological data collected by a Sensor Web, developed at NASA/JPL [11,2], deployed at the Sevilleta LTER site. A Sensor Web is a spatially distributed macro instrument, where every component sensor node (or "pod") shares its readings, at each measurement cycle, with all other pods in the system. The Sensor Web is designed to maintain synchronicity among all component pods which makes it ideal for the type of correlated statistical analysis proposed in the previous section.

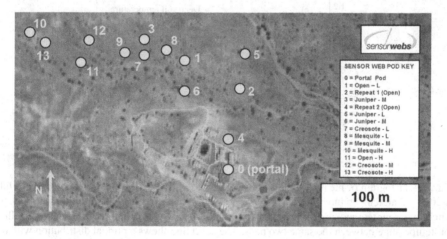

Fig. 2. Aerial photograph showing the Sensor Web layout at the Sevilleta LTER site. Fourteen sensor pods are distributed over a range of a few hundred meters to measure microclimate effects of the surrounding arid land plants. At regular time intervals the pods transmit data wirelessly to nearby pods. Sensor measurements eventually reach pod 0 where they are recorded.

The Sensor Web was initially deployed at the Sevilleta LTER site in 2003 as part of an ongoing effort to measure canopy microclimate effects of three arid land plant species: *Juniperus monosperma* (one-seeded juniper), *Larrea tridentata* (creosote bush), and *Prosopis glandulosa var. torreyana* (honey mesquite) [13]. The deployed Sensor Web consists of 14 sensor pods (see Fig. 2) which measure temperature, humidity, light flux, soil temperature, and soil moisture and transmit the data wirelessly to nearby pods.

The method for inferring missing readings, presented in the previous section, was tested by comparing inferred values to actual measurements. In this example, see Figs. 2 and 3, we selected an environmental variable (air temperature), a pod (pod 5), a set of neighbors (pods 8, 9, 11, 12, and 13), and a period of time (the first 2 days of July, 2005). We used the parametric version of the method [Equations (6) and (7)] because the distributions of differences are approximately normal (e.g., see Fig. 3). Figure 4(a) shows the inferred and actual readings for pod 5. The average error over the time period was 0.717 degrees Celsius.

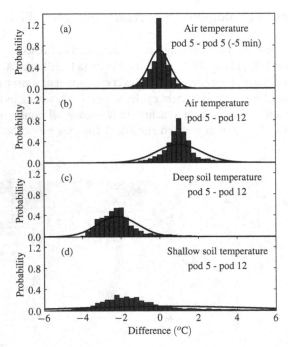

Fig. 3. A histogram of measurement differences recorded at the Sevilleta LTER site during July of 2004. (a) air temperature differences between pod 5 its previous reading (5 minutes earlier), (b) synchronous air temperature differences between pod 5 and pod 12, and (c) deep and (d) shallow soil temperature between the same two pods. The solid line shows a normal distribution with the same mean and variance as the data.

Because nodes have different placements, corresponding to distinct micro-climates, the distributions of differences are still weakly non-stationary. During warmer parts of the day, the more exposed nodes are warmer, but during cooler parts of the day (e.g. at night) the the more exposed nodes are cooler. Under these non-stationary conditions the average measurement error can be reduced by using Eqs. (6) and (7) with the appropriate value of K_t that optimizes the learning rate. Figure 4(c) shows the average error

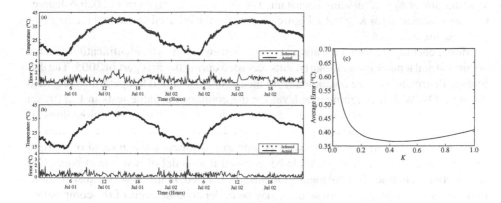

Fig. 4. Actual versus inferred air temperatures at sensor pod 5 for measurements taken in July 2005. The inferred measurements were computed using Eq. (11), with the average estimated via Eq. (6) with (a) $K_t = 1/t$, (b) $K_t = K = 0.46$. (c) The average error between the actual and inferred air temperature data as a function of the learning rate, K. The average error is computed using the entire two-day period of measurements. The minimum average error of 0.366 degrees Celsius is obtained for $K = 0.46$.

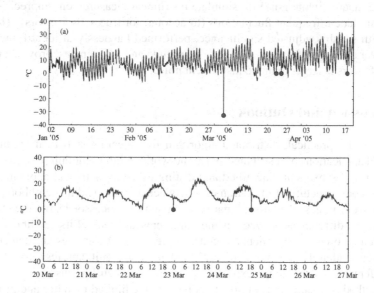

Fig. 5. Detected anomalies (marked by circles) in the pod 13 air temperature measurements for a period at the beginning of 2005. To detect the anomalies difference distributions for pod 13 (time difference) and pods 5, 11, 12, and 13 (space differences) were recorded for all of 2004 and the significance was computed using the combined p-value test of Eq. (9) with $\alpha = 0.005$. The method clearly captures the anomalies, as seen in (a), including some that are within the range of valid measurements. The two anomalies on March 23 and 24, shown in more detail in (b) are near zero degrees which is a common nighttime low temperature during that time of year.

as a function of $K_t = K$, assumed constant. The minimum average error of 0.366 degrees Celsius is achieved for $K = 0.46$. Figure 4(b) shows the inferred and actual readings for pod 5, using $K = 0.46$.

More generally we tested the anomaly detection and event type identification scheme on air and soil temperature measurements recorded during the first part of 2005. The difference distributions were computed from measurements of pod temperatures recorded during 2004. With the significance level for the combined p-value tests in Eq. (9) set to $\alpha = 0.005$, the method detects measurements that appear likely to be anomalous, as shown in Fig. 5, with no obvious false positives. Increasing α leads to the detection of more events, which may in some cases be due to instrument noise instead of outright failure. These effects can in principle be assessed if a model of instrument noise, and how it couples to true physical measurements, is provided. In this case, the distributions of differences between any two sensors may be understood in terms of the composition of failures, instrument noise, and physical measurements. Prior knowledge, or estimation, of the former may allow their subtraction from truly physical data streams. Here we have shown that, even in the absence of this knowledge, failures and instrument noise can be excluded from recordings and automatically corrected for at a chosen level of significance.

To understand these effects more clearly we have also applied our procedure to synthetically generated data containing diurnal and seasonal cycles, and with added small amplitude random white noise (to simulate instrument measurement imprecision) and larger amplitude infrequent fluctuations (to account for true sensor errors). The algorithm, with suitably adjusted significance, performed flawlessly at identifying sensor errors, over a variety of noise and failure amplitude and frequencies, provided the amplitude of errors is larger than the instrument noise.

6   Discussion and Outlook

We presented a practical, distributed algorithm for detecting statistical anomalies in ecological applications of distributed sensor networks. Both point failures and common mode events of sensors are identified and distinguished as statistical anomalies in the spatio-temporal structure of measurements between a sensor and its neighbors. Specifically, to avoid issues of non-stationarity, each sensor-processor learns the statistical distribution of differences between its measurements and each of its neighbors, as well as between its own measurements at consecutive times. Anomalies are detected, and their structure identified, in terms of statistical p-value significant tests for new measured differences relative to the expectations from these distributions.

The method is intentionally simple to cope with the limited memory and processing capabilities that characterize current sensor network technology. For this reason there are several directions for improvement. First, the operation of differencing, aimed here at factoring out the effects of common diurnal and seasonal drivers and reducing the size of the estimation space, can be achieved in principle by more sophisticated and accurate methods that are inspired by similar problems in image processing [20], signal processing [21] or component decomposition [29]. Methods for meta-analysis [30] to combine a variety of statistical tests can also be constructed to take into account external information about sensor or environmental specificities.

While these are interesting directions for future research we also emphasize that, for the empirical environmental data streams discussed above, the methods developed here suffice. They have the added bonus of being simple and implementable in sensors with very small amounts of memory and processing. The consideration of further constraints such as hard energy limitations, specific network and routing geometries, etc., is not necessary for most practical ecological distributed sensing problems. Instead the real challenge typical of ecological applications (and shared by others that measure physical and/or social environments) is the unpredictable, non-stationary nature of data streams and the fact that measurements tend to relate only indirectly to the hypotheses of interest. These issues place the emphasis on methods that use the rich spatiotemporal structure collected by networks of sensors to provide reliable statistical inference and to identify multi-variable event structures that may allow the testing of high level hypotheses. We believe that differencing, broadly understood, combined with sequential real-time estimation and meta-analysis of simultaneous statistical tests are important ingredients of any method concerned with automatic event detection and error correction in distributed sensor networks.

From the practical point of view, we have also shown that the combination of these ingredients, when compared to an alternative method based on Bayesian classifiers, leads to algorithms that are more storage-efficient, learn faster, and are more robust to non-stationary phenomena. In addition, the storage, processing, and communication requirements are such that it can be implemented in a distributed fashion, on each of the nodes in the network, thus reducing remote communication. Because of these qualities, this class of algorithms can provide data quality assurance for current generation of wireless sensor networks, such as the Sensor Web deployed at the Sevilleta LTER site. In the process of learning distributions of differences for data quality assurance, the algorithm also produces statistics that compare different microclimate environments [13], to each other and to control experiments, which are of immediate scientific ecological interest.

Acknowledgments

This work was supported in part by a DCI Postdoctoral Fellowship to L. B. Larkey. We thank S. Collins and R. Brown at Sevilleta LTER, and K. Delin at NASA JPL, for enabling and encouraging this work, and R. Nemzek at LANL for discussions. The image in Fig. 2 was taken by the USDA-ARS Remote Sensing Research Unit, Subtropical Agricultural Research Laboratory, Weslaco, TX.

References

1. Szewczyk, R., Osterweil, E., Polastre, J., Hamilton, M., Mainwaring, A., Estrin, D.: Habitat monitoring with sensor networks. Communications of the ACM 47(6), 34–40 (2004)
2. Delin, K.A.: Sensor Webs in the wild. In: Bulusu, N., Jha, S. (eds.) Wireless Sensor Networks: A Systems Perspective. Artech House (2005)

3. Marzullo, K.: Tolerating failures of continuous-valued sensors. ACM Trans. Comput. Syst. 8(4), 284–304 (1990)
4. Elnahrawy, E., Nath, B.: Cleaning and querying noisy sensors. In: Proceedings of the Second ACM International Workshop on Wireless Sensor Networks and Applications (September 2003)
5. Bychkovskiy, V., Megerian, S., Estrin, D., Potkonjak, M.: A collaborative approach to in-place sensor calibration. In: Zhao, F., Guibas, L.J. (eds.) IPSN 2003. LNCS, vol. 2634, pp. 301–316. Springer, Heidelberg (2003)
6. Sharma, A., Leana Golubchik, R.G.: On the prevalence of sensor faults in real world deploy-ments. In: Proceedings of the IEEE Conference on Sensor, Mesh and Ad Hoc Communica-tions and Networks (SECON) (June 2007)
7. Jeffery, S.R., Alonso, G., Franklin, M.J., Hong, W., Widom, J.: Declarative support for sensor data cleaning. In: Fishkin, K.P., Schiele, B., Nixon, P., Quigley, A. (eds.) PERVASIVE 2006. LNCS, vol. 3968, pp. 83–100. Springer, Heidelberg (2006)
8. Estrin, D., Culler, D., Pister, K., Sukhatme, G.: Connecting the physical world with pervasive networks. IEEE Pervasive Computing 1(1), 59–69 (2002)
9. Tulone, D., Madden, S.: An energy-efficient querying framework in sensor networks for detecting node similarities. In: MSWiM'06 (2006)
10. Delin, K.A.: The Sensor Web: A macro-instrument for coordinated sensing. Sensors 2, 270–285 (2002)
11. Delin, K.A., Jackson, S.P., Johnson, D.W., Burleigh, S.C., Woodrow, R.R., McAuley, J.M., Dohm, J.M., Ip, F., Ferre, T.P.A., Rucker, D.F., Baker, V.R.: Environmental studies with the Sensor Web: Principles and practice. Sensors 5, 103–117 (2005)
12. Meguerdichian, S., Slijepcevic, S., Karayan, V., Potkonjak, M.: Localized algorithms in wire-less ad-hoc networks: Location discovery and sensor exposure. In: Proceedings of MobiHOC 2001, Long Beach, CA, pp. 106–116 (2001)
13. Collins, S.L., Bettencourt, L.M.A., Hagberg, A., Brown, R.F., Moore, D.I., Delin, K.A.: New opportunities in ecological sensing using wireless sensor networks. Frontiers in Ecology 4(8), 402–407 (2006)
14. Szewczyk, R., Polastre, J., Mainwaring, A., Culler, D.: Lessons from a sensor network expe-dition. In: Karl, H., Wolisz, A., Willig, A. (eds.) Wireless Sensor Networks. LNCS, vol. 2920, Springer, Heidelberg (2004)
15. Ramanathan, N., Balzano, L., Burt, M., Estrin, D., Harmon, T., Harvey, C., Jay, J., Kohler, E., Rothenberg, S., Srivastava, M.: Rapid deployment with confidence:calibration and fault detection in environmental sensor networks. Technical Report 62, CENS, UCLA (2006)
16. Werner-Allen, G., Lorincz, K., Johnson, J., Lees, J., Welsh, M.: Fidelity and yield in a vol-cano monitoring sensor network. In: Proceedings of the 7th USENIX Symposium on Oper-ating Symposium (OSDI 2006) (2006)
17. Deshpande, A., Guestrin, C., Madden, S.R., Hellerstein, J.M., Hong, W.: Model-driven data acquisition in sensor networks. In: 30th International Conference on Very Large Data Bases, pp. 588–599 (2004)
18. Liu, K., Sayeed, A.: Asymptotically optimal decentralized type-based detection in wireless sensor networks. In: Acoustics, Speech, and Signal Processing, IEEE International Confer-ence (ICASSP '04), vol. 3, pp. 873–876 (2004)
19. Gupta, H., Navda, V., Das, S.R., Chowdhary, V.: Efficient gathering of correlated data in sensor networks. In: MobiHoc '05: Proceedings of the 6th ACM international symposium on Mobile ad hoc networking and computing, pp. 402–413. ACM Press, New York (2005)
20. Radke, R.J., Andra, S., Al-Kofahi, O., Roysam, B.: Image change detection algorithms: A systematic survey. IEEE Trans. on Image Proc. vol. 14(3) (2005)
21. Markou, M., Singh, S.: Novelty detection: A review - part 1: Statistical approaches. Signal Process. 83(12), 2481–2497 (2003)

22. Gelman, A., Carlin, J.B., Stern, H.S., Rubin, D.B.: Bayesian Data Analysis, 2nd edn. CRC Press, Boca Raton (2003)
23. Elnahrawy, E., Nath, B.: Context-aware sensors. In: Karl, H., Wolisz, A., Willig, A. (eds.) Wireless Sensor Networks. LNCS, vol. 2920, pp. 77–93. Springer, Heidelberg (2004)
24. DeGroot, M.H.: Optimal Statistical Decisions. Wiley, Chichester (2004)
25. Maybeck, P.S.: Stochastic Models, Estimation, and Control. In: Mathematics in Science and Engineering, vol. 141, Academic Press, San Diego (1979)
26. Duda, R.O., Hart, P.E., Stork, D.G.: Pattern Classification, 2nd edn. Wiley, Chichester (2000)
27. Rice, W.R.: A consensus combined p-value test and the family-wide significance of component tests. Biometrics 46(2), 303–308 (1990)
28. Folks, L.J.: Combination of independent tests. In: Krishnaiah, P.R., Sen, P.K. (eds.) Handbook of Statistics 4. Nonparametric Methods, North Holland, New York (1984)
29. Lakhina, A., Crovella, M., Diot, C.: Diagnosing network-wide traffic anomalies. SIGCOMM Comput. Commun. Rev. 34(4), 219–230 (2004)
30. Hedges, L.V., Olkin, I.: Statistical Method for Meta-Analysis. Academic Press, San Diego (1985)

Image Change Detection Using Wireless Sensor Networks

SreeRamya Yelisetty and Kamesh R. Namuduri

Department of Electrical and Computer Engineering,Wichita State University,
Wichita, KS 67260

Abstract. Change detection in images is of great interest due to its relevance in applications such as video surveillance. This paper presents the underlying theoretical problem of image change detection using wireless sensor network. The proposed system consists of multiple image sensors which make local decisions independently and send them to the base station which finally makes a global decision and declares whether a significant change has occurred or not.

1 Introduction

One of the main applications of image change detection is video surveillance. There is a great need for automated video surveillance system in commercial, defence and military applications. Existing video surveillance systems need continuous human monitoring to alert if any unusual events happen in the scene that is being monitored. In hostile environments such as a battle field, it is hard to deploy traditional video surveillance systems due to resource constraints. There comes the need for wireless sensor networks due to their attractive features such as rapid deployment, self organization and fault tolerance.

A wireless sensor network consists of low cost and low energy sensors which are deployed in a region of interest to observe a phenomenon and send the observations to a fusion center or a base station which makes a global assessment. The observations could be in terms of sensor readings such as light intensity, temperature, pressure, sound intensity, images and so on. System resources such as bandwidth and energy are very limited in a wireless sensor network. Hence it is very essential to limit the communication between the sensors and the fusion center as much as possible.

A wireless image sensor network consists of image sensors or cameras as nodes. The main objective of the wireless image sensor network is to monitor and detect changes in a given region. The image sensors capture images at fixed intervals of time and relay them to the base station which makes the final decision. Instead of sending raw image data from the sensors to the fusion center, the sensor nodes are allowed to carry out simple computations. These local computations help the sensors to carry out local decisions. The local decisions at each sensor are of the form of "1" or "0" for a change and no change respectively. Hence by sending binary decisions to the fusion center, the data transmission between the sensor

J. Aspnes et al. (Eds.): DCOSS 2007, LNCS 4549, pp. 240–252, 2007.

nodes and the base station is reduced to a great extent. The local decisions are relayed to a fusion center which makes a final decision by aggregation. This aggregation of decisions from all sensors and making a global decision is termed as distributed detection or decision fusion [1], [2].

The problem of distributed and decentralized detection has been investigated in the literature. In the papers [3], [4], the authors consider a distributed sensor network which is subjected to power constraint. The local processing at the sensors involves amplifying and forwarding the data to the fusion center. The decision rule at the fusion center is derived using Neyman-Pearson and Bayesian criterion. Decentralized detection under dependent observations has been studied in [4] and decentralized detection under conditionally independent observations has been studied in [3]. The evolution of visual sensor networks, wireless radio technologies and embedded system platforms in the recent technology defines a new set of applications along with few challenges. Due to their wide application range such as video surveillance, remote sensing, tracking, face recognition and so on, visual sensor networks have become very important in recent years. Hierarchial networks or multi-tier systems [5] are the recent research topics in this area. Multi-tier system deals with a network consisting of multiple tiers, where each tier constitutes of different types of sensors. For example, a two tier system consists of low resolution camera sensors in the lower tier and high resolution camera sensors in the higher tier. By using this multi-tier system we can achieve several benefits such as low cost, high reliability, high functionality and high coverage.

In general, visual sensors relatively consume high power. In [5], the power utilization in different types of sensors is presented. In [6], the authors have presented different modes of operation of visual sensor networks which enhances their life time and maximizes the probability of detecting an event. One good example of low resolution camera sensors is a Cyclops camera [7]. The cyclops camera is shown in Fig.1.

Fig. 1. Cyclops Camera designed jointly by Agilent Laboratories and the Center for Embedded Network Sensing at UCLA

There are few processing steps that need to be performed by a local sensor prior to making a local decision. These processing steps involves the implementation of image change detection algorithms at the sensor level. Image change detection typically requires image differencing or image ratioing etc [8]. The image change can be caused due to significant reasons such as appearance and disappearance of objects, motion of objects and some insignificant changes caused due to sensor noise, camera motion, atmospheric absorption and so on [8].

In [9], the authors developed a change detection algorithm which uses image difference with two thresholds(one low and the other high). If the pixel intensity in the corresponding difference image, is greater than the higher threshold, then the corresponding pixel is categorized in a change class. On the other hand, if the difference pixel intensity is lower than the lower threshold, then the corresponding pixel is categorized in an unchanged class. The remaining pixels whose difference intensity levels are between these two thresholds are allowed for further processing where the spatial-contextual information based on Markov random field model is considered. In [10], the authors tested the performance of eight different thresholding algorithms.

In this paper, we propose the underlying theoretical problem of image change detection in a wireless sensor network. The system makes use of four thresholds to detect local and global changes in the area being monitored. Two thresholds defined at the sensor level help the sensor make a local decision and the remaining two thresholds are defined at the system level which help the fusion center make a global decision. The challenge is to distinguish between significant and insignificant changes in the area being monitored.

The organization of the paper is as follows. Section II describes the processing at each sensor and Section III describes, processing at the fusion center. Section IV deals with performance analysis of the system followed by the results. Conclusions and future work are discussed in Section V.

2   Processing at Sensors

Consider a scenario of a wireless sensor network which is a collection of m image sensors observing a particular scene. The main objective of the sensor network is to detect significant change in the scene being monitored. Each image sensor is allowed to take images at fixed intervals of time. Let $D(\mathbf{x}_i) = |f_1(\mathbf{x}_i) - f_2(\mathbf{x}_i)|$ represent difference pixel where $\mathbf{x}_i$ corresponds to a pixel coordinate and $f_1(\mathbf{x}_i)$ and $f_2(\mathbf{x}_i)$ represents the pixels at location $\mathbf{x}_i$ in two different images taken at times t_1 and t_2 respectively. Here i=1,2,..n where n corresponds to size of the image. The processing flow at each sensor is illustrated in Fig.2. The set of difference pixels between two different images forms a difference image. This difference image can be considered as an instance of a random vector with each element or pixel corresponding to an instance of a random variable. A pixel in the difference image is considered as an active pixel if its intensity exceeds a certain threshold, otherwise it is considered as an inactive pixel. This threshold T_1 is computed based on a binary hypothesis test which results in 1/0 decisions.

Suppose $\delta(D(\mathbf{x}_i))$ represents the decision at pixel $\mathbf{x}_i$, then this decision rule can be expressed as follows,

$$\delta(D(\mathbf{x}_i)) = \begin{cases} 1, \text{ if } \mathbf{x}_i \text{ is active and } D(\mathbf{x}_i) > T_1; \\ 0, \text{ if } \mathbf{x}_i \text{ is inactive and } D(\mathbf{x}_i) < T_1. \end{cases} \tag{1}$$

Let $p(D(\mathbf{x}_i)|H_0)$ and $p(D(\mathbf{x}_i)|H_1)$ represent the density functions of inactive pixels(H_0) and active pixels(H_1) respectively. In this paper, we assume that both the densities follow Gaussian distributions which are defined as follows,

$$p(D(\mathbf{x}_i)|H_0) = \frac{1}{\sqrt{2\pi\sigma_0^2}} \exp\left\{ \frac{-(D(\mathbf{x}_i) - \mu_0)^2}{2\sigma_0^2} \right\} \tag{2}$$

$$p(D(\mathbf{x}_i)|H_1) = \frac{1}{\sqrt{2\pi\sigma_1^2}} \exp\left\{ \frac{-(D(\mathbf{x}_i) - \mu_1)^2}{2\sigma_1^2} \right\} \tag{3}$$

where μ_0, μ_1 and σ_0^2, σ_1^2 denote the means and variances for the corresponding density functions respectively.

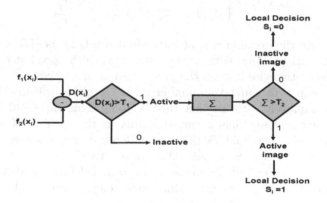

Fig. 2. Information processing flow at individual sensors. In the first stage, the difference $D(\mathbf{x}_i)$ at each pixel in two different images of the same scene is computed and compared with a threshold T_1. If the difference is greater than T_1, 1 is generated, else 0 is generated. The next step is to count the number of "1" 's or active pixels and compare it with another threshold T_2. If the count is greater than the T_2, 1 is generated signifying an active difference image, else 0 is generated which signifies an inactive difference image. These local decisions S_i in the form of 1/0 are transmitted via a wireless channel to the base station.

Based on (2) and (3), likelihood ratio can be defined as shown below,

$$L(D(\mathbf{x}_i)) = \frac{p(D(\mathbf{x}_i)|H_1)}{p(D(\mathbf{x}_i)|H_0)} = e^{\frac{\mu_1 - \mu_0}{\sigma^2}\left\{ D(\mathbf{x}_i) - \frac{\mu_0 + \mu_1}{2} \right\}} \tag{4}$$

assuming $\sigma_0^2 = \sigma_1^2 = \sigma^2$.

Now, the likelihood ratio needs to be compared with a threshold T_1 to come up with a decision rule. The selection of T_1 decides whether a change at a pixel is significant or not. The threshold can be fixed using Baye's criterion or Neyman-Pearson criterion or Minimax criterion. In this paper, we make use of Baye's criterion to compute the threshold T_1 which is defined as follows,

$$T_1 = \frac{p(H_0)(C_{10} - C_{00})}{p(H_1)(C_{01} - C_{11})}, \tag{5}$$

where $p(H_0)$ and $p(H_1)$ are the a priori probabilities that are independent of the observations. The elements C_{10}, C_{00}, C_{01}, C_{11} form a cost matrix. The elements C_{00} and C_{11} represents the costs incurred in taking correct decisions, i.e, costs incurred in stating inactive and active pixels as inactive and active pixels respectively. Similarly the costs C_{10} and C_{01} represents the costs incurred in taking wrong decisions, i.e, costs incurred in stating inactive and active pixels as active and inactive pixels respectively.

Based on (2), (3), (4) and (5), the decision rule can be simplified as follows,

$$\delta(D(\mathbf{x}_i)) = \begin{cases} 1, D(\mathbf{x}_i) \geqslant \frac{(\mu_0 + \mu_1)}{2} + \frac{\sigma^2 log(T_1)}{(\mu_1 - \mu_0)} ; \\ 0, D(\mathbf{x}_i) < \frac{(\mu_0 + \mu_1)}{2} + \frac{\sigma^2 log(T_1)}{(\mu_1 - \mu_0)}. \end{cases} \tag{6}$$

Let us denote the decision rule at individual pixels $\mathbf{x}_i$ as $\delta(D(\mathbf{x}_i)) = \{0, 1\}$. The decision statistic $(D(\mathbf{x}_i))$ takes the value 1 when the pixel at location $\mathbf{x}_i$ is an active pixel, otherwise 0 when the pixel is an inactive pixel.

The next step is to count the number of active pixels within the difference image and compare it with another threshold T_2. This threshold T_2 selection decides whether a change has occurred locally. If the number of active pixels is greater than the threshold T_2 then the difference image is considered as an active image, otherwise it is considered as an inactive image.

Let us assume the size of the image to be n and define a statistic λ_1 which describes sum of active pixels in the difference image as given below [1], [2],

$$\lambda_1 = \sum_{i=1}^{n} \delta(D(\mathbf{x}_i)). \tag{7}$$

Then, the probabilities for active image, inactive image, false alarm and miss can be written as shown below [11],

$$P_{ii} = P\{\lambda_1 < T_2 | H_0\}, \tag{8}$$

$$P_{ai} = P\{\lambda_1 \geq T_2 | H_1\}, \tag{9}$$

$$P_{fa} = P\{\lambda_1 \geq T_2 | H_0\}, \tag{10}$$

$$P_{miss} = P\{\lambda_1 < T_2 | H_1\}. \tag{11}$$

It can be shown that λ_1 follows a Binomial distribution. Let the probabilities of active pixels and inactive pixels in a difference image be represented as p_{ap} and p_{ip} respectively. Therefore for a given threshold T_2, the probabilities can be calculated as follows,

$$P_{ii} = \sum_{i=0}^{T_2} \binom{n}{i} p_{ip}^i (1 - p_{ip})^{(n-i)}, \tag{12}$$

$$P_{ai} = \sum_{i=T_2}^{n} \binom{n}{i} p_{ap}^i (1 - p_{ap})^{(n-i)}, \tag{13}$$

$$P_{fa} = 1 - P_{ii} = 1 - \sum_{i=0}^{T_2} \binom{n}{i} p_{ip}^i (1 - p_{ip})^{(n-i)}, \tag{14}$$

$$P_{miss} = 1 - P_{ai} = 1 - \sum_{i=T_2}^{n} \binom{n}{i} p_{ap}^i (1 - p_{ap})^{(n-i)}. \tag{15}$$

When n is large enough, the probabilities in (12), (13), (14) and (15) can be calculated by using Laplace-Demoivre approximation [12],

$$P_{ii} \simeq Q\left(\frac{-T_2 p_{ip}}{\sqrt{n p_{ip}(1-p_{ip})}} \right) \tag{16}$$

$$P_{ai} \simeq Q\left(\frac{T_2 - n p_{ap}}{\sqrt{n p_{ap}(1-p_{ap})}} \right) \tag{17}$$

$$P_{fa} \simeq 1 - Q\left(\frac{-T_2 p_{ip}}{\sqrt{n p_{ip}(1-p_{ip})}} \right) \tag{18}$$

$$P_{miss} \simeq 1 - Q\left(\frac{T_2 - n p_{ap}}{\sqrt{n p_{ap}(1-p_{ap})}} \right) \tag{19}$$

The threshold T_2 can be fixed in the same way as that of T_1 based on Baye's criterion, as given below,

$$T_2 = \frac{p(H_0)(C_{10} - C_{00})}{p(H_1)(C_{01} - C_{11})}, \tag{20}$$

In this scenario, C_{00} and C_{11} represent the costs incurred by making correct decisions, that means stating active and inactive images as active and inactive images respectively. The elements C_{10} and C_{01} of the cost matrix represent the costs incurred by making wrong decisions, i.e, stating active image and inactive image as inactive and active images respectively.

Based on these calculations, if the number of active pixels exceeds the threshold T_2, then there is a significant change in the scene and hence the corresponding image is considered as an active image, otherwise an inactive image. This decision test is a final test made at the sensor level which tells whether a local change at the sensor has occurred or not. Let us assume the local decisions from all the sensors as S_i where i=1,2,...m. These local decisions S_i's from individual sensors are relayed in the form of "1" or "0" to the base station through a wireless channel. Hence by sending binary decisions, the data transmission between the sensor nodes and the base station is reduced to a great extent. The local decisions S_i can now be written as follows,

$$S_i = \begin{cases} 1, \text{If } \lambda_1 \geq T_2 \text{ consider as an active image;} \\ 0, \text{If } \lambda_1 < T_2 \text{ consider as an inactive image.} \end{cases} \tag{21}$$

3   Processing at Fusion Center

Let us assume that there are m sensors deployed in the region of interest. Let all the m sensors transmit their local decisions S_i's to the fusion center through an additive white Gaussian channel (AWGN) as depicted in Fig.3. The fusion center receives noisy versions Y_i's of decisions S_i's sent by the sensors. This can be represented as,

$$Y_i = S_i + N_i, \tag{22}$$

where $N_i \sim \mathcal{N}(0, \sigma_n^2)$ and $i = \{1, 2, ...m\}$. In order to detect incoming bits at the base station, a hypothesis test similar to the one used at sensor level can be used at the base station also.

Let $p(Y_i|H_0)$ and $p(Y_i|H_1)$ represent the density functions of receiving bit "0"(H_0) and receiving bit "1"(H_1) respectively. In this paper, we assume that both the densities follow Gaussian distributions [13] which are defined as follows,

$$p(Y_i|H_0) = \frac{1}{\sqrt{2\pi\sigma_n^2}} \exp\left\{\frac{-Y_i^2}{2\sigma_n^2}\right\}, \tag{23}$$

$$p(Y_i|H_1) = \frac{1}{\sqrt{2\pi\sigma_n^2}} \exp\left\{\frac{-(Y_i - 1)^2}{2\sigma_n^2}\right\}. \tag{24}$$

If $p(H_0|Y_i)$ and $p(H_1|Y_i)$ denote the a posteriori probabilities, we define the decision rule that maximizes the probability of correct decision, and such a decision can be stated as,

$$\frac{p(H_1|Y_i)}{p(H_0|Y_i)} \underset{H_0}{\overset{H_1}{\gtrless}} 1 \tag{25}$$

The above equation is called as maximum a posteriori (MAP) criterion. Using Baye's rule in (25), the decision rule can be rewritten as follows,

$$\frac{p(H_1)p(Y_i|H_1)}{p(H_0)p(Y_i|H_0)} \overset{H_1}{\underset{H_0}{\gtrless}} 1 \tag{26}$$

or

$$\frac{p(Y_i|H_1)}{p(Y_i|H_0)} \overset{H_1}{\underset{H_0}{\gtrless}} \frac{p(H_0)}{p(H_1)} \tag{27}$$

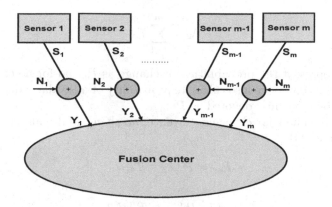

Fig. 3. Information processing flow at fusion center. The local decisions from sensors S_i are relayed to the base station via a wireless channel which is prone to additive white gaussian noise N_i. The noisy versions of the local decisions are represented by Y_i. Before the fusion center makes a global decision, MAP rule is used to detect the incoming bits. With these detected bits, a fusion rule is computed to make a global decision, whether a change has taken place or not.

Based on (23) and (24), the likelihood ratio can be computed as follows

$$L(Y_i) = \frac{p(Y_i|H_1)}{p(Y_i|H_0)} = e^{(2Y_i-1)/2\sigma_n^2}, \tag{28}$$

where $L(Y_i)$ is called likelihood statistic. The MAP decision rule consists of comparing this ratio with the constant $p(H_0)/p(H_1)$ which is called decision threshold.

$$T_3 = \frac{p(H_0)}{p(H_1)} \tag{29}$$

Based on (28) and (29), will lead to the formulation of a decision rule that helps in making the final decision, that decides if a global change took place in the scene being monitored. The effect of a binary hypothesis decision rule is to divide the observation space into two regions R_0 and R_1 where, $R_0 = (-\infty, T_3)$

and $R_1 = (T_3, \infty)$. If $Y_i \in R_0$, then a decision is made against H_0 which implies bit "0" is transmitted. Similarly if $Y_i \in R_1$, then a decision is made against H_1, i.e, bit "1" is transmitted.

$$\text{If } Y_i \in \begin{cases} R_0 , & 0 \text{ is transmitted,} \\ R_1 , & 1 \text{ is transmitted.} \end{cases} \tag{30}$$

Once the local decisions are detected at the fusion center, a fusion rule can be derived which decides whether a global change has occurred or not. This is the final decision that is made at the system level. A statistic λ_2 is defined which counts number of "1"'s received at the fusion center as follows,

$$\lambda_2 = \sum_{i=1}^{m} Y_i. \tag{31}$$

Let us represent the probabilities for change as P_c and for no change as P_{nc}. At the fusion center, we also calculate probability of false alarm denoted as P_{ffa} and probability of miss denoted as P_{fmiss}.

The probabilities for change occurrence, no change, false alarm and miss can be written as [11],

$$P_{nc} = P\{\lambda_2 < T_4 | H_0\}, \tag{32}$$

$$P_c = P\{\lambda_2 \geq T_4 | H_1\}, \tag{33}$$

$$P_{ffa} = P\{\lambda_2 \geq T_4 | H_0\}, \tag{34}$$

$$P_{fmiss} = P\{\lambda_2 < T_4 | H_1\}. \tag{35}$$

It can be shown that the sum follows Binomial distributions [1]. Let the probabilities of receiving bit "1" and bit "0" at the fusion center be represented as p_1 and p_0 respectively. Therefore for a given threshold T_4, the probabilities can be calculated as follows,

$$P_{nc} = \sum_{i=0}^{T_4} \binom{m}{i} p_0^i (1 - p_0)^{(m-i)}, \tag{36}$$

$$P_c = \sum_{i=T_4}^{m} \binom{m}{i} p_1^i (1 - p_1)^{(m-i)}, \tag{37}$$

$$P_{ffa} = 1 - P_{nc} = 1 - \sum_{i=0}^{T_4} \binom{m}{i} p_0^i (1 - p_0)^{(m-i)}, \tag{38}$$

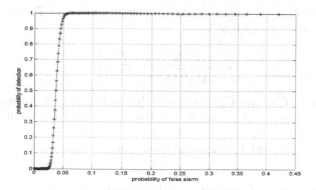

Fig. 4. ROC curve at sensor

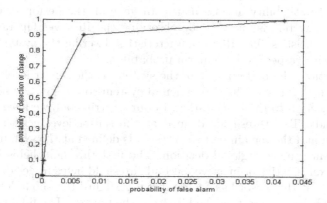

Fig. 5. ROC curve at fusion center

$$P_{fmiss} = 1 - P_c = 1 - \sum_{i=T_4}^{m} \binom{m}{i} p_1^i (1-p_1)^{(m-i)}. \tag{39}$$

When m is large enough, the probabilities in (36), (37), (38) and (39) can be calculated by using Laplace-Demoivre approximation [12],

$$P_{nc} \simeq Q\left(\frac{-T_4 p_0}{\sqrt{m p_0 (1-p_0)}}\right) \tag{40}$$

$$P_c \simeq Q\left(\frac{T_4 - m p_1}{\sqrt{m p_1 (1-p_1)}}\right) \tag{41}$$

$$P_{ffa} \simeq 1 - Q\left(\frac{-T_4 p_0}{\sqrt{m p_0 (1-p_0)}}\right) \tag{42}$$

$$P_{fmiss} \simeq 1 - Q\left(\frac{T_4 - m p_1}{\sqrt{m p_1 (1-p_1)}}\right) \tag{43}$$

The threshold T_4 can be fixed in the same way as that of T_1 and T_2 based on Baye's criterion, as given below,

$$T_4 = \frac{p(H_0)(C_{10} - C_{00})}{p(H_1)(C_{01} - C_{11})}, \tag{44}$$

In this scenario, C_{00} and C_{11} represent the costs incurred by making correct decisions. The elements C_{10} and C_{01} of the cost matrix represent the costs incurred by making wrong decisions. Based on these calculations, if number of "1" 's exceeds the threshold T_4, then a global change is declared.

4   Performance Analysis and Results

The performance metrics of a decision support system include probability of detection and probability of false alarm. In general, the performance of a decision support system is analyzed by observing the Receiver Operating Curve (ROC) characteristics. The ROC characteristics describe the system detection probability with respect to false alarm probability.

In this section, the performance of the system at the sensor level and at the system level is discussed. The experimental system consists of five sensors with wireless connection to the base station. In our experiments, we have considered four thresholds. Two thresholds defined at the sensor level help in making a local decision and the remaining two thresholds defined at the system level help the fusion center make a global decision. The first threshold helps in deciding whether a pixel is active or inactive. Second threshold helps in deciding whether a local change at the sensor has occurred or not. Third threshold defined at the fusion center helps in detecting the bits from the sensors. The final threshold at the fusion center helps in making a global decision to decide whether a change has occurred or not. The set of images used for simulations are shown in Figure 6.

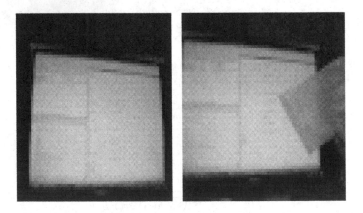

Fig. 6. Test images taken from cyclops camera

4.1   Performance at the Sensor

The performance analysis at the sensor level is illustrated in Figure 4. In this figure, the probability of detection is plotted against the probability of false alarm. The data for the plot is obtained by varying T1 and setting T_2 to 300. The size of the test images is $64X64$. The results shown in Figure 4 suggest that the system is very sensitive to the threshold.

4.2   Performance at the Fusion Center

The performance analysis at the fusion center is illustrated in Figure 5. In our experiments, the channel is considered to be additive white Gaussian (AWGN). Five sensors are considered in the system. The threshold T_4, which helps in making the final decision for a change is varied between (1,5). The number of detections or (1's) from the sensors are added and compared with this threshold T_4. From Figure 5, we observe that the system is able to detect a change with a high probability of detection and less number of false alarms.

5   Conclusions and Future Work

In this paper, we proposed a solution to image change detection problem in a wireless sensor network which makes use of four thresholds in order to make local and global changes in the area being monitored. It is assumed that an additive white Gaussian noise channel corrupts the local decisions transmitted from the sensors. In future, a system can be developed for real world applications with more realistic models of channel noise.

Acknowledgements

This work was carried out under the Kansas Space Grant Consortium grant funded by the Exploration Systems Mission Directorate (ESMD) program at NASA.

References

1. Niu, R., Varshney, P.K., Moore, M., Klamer, D.: Decision fusion in a wireless sensor network with a large number of sensors. In: Proc. 7th International. Conference on Information Fusion, Stockholm, Sweden (June 2004)
2. Niu, R., Varshney, P.K.: Distributed detection and fusion in a large wireless sensor network of random size. EURASIP Journal on Wireless Communications and Networking
3. Jayaweera, S.K.: Decentralized detection of stochastic signals in power-constrained sensor networks. In: IEEE Workshop on Signal Processing Advances in Wireless Communications (SPAWC)

4. Chamberland, J.-F., Veeravalli, V.V.: Decentralized detection in wireless sensor networks with dependent observations. In: Proc. 2nd Intl. Conf. on Computing, Commun. and Contrl Technologies (CCCT04)
5. Kulkarni, P., Ganesan, D., Shenoy, P.: The case for multi-tier camera sensor networks. In: Proceedings of the international workshop on Network and operating systems support for digital audio and video, pp. 141–146
6. Margi, C.B., Manduchi, R., Obraczka, K.: Energy consumption tradeoffs in visual sensor networks. In: Proceedings of 24th Brazilian Symposium on Computer Networks (SBRC 2006) (June 2006)
7. Rahimi, M., Baer, R.: Cyclops: Image sensing and interpretation in wireless sensor networks, reference manual,
   http://www.cens.ucla.edu/~mhr/cyclops/cyclops.pdf
8. Radke, R.J., Andra, S., Al-Kofahi, O., Roysam, B.: Image change detection algorthims: A systematic survey. IEEE Trans.on Image Processing (March 2005)
9. Bruzzone, L., Prieto, D.: Automatic analysis of the difference image for unsupervised change detection. IEEE Trans. Geoscience and Remote Sensing 38, 1171–1182 (2000)
10. Paul, E.I., Rosin, L.: Evaluation of global image thresholding for change detection. Pattern Recognition Letters 24, 2345–2356 (2003)
11. Duda, R.O., Hart, P.E., Stork, D.G.: Pattern Classification. Wiley, USA (2001)
12. Papoulis, A., Pillai, S.: Probability, Random Variable and Stochastic Processes. McGraw-Hill, New York
13. Sam, S.K., Breipohl, A.M.: Random Signals Detection, Estimation and Data Analysis. John Wiley, New York (1988)

Near Optimal Sensor Selection in the COlumbia RIvEr (CORIE) Observation Network for Data Assimilation Using Genetic Algorithms

Thanh Dang[1], Sergey Frolov[2], Nirupama Bulusu[1], Wu-chi Feng[1], and António Baptista[2]

[1] Portland State University, Portland OR, USA
{dangtx,nbulusu,wuchi}@cs.pdx.edu
[2] Oregon Health and Science University, Beaverton, OR, USA
{frolovs,baptista}@ccalmr.ogi.edu

Abstract. CORIE is a pilot environmental observation and forecasting system (EOFS) for the Columbia River. The goal of CORIE is to characterize and predict complex circulation and mixing processes in a system encompassing the lower river, the estuary, and the near-ocean using a multi-scale data assimilation model.

The challenge for scientists is to maintain the accuracy of their modeling system while minimizing resource usage. In this paper, we first propose a metric for characterizing the error in the CORIE data assimilation model and study the impact of the number of sensors on the error reduction. Second, we propose a genetic algorithm to compute the optimal configuration of sensors that reduces the number of sensors to the minimum required while maintaining a similar level of error in the data assimilation model. We verify the results of our algorithm with 30 runs of the data assimilation model. Each run uses data collected and estimated over a two-day period. We can reduce the sensing resource usage by 26.5% while achieving comparable error in data assimilation. As a result, we can potentially save 40 thousand dollars in initial expenses and 10 thousand dollars in maintenance expense per year.

This algorithm can be used to guide operation of the existing observation network, as well as to guide deployment of future sensor stations. The novelty of our approach is that our problem formulation of network configuration is influenced by the data assimilation framework which is more meaningful to domain scientists, rather than using abstract sensing models.

Keywords: Sensor selection, network configuration, coastal monitoring, data assimilation, genetic algorithm.

1 Introduction

Earth and ocean sciences confront great opportunities and challenges in understanding the complex behaviors of large-scale physical systems with next generation sensing systems [6]. Modeling the behavior of the oceans and river estuaries

J. Aspnes et al. (Eds.): DCOSS 2007, LNCS 4549, pp. 253–266, 2007.

is a challenging but important research field. In order to understand the state of the physical process, sensors are deployed in the environment to collect data for the modeling process. Ideally, a highly dense network of sensors will enable the collection of fine-grained information about the physical system under observation. However, for systems that operate over a large geographical region, such a deployment of sensors is infeasible. Hence, most of the existing large-scale sensing systems only deploy a sparse network of sensors and use advanced numerical methods in estimating and modeling the physical processes.

Figure 1 shows CORIE, an observation network that monitors the Columbia River estuary and the Eastern North Pacific ocean. CORIE integrates a real-time sensor network, a data management system and advanced numerical models. The goal of CORIE is to characterize and predict complex circulation and mixing processes in a system encompassing the lower river, the estuary and the near-ocean. The CORIE observation network includes an extensive array of 24 stations in the Columbia River estuary and the nearby coastal ocean. At each station, variable combinations of in-situ sensors measure one or more physical properties of water or atmosphere. Water temperature, salinity, and water levels are measured at most stations. Profiles of velocity and acoustic backscatter are measured at three stations.

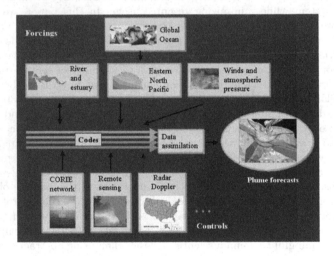

Fig. 1. CORIE Data Assimilation Architecture. Source: The CORIE project website.

Data assimilation combines observational data with numerical data models to produce an estimated system state for the physical process. Data assimilation plays an important role in predicting the state of the dynamic physical process such as estuary circulation, weather and climate changes. Unlike low-powered wireless sensors such as the popular Crossbow motes which are tiny and cheap, the sensor stations in ocean monitoring are usually very expensive to deploy and operate. Such stations typically have a number of sensors between the surface and the anchor, providing a vertical array of sensors. Measurement data from the observation network directly impacts the accuracy of the estimated system. Hence, finding a suitable network configuration is an important problem in deploying and operating an observation network because it can help reduce the resource usage while maintaining or improving the estimation accuracy.

In this paper, we first propose a metric for characterizing the error in the CORIE data assimilation model and study the impact of the number of sensors on the error reduction. Second, we propose a genetic algorithm to compute the near optimal configuration of sensors that reduces the number of sensors to the minimum required while maintaining a similar level of error in the data assimilation model.

This problem is relevant to the sensor network research community because the deployment of the existing observation network was based on an intuition of the underlying physical process, with little knowledge about how sensor placement would affect the resulting data models. Hence, a solution to the problem will help conserve resources by using fewer sensors.

The problems in oceanography have their own distinct challenges. First, the physical model is extremely complex. Unlike models in previous work [1, 8, 20, 11, 12], we address a complex 3D circulation and mixing processes in a system encompassing the lower river and the estuary. This has been formally recognized as a challenging task in ocean modeling [10]. Second, the computation is very expensive due to its large state space size. For example, the state space size of the CORIE model is 878,850. Given larger memory and computing platforms, this number would increase with new resources by increasing spatial and temporal resolution. Third, the observation model incorporates multiple sensing modalities such as salinity, temperature, elevation, and velocities. Each sensing modality provides different information about the observed environment. Therefore, the solution must take into account not only the correct set of sensors but also the correct type of sensors to ensure good estimation results. In addition, the solution must be model independent so that it can be used in other environmental monitoring deployments provided that they use the same data assimilation framework. Fortunately, the framework we use is a state-of-the-art data assimilation and estimation system [10]. Finally, selecting an optimal sequence of sets has already been shown to be NP-hard in many settings [1]. Therefore, we must consider not just only polynomial class solutions but also how much time it actually takes to converge to an acceptable result.

We present a method to partially address the problem of finding near optimal network configuration for an observation network, which uses a data assimilation framework based on a sigma-point Kalman filter [10]. The main contributions of this paper are:

– We formulate the problem of optimizing network configuration based upon data assimilation (Section 3) and apply it to an ocean modeling application.
– We propose a framework that uses genetic algorithms to partially address the problem of selecting a suitable subset of sensors (Section 4).

– We evaluate the approach on data from the CORIE observation network (Section 5) and demonstrate that we can reduce the use of sensing resources by 26.5% and operating expenses by $10,000 a year while maintaining a similar level of estimation accuracy.

2 Data Assimilation Overview

This section provides a brief overview of the CORIE data assimilation framework used in the formulation of the sensor selection problem. While we describe only the data assimilation framework used in the CORIE project [10], we do not imply that the problem is only applicable for this specific framework. In fact, the problem is suitable for any situation provided that the error of the estimation can be calculated.

2.1 CORIE Data Assimilation Framework

The complete data assimilation framework, proposed and implemented by Frolov *et al.* [10], is complex and draws upon several disciplines including numerical analysis, machine learning, and estimation theory. We provide a brief overview of CORIE here and refer readers to [10] for more details. Figure 1 shows the high level components of the CORIE modeling system. It integrates model and field controls. The main purpose of CORIE is to simulate 3D circulation in the region that lies between the Columbia River estuary and near ocean but also extends further inland in Oregon, USA to the Eastern North Pacific. CORIE performs multiple tasks and provides the following: short term forecasts, actual past conditions, characteristic climatology conditions, and scenario conditions.

In order to accomplish these tasks, one of the key components is data assimilation which integrates observational data from sensors into a non-linear ocean model. The model integrates information from the CORIE network, Doppler radar and remote sensing with forcings from the river, estuary, winds, atmosphere, and ocean to predict the behavior of the underlying physical processes. The work of Frolov *et al.* [10] proposes and implements a fast framework with model surrogates for data assimilation, illustrated in Figure. 2. The data assimilation framework includes two main components. The first component is *off-line learning* illustrated on the left block in Figure. 2. Its main purpose is to train a model surrogate, which is an equivalent model in the reduced space. In order to do that, the original system state of 878,850 variables is reduced to 60 variables using principle component analysis, a popular method for extracting patterns and compressing data [17], based on the singular value decomposition (SVD) algorithm. The model surrogate is trained using a recurrent neural network [15]. All training is carried out off-line using an existing database of model hindcasts generated by the traditional circulation model [18]. The second component is *data assimilation* illustrated on the right block in Figure. 2. The core of the assimilation algorithm is the *sigma-point kalman filter* [19]. The filter estimates the state of the dynamic system using the model surrogate and measurements

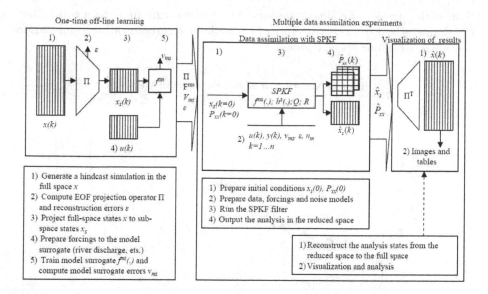

Fig. 2. CORIE Data assimilation framework — Reproduced with permission from Frolov *et al.* [10]

from sensors. An existing framework such as data assimilation using ensemble Kalman filter [9] is computationally very expensive, limiting its use. In contrast, the model surrogate framework performs 1000 times faster than existing frameworks and can significantly increase the estimation error reduction (defined in Equation 2) on the given measurement set [10].

Clearly, this framework provides a significant contribution in improving the estimation of the ocean model. Nevertheless, a trivial observation is that the configuration of sensors including not only sensor location but also sensor type plays a very important role in providing better estimation. Hence, there are two problems:

- What is the configuration of sensors that would maximize the estimation accuracy given an observation network?
- What is a subset of sensors to achieve or maintain a certain estimation accuracy?

These have been proven to be hard problems [1]. In our work, we only address a part of the latter problem, which we describe formally in the next section.

3   Problem Formulation

Network configuration refers to the number of sensors, their type and their locations. The configuration of sensors plays an important role in estimation in general. For example, an object's location in two dimensional space can be better estimated from range measurements with a triplet of non-collinear sensors than with a triplet of collinear sensors. The temperature in a room can be better characterized from sensors spread throughout the room rather than sensors concentrated in one specific area.

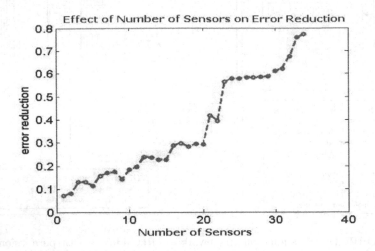

Fig. 3. Number of sensors versus estimation error reduction

For example, we have run data assimilations on an increasing number of sensors in CORIE network. Figure 3 shows the estimation error reduction (defined in Equation 2) versus the number of sensors used in CORIE network. As we can see, some sensors have more impact on error reduction than others [1]. In addition, some configurations of sensors might have lower error reduction even though they have more sensors than other configurations. Therefore, finding a minimum set of sensors that can provide the most information is an interesting problem, the *network configuration problem*. Formally, it can be stated as the following optimization problem:

$$min|S| \ subject \ to \ D(S) \leq \varepsilon \ and \ S \subseteq A \qquad (1)$$

where A is the set of all sensor information $A = \{s_1, s_2, ..., s_n\}$ and $s_i = (type, x, y, z, \delta)$ in which *type* is the sensor type, which can be temperature, salinity, or elevation. (x, y, z) is the sensor's location. δ is the standard deviation in the sensor reading obtained by calibration. ε is the threshold error. $D(S)$ is the simplified form of the function of error reduction of the data assimilation. In other words, it is the cost function to be optimized. $D(S)$ can be calculated as the estimation error reduction as follows:

$$error\_reduction = 1 - \frac{sum[(xs\_twin - xs\_data).^2]}{sum[(xs\_twin - xs\_free).^2]} \qquad (2)$$

---

[1] We compare the error reduction in data assimilation using the observational data relative to relying on the numerical model alone.

where xs_twin is the true system state, xs_data is the estimated system state, xs_free is the simulated system state, i.e., the estimated system state without considering sensor measurements. The notation $(.)^2$ denotes the vector of squared elements. We can consider $sum[(xs\_twin - xs\_data).^2]$ as the squared error of the data assimilation using sensor measurements and $sum[(xs\_twin - xs\_free).^2]$ as the squared error of the data assimilation without sensor measurements. Hence, Equation 2 shows how much error the data assimilation can reduce when it uses additional sensor measurements.

A similar derived optimization problem can be formulated as follows:

$$max|D(S)| \ subject \ to \ S \subseteq A \ and \ |S| = n \ (n \leq |A|) \tag{3}$$

to find a configuration of the network that maximizes the error reduction $D(S)$ of the data assimilation.

There are several parameters to be considered here. The first parameter is a sensor's *type*. Intuitively, sensors of different types may provide a better data set for data assimilation than sensors of a single modality e.g. temperature sensors. The second parameter is the sensor *location*. It is important that sensors should be deployed in critical locations such that they together report data representing the underlying physical process. The final parameter is the *number* of sensors which is our optimization objective. Unfortunately, the complete problem is very difficult to solve due to the fact that selecting an optimal sequence of sets is NP-hard [1] and the behavior of function $D(S)$ is unknown. Therefore, our work can only address a part of the problem where the sensor locations are fixed. Hence, the problem becomes a *sensor selection* problem. The next section presents our approach to solve this problem.

4   Sensor Selection Using Genetic Algorithm

The key idea in our proposed solution is to apply genetic algorithms to search for an acceptable sensor set. We consider genetic algorithms (GAs) for this problem because they have been applied successfully to a variety of optimization problems, and especially for optimizing the topology and learning parameters for artificial neural networks [15]. GAs can search for the optimal solution by observing the behavior of the system without actually knowing how the system works. GAs can optimize cost functions with multiple minima without numerical gradients for the cost functions. Hence, it is well suited for our sensor selection problem because we have little prior knowledge about the relationship between the error reduction and the configuration of sensors. For a complete discussion on genetic algorithms, please see [15].

The search for an appropriate configuration begins with a collection of initial configurations. Members of the current population are used to generate the next generation population by means of operations such as random mutation and crossover, which are patterned after processes in biological evolution. At each step, the configurations in the current population are evaluated by the reduction of error after data assimilation. Those with the highest error reduction are selected probabilistically as seeds to produce the next set of configurations.

4.1 Representing the Network Configuration

We employ a standard hypothesis bit-string (or chromosome) representation that is often used in GAs. The advantage of this representation is that it can be easily manipulated by genetic operators such as crossover and mutation.

Since we are only optimizing the number of sensors in the network, the network configuration can be represented by an n-bit string $\underbrace{10111...1}_{n}$

where n is the total number of sensors in the network.

0 means that the sensor is not used.

1 means that the sensor is used in the configuration.

For example: 10101 is a configuration of a network of 5 sensors in which the 1st, 3rd, and 5th sensors are used while 2nd and 4th sensors are not used.

4.2 Fitness Function and Selection

The fitness function defines the criterion to rank the configurations for the purpose of selection. In our problem settings, the most appropriate criterion is the error reduction in data assimilation. Hence, the fitness function calculates the error reduction in the data assimilation using that configuration.

There are several popular selection methods such as fitness proportionate selection, tournament selection, and rank selection. Each selection method has its own advantages and disadvantages [15]. In our approach, we use the tournament selection method which runs a competition among a few individuals selected randomly and select ones with the best fitness. The tournament selection method often yields a more diverse population than other methods. Hence, a broader range of configurations can be considered during training.

4.3 Crossover and Mutation

We use standard settings for the crossover and mutation functions. We use scattered crossover as the operator instead of single point or intermediate crossover because it maximizes the information exchange among individuals. We use the gaussian mutation strategy because it is popular and standard in GAs.

5 Experimental Results

This section describes the experiments conducted to evaluate if GAs can produce a good set of sensors. The hypothesis we propose and test is that the sensor set found by a GA can save significant resources while maintaining a level of estimation accuracy similar to the current observation network.

The metric we used is the same as the cost function in Equation 2 because it is the optimization criterion. The evaluation of the sensor set is based on the error reduction in the data assimilation using the data from this configuration.

$$error\_reduction = 1 - \frac{sum[(xs\_twin - xs\_data).^2]}{sum[(xs\_twin - xs\_free).^2]} \qquad (4)$$

The *error_reduction* lies between 0 to 1 because $sum[(xs\_twin - xs\_data).^2]$ is smaller than $sum[(xs\_twin - xs\_free).^2]$ as the measurements are incorporated in the estimation of xs_data. Ideally, the higher the *error_reduction* is, the better the set of sensors.

Another metric that we consider is the cost of sensor equipment, deployment, and maintenance, that we can save by reducing the number of fully operational sensors in CORIE. We assume the costs for different sensors are the same. In practice, this is not true. Deployment and operation costs for sensors depend on sensor location and sensor type. However, we simplify the model to give an idea of how much money we can save by selectively reducing the number of sensors as follows:

$$cost\_reduction = (eq\_cost + dep\_cost + mnt\_cost) * num\_sensor \qquad (5)$$

where num_sensor is the number of sensors we can remove from CORIE. The average equipment cost, eq_cost, is approximately \$4,500 per sensor. The deployment cost, dep_cost, is about \$500 per sensor. The maintenance cost, mnt_cost is about \$1000 per sensor. These are derived from actual costs in CORIE. We do not take into account the cost to deploy the station and the power and communication system because we can use one station for several sensors.

5.1   Experimental Design

We conduct the experiments using data from the CORIE observation network. The network consists of 23 stations with 34 sensors deployed in the Columbia river estuary.

Due to the fact that we never know the true state of the dynamic system, we set up twin experiments that use the real data to estimate the true state of the system and use this estimated true state to simulate the measurements for the data assimilation.

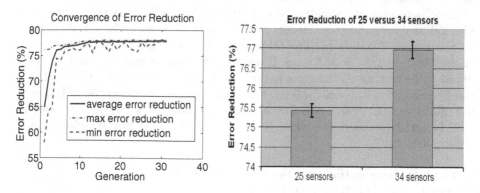

Fig. 4. Results: a) The error reduction converges and reaches a stable state after 10 generations. b) The error reduction of the data assimilation using 25 GA-selected sensors is only 1.55% smaller than using all 34 sensors.

We use a separate hindcast data xs_twin and consider it as the true state. xs_twin is then used to simulate the observations from the sensor network. The measurements are used as the input to the data assimilation. The output of the data assimilation is xs_data, which is the estimated system state. On the other hand, by using the model only, we also simulate xs_free as the simulated system state. xs_free is obtained without data assimilation. Readers should distinguish between the process model, which is known and used to simulate xs_free and the error reduction model, which we have little understanding about. The settings for GA are listed below.

- Hypothesis representation: 34-bit chromosome
- population size: 20
- crossover rate: 0.8
- crossover operator: scattered
- mutation strategy : gaussian
- selection method: tournament
- number of generation: 30
- fitness function: average error reduction of 5 runs
- runtime: 35 days

Due to limited processing capability, we do not set the threshold error reduction ε to find the configuration. Instead, we observe the best configuration after 30 generations.

5.2 Results and Analysis

The experiments finished after 35 days with the error reduction convergence. One might wonder about the experiment run time. As we mentioned earlier, the 3D circulation model state size is 878,850 — 8370 grid points × [(1 salinity + 1 temperature + 2 velocities) × 26 levels + 1 elevation]. Although the data assimilation is operated in the reduced space of 60 variables, the evaluation of error reduction of individual sensing type must be done in the full space of 878,850 at each time step. Hence, one complete data assimilation alone takes 20 minutes on 2-day data. The total time to finish the GA can be estimated as $20 \times 30 \times 5 \times 0.3/24 = 37.5$ days. However, this number can be significantly reduced by leveraging the inherent parallelism in GA. For example, the total time can be reduced to one week if 5 machines are used for the experiment. However, this motivates the design of a new algorithm to make GA parallelism possible. This is not the focus of our work. However, there exist several popular ways to accomplish it [15].

The best configuration after the 30th generation was:
1111101001111111100111100101110111.

This means that 9 sensors or 26.5% resources are not used. We verify this configuration by the second experiment, in which we run data assimilation 30 times for 2-day data. The error reduction achieved was 75.42%. This is only 1.55% lower than that using all 34 sensors as shown in Figure 4.

If the difference in the error reduction is negligible, it means that we can save 9 sensors. According to the estimated costs for initial equipment, deployment and maintenance, we can save around 40 thousand dollars of initial expense and 10 thousands dollars for maintenance per year.

6 Related Work

The problem of network configuration or sensor selection has attracted significant interest in the sensor networks research community. Several papers [22] [13] [20] [1] try to address the problem for varying classes of sensors, network scale and the underlying physical process that the network is monitoring.

In one of the earliest works on the sensor coverage problem, Megeurdicherian *et al.* [14] proposed a solution that given the knowledge of existing sensor positions uses Voronoi diagrams to compute the maximal breach paths in the sensor field and find gaps in coverage, where additional sensors can be deployed. Similarly, Wang *et al.* [21] proposed a solution to network coverage by integrating sensing and connectivity constraints. The limitation of their work is that they use a simple signal attenuation model for a particular sensing modality to evaluate the utility of each sensor, rather than considering the complete data assimilation process.

Willett *et al.* [22] proposed an adaptive sampling scheme called Backcasting. They try to address a similar problem to ours, which is to minimize the number of active sensors while maintaining high accuracy. However, they assume a dense uniform distribution of sensors and eliminate sensors by considering the correlation of the environment estimated from a fusion center. The context in coastal modeling is slightly different w there are only a few expensive sensor stations deployed in a very large geographical area. Hence, the assumptions are no longer valid.

One direction in solving the sensor selection problem for target tracking tries to use concepts from information theory [8] [13] [20]. Ertin *et al.* [8] and Liu *et al.* [13] consider the *mutual information* between the predicted sensor observation and the current target location distribution as the criterion for selecting sensors. This approach works because mutual information actually represents the reduction in the uncertainty of one random variable to the knowledge of the other [4]. Wang *et al.* [20] overcome the expensive computation of mutual information by introducing an *entropy-based* approach. The authors claim that the difference between the entropy of the probability distribution of the sensor view and the entropy of the sensing model for a true target is strongly related to the mutual information. Hence, this information can be used to sort sensors more quickly while still maintaining similar results. While these attempts show very interesting findings, they are formulated for target tracking and localization problems. It is unclear how the approaches can be applied to the ocean monitoring problem. In addition, the approaches implicitly assume that a greedy selection of the set of the most informative sensors provides the most information. However, as we observed in our data assimilation problem, this is not always the case.

The work of Krause *et al.* [12] and Bian *et al.* [1] formulate the problem as a form of optimization with some cost function, utility function or sensing quality, subject to constraints such as energy consumption [1] or communication cost [12]. We found that they are very close to our problem theoretically. However, their problem context is different from ours because we do not have any constraints on energy or communication cost. All sensor stations in CORIE are wired with power and data cables. Another difference is that our problem addresses a very large and complex geographical region, the Columbia river estuary. Therefore, determining the super modular utility function [1] or predicting sensing quality [12] is infeasible.

There are various works attempting to solve related problems such as adaptive sampling for localized phenomena [7], and sensor deployments that differ from our work in that they optimize certain specific criteria [16] [2] [5]. The problems are related but different to ours. Therefore, most of their approaches are not applicable to the problem we address.

Finally, there are attempts to use genetic algorithms to select sensor parameters [3] or select noisy sensor data [11]. Our work is different in that we show that we can use genetic algorithms combined with data assimilation for applications in ocean observation and coastal monitoring.

7   Future Work

Genetic algorithms do not use knowledge about the relationship between sensor configuration and monitoring precision. However, detailed investigation about the physical process model may help in better explaining the relationship between sensor configuration and monitoring precision. We also would like to assess the effectiveness of the genetic algorithm results with monthly, seasonal, and yearly environmental changes.

As mentioned before, finding an optimal configuration of sensors is an unsolved hard problem. As future work, we would like to investigate optimization algorithms that take into account not only the number of sensors but also the sensor type and sensor location to determine an optimal network configuration. We would also like to try our framework with other modeling and sensing systems such as atmospheric sensing besides the Columbia river estuary system to ensure the usefulness of our approach in practice. Finally, in sparse wide-area observation networks such as CORIE, the long term data collection from static stations is often augmented with opportunistic data collection from mobile stations. In the CORIE project, Clatsop Community College's M/V Forerunner serves as a mobile station of opportunity, and several cruises have been conducted over the years. As part of the CORIE project, we are currently investigating how to guide the trajectory of these vehicle cruises to optimize the observation process.

8   Conclusion

CORIE is a pilot environmental observation and forecasting system for the Columbia River. The CORIE observation network differs from low-power, dense wireless sensor networks in one aspect - sensor stations are sparse and expensive to deploy and maintain. The challenge for scientists is to maintain the accuracy of their modeling system while reducing the use of expensive resources.

We showed that genetic algorithms can aid in optimizing the configuration of the CORIE observation network. Specifically, we were able to reduce the number of observation stations without compromising the accuracy of the state estimate; leading to potential savings in the deployment and maintenance cost for the observatory. The novelty of this paper is that our problem formulation of sensor selection is influenced by the data assimilation framework which is more meaningful to domain scientists, rather than by abstract sensing models. Our approach and algorithm are simple and potentially generalizable to other wide area environmental sensing systems.

Acknowledgements. We would like to thank Michael Wilkin for providing information on the costs and operating expenses of equipment in the CORIE project. We also would like to thank Dave Maier, Todd Leen, and Eric Wan for useful discussions during the project. The research described in this paper was supported by National Science Foundation grants NSF 05-14818 and 01-21475.

References

1. Bian, F., Kempe, D., Govindan, R.: Utility based sensor selection. In: Proceedings of the Fifth International Conference on Information Processing in Sensor Networks (IPSN 06), pp. 11–18, Nashville, Tennessee (April 2006)
2. Bredin, J.L., Demaine, E.D., Hajiaghayi, M., Rus, D.: Deploying sensor networks with guaranteed capacity and fault tolerance. In: Proceedings of the 6th ACM international symposium on Mobile ad hoc networking and computing, pp. 309 – 319, Urbana-Champaign, Illinois (May 2005)
3. Corcoran, P., Anglesea, J., Elshaw, M.: The application of genetic algorithms to sensor parameter selection for multisensor array configuration. Sensors and Actuators A Physical 76, 57–66 (1999)
4. Cover, T.M., Thomas, J.A.: Elements of Information Theory. John Wiley and Sons, New York (1991)
5. Cristescu, R., Vetterli, M.: On the optimal density for real-time data gathering of spatio-temporal processes in sensor networks. In: Proceedings of the Fourth International Symposium on Information Processin In Sensor Networks, pp. 159–164, Los Angeles, California (April 2005)
6. Delaney, J.: Keynote: Next-generation earth and ocean sciences: Opportunities and challenges. In: Proceedings of the 3rd ACM Conference on Embedded Networked Sensor Systems (SenSys), San Diego, California (November 2005)
7. Ermis, E.B., Saligrama, V.: Adaptive statistical sampling methods for decentralized estimation and detection of localized phenomena. In: Proceedings of the Fourth International Symposium on Information Processing in Sensor Networks (IPSN 05), pp. 143–150, Los Angeles, California (April 2005)

8. Ertin, E., Fisher, J.W., Potter, L.C.: Maximum mutual information principle for dynamic sensor query problems. In: Zhao, F., Guibas, L.J. (eds.) IPSN 2003. LNCS, vol. 2634, pp. 405–416. Springer, Heidelberg (2003)

9. Evensen, G.: Sequential data assimilation with a nonlinear quasi-geostrophic model using monte carlo methods to forecast error statistics. Journal of Geophysical Research, vol. C5(10) (1999)

10. Frolov, S., Baptista, A., Lu, Z., van der Merwe, R., Leen, T.: Fast data assimilation with model surrogates: Application to circulation in a highly stratified estury. In: Submission to Ocean Modeling

11. Khan, A.A., Zohdy, M.A.: A genetic algorithm for selection of noisy sensor data in multisensor data fusion. In: Proceedings of American Control Conference, pp. 2256–2262, Albuquerque, NM (June 1997)

12. Krause, A., Guestrin, C., Gupta, A., Kleinberg, J.M.: Near-optimal sensor placements: Maximizing information while minimizing communication cost. In: Proceedings of the Fifth International Conference on Information Processing in Sensor Networks (IPSN 06), pp. 2–10, Nashville, Tennessee (April 2006)

13. Liu, J., Reich, J., Zhao, F.: Collaborative in-network processing for target tracking. EURASIP Journal on Applied Signal Processing, vol. 4 (2002)

14. Meguerdichian, S., Koushanfar, F., Potkonjak, M., Srivastava, M.B.: Coverage problems in wireless ad-hoc sensor networks. In: Proceedings of the Conference on Computer Communications 2001 (INFOCOM 2001), pp. 1380–1387. Anchorage, Alaska (April 2001)

15. Michell, T.M.: Machine Learning. Mc Graw Hill, New York (1997)

16. Ray, S., Lai, W., Paschalidis, I.C.: Deployment optimization of sensornet-based stochastic location-detection systems. In: Proceedings of the Conference on Computer Communications (INFOCOM 2005), Miami, Florida (March 2005)

17. Smith. L.I.: A tutorial on principle component analysis (February 2007) http://kybele.psych.cornell.edu/Ēedelman/Psych-465-Spring-2003/PCA-tutoria 1.pdf

18. van der Merwe, R., Leen, T., Lu, Z., Frolov, S., Baptista, A.M.: Fast neural network surrogates for very high dimensional physics-based models in computational oceanography. Neural Computation (To appear, 2007)

19. van der Merwe, R., Wan, E.A.: Sigma-point kalman filters for probabilistic inference in dynamic state-space models. In: Proceedings of the Workshop on Advances in Machine Learning, Montreal, Canada (June 2003)

20. Wang, H., Yao, K., Pottie, G., Estrin, D.: Entropy-based sensor selection heuristic for target localization. In: Proceedings of the third international symposium on Information processing in sensor networks (IPSN 04), pp. 36–45, Berkeley, California, USA (April 2004)

21. Wang, X., Xing, G., Zhang, Y., Lu, C., Pless, R., Gill, C.: Integrated coverage and connectivity configuration in wireless sensor networks. In: Proceedings of the 1st international conference on Embedded networked sensor systems (Sensys), pp. 28–39. ACM Press, New York (2003)

22. willett, R., Martin, A., Nowak, R.: Backcasting: Adaptive sampling for sensor networks. In: Proceedings of the Fifth International Conference on Information Processing in Sensor Networks (IPSN 06), pp. 36–45, Nashville, Tennessee (April 2006)

Data Salmon: A Greedy Mobile Basestation Protocol for Efficient Data Collection in Wireless Sensor Networks

Murat Demirbas[1], Onur Soysal[1], and Ali Şaman Tosun[2,*]

[1] Dept. of Computer Science & Engineering, University at Buffalo, Suny
{demirbas, osoysal}@cse.buffalo.edu
[2] Dept. of Computer Science, The University of Texas at San Antonio
tosun@cs.utsa.edu

Abstract. Our work addresses the spatiotemporally varying nature of data traffic in environmental monitoring and surveillance applications. By employing a network-controlled mobile basestation (MB), we present a simple energy-efficient data collection protocol for wireless sensor networks (WSNs). In contrast to the existing MB-based solutions where WSN nodes buffer data passively until visited by an MB, our protocol maintains an always-on multihop connectivity to the MB by means of an efficient distributed tracking mechanism. This allows the nodes to forward their data in a timely fashion, avoiding latencies due to long-term buffering. Our protocol progressively relocates the MB closer to the regions that produce higher data rates and reduces the average weighted multihop traffic, enabling energy savings. Using the convexity of the cost function, we prove that our local and greedy protocol is in fact optimal.

1 Introduction

A wireless sensor network (WSN) consists of potentially hundreds of sensor nodes and is deployed in an ad hoc manner for collecting data from a region of interest over a period of time [1,2]. Even though the technology is new, WSNs received an enthusiastic reception in the science community as WSNs enable precise and fine-grain monitoring of a large region in real-time. Some examples of successful large-scale deployments of WSNs to date are in the context of ecology monitoring (monitoring of micro-climate forming in redwood forests [3]), habitat monitoring (monitoring of nesting behavior of seabirds [4]), and military surveillance (detection and classification of an intruder as a civilian, soldier, car, or SUV [5,6]).

In traditional WSN deployments (including all of the above deployments), the data collection is achieved by using a multihop data forwarding mechanism toward a static basestation (SB), which has the computational power to store and process all the collected data. A major shortcoming of this approach is that it neglects the *spatiotemporal nature of data generation in the WSN*. That the WSN data generation rates are local both in time and space has been observed

---

* Partially supported by Center for Infrastructure Assurance and Security at UTSA.

J. Aspnes et al. (Eds.): DCOSS 2007, LNCS 4549, pp. 267–280, 2007.

Table 1. Comparison of MB protocols

	Mobility	Buffering	Latency	Energy consumed
Data mules	random	long-term	high	low
Predictable MB	periodic	long-term	high	low
MES	self-controlled	long-term	medium	low
Data salmon	network-controlled	short-term	low	medium

in several WSN deployments. Investigations of natural phenomena in forest environments have validated spatiotemporal distribution of solar illumination, temperature, and humidity [3, 7, 8]. This effect is especially prevalent in intrusion detection applications [5, 6].

As the observed environmental phenomena changes inevitably with time, the performance of data collection in the WSN suffer using the SB approach. In typical WSN applications most of the time the network remains idle, and when there is an interesting event (such as a rapid change in ambient features or detection of an intruder), a bursty generation of traffic occurs at a region of the network for some time. Fixing the location of the SB as the center or one corner of the network penalizes these bursts of data generation. Multihop relaying of this high data-rate traffic from the originating region toward the SB results in the depletion of energy at the relaying nodes for the duration of the traffic. Also collisions and message losses occur as this high data-rate traffic contends with itself over multiple hops. Clustering and aggregation techniques [9, 10] help to alleviate these contention and surge conditions, but they fail to address the root cause of the problem.

In order to address the drawbacks of the SB approach, there has been a flurry of work on employing a mobile basestation (MB) for data collection. The *data mules* [11] work exploit random movement of MBs to opportunistically collect data from a sparse WSN. Here, the nodes buffer all their data locally, and upload the data only when the MB arrives within direct communication distance. Although this approach is energy-efficient (in that the nodes do not engage in multihop data forwarding), the tradeoff is the very high latency and buffering costs. A similar approach is investigated in the context of predictable movement of the MB. Here sensors are assumed to know the trajectory of the MB and predict when the data transfer will occur accordingly. This work also shares similar drawbacks as the data mules work: since the data rate may be varying in time among the regions, buffer overflows may occur due to high data-rate traffic.

Mobile element scheduling (MES) work [12] considers controlled mobility of the MB in order to reduce latency and serve the varying data-rates in the WSNs effectively. The MES work shows that the problem of planning a path for the MB to visit the nodes before their buffers overflow is NP-complete. Although some heuristic based solutions are proposed to address this problem [12, 13, 14], these solutions ignore the problem of communicating the status of the spatiotemporally varying data-rates to the MB.

Contributions. Our work addresses the spatiotemporally varying nature of data traffic in environmental monitoring and surveillance applications. In contrast to previous MB-based solutions, we avoid indefinite buffering of data at the sensor nodes and let them forward the data toward the MB to avoid any latencies. In order to reduce the energy-consumption due to multihop data forwarding, our MB protocol, namely the Data Salmon, progressively relocates the MB to minimize the average weighted-multihop distance from the data producing nodes to the MB. To this end, our protocol directs the MB closer to the regions that produce higher data rates so that most of the traffic arrives to the MB via a small number of hops. Intuitively speaking our MB always tends toward the center of mass of the network based on the data rate distribution.

Secondly, we prove that it suffices to design a local and greedy protocol to achieve an optimal relocation of the MB, minimizing the average weighted-multihop distance in the WSN. This proof involves showing that the cost function is convex and a local minima is a global minima. Our Data Salmon protocol exploits this result in that at each position in the network it decides on the next position via a simple greedy decision, using only the information available at that node. Our protocol relocates the MB toward the edge with the largest flow.[1] Another implication of such a greedy approach is that it is easy to parallelize the solution via divide and conquer technique: Adding more MBs to the network is easy since the MBs do not need to coordinate, yet each by optimizing its own gain implicitly cooperates to achieve a desirable global behavior. We present an extension of our Data Salmon protocol to multiple MBs along these lines.

Thirdly, our work demonstrates a synergistic cooperation of the MB and the underlying WSN for achieving efficient and effective data collection. In our protocol, the MB uses an underlying spanning tree structure to receive the data and to decide which direction to move on this backbone tree. In return when the MB moves along one edge of the tree, it updates the direction of the edge to point to its new location to ensure that a dynamic tree is always rooted at the MB. This way it is possible to keep the MB always reachable from the backbone tree structure, and the movement of MB also becomes relatively simple (by following one edge on the backbone tree). Although previous work assumed that the data-rates in the WSN is known and fixed [12], our protocol addresses this problem explicitly and discovers the current data-rates on-the-fly by means of this distributed dynamic tree structure.

Finally, we simulate our Data Salmon protocol using real WSN data (collected from an intrusion detection application) and some synthetic data. Via these simulation results, we compare the improvements gained by using Data Salmon over using SB under various configurations.

Applications. Since the Data Salmon protocol action for the MB is simple and the MB is virtually controlled by the network, our protocol does not require a fully-autonomous robot to implement the MB. Thus, it is practical to implement and deploy Data Salmon in real-world environmental monitoring and

---

[1] Due to this greedy behavior to move toward the largest flow, we name our protocol after the Salmon fish which swim upstream to lay eggs.

surveillance applications using semi-autonomous MBs. A suspended cableway infrastructure for MB mobility would provide a suitable framework for the deployment of the Data Salmon protocol. For example, the Networked Infomechanical Systems (NIMS) architecture [7] successfully avoids surface-based obstacles found in natural environments by employing a horizontally mobile node suspended via an aerial cable, and achieves adaptive sampling and effective solar radiation mapping in microclimate monitoring applications. Another example of such a system is the SkyCam platform [15], which is suitable for intrusion detection and surveillance applications.

In our model, we have not included the energy required for moving the MB. The reason behind our willingness to generously tradeoff the energy required for relocating the MB with the energy gain in data collection is that it is much easier to replenish and maintain the batteries of one MB than those of the sensor nodes in the entire network. As it was observed through the NIMS deployment [16], by using a solar panel attached to the mobile node it is possible to harvest an average of 250 Watt hours of energy per day and sustain the mobility of the node. Such an alternative energy source creates a virtual flow of energy into the system, hence, the WSN lifetime is also elongated. Another benefit a network controlled MB provides is the increased traffic capacity and network throughput as mentioned in [17].

2   Model

We consider a dense, connected, multihop WSN. The sensor nodes are static after the initial deployment. There is a distinguished MB in the network whose current location (the node it resides on) is denoted by m.

We assume that a spanning tree structure is overlaid over the WSN during the network initialization phase. To reduce the height of the tree, the tree root (denoted as *root*) may be a node in the center of the network. By using a flooding protocol initiated by the *root* it is easy to construct this backbone tree structure [5, 6]. We denote the set of neighbors of node i on the tree as $N(i)$, and use $d(i, m)$ to denote the hop distance over the tree structure between a node i and the MB at node m. We denote the data rate generated by a node i at a given time as w_i. For a node m we define $M(m)$, the cost of forwarding all the data to m from the entire network, as $M(m) = \sum_i w_i * d(i, m)$. Then the problem of finding the optimal location for the MB reduces to finding a node m^* with minimal $M(m)$.

3   The Data Salmon Protocol

After discussing how we maintain a dynamic tree rooted at the MB, we give our greedy MB relocation protocol and prove its optimality.

3.1   The Dynamic Tree Maintenance Protocol

Keeping the MB always reachable from the backbone tree structure is essential to guarantee always-on data forwarding to the MB. In order to maintain a dynamic tree that is always rooted at the MB over the static backbone tree, we adopt the distributed arrow protocol [18].

After the backbone tree structure is set up as discussed in the Model section, we assume that the tree edges all point to the MB initially. As the MB moves over one of the tree edges, the arrow protocol prescribes flipping the direction of the edge. This way the tree is always rooted at the MB. By locally updating a tree edge, a dynamic tree rooted at MB can be thus maintained over the static backbone tree.

Of course, embedding a tree constrains how nodes can forward the traffic to the MB. For example, shortest path forwarding may not be achievable for some nodes as they are constrained to follow the tree while forwarding data to the MB. However using a backbone tree for forwarding of the traffic reduces the tracking cost of the MB drastically: In our scheme as the MB moves only one edge needs to be updated. Had we not used a tree backbone for data forwarding toward the MB, the tracking of the MB would incur an expensive (nonlocal) communication cost for updating the tracking structure as the MB relocates. Investigating update-efficient and local tracking structures is an active topic of research, and we give some pointers to this work in Section 5.

3.2   The Greedy Data Salmon Protocol

Our Data Salmon protocol for the MB runs on top of the dynamic tree structure, and uses the incoming data rates from neighboring nodes for deciding which neighbor to move the MB next. For each neighbor i of the current node m, we denote the forwarded data rate from i with ε_i. Note that, ε_i corresponds to the cumulative weights of all nodes in the subtree rooted at i. We denote the total data rate in the WSN with ε, which is calculated at m as $(\sum_{i \in N(m)} \varepsilon_i) + w_m$.

To minimize the cost function M, it is natural for the MB to move toward a neighbor i with a lower cost function $M(i)$. We prove in Theorem 1 that in fact such a neighbor i is unique since the ε_i is at least more than half of the total data rate ε in the WSN if and only if $M(i) < M(m)$.

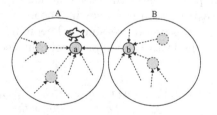

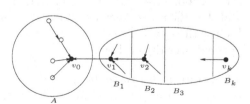

Fig. 1. Conceptual representation for proof of Theorem 1

Fig. 2. Visual Representation of Theorem 2

Theorem 1. *Let MB be at node v_a, and $v_b \in N(v_a)$, then $M(v_b) < M(v_a) \iff \varepsilon_{v_b} > \frac{\varepsilon}{2}$.*

Proof. Consider Figure 1, where MB is at v_a. If MB is moved from v_a to v_b, since this graph is a tree, all data generated at nodes in set A has to be forwarded through edge (v_b, v_a) toward v_b. That is, the distance to the MB increases by 1 for all nodes in set A, and the distance decreases by 1 for all nodes in set B. Thus, we can write the following:

$$M(v_b) = M(v_a) + \varepsilon_A - \varepsilon_B \tag{1}$$

Since $A \cup B$ contains all the nodes, the following also holds:

$$\varepsilon_A + \varepsilon_B = \varepsilon \tag{2}$$

Case ($\Longrightarrow$): Using the assumption $M(v_b) < M(v_a)$, from (1) we can write, $\varepsilon_A - \varepsilon_B = M(v_b) - M(v_a) < 0$. So subtracting (2) from this term we can write $-2\varepsilon_B < -\varepsilon$ and thus conclude $\varepsilon_B > \frac{\varepsilon}{2}$.

Case ($\Longleftarrow$): Using the assumption $\varepsilon_{v_b} > \frac{\varepsilon}{2}$, from equation (2) we have $\varepsilon_A < \frac{\varepsilon}{2} < \varepsilon_B$. This entails $M(v_b) - M(v_a) = \varepsilon_A - \varepsilon_B < 0$. So $M(v_b)$ has smaller cost. $\square$

Algorithm 1. MB control action at m

1: $\varepsilon \leftarrow \left(\sum_{i \in N(m)} \varepsilon_i \right) + w_m$
2: **if** $\exists i \in N(m) : \varepsilon_i > \frac{\varepsilon}{2}$ **then**
3:     move to i
4: **end if**
5: // else stay at m, since m is optimal

Therefore, the MB control action at node m is given as in Algorithm 1. We prove the optimality of this protocol in the next section.

3.3   Proof of Optimality

Our optimality discussion depends on some properties of the cost function over the backbone tree. We first show that the cost function is convex in Theorem 2, and that the rate of increase of the cost function is non-decreasing in Theorem 3. We use these two properties to conclude that the Data Salmon protocol is indeed optimal over the backbone tree.

Theorem 2. *Let v_0 be an optimal location for MB. Consider a path $v_0, v_1, \ldots, v_k$ over the backbone tree. $M(v_0) \leq M(v_1) \leq \ldots \leq M(v_k)$ holds for the path.*

Proof. Consider Figure 2. Since v_0 is an optimal location of MB, $M(v_0) \leq M(v_1)$, and this proves the first inequality. If the MB is moved from v_0 to v_1, the distance increases by 1 for all nodes in set A, and the distance decreases by 1 for all nodes in the set $B = \cup_{i=1}^{k} B_i$. So,

$$M(v_1) = M(v_0) + \sum_{i \in A} w_i - \sum_{i \in B} w_i \qquad (3)$$

Since we have $M(v_0) \le M(v_1)$ we get

$$\sum_{i \in A} w_i - \sum_{i \in B} w_i \ge 0. \qquad (4)$$

If the MB is moved from v_1 to v_2, similarly, the distance increases by 1 for all nodes in set $A \cup B_1$ and the distance decreases by 1 for all nodes in the set $B - B_1$. We can write $M(v_2)$ using $M(v_1)$ as follows:

$$M(v_2) = M(v_1) + \sum_{i \in A \cup B_1} w_i - \sum_{i \in B - B_1} w_i \qquad (5)$$

This can be rewritten as

$$M(v_2) = M(v_1) + \sum_{i \in A} w_i + \sum_{i \in B_1} w_i - \sum_{i \in B} w_i + \sum_{i \in B_1} w_i$$

$$M(v_2) = M(v_1) + \sum_{i \in A} w_i - \sum_{i \in B} w_i + 2 \sum_{i \in B_1} w_i$$

Since all weights are non-negative, the last term is non-negative. First two terms are shown to be non-negative in equation 4, so we get $M(v_2) \ge M(v_1)$. This can be generalized to the following using the same approach:

$$M(v_k) = M(v_{k-1}) + \sum_{i \in A} w_i - \sum_{i \in B} w_i + 2 \sum_{j=1}^{k-1} \sum_{i \in B_j} w_i \qquad (6)$$

Thus, we have $M(v_0) \le M(v_1) \le \ldots \le M(v_k)$. $\qquad\square$

Since we use this result later we introduce ε_S, which corresponds to the sum of weights of all members of set S, as $\varepsilon_S = \sum_{i \in S} w_i$. Hence, equation (6) can be rewritten as:

$$M(v_k) = M(v_{k-1}) + \varepsilon_A - \varepsilon_B + 2 \sum_{j=1}^{k-1} \varepsilon_{B_j} \qquad (7)$$

Theorem 2 shows that the cost function is convex, but in order to guarantee that there are no oscillations in the MB control protocol we need to show that the rate of increase is also non-decreasing.

Theorem 3. *Let v_0 be the optimal location of MB. Over the backbone tree consider a path $v_0, v_1, \ldots, v_a, \ldots, v_b, \ldots, v_k$, where $0 < a < b \le k$.*
$M(v_a) - M(v_{a-1}) \le M(v_b) - M(v_{b-1})$ *holds for the path.*

Proof. By using equation (7), we get the following:

$$M(v_a) - M(v_{a-1}) = \varepsilon_A - \varepsilon_B + 2\sum_{j=1}^{a-1} \varepsilon_{B_j}$$

$$M(v_b) - M(v_{b-1}) = \varepsilon_A - \varepsilon_B + 2\sum_{j=1}^{b-1} \varepsilon_{B_j} \qquad (8)$$

$$[M(v_{b+1}) - M(v_b)] - [M(v_{a+1}) - M(v_a)] = 2\sum_{j=1}^{b-1} \varepsilon_{B_j} - 2\sum_{j=1}^{a-1} \varepsilon_{B_j} \qquad (9)$$

Since $b > a$, and all weights are non-negative, we can rewrite above as:

$$[M(v_{b+1}) - M(v_b)] - [M(v_{a+1}) - M(v_a)] = 2\sum_{j=a}^{b-1} \varepsilon_{B_j} \geq 0 \qquad \square$$

As a corollary to these theorems, we observe that when the data rates are stable for a sufficient enough period, the MB progressively relocates to the optimal location in the WSN. Our corollary follows by using $M(m)$ as the variant function. From Theorem 2, we know that there are no local optimum points, and $M(m)$ is non-increasing toward the direction of the optimal location. Furthermore, Theorem 3 states that equality among $M(m)$ values is only possible between optimal nodes, so at any suboptimal node we are guaranteed to have a neighbor with lower cost. The decrease of $M(m)$ is bounded below by $M(m^*)$, so the MB eventually reaches and comes to a rest at an optimal location m^*.

4 Simulation Results

In order to evaluate the performance of our protocol, we use real-world WSN deployment data from the "Catch Me if You Can" project [19]. This project implements a multiple-pursuer, multiple-evader tracking application by utilizing the WSN to help the pursuers in protecting an asset from the evaders. Figure 3 shows the topology of the 60 nodes deployed for this project. The distance between any neighboring nodes in the topology is 10 meters.

The *Catch Me if You Can* project collected data sets for over 50 experiments. Each data set contains onsets and offsets of detection for the nodes: during these detection periods, the sensor nodes generate detection data. In order to simulate our Data Salmon protocol for collecting the generated detection data, we overlay a randomly generated backbone tree over the WSN as shown in Figure 3. This way, we calculate the approximate energy consumption for the SB and MB approaches using the durations of detections and the distances on the backbone tree.

The introduced locomotion model for the MB is a high-level abstraction of the mobile platform details. We assume that the MB moves following the tree

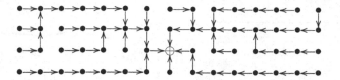

Fig. 3. The topology of sensors in *Catch Me if You Can* experiments. Dots denote the sensors. Embedded random backbone tree on the topology is shown by arrows and the *root* is indicated with a plus.

edges, with constant speed. The MB only makes decisions at nodes, so it can not change direction during transitions between nodes.

We developed a Java application named *SalmonSim*[2] to interpret and emulate the data sets from the *Catch Me if You Can* experiments. The Java application uses constant time steps to measure the performance. During these time steps, the MB is simulated and the events from experiment logs are emulated. The Java application also provides a user interface to visualize the progress of the MB running the Data Salmon protocol. Using this simulator, we compare the performance of the SB positioned in the *root* of the tree with the MB using the Data Salmon protocol. For the comparisons, we use the same cost metric defined in the Section 3.

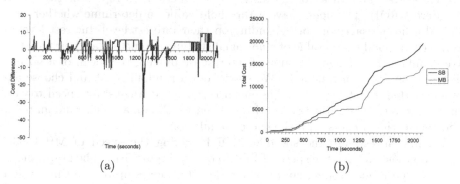

(a)   (b)

Fig. 4. (a) Difference between costs of SB and MB at any given time in a reference data set. (b) Total costs of SB and MB in a reference data set.

We first compare the performance of the Data Salmon protocol with the SB using data from a reference data set. Figure 4(a) shows the instantaneous cost difference between the SB and the MB approaches. In the figure, the areas below the $y = 0$ baseline show that the MB may become disadvantageous (albeit, briefly) due to some abrupt changes in data rate. Since the cost difference stay above the $y = 0$ baseline most of the time, we observe that the cost of MB is less

---

[2] An applet version of the simulator is available in
   http://www.cse.buffalo.edu/~osoysal/salmonSim/.

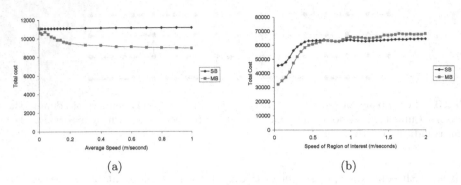

Fig. 5. (a) Total cost with respect to varying MB speed (b) Total cost with respect to varying region of interest speed

than that of SB. The cumulative of these differences, which gives us the total energy costs of each approach, are graphed in Figure 4(b).

Secondly, we investigate the effect of the MB speed. For chosen speed values the average of total cost for all data sets is shown in Figure 5(a). This graph shows that even with low speeds the MB approach can outperform the SB.

Since the emulation dataset does not lend itself well for controlling the data generation, we devised a second set of experiments using a synthetic data set. We chose the value of a normal distribution function, to represent the region of interest(ROI). The nodes have a threshold value to determine whether they should send messages or not depending on their interest level. Interest level of nodes correspond to the value of a normal distribution function at their position. We simulate the change in region of interest by moving the mean of a normal distribution function randomly. We start from a random point and choose another random point for mean. The mean is moved at a constant speed toward this random point until it reaches there. Then we choose another random point and repeat this process until the end of simulations.

We first investigate the speed of ROI by fixing the speed of MB to 0.4 m/second and varying the speed of ROI to obtain Figure 5(b). The graph shows that MB performs better than SB when the ROI moves up to two times faster than MB. After this point, MB cannot keep up with the sudden changes in ROI and falls beyond the SB case. Still the difference in performance is much less than the cases where ROI moves slowly.

We also replicated the experiments in Figure 4(a), Figure 4(b) and Figure 5(a) and obtained similar results. Synthetic results are more regular than regular data set which can be explained by the effect of the normal function used in modeling activation.

The Data Salmon protocol banks on the spatio-temporal nature of the data in WSNs. Through our experiments, we have seen that even for the rare cases this locality assumption does not hold, our protocol performs at least as good as the SB approach most of the time. For example, if the data rate is uniform throughout the network, our Data Salmon protocol fixes the location of the

MB in an optimal location in the center of the WSN and acts more like an SB approach. The Data Salmon performs worse than the SB approach only in extremely pathological cases, where an adversary lures the MB toward one corner of the network only to follow it up with a large but short-lived surge of data from the opposite corner.

5   Discussion

Fault-tolerance. The backbone tree structure we use does not provide any redundancy to the face of node failures. When one node goes down, the result is a partitioned network. Fortunately, there has been a lot of work on self-stabilizing tree maintenance protocols [20, 21, 22] that enable the tree to recover itself upon node failure or corruption of the pointer structure at the nodes. (This is, of course, provided that the network is not physically partitioned.)

After the backbone tree is fixed as discussed above, we also need to consider the recovery of the distributed-arrow protocol. In [23], it is shown that by adding some self-stabilizing actions, it is possible to achieve self-stabilization of the distributed arrow protocol in a local and efficient way. Finally, when the underlying static tree and the distributed-arrow protocol stabilizes, the Data Salmon protocol stabilize trivially by virtue of being stateless.

Another issue for fault-tolerance is the collision-free collection of data packets. Since the tree structure is commonly used for data collection, there has been several work on collision-avoidance protocols for tree structures [24, 25].

Load-balancing. The static backbone tree imposes a strain on the static *root* and core of the tree as there is always considerable amount of traffic routed through the core toward the MB. Improving load-balancing in the network and reducing the hot-spot in the core would help elongate the network lifetime. Existing work in reducing uneven energy consumption in WSNs by using a MB [26, 27] show that the optimum movement strategy for the MB is to follow the periphery of the network when the deployment area is circular. However, since these work assume uniform data generation by the sensor nodes every time unit, and reducing the average weighted multihop in the face of varying traffic is not a goal, these work are inapplicable in our context.

Modifying the backbone tree as the MB relocates may help reduce hot-spots in the tree. It is important to keep such modifications to be as localized around the MB as possible in order not to introduce excessive communication, hence, excessive energy-consumption into the WSN. A relatively local tree reconfiguration algorithm for bounded-length (4-5 hops) trees is presented in [28]. However, local reconfigurations alone are insufficient for maintaining a globally desirable tracking tree structure, and in the worst case local modifications may—over time—result in pathological cases where the height of the tree can be several orders of magnitude larger than the diameter of the network.

Relaxing the backbone tree structure by replacing it with a more permissive and load-balanced topology, such as a grid topology, may alleviate the hot-spot issue, as this allows multiple forwarding paths between any two points in the structure. Unfortunately, such a replacement introduces the problem of efficient tracking of the MB over the structure. Except for simple structures, such as a linear topology or a tree structure—as in our case—, designing update-efficient and local tracking protocols is a challenging problem. A tracking protocol for grid topology is investigated in [29] and several tracking protocols for more general network topologies are proposed in the literature [21, 30]. However, when adopting such an approach, it is unclear whether the overhead involved in tracking would be commensurate with the gains achieved from using a MB.

Multiple MB extension. An implication of our greedy protocol is that it is easy to parallelize the solution via divide and conquer: Adding more MBs to the network is easy since the MBs do not need to coordinate, yet each by optimizing its own gain implicitly cooperates to achieve a desirable global behavior. As a demonstration of this claim, we present a simple scheme to extend our Data Salmon protocol to support multiple MBs. This scheme is based on the observation that when there are multiple MBs on the backbone tree, the arrow protocol maintains a dynamic directed acyclic graph (DAG) structure with multiple sinks instead of a tree structure with one root. A DAG structure implies that some nodes in the backbone tree now have multiple outgoing edges. Our modification, then, is to divide the incoming traffic at a node in an equal manner among the outgoing edges of the node. The MBs decide on their relocation in a local, greedy manner as before and, as before, an edge direction is reversed when a MB traverses the edge. This simple protocol leaves it solely to the discretion of the MBs to sort out how to share the network traffic and is not optimal. Devising optimal solutions for the multiple MB case is part of our ongoing work.

6   Concluding Remarks

We presented a simple, low-latency, and energy-efficient protocol for data collection in WSNs using a network controlled MB. In contrast to the existing MB-based solutions where WSN nodes buffer data passively until visited by an MB, our protocol overlays a spanning backbone tree and maintains an always-on multihop connectivity to the MB by employing the distributed-arrow tracking protocol on top of this tree. This enables the nodes to forward their data to the MB anytime, in a timely, and efficient fashion avoiding latencies due to long-term buffering. Our protocol achieves energy-efficiency for the WSN by greedily relocating the MB toward the direction of the tree that produce higher data rates and, hence, reducing the average weighted multihop traffic. Using the convexity of the cost function in this problem, we were able to prove that our local greedy protocol also optimizes the network-wide energy-efficiency metrics. An implication of this local, greedy, and optimal protocol is that it is easy to parallelize the data collection by adding more MBs to the WSN.

Devising and proving optimal solutions for the multiple MB case is part of our ongoing work. In future work we will focus on relaxing the underlying static backbone tree structure by replacing it with a less restrictive variant, such as a grid structure. Another extension we will pursue for alleviating the hot-spots problem is the inclusion of the energy constraints and the remaining lifetime of nodes (in addition to the data rates) for the calculation of the cost function.

References

1. Akyildiz, I.F., Su, W., Sankarasubramaniam, Y., Cayirci, E.: A survey on sensor networks. IEEE Communications Magazine 38, 393–422 (2002)
2. Estrin, D., Govindan, R., Heidemann, J.S., Kumar, S.: Next century challenges: Scalable coordination in sensor networks. In Mobile Computing and Networking, pp. 263–270 (1999)
3. Tolle, G., Polastre, J., Szewczyk, R., Turner, N., Tu, K., Buonadonna, P., Burgess, S., Gay, D., Hong, W., Dawson, T., Culler, D.: A macroscope in the redwoods. In: Proceedings of the Third ACM Conference on Embedded Networked Sensor Systems, pp. 51–63 (2005)
4. Mainwaring, A., Polastre, J., Szewczyk, R., Culler, D., Anderson, J.: Wireless sensor networks for habitat monitoring. In: ACM Int. Workshop on Wireless Sensor Networks and Applications, pp. 88–97 (2002)
5. Arora, A., et al.: A line in the sand: A wireless sensor network for target detection, classification, and tracking. Computer Networks (Elsevier) 46(5), 605–634 (2004)
6. Arora, A., et al.: Exscal: Elements of an extreme scale wireless sensor network. In: 11th IEEE International Conference on Embedded and Real-Time Computing Systems and Applications, pp. 102–108 (2005)
7. Batalin, M., Rahimi, M., Yu, Y., Liu, D., Kansal, A., Sukhatme, G., Kaiser, W., Hansen, M., Pottie, G., Srivastava, M., Estrin, D.: Call and response: experiments in sampling the environment. In: Proceedings of the 2nd international conference on Embedded networked sensor systems, pp. 25–38 (2004)
8. Yu, Y., Ganesan, D., Girod, L., Estrin, D., Govindan, R.: Synthetic data generation to support irregular sampling in sensor networks. In: Geo Sensor Networks, Taylor and Francis Publishers (October 2003)
9. Pattem, S., Krishnamachari, B., Govindan, R.: The impact of spatial correlation on routing with compression in wireless sensor networks. In: Proceedings of the third int. symposium on Information processing in sensor networks, pp. 28–35 (2004)
10. Krishnamachari, B., Estrin, D., Wicker, S.B.: The impact of data aggregation in wireless sensor networks. In: Proceedings of the 22nd International Conference on Distributed Computing Systems, pp. 575–578 (2002)
11. Shah, R.C., Roy, S., Jain, S., Brunette, W.: Data mules: modeling a three-tier architecture for sparse sensor networks. In: Proceedings of the First IEEE International Workshop on Sensor Network Protocols and Applications, pp. 30–41 (2003)
12. Somasundara, A., Ramamoorthy, A., Srivastava, M.: Mobile element scheduling for efficient data collection in wireless sensor networks with dynamic deadlines. In: Proceedings of the 25th IEEE International Real-Time Systems Symposium, pp. 296–305 (2004)
13. Gu, Y., Bozdag, D., Ekici, E., Ozguner, F., Lee, C.: Partitioning based mobile element scheduling in wireless sensor networks. In: IEEE SECON, pp. 386–395 (2005)

14. Zhao, W., Ammar, M.: Message ferrying: Proactive routing in highly-partitioned wireless ad hoc networks. In: Proceedings of the The Ninth IEEE Workshop on Future Trends of Distributed Computing Systems, pp. 308–314 (2003)
15. Cone, L.L.: Skycam: An aerial robotic camera system. Byte. 10, 122–132 (1985)
16. Pon, R., Batalin, M., Gordon, J., Kansal, A., Liu, D., Rahimi, M., Shirachi, L., Yu, Y., Hansen, M., Kaiser, W., Srivastava, M., Sukhatme, G., Estrin, D.: Networked infomechanical systems: a mobile embedded networked sensor platform. In Information Processing in Sensor Networks, pp. 376–381 (2005)
17. Kansal, A., Rahimi, M., Estrin, D., Kaiser, W.J., Pottie, G., Srivastava, M.: Controlled mobility for sustainable wireless sensor networks. In Sensor and Ad Hoc Communications and Networks, pp. 1–6 (2004)
18. Demmer, M.J., Herlihy, M.: The arrow distributed directory protocol. In: Proceedings of the 12th International Symposium on Distributed Computing, pp. 119–133 (1998)
19. Cao, H., Ertin, E., Kulathumani, V., Sridharan, M., Arora, A.: Differential games in large-scale sensor-actuator networks. In: Proceedings of the fifth international conference on Information processing in sensor networks, pp. 77–84 (2006)
20. Dolev, S.: Self-Stabilization. MIT Press, Cambridge (2000)
21. Demirbas, M., Arora, A., Gouda, M.: Pursuer-evader tracking in sensor networks. In: Sensor Network Operations, IEEE Press, New York (2006)
22. Chen, N., Huang, S.: A self-stabilizing algorithm for constructing spanning trees. Information Processing Letters (IPL) 39, 147–151 (1991)
23. Herlihy, M., Tirthapura, S.: Self-stabilizing distributed queueing. In: Proceedings of 15th International Symposium on Distributed Computing, pp. 209–219 (October 2001)
24. Woo, A., Culler, D.E.: A transmission control scheme for media access in sensor networks. In: Proceedings of the 7th annual international conference on Mobile computing and networking, pp. 221–235 (2001)
25. Kulkarni, S.S., Arumugam, M.: Ss-tdma: A self-stabilizing mac for sensor networks. In: Sensor Network Operations, IEEE Press, Los Alamitos (2005)
26. Gandham, S., Dawande, M., Prakash, R., Venkatesan, S.: Energy-efficient schemes for wireless sensor networks with multiple mobile base stations. In: Proceedings of IEEE GLOBECOM, pp. 377–381 (2003)
27. Wang, Z., Basagni, S., Melachrinoudis, E., Petrioli, C.: Exploiting sink mobility for maximizing sensor networks lifetime. In: Proceedings of the 38th Annual Hawaii International Conference on System Sciences, pp. 287a–287a (2005)
28. Zhang, W., Cao, G.: Dctc: Dynamic convoy tree-based collaboration for target tracking in sensor networks. IEEE Transactions on Wireless Communication 3(5), 1689–1701 (2004)
29. Ye, F., Luo, H., Cheng, J., Lu, S., Zhang, L.: A two-tier data dissemination model for large-scale wireless sensor networks. In: Proceedings of the 8th annual international conference on Mobile computing and networking, pp. 148–159 (2002)
30. Demirbas, M., Arora, A., Nolte, T., Lynch, N.: A hierarchy-based fault-local stabilizing algorithm for tracking in sensor networks. In: Higashino, T. (ed.) OPODIS 2004. LNCS, vol. 3544, pp. 299–315. Springer, Heidelberg (2005)

SDIP³: Structured and Dynamic Information Push and Pull Protocols for Distributed Sensor Networks

Ying Zhang and Qingfeng Huang

Palo Alto Research Center Inc.,
3333 Coyote Hill Rd,
Palo Alto, CA 94304, USA
{yzhang,qhuang}@parc.com
http://www.parc.com/yzhang,
http://www.parc.com/qhuang

Abstract. We propose and study a class of structured and dynamic information push and pull protocols for wireless sensor networks. For structured information dissemination, our study focuses on the impact of various information demand characteristics on dissemination along some type of backbone structures. Our exploration of dynamic information push and pull focuses on finding optimal strategies in a distributed manner without prior knowledge of information demand characteristics and/or with heterogeneous query distributions. Our theoretical analysis uses a simple grid structure, but the protocol is applicable to arbitrary network topologies. A distributed traffic information system is used as the context of study and the simulation study uses a microscopic traffic simulator to demonstrate some of the ideas discussed in the paper.

Keywords: Distributed algorithms for collaborative information processing, Communication and processing primitives.

1 Introduction

The goal of this work is to advance the understanding of optimal information push and pull strategies for wireless sensor networks, with and without the knowledge about the spatial and temporal characteristics of information demand. Since data gathering and provision are sensor networks' main functions, efficient information gathering protocols and information dissemination/query/discovery protocols ([1] [2] [3] [4] [5]) are essential in building efficient sensor network applications. In this paper we focus on distributed information dissemination and discovery protocols.

This work is directly motivated by the following work and observations:

- Oftentimes query interest and frequency of an application exhibits location-varying characteristics. For instance, in an intelligent transportation system scenario, a vehicle could be more interested in traffic information on its

J. Aspnes et al. (Eds.): DCOSS 2007, LNCS 4549, pp. 281–294, 2007.

potential paths to the destination and a piece of information about traffic jam could be queried more frequently from those roads closer by than those further away.

- Previous work [2][6] have shown that knowledge about query frequency and data event frequency can be used to create efficient push-pull balanced information discovery and dissemination protocols. However, in these earlier work, query frequency is uniform or homogeneous across nodes/space;
- If the non-uniform spatial distribution of information demand is known, one maybe able to further optimize push-pull strategies;
- When good estimation about spatial distribution of query frequency does not exist a priori, information dissemination structure should try to adapt to ongoing event and query distribution patterns for improved efficiency.

As a result, our investigation starts from exploring how to use potential knowledge about spatial distribution of information demand, and investigating what to do when such information is not available.

The contributions of this paper include:

- A class of structured information push-pull protocols for applications that have spatially non-uniform information demand;
- Theoretical results for the performance of these protocols regarding optimal push-pull balance, given a variety of spatial distribution of query frequencies;
- A distributed dynamic push-and-pull protocol for scenarios where good estimation about spatial distribution of query frequency is not known a priori, and/or the query distribution is heterogeneous.

Comparing to related work in information dissemination area, we address the optimization perspective with the consideration of query frequency and distribution which is rather unique to our knowledge. Much related research on distributed query efficiency [6] [7] has been done in the last couple of years, including "comb-needle" [2], DIM [3], GHT [4] and DIMENSIONS [8]. However, all these related work only explicitly or implicitly focused spatially uniform queries. The same if true for another related line of work points to potential of reducing dissemination/query cost by taking into account application semantics [5,9,10] and the study of fundamental scaling laws on energy-efficient storage and querying such as [11]. Moreover, related work on tracking (STALK [12], LLS [13]) explored multi-resolution structures; and recent work on distance-sensitive information services ([14], Trail [15], Glance [16]) studied optimal structures for information dissemination. However, none of them has considered query distribution or frequencies either.

The rest of this paper is organized as follows. Section 2 provides the problem formulation. Section 3 presents some theoretical results on a grid structure. Section 4 describes the dynamic information push and pull protocols. Section 5 focuses on the traffic network application with simulation validation, followed by conclusions in Section 6.

2   Problem Formulation

Systematic analysis of the push-pull strategies relies on the underlying system, protocol, event generation and demand distribution models. In this section we define these models in our study.

2.1   System Model

We consider two topologies in our model of the distributed information system: a *information relevance network topology* and a *communication network topology*. In the example of distributed traffic information system, we represent each road segment as a node in the information relevance network, node X has a directed link to node Y if and only if the segment represented by Y is connected to the segment represented by X in the road traffic topology. Each node in the communication network represent a communication interface node and a link between two nodes represent their ability to communicate directly. Fig. 1 shows an example of a road configuration with traffic sensors distributed on road segments. Fig. 2 only shows one intersection in Fig. 1 with detailed road connectivity, which determines the connectivity in the corresponding information relevance network. $L(i, j)$ is a notation for a node in the information network, representing the road segment from point i to point j. $L(i, j) \rightarrow L(j, k)$ if there is legal turn at j. The example of communication network is shown on the right of Fig. 1 where an arrow from node i to node j indicates that node j can hear from node i. Note that although node 33 has no road connection to node 47, it can communicate with it.

Fig. 1. A traffic information network, circles indicate locations of sensor nodes

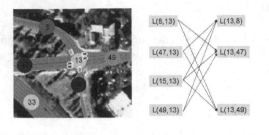

Fig. 2. Left: Traffic/road network abstraction; Right: Communication structure

For simplicity, in this paper we assume both the information network and the communication network are fixed.

2.2 Push and Pull Model

Our structured information dissemination strategy combines pro-active push and reactive pull. The push and pull model is formally defined as follows.

Push. The structured information push model is specified by: $\langle m, R \rangle$ where topological constraint R defines the recipient set of the pro-active information push, m is a message that can be further defined as a 4-tuple: $m = \langle s_i, i, t, \tau \rangle$ where

- s_i is the ID of the information source,
- i is the information of that source,
- t is the time that the information is obtained,
- τ is the expiration time. When $\tau = \infty$, it is event driven, i.e., it will not be expired until the next publication.

One simple example constraint of R is the maximum "distance" the information should be pushed. Interestingly, for a traffic network, a meaningful distance may be represented by time, e.g., T seconds away (given current, historical or nominal vehicle speed information on relevant roads) to the source road segment.

More complicated constraint for R could represent fine grain topological structures that we may call "push backbone". Fig. 3 is an example of such push backbones. In this example, the source of information is located at the center, the structure is sparser when further away from the source. As one can see later, this kind of multi-resolution push structure would help increase the efficiency of information dissemination when information demand density have a distance varying distribution.

A push process can be carried out as follows. For simplicity in description, let us assume that the nodes within constraint R of a source are directly or

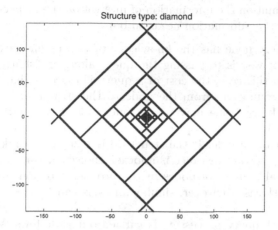

Fig. 3. A multi-resolution diamond structure

transitively connected to the source. Each node use a list to keep track of the most recent updates it has heard about. When a node n receives a new piece of information $\langle s_i, i, t, \tau, R \rangle$, it checks its constraints in R: if R is not satisfied, or it is expired (i.e., $t + \tau$ is smaller than the current time), then the message is dropped, otherwise rebroadcast $\langle s_i, i, t, \tau, R \rangle$ and:

- if s_i does not exist in the list, add the entry with $\langle s_i, i, t, \tau \rangle$,
- if s_i exists in the list with smaller t, update the entry with $\langle s_i, i, t, \tau \rangle$.

Note that as we assume the recipient set R and the source are self-connected without other relay nodes, this simple implementation guarantees that all recipients within the expiration time bound will receive the information. If the communication links are not reliable, one can increase the thickness of the lines in the structure [2] to achieve the desired level of reliability. Note that in this implementation, nodes not in R may also overhear the information, in that case they shall stored the information in their lists as well if memory is less expensive than communication cost.

In our cost model, we assume each transmission costs one unit. When R is connected, the cost of dissemination is bounded by the number of nodes in R, i.e., each node in R transmits at most once. If a routing tree such as a connected minimum dominating set tree is built a priori, the number of transmissions can be less than the number of nodes in R, since leaf nodes do not need to transmit. For simplicity in our analysis, we assume the cost of dissemination be the number of nodes in R.

Pull. The query we are concerned with is specified by a triple: $\langle n_q, s_i, c \rangle$ where

- n_q is the ID of the node generating the query,
- s_i is the source of the information that is queried,

- c is a constraint on the information of interest on the respective source, (e.g, how "fresh" the information needs to be).

If we assume every node has the knowledge of the dissemination structure R, the most efficient way is to passing the query along the shortest hop-count to the structure. In this case, the cost of a query is twice of that hop-count since a query goes a round trip from the node to the structure. If the structure is not completely known, one may pass a query along a "needle" that may hit the structure within certain hop-counts, and the cost of a query is proportional to the length of the needle. If the structure is completely unknown, a limited broadcast within certain distance is another choice. In this case, the cost of a query is the total number of nodes in the broadcast region. In this paper, we focus on the first case, however, similar analysis can be easily carried out for other cases.

The in-network query processing is carried out as follows. When a network node receives a query $\langle n, s_i, c \rangle$, it checks to see if it has relevant data about source s_i that satisfies the constraint c in the source list. If the information does not exist locally, the query will be forwarded towards the backbone. Along the way to its destination, as soon as the query hits a node that has the piece of information, a reply $\langle s_i, i, \min(\mathbf{t}, t + \tau), 0 \rangle$, where $\mathbf{t}$ is the current time, is generated and propagated along the shortest path from this node to the origin of the query. Each node that receives or overhears the reply will update its source list with the information.

2.3   Event and Demand Distribution Models

We assume source s_i have event generation frequency f_i. The information demand frequency for s_i may be specified by a function $\mathbf{f_q} : X \to \mathcal{R}^+$, where X is a set of locations relative to the source of information s_i, $\mathcal{R}^+$ is the set of non-negative real numbers, and the value $\mathbf{f_q}(\bar{x})$ represents query frequency about information s_i from location $\bar{x}$. For simplicity, we start from a omni-directional version where query frequency is only related to the information distance from the query node to the source, i.e., $\mathbf{f_q} = \mathbf{f_q}(|\bar{x}|)$. We will use three types of frequency models:

- **uniform:** the frequencies at every location are a constant, i.e., $\mathbf{f_q}(d) = f_q$;
- **linear:** the frequencies are decreasing linearly with respect to the distance, i.e., $\mathbf{f_q}(d) = f_q(1 - kd)$ where $k > 0$ is a constant;
- **power:** the frequencies are inversely related to the distance, i.e., $\mathbf{f_q}(d) = f_q/d^\alpha$, where $\alpha > 0$ is a constant.

All three forms can be written as $\mathbf{f_q}(d) = f_q q(d)$ where f_q is a constant, and $q(d)$ is a function of d.

2.4   Formulation of the Optimization Problem

Our objective is, given the knowledge about the demand distribution and event generation frequency, find a push backbone structure R that minimize the overall communication cost. Let $C_{push}(R)$ be the cost of propagating one data packet from the source node on R; let $C_{pull}(R, \bar{x})$ be the query cost for a node located at $\bar{x}$ to find out the information from R. Note that this query cost depends on the push backbone structure R. The denser the push structure, the smaller the average query cost. With these definitions, the total push-pull cost rate (number of data packet per second) related to information s_i can be expressed as follows:

$$Cost(R) = f_i C_{push}(R) + \int_X C_{pull}(R, \bar{x}) \mathbf{f_q}(\bar{x}) d\bar{x} \qquad (1)$$

The optimization problem can be formally expressed as: find an optimal push structure R_{opt} which minimizes the cost $Cost(R)$, i.e.

$$R_{opt} = \min arg_R Cost(R) \qquad (2)$$

Since there is no obvious way to express in analytical form for the "structure" R, in this paper, we take the approach of exploring the optimal parameters within a specific class of structures.

3   Theoretical Analysis

In this section, we will do some preliminary analysis of the structures described in the previous section, and find out the closed-form solutions of the optimal parameters for the structures, with respect to three different demand distribution models: **uniform**, **linear** and **power**, using the following two structures. Let $\bar{x} = (x, y)$:

- **disk:** the only constraint of this structure is the push scope R_i, i.e., $d \le R_i$ where $d = |x| + |y|$ for a grid network;
- **diamond** (Fig. 3): let r be the size of the base diamond, $b_+ = sign(x + y)\lceil r^{\lfloor \log_r |x+y| \rfloor} \rceil$, $b_- = sign(x - y)\lceil r^{\lfloor \log_r |x-y| \rfloor} \rceil$, $c_+ = \lceil r^{\lfloor \log_r |x+y| \rfloor + 1} \rceil$, and $c_- = \lceil r^{\lfloor \log_r |x-y| \rfloor + 1} \rceil$, the diamond structure can be represented by:

$$|b_+ - (x + y)| \le 1 \wedge |x + y| \le c_+ \vee |b_- - (x - y)| \le 1 \wedge |x - y| \le c_-.$$

Let the information generation and query frequencies be f_i and $\mathbf{f_q}$, respectively. Let $\mathbf{f_q}(d) = f_q q(d)$. Let R_i and R_q be the push and pull scopes, respectively. C_{push} is equal to the total number of nodes in R. $C_{pull}(R, \bar{x})$ is the number of transmissions of a query at $\bar{x}$, which is twice of the distance from $\bar{x}$ to the structure R. Given R is omni-directional, we have $C_{pull}(R, \bar{x}) = C_{pull}(d)$ where $d = |\bar{x}|$. The total cost for information push is $C_{push} = f_i C_{push}$ and the total cost of information pull is $C_{pull} = f_q C_p$ where $C_p = \Sigma_{d=0}^{R_q} q(d) C_{pull}(d) 4d$. Here $4d$ is the number of nodes at distance d for a grid network. For each case, we estimate C_{push} and C_{pull} and then find the optimal parameter by solving equation $\frac{\partial(C_{pull} + C_{push})}{\partial p} = 0$ where p is a parameter of the structure R.

3.1   Optimal Push Scopes for Disk Structures

For a disk structure with push scope R_i, it is easy to see that the total number of nodes in R is about $C_{push} = 2R_i^2$, and $C_{pull}(d) = 0$ if $d \le R_i$ and $C_{pull}(d) = 2(d - R_i)$ otherwise, where the factor 2 comes from the fact that query is a process involving a round trip.

Uniform $(q(d) = 1)$

$$C_p = \sum_{d=R_i}^{R_q} 2(d - R_i)4d = 8 \sum_{d=R_i}^{R_q} d(d - R_i) \sim 8((R_q^3 - R_i^3)/3 - R_i(R_q^2 - R_i^2)/2)$$

Let $\frac{\partial(C_{pull} + C_{push})}{\partial R_i} = 0$, we have an optimal push and pull balance scope condition, by $f_i R_i = f_q(R_q^2 - R_i^2)$ We can see that if f_q or f_q/f_i approaches 0, R_i should approach 0, i.e., no push is needed. On the other hand, if f_i or f_i/f_q approaches 0, $R_i \to R_q$, i.e., push as far as the query scope.

Linear $(q(d) = 1 - d/R_q)$

$$C_p = \sum_{d=R_i}^{R_q} 2(d - R_i)4d(1 - d/R_q) = 8 \sum_{d=R_i}^{R_q} d(d - R_i) - \sum_{d=R_i}^{R_q} d^2(d - R_i)/R_q$$
$$\sim 8((R_q^3 - R_i^3)/3 - R_i(R_q^2 - R_i^2)/2) - 8((R_q^4 - R_i^4)/4 - R_i(R_q^3 - R_i^3)/3)/R_q$$

Let $\frac{\partial(C_{pull} + C_{push})}{\partial R_i} = 0$, we have an optimal push and pull balance scope condition, by $f_i R_i = f_q(R_q^2 - R_i^2) - \frac{2f_q}{3R_q}(R_q^3 - R_i^3)$ It is easy to see that $R_i^{linear} < R_i^{uniform}$.

Power $(q(d) = 1/d)$

$$C_p = \sum_{d=R_i}^{R_q} 2(d - R_i)4d/d = 8 \sum_{d=R_i}^{R_q} (d - R_i) \sim 8((R_q^2 + R_i^2)/2 - R_i R_q)$$

Let $\frac{\partial(C_{pull} + C_{push})}{\partial R_i} = 0$, we have an optimal push and pull balance scope condition, by $f_i R_i = 2f_q(R_q - R_i)$ i.e. $R_i = \frac{R_q}{1 + f_i/(2f_q)}$. It is easy to see that $R_i^{power} < R_i^{uniform}$.

3.2   Optimal Diamond Size

In this case, we assume the push scope is as far as the query scope, i.e., $R = R_i = R_q$, and find out the optimal spacing parameter given a diamond structure (Fig. 3). Note that the line thickness can vary; in most cases, it has to be at least 2 for guarantee the connectivity of the structure. In the analysis below, we use β for the line thickness in the structure. Given the the base diamond size r, the cost of push for each dissemination is

$$C_{push} = \beta \sum_{k=0}^{\lfloor log_r R \rfloor} 4r^k \sim \beta 4R/(r-1)$$

Here we use $r^{\lfloor log_r R \rfloor} \sim R$ to simplify the analysis. The spacing between two lines at distance d is $r^{\lfloor log_r d \rfloor + 1} - r^{\lfloor log_r d \rfloor} \sim (r-1)d$. The average pull cost for a query $C_{pull}(d)$ is about $2(L/4) = L/2$ where $L = (r-1)d$. Note that for this structure the pull cost is increasing with respect to the distance.

Uniform ($q(d) = 1$)

$$C_p = \sum_{d=0}^{d=R} 4d(r-1)d/2 \sim \frac{2}{3}R^3(r-1)$$

$\frac{\partial(C_{pull}+C_{push})}{\partial(r-1)} = 0$, yields an optimal diamond size, by $6\beta f_i = f_q R^2 (r-1)^2$, i.e.
$r - 1 = \frac{\sqrt{6\beta f_i/f_q}}{R}$.

Linear ($q(d) = 1 - d/R$)

$$C_p = \sum_{d=0}^{d=R} 2d^2(r-1)(1 - d/R) \sim \frac{2}{3}R^3(r-1) - \frac{1}{2}R^3(r-1) \sim \frac{1}{6}R^3(r-1)$$

$\frac{\partial(C_{pull}+C_{push})}{\partial(r-1)} = 0$, yields $24\beta f_i = f_q R^2 (r-1)^2$, i.e., $r - 1 = 2\frac{\sqrt{6\beta f_i/f_q}}{R}$.

Power ($q(d) = 1/d$)

$$C_p = \sum_{d=0}^{d=R} 2d^2(r-1)/d \sim R^2(r-1)$$

$\frac{\partial(C_{pull}+C_{push})}{\partial(r-1)} = 0$, yields $4\beta f_i = f_q R(r-1)^2$, i.e., $r - 1 = 2\sqrt{\beta \frac{f_i/f_q}{R}}$.

It is easy to see that $r^{linear} \sim 2r^{uniform}$ and $r^{power} \sim \sqrt{\frac{R}{6}} r^{linear}$.

3.3 Push Scope Verification

We use *load factor* as a performance metric, which is the total amount of pushes and pulls for a piece of information, normalized by $n(f_q + f_i)$ where n is the number of nodes involved.

In the study, we assume the pull scope to be 150 hop-count to a source in a grid network. Let $f_i = 1$ without lose of generality, and set f_q 0.01 and 0.05

for **uniform**, and 0.05 and 0.1 for **linear**, and 0.1 and 0.2 for **power** models, respectively.

The load factor is computed using Eq. 1, where $C_{push}(R)$ is the number of nodes in R and $C_{pull}(R, \bar{x})$ is $2h$ where h is the minimum hop-count from $\bar{x}$ to structure R.

Fig. 4 show the load factors using the **disk** structure, for **uniform** and **power** query models, respectively.

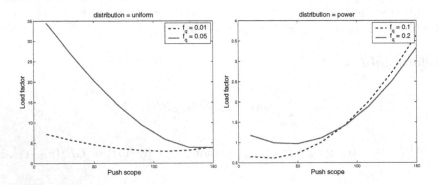

Fig. 4. Load factors for (Left) the uniform and (Right) the power query models

From the theoretical analysis, we know $f_i R_i = f_q(R_q^2 - R_i^2)$ for the uniform query model. Given $f_i = 1$ and $f_q = 0.05$, and $R_q = 150$, we obtain $R_i = 108$ which is consistent to the left curve in Fig. 4. Similarly, for $f_q = 0.2$, we have optimal $R_i = R_q/(1 + f_i/(2f_q)) = 43$ for the power model, which is consistent to the right curve in Fig. 4.

4   Dynamic Balancing of Push and Pull

In the previous section, we see that the average-case optimal values of a information push structure depends on two key factors: the rate of information generation and the rate of request for the information. Once one collects the historical information regarding these two rates in the information network, one can determine the average-case optimal push parameters for expected rates of information generation and request.

This average-case optimality can be further improved by capturing and responding to the fluctuations in the information generation and request rates in real-time. We propose a distributed dynamic optimization scheme called *micro-balancing* to achieve this goal. Given the recipient set R, the micro-balancing scheme reactively changes R to minimize the total cost of information sharing. The rest of this section describes this mechanism. Such a scheme can be also applied to cases where the query distribution is not omni-directional or unknown at all.

4.1 Boundary States

We use two boolean values to denote the boundary of push dissemination. Assume the push structure forms a spanning tree rooted at the source. A node in the structure is in state 1 if it needs to forward the information and in state 0 if it does not need to forward the information. The set of nodes in state 1 is called *active* nodes. A node in the structure is in boundary state 0 if it is in 0 but its parent is 1. A node in the structure is in boundary state 1 if it is in 1 but all its children in the structure are in 0.

For micro-balancing, the push boundary can be changed dynamically according to the push and pull rates. To obtain the push and pull rates, nodes in both boundary states count the number of push events and the number of queries. Let N_q be the number of different queries a node receives, and N_p be the number of different "pushed" information a node receives. It is called *information-dominate* if $\frac{N_p}{N_q} > 2 + \sigma$ and *query-dominate* if $\frac{N_p}{N_q} < 2 - \sigma$ where σ is a small value parameter for the purpose of reducing the oscillation in state-change condition boundary to reduce potential protocol overhead. The insight that leads to the inequality is a simple one: a query and a reply need two uses of the wireless medium and a push only need one local broadcast. To make the switch between these two conditions more stable for small N_q and N_p, we compute the inequalities only after $N_q + N_p > \Delta$ for some constant Δ.

4.2 Boundary State Transition

The rules for state transition are as follows:

- $0 \rightarrow 1$: if it becomes query-dominate,
- $1 \rightarrow 0$: if it becomes information-dominate

Whenever a node in the boundary state changes its state, its neighbor's state will be changed. For example, if a node changes to state 1, all its children become boundary state 0; if a node changes to 0, its parent may be a boundary node 1. When a node enters the boundary state, both N_p and N_q are reset to 0. It is easy to see that the state transition guarantees that the set of active nodes are connected.

Note that micro-balancing will not change the push structure, but it changes the active nodes in the structure, forming a more efficient push structure that is optimal for an unpredictable query model.

The push operation is the same as before except each node at the structure need to check if it is active or not. Only active nodes participate in the push.

5 Application: Distributed Traffic Information Networks

5.1 Traffic Network Simulations

We have implemented and simulated the push-pull protocols using a commercial on-the-shelf microscopic traffic simulator Paramics [17]. Fig. 1 shows a traffic network in Paramics Simulator (demo5 in Paramics software distribution) that we use for testing our protocol. The area of the network is 1466.6m × 1102.3m. This network has 58 intersections and 118 road segments. Our protocol and event/query generations model are implemented as a plug-in to the simulator. We assume each road segment generate some events with probability p_e at every second, and each car at a road segment with a probability p_q queries another road segment that it may travel within a time bound T_q. Parameters p_e and p_q are used to control the event and query frequencies. The communication range is set to be 300m. The simulation data are collected within a time window $[10s, 1010s]$. We record the total number of transmissions for pull and push at each node. The performance metrics is the total number of transmissions, i.e., the sum of pushes and pulls over the whole network.

A shortest path routing table is built up at the initialization stage for query propagation. For the static push protocol, any piece of new information will travel up to nodes that are within T_i seconds away from the source along the road network, where T_i is a parameter set at initialization. For the dynamic push, we set σ be 0.1 and Δ be 5. Since the network is small, we set $T_q = 100$ which effectively includes all the links when querying at any node. We also assume information to be event-driven, i.e., $\tau = \infty$. And query constraint c is a timeout that equals to the travel time to the location of the information. Note that in this simulation, we did not use linear or power query models since the size of the traffic network is too small. A larger size simulation will be conducted in the future.

Our implementation also includes visualization of communication. In Fig. 1, for example, colors of circles indicate push/pull ratios, where red is pull dominated and blue is push dominated.

5.2   Simulation Results

The purpose of the simulation is to verify our theoretical analysis about push and pull scope, and show the effectiveness of the dynamic push and pull protocol. Three cases are studied:

  I. Information dominate ($p_e = 0.01, p_q = 0.03$): in this case push scope should be small;
 II. Query dominate ($p_e = 0.001, p_q = 0.3$): in this case push scope should be as large as pull scope;
III. Neither ($p_e = 0.01, p_q = 0.3$): in this case push scope should be somewhere in the middle.

First we test the optimal scopes for the static push protocol. We set the push scope to be small $T_i = 5$, median $T_i = 15$, and large $T_i = 30$, for each case. A total of nine sets of data are collected and compared. Figure 5 Left shows the results. We observe that number of pushes increases and number of pulls decreases with the increase of the push scope, and for case I, push scope small is most effective, for case II, push scope large is most effective, and for case III the ideal push scope is somewhere in the middle.

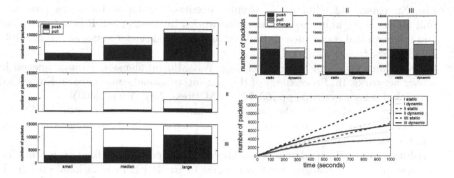

Fig. 5. Left: Number of transmissions for three cases and three push scope settings, Right: Static vs. dynamic push strategies

We also tested the effectiveness of the dynamic push protocol. We set the initial push scope $T_i = 15$ for each case, then collect data for both static and dynamic push strategies. A total of six sets of data are collected and compared. Figure 5 Right shows the results. Note that in addition to pushes and pulls, the dynamic protocol introduces the extra state change cost, each of which costs one transmission.

We observe that in all these cases the dynamic protocol reduces the total cost by about 50%. There are two reasons that the dynamic protocol outperforms the static one: (1) when the pull frequencies are unknown, the dynamic protocol can adjust to the push scope to balance the demands and changes, and (2) when the pull frequencies are not uniformly distributed over all directions, the dynamic protocol can adjust the push scope locally at different directions.

6   Conclusions

In this paper we proposed a class of the structured information push and pull protocols. We first analyzed these protocols using various query models and then presented dynamic micro-balancing strategies when the query models are unknown. We also applied this class of protocols to the distributed traffic information system and demonstrated the benefit of push-pull micro-balancing.

References

1. Intanagonwiwat, C., Govindan, R., Estrin, D.: Directed diffusion: a scalable and robust communication paradigm for sensor networks. In: Proceedings of the Fifth Annual ACM/IEEE International Conference on Mobile Computing and Networking, pp. 56–67 (2000) [Online]. Available `citeseer.ist.psu.edu/intanagonwiwat00directed.html`

2. Liu, X., Huang, Q., Zhang, Y.: Combs, needles, haystacks: balancing push and pull for discovery in large-scale sensor networks. In: SenSys '04: Proceedings of the 2nd international conference on Embedded networked sensor systems, pp. 122–133. ACM Press, New York (2004)
3. Li, X., Kim, Y.J., Govindan, R., Hong, W.: Multi-dimensional range queries in sensor networks. In: Proceedings of the first international conference on Embedded networked sensor systems, pp. 63–75. ACM Press, New York (2003)
4. Ratnasamy, S., Karp, B., Shenker, S., Estrin, D., Govindan, R., Yin, L., Yu, F.: Data-centric storage in sensornets with GHT, a geographic hash table. Mob. Netw. Appl. 8(4), 427–442 (2003)
5. Krishnamachari, B., Heidemann, J.: Application-specific modelling of information routing in sensor networks. In: Proceedings of the IEEE International on Performance, Computing, and Communications Conference, pp. 717–722. IEEE, Phoenix, Arizona, USA (2004)
6. Kapadia, S., Krishnamachari, B.: Comparative analysis of push-pull query strategies for wireless sensor networks. In: Gibbons, P.B., Abdelzaher, T., Aspnes, J., Rao, R. (eds.) DCOSS 2006. LNCS, vol. 4026, Springer, Heidelberg (2006)
7. Madden, S., Franklin, M.J., Hellerstein, J.M., Hong, W.: The design of an acquisitional query processor for sensor networks. In: Proceedings of the 2003 ACM SIGMOD international conference on on Management of data, pp. 491–502. ACM Press, New York (2003)
8. Ganesan, D., Estrin, D., Heidemann, J.: DIMENSIONS: Why do we need a new data handling architecture for sensor networks? In: Proceedings of the First Workshop on Hot Topics In Networks (HotNets-I), Princeton, NJ (October 2002)
9. Heidemann, J., Silva, F., Estrin, D.: Matching data dissemination algorithms to application requirements. In: Proceedings of the first international conference on Embedded networked sensor systems, pp. 218–229. ACM Press, New York (2003)
10. Sadagopan, N., Krishnamachari, B., Helmy, A.: Active query forwarding in sensor networks. Elsevier Journal of Ad Hoc Networks (2003)
11. Ahn, J., Krishnamachari, B.: Fundamental scaling laws for energy-efficient storage and querying in wireless sensor networks. In: Proceedings of International Symposium on Mobile Ad Hoc Networking and Computing (MobiHoc) (May 2006)
12. Demirbas, M., Arora, A., Nolte, T., Lynch
13. Abraham, I., Dolev, D., Malkhi, D.: Lls: a locality aware location service for mobile ad hoc networks. In: Workshop on Discrete Algothrithms and Methods for MOBILE Computing and Communications (2004)
14. Funke, S., Guibas, L.J., Nguyen, A., Wang, Y.: Distance-sensitive routing and information brokerage in sensor networks. In: Gibbons, P.B., Abdelzaher, T., Aspnes, J., Rao, R. (eds.) DCOSS 2006. LNCS, vol. 4026, pp. 234–251. Springer, Heidelberg (2006)
15. Kulathumani, V., Arora, A., Demirbas, M., Sridharan, M
16. Demirbas, M., Arora, A., Kulathumani, V
17. Paramics website. [Online]. Available: http://www.paramics-online.com/

Efficient Computation of Minimum Exposure Paths in a Sensor Network Field

Hristo N. Djidjev[1,*]

Los Alamos National Laboratory, Los Alamos, NM 87545

Abstract. The exposure of a path p is a measure of the likelihood that an object traveling along p is detected by a network of sensors and it is formally defined as an integral over all points x of p of the sensibility (the strength of the signal coming from x) times the element of path length. The minimum exposure path (MEP) problem is, given a pair of points x and y inside a sensor field, find a path between x and y of a minimum exposure. In this paper we introduce the first rigorous treatment of the problem, designing an approximation algorithm for the MEP problem with guaranteed performance characteristics. Given a convex polygon P of size n with $O(n)$ sensors inside it and any real number $\varepsilon > 0$, our algorithm finds a path in P whose exposure is within an $1 + \varepsilon$ factor of the exposure of the MEP, in time $O(n/\varepsilon^2 \psi)$, where ψ is a topological characteristic of the field. We also describe a framework for a faster implementation of our algorithm, which reduces the time by a factor of approximately $\Theta(1/\varepsilon)$, by keeping the same approximation ratio.

1 Introduction

Wireless sensor networks have been attracting the interest of computer scientists and engineers recently due to their potential to impact our everyday lives and because of their numerous applications in areas such as health care, national security, inventory tracking, surveillance, and environmental monitoring.

One of the fundamental issues in sensor networks is related to analyzing the *coverage*, or how well a network of sensors monitors the physical space for an intrusion. The coverage is a measure of the quality of service (QoS) of the sensing function and has been studied by several authors (see [5] for a recent survey of results). The concept of coverage was introduced by Gage [7], who studied it in relation to multi-robot systems. He defined three classes of coverage problems: *blanket coverage* (also known as *area coverage*), where the goal is to achieve a static arrangement of sensing elements that maximizes the detection rate of targets appearing in the region, *sweep coverage*, where the goal is to move a number of sensors across the region as to maximize the probability of detecting a target, and *barrier coverage*, where the objective is to protect the region from undetected penetration.

---

* This work has been supported by the Department of Energy under contract W-705-ENG-36.

J. Aspnes et al. (Eds.): DCOSS 2007, LNCS 4549, pp. 295–308, 2007.

The first model proposed for the barrier coverage problem is due to Meguerdichian et al. [12], who defined the *maximum breach path problem* as a problem of the following type: given a sensor field with known locations of the sensors, find a path such that the distance from any point on the path to the closest sensor is maximized. Meguerdichian et al. solve the maximum breach path problem by using the fact that there is always a maximum breach path that goes along the edges of the Voronoi diagram [4] computed for the set of sensor locations. This concept was further developed by Meguerdichian et al. [11] and by Veltri et al. [14], who define the *exposure* of a path p as an integral over all points x of p of the ability of sensing (detecting) x, which ability is given as a function that depends on the distance between x and the closest sensor, as well as the sensing time.

The *minimum exposure path* (MEP) problem is, given a sensor field and a pair of points x and y inside it, find a path between x and y of a minimum exposure. Meguerdichian et al. [11] give an exact formula for computing the MEP between two points at equal distances from the sensor in a single-sensor field and a sensing function $\gamma/d(s,x)$, where γ is a constant and $d(s,x)$ is the Euclidean distance between the sensor location s and the point x on the path. Although there is no formal proof published, the MEP problem is believed to be unsolvable in the case of multi-sensor fields. In order to solve it approximately, in [11] the region is covered by an $k \times k$ grid of points, each point is connected by edges to l other points following a given pattern, edge weights are assigned equal to the approximated exposures computed using numerical integration techniques, and finally a single pair shortest path problem is solved on the resulting graph. There is no analysis in [11] of how closely the path constructed by their algorithm approximates the MEP or of the time complexity of the algorithm, although it is easy to see that the size of the graph they construct is $\Omega(k^2 l)$ and that it is dependent on the area of the region and is independent of the number of the sensors. Veltri et al. [14] find a partial solution to the problem of exactly computing the exposure between two points in a single-sensor field and describe, for the case of many-sensor fields, a heuristic message-passing distributed algorithm that allows the sensor network to estimate a minimum exposure without knowing the network topology. Distributed algorithms for coverage problems were also studied by Li et al. in [10], for the problem of finding a path with maximum observability, and by Huang et al. in [9], who consider a dynamic version of the maximum breach and maximum observability path problems, where the topology of the sensor network may change due to new sensors being inserted, relocated, or deleted from the network.

In this paper we describe an approximation algorithm for solving the MEP problem. Our algorithm takes as input a description of a sensor field consisting of n sensors positioned inside an $O(n)$ vertex simple polygon P, two points x and y in P, and any number $\varepsilon > 0$. It returns a path between x and y in P whose exposure is within $1 + \varepsilon$ factor of the exposure of the MEP. The algorithm is based on an analysis of the properties of MEPs and on a discretization of the region by constructing a Voronoi diagram and defining additional points and edges. Then a shortest path problem is solved on the resulting graph and the resulting shortest path is used as an approximation to the MEP. The time of the algorithm is $O(n/\varepsilon^2\psi)$, where ψ is a topological characteristic of the field. We also describe a faster version of the algorithm that improves the computation time by a factor of roughly $\Theta(1/\varepsilon)$.

The main contributions of this paper are the following: (i) We find an *exact* solution for the MEP problem in a single-sensor field – the previous solution [14] was valid only in special cases; (ii) We develop the first approximation algorithm for the MEP problem in a multi-sensor field with theoretically guaranteed running time and approximation ratio; (iii) We develop a theoretical framework that can be applied for designing approximation algorithms for related minimum exposure and coverage problems; (iv) Our algorithm is much faster and uses much less memory than the previous algorithms [11], [14] since the latter create a 2-D mesh of points covering the entire region, while we only place additional points on the edges.

The paper is organized as follows. In Section 2 we formally introduce the MEP problem and give some definition. In Section 3 we study MEPs in sensor fields of a single sensor and in Section 4 we study the multiple-sensor case. In Section 5 we describe our approximation algorithms for computing a minimum exposure path and in the last section we conclude with a list of open problems and ongoing work.

2   Preliminaries and Problem Formulation

We consider a connected region P in the plane of *bounded aspect ratio*, e.g., such that the ratio of the square root of the area of P and the perimeter of P is bounded. We have n identical sensors located at points $l_1, \ldots, l_n$ in P monitoring for a target in P. Each target emits a signal that the sensors try to detect. The strength of that signal diminishes with the distance traveled. The probability that a target will be detected depends also on parameters such as the energy emitted by the target, the nature of the signal, the sensitivity of the sensors, and the noise in the environment. Adopting a widely used sensibility model [8,11,14,6], we assume that the signal energy of a target at point x detected by a single sensor at point l is

$$S(l, x) = \frac{\gamma}{d(l, x)^\mu}, \tag{1}$$

where $d(l, x)$ is the Euclidian distance between l and x and γ and μ are constants. Depending on the technology and the environment, the value of μ, called *sensibility exponent*, is typically between 1 and 5.

A *sensor field* F is defined as a 3-tuple $F = (P, L, S)$, where P is a connected region in the plane, $L = \{l_1, \ldots, l_n\}$ is the set of sensor locations, and the function S, called *sensibility* of F, is defined by (1).

For the case of multiple sensors, the notion of *sensor field intensity* for a given point x in the sensor field F has been introduced in [11] in order to measure the likelihood that a target on x will be detected by any of the sensors. There are two basic variations of the model. In the *all-sensor intensity* model, the *intensity* at point x, denoted by $I_A(F, x)$, is defined as a sum of the sensibilities of individual sensors, e.g., $I_A(F, x) = \sum_{i=1}^{n} S(l_i, x)$. The all sensor intensity model reflects more accurately the capability of the sensors to detect a target, but it has also a number of weaknesses: (i) it assumes that all sensors are active during most of the time, which would be energy inefficient; (ii) it presents greater communication and data fusion challenges; (iii) the collection of data from weak sources increases the total noise-to-signal ratio.

In the *closest-sensor field intensity* model the *intensity* at a point x, denoted by $I_C(F, x)$, is defined as $I_C(F, x) = S(l_i, x)$, where l_i is the closest sensor to x.

Let p be a path given as $p = \{p(t) \in P \mid t \in [t_1, t_2]\}$, where $[t_1, t_2]$ is a given interval and $p(t)$ is a continuous function differentiable everywhere in $[t_1, t_2]$ except for a finite number of point. The *exposure* of p with regard to intensity model I and field F is defined [14,11] as

$$\exp(p, I, F) = \int_{t_1}^{t_2} I(F, p(t)) \left| \frac{dp(t)}{dt} \right| dt, \tag{2}$$

where $I(F, x)$ is either $I_A(F, x)$ or $I_C(F, x)$ and $|dp(t)/dt|$ is the element of arc length. In the rest of this paper we assume that $I(F, x) = I_C(F, x)$. The definition of exposure accounts for the fact that the probability for a target traveling at a constant speed along the path p to be detected by a sensor is proportional to the intensity of the field along p and the length of the path. A *minimum exposure path* $MEP(x, y, F)$ between x and y is defined as a path between x and y in P with a minimum exposure.

The *minimum exposure path problem* is, given a sensor field $F = (P, L, S)$ and a pair of points x and y, find $MEP(x, y, F)$. In order to simplify the notations, we use $\exp(p)$ or $\exp(p, S)$ instead of $\exp(p, I, F)$ and $MEP(x, y)$ instead of $MEP(x, y, F)$, when P, L, F, and/or I are clear from the context.

We end this section with several definitions from graph theory. A *graph* G is a pair of two sets denoted by $V(G)$ and $E(G)$, where $V(G)$ is the set of the *vertices* and $E(G)$ is the set of the *edges* of G, where each edge is a pair (v, w) of vertices. A *path* p in G is a sequence $v_0, \ldots, v_k$ of vertices, where $(v_{i-1}, v_i) \in E(G)$ for $i = 1, \ldots, k$. If $k = 0$ then p is a *null* path. If there are weights associated with the edges of G, then the *length* of p is defined as the sum of the weights of all edges (v_{i-1}, v_i). Given two vertices $v, w \in V(G)$, the *distance* between v and w is the minimum length of any path between v and w (infinity, if there is no such path). The *shortest path problem* is, given v and w, find a shortest path between v and w.

3   Single-Sensor Fields

Next, we will study the MEP problem in the case of a single sensor. Without loss of generality, in the rest of this paper we assume that $\gamma = 1$, where γ is the constant from (1). (Changing γ scales the exposures of all paths by the same factor and hence preserves the minimum exposure paths.)

Case A: Unrestricted region. We will start by considering the case of an unbounded region, e.g., where P is the entire plane. We use polar coordinates to represent each point q as a pair (ρ, α), where ρ is the distance between q and the origin O (which we choose to be the sensor location) and $\alpha \in [0, 2\pi)$ is the angle between the polar axis and $\overrightarrow{Oq}$. The exposure of a path p with endpoints $x(\rho_0, 0)$ and $y(\rho_\alpha, \alpha)$ in polar coordinates given as $p = \{(\rho(\theta), \theta) \mid \theta \in [0, \alpha]\}$ can be written as

$$\exp(p, d^{-\mu-1}) = \int_0^\alpha \rho(\theta)^{-\mu-1} \sqrt{\rho(\theta)^2 + \rho'(\theta)^2} \, d\theta. \tag{3}$$

Using the *Beltrami identity* [15], we can find that if ρ is a nonnegative function defined in the interval $[0, \alpha]$ that minimizes the integral (3), then

$$\rho(\theta) = \begin{cases} \rho_0 \left(\dfrac{\rho_\alpha}{\rho_0} \right)^{\frac{\theta}{\alpha}} & \text{if } \mu = 0; \tag{4} \\[2ex] \left(\dfrac{\rho_0^\mu \sin(\mu\alpha - \mu\theta) + \rho_\alpha^\mu \sin(\mu\theta)}{\sin(\mu\alpha)} \right)^{1/\mu} & \text{if } \mu \neq 0. \tag{5} \end{cases}$$

Formulas (4)–(5) were derived in [14] using the Euler-Lagrange differential equation, but were not analyzed whether they correspond to a minimum of (3). But since (4)-(5) are only necessary conditions, one needs to additionally check whether a function ρ satisfying (4) or (5) for a particular set of values for μ, ρ_0, ρ_α, and α is a minimum or an inflexion point. (Clearly, ρ from (4) or (5) can not be a maximum since (3) is unbounded from above for $\mu \geq 0$ – it tends to infinity when $\rho \to 0$.) Consider the following two cases.

Case 1: $\mu = 0$. Let ϕ_M be the set of all nonnegative continuous functions defined in $(0, \alpha]$ and upper bounded by M. The integral (3) is unbounded from above (for any $\mu \leq 0$) as it tends to infinity when $\rho \to \infty$. Therefore, for some M sufficiently large, the exact lower bound of (3) for all functions in $(0, \alpha]$ will be the same as the exact lower bound of (3) restricted to the set of functions from ϕ_M. But ϕ_M is a compact set and, hence, the exact lower bound over ϕ_M (and therefore over all functions in $(0, \alpha]$) is reached for some function $\tilde{\rho}$ from ϕ_M. Since there is a unique function satisfying the necessary condition (4), then $\tilde{\rho}$ should be the function defined by (4). Hence (4) does define a MEP between x and y (it is not an inflexion point).

In order to compute the exposure of that path, substitute the expressions for ρ from (4) into the exposure expression (3), resulting in

$$\text{minExp}(x, y, d^{-1}) = \int_0^\alpha \sqrt{1 + \frac{\ln^2(\rho_\alpha/\rho_0)}{\alpha^2}} \, d\theta = \sqrt{\alpha^2 + \ln^2(\rho_\alpha/\rho_0)}. \tag{6}$$

Case 2: $\mu > 0$. In this case the function (5) may or may not represent a minimum, depending on the values of α, ρ_0, and ρ_α. For instance, if $\alpha = \pi/\mu$, the path

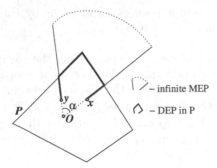

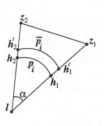

Fig. 1. A MEP in an infinite region may be infinite. In a polygon, such a path is projected into a DEP.

Fig. 2. Illustration to the proof of Lemma 7

given by (5) is not defined. If $\rho_0 = \rho_a$ and $\alpha \to (\pi/\mu)^+$, ρ is unbounded from above and hence ρ would not define an optimal path for α close enough to π/μ. If $\rho_0 = \rho_a$ and $\alpha \to (\pi/\mu)^-$, ρ is unbounded from below (and, in particular, gets negative values). In these cases (5) does not correspond to a solution of the optimization problem and the minimum of (3) is not reached for any (finite) function ρ. However, as we show next, a MEP always exists if the region P is bounded.

Case B: Minimum exposure paths in a polygonal region P. Intuitively, if we consider paths in the entire plane in the case where (5) corresponds to an inflexion point, the MEP from x to y will follow the ray from x in the direction $\overrightarrow{Ox}$ to infinity, then move along an infinite circle to align with the line yO (the exposure along that semicircle will be zero), and finally move in the direction of $\overrightarrow{yO}$ to point y (Figure 1). Although this path is of infinite length, its exposure if finite; the exposure of the path is $\mathrm{minExp}_1(x, y, d^{-\mu-1}) = (\rho_0^{-\mu} + \rho_\alpha^{-\mu})/\mu$.

In a polygonal region, the portion of the path described above that is outside P is replaced by the path of lower exposure among the two paths along the boundary of P connecting the same endpoints. We will refer to the latter path as the *direct escape path* (DEP). As a DEP in P is a chain of straightline segments, we will need a formula for the exposure along a single such segment. If the segment xy belongs to a line containing point O, then the exposure along the DEP p between the points $x(\rho_1, 0)$ and $y(\rho_2, 0)$, $0 < \rho_1 \le \rho_2$, which we denote by $minExp_1$, can be computed by the formula

$$\mathrm{minExp}_1(x, y, d^{-\mu-1}) = \exp(p, d^{-\mu-1}) = \begin{cases} \ln \rho_2 - \ln \rho_1 & \text{if } \mu = 0 \quad (7) \\ \dfrac{1}{\mu}(\rho_1^{-\mu} - \rho_2^{-\mu}) & \text{if } \mu > 0. \quad (8) \end{cases}$$

Otherwise, the exposure $\mathrm{minExp}_1(x, y, d^{-\mu-1})$ along the segment (DEP) p between points $x(\rho_0, 0)$ and y such that $\angle Oyx = \pi/2$ and $\angle xOy = \alpha$ is

$$\exp(p, d^{-\mu-1}) = \begin{cases} \ln(\sec\alpha + \tan\alpha) & \text{if } \mu = 0 \\ \rho_0^{-\mu} \, _2F_1\left(\frac{1}{2}, \frac{\mu+1}{2}; \frac{3}{2}; -\tan^2\alpha\right)\tan(\alpha) & \text{if } \mu > 0, \text{ (9)} \end{cases}$$

where $_2F_1$ is the Gaussian hypergeometric function [1]. A segment xy for which $\angle Oyx \neq \pi/2$ can always be represented as a sum or a difference of segments of the above type.

Next, the exposure of the path p defined by (5) is

$$\mathrm{minExp}_2(x, y, d^{-\mu-1}) = \frac{\sin(\mu\alpha)(\tan(\mu\alpha - c_2) + \tan c_2)}{\mu\sqrt{\rho_0^{2\mu} + \rho_\alpha^{2\mu} - 2\rho_0^\mu \rho_\alpha^\mu \cos(\mu\alpha)}}, \qquad (10)$$

where $\tan(c_2) = \frac{\rho_\alpha^\mu - \rho_0^\mu \cos(\mu\alpha)}{\rho_0^\mu \sin(\mu\alpha)}$.

Finally, the minimum exposure between x and y is determined as

$$\mathrm{minExp}(x, y, d^{-\mu}) = \min\{\mathrm{minExp}_1(x, y, d^{-\mu}), \mathrm{minExp}_2(x, y, d^{-\mu})\}.$$

The path corresponding to the smaller of the two exposures is the MEP.

4   Multiple-Sensor Fields

Here we consider the case of a sensor field $F = (P, L, S)$ with an arbitrary number n of sensors, where $L = \{l_1, \ldots, l_n\}$ is the set of the locations of the sensors and P is a convex polygon of size $O(n)$ that contains all points l_i.

We construct a *Voronoi diagram* $\mathrm{Vor}(L)$ for L in P, which is a tessellation of P into n convex polygons $C_1, \ldots, C_n$, which we call *cells* of $\mathrm{Vor}(L)$, such that C_i is the set of all points that are closer to l_i than to any other point from L. The Voronoi diagram can be constructed in $O(n \log n)$ time. (See [4] for more information about Voronoi diagrams.) Denote by $V(C_i)$ and $E(C_i)$ the sets of the vertices and the edges of C_i, respectively.

Next we analyze the structure of the MEP between any two points x and y from P. First we consider the case where x and y belong to the same cell C_i.

Lemma 1. *Define sensor fields* $F_1 = (C_i, L, S)$ *and* $F_2 = (P', L, S)$, *where* $|L| = 1$ *and* P' *is the entire plane and let* x *and* y *be points from* C_i. *Then*

  (a) *the minimum exposure path* $p = \mathrm{MEP}(x, y, F_1)$ *either contains a point from the boundary of* C_i, *or* $p = \mathrm{MEP}(x, y, F_2)$;
  (b) *the intersection between* p *and any edge of* P *is either empty or a single segment.*

Proof. (a) If p is disjoint with the boundary of C_i, then p is a stationary point for (3) and hence it can be determined by the method discussed in Section 3 Case A.

(b) Assume that claim (b) does not hold. Then there will exist a subpath p' of p that joins a pair of points x_1, x_2 on an edge of C_i and whose interior is entirely inside C_i. Then, by (a), $p' = \text{MEP}(x_1, x_2, F_2)$. Consider the path p'' with endpoints x_1, x_2 that is symmetrical to p' with respect to the line $x_1 x_2$. Then p'' will have a smaller intensity and the same element of arc length compared to p' and hence, by (2), a smaller exposure, which is a contradiction.   □

Next we consider the case where x and y can belong to different cells of $\text{Vor}(L)$.

Lemma 2. *Given two points $x \in C_i$ and $y \in C_j$, $i \neq j$, $\text{MEP}(x, y, F)$ consists of a sequence of subpath each of them of one of the following types:*

(i) *a MEP from x to a point on an edge of C_i or from a point on an edge of C_j to y;*

(ii) *a MEP between points belonging to two different edges of the same cell of $\text{Vor}(L)$;*

(iii) *a segment on an edge of $\text{Vor}(L)$.*

Proof. Follows from the discussion in Section 3 and Lemma 1.   □

5   An Approximation Algorithm for Constructing MEPs

Next we describe and analyze an algorithm that computes an approximation of the MEP between a pair of points x and y. In the algorithm, we first discretize the region by triangulating it and creating a set S of additional points called *Steiner points* (SPs) on the edges of the triangulation. This is similar to the discretization schemes used for solving shortest path problems on weighted polyhedral surfaces, e.g., [3]. Then we define a graph with a vertex set $\{x\} \cup \{y\} \cup \text{Vor}(L) \cup S$ and an edge between any pair of vertices belonging to the same triangle. We define a weight on each edge (v, w) equal to the exposure either along the minimum exposure path between v and w in the triangle containing v and w, if v and w belong to different edges, or along the edge containing v and w, otherwise (see Lemma 2). Then the algorithm finds the shortest path in the resulting graph between x and y using a modification of Dijkstra's algorithm. Next we describe the steps in more detail and analyze the accuracy and the efficiency of the algorithm.

5.1   Defining Steiner Points

First we will define "empty" regions around each sensor location l_i that will contain no SPs. The rationale is to limit the number of SPs we have to define in each Voronoi cell C_i, because the number of SPs needed to achieve a given approximation ratio increases when the distance to l_i decreases. We need the following properties.

Lemma 3. *Let F_1 and F_2 be fields with sensibility exponents $\mu_1 = 0$ and $\mu_2 > \mu_1$, respectively. Given two points $x_1(r, \alpha_1)$ and $x_2(r, \alpha_2)$ belonging to cell $C_i \in \text{Vor}(L)$, let*

$$p_1 = \text{MEP}(x_1, x_2, F_1) = \{(\rho_1(\theta), \theta) \mid \theta \in [0, \alpha]\}$$

and

$$p_2 = \text{MEP}(x_1, x_2, F_2) = \{(\rho_2(\theta), \theta) \mid \theta \in [0, \alpha]\}.$$

Then $\rho_1(\theta) \leq \rho_2(\theta)$ for all $\theta \in (\alpha_1, \alpha_2)$.

Let F be a field with a scalability exponent $\mu \geq 0$, and let $d_i = \min\{d(l_i, z) \mid z \in E(C_i)\}$. Define a circle κ_i with center l_i and radius d_i.

Lemma 4. *Any MEP in F with both endpoints on $E(C_i)$ contains no points from the inside of κ_i.*

Proof. Suppose p is a MEP for a sensor field with sensibility exponent μ that contains a point from the inside of κ_i. Then p contains a subpath p_1 with endpoints, say, a and b on κ_i and the rest of p_1 in the interior of κ_i. Consider the following two cases:

(i) $\mu = 0$. By (4), the portion p_1' on κ_i between a and b, being a minimum exposure path, has a smaller exposure than p_1. Replacing p_1 by p_1' in p results in a path with a smaller exposure than p, a contradiction.

(ii) $\mu > 0$. Combine the proof of case (i) with Lemma 3. $\square$

Next we define $\mathcal{S}$, the set of Steiner points for $\text{Vor}(L)$. For each $C_i \in \text{Vor}(L)$, triangulate C_i by adding straightline segments joining l_i to each vertex of $V(C_i)$. Let T_i be the resulting set of triangles for C_i, let $\mathcal{T}$ be the resulting triangulation of $\text{Vor}(L)$, and let $t \in T_i$ for some i. Let l_i, a, b be the vertices of t and $d(l_i, a) \geq d(l_i, b)$. Call (l_i, a) and (l_i, b) *new* edges and call (a, b) an *old* edge. Let $l = d(l_i, a)$ and let $\varepsilon > 0$. Define a set of Steiner points $s_0, s_1, \ldots, s_\lambda$ on the segment $\overline{l_i x}$ such that

$$d(l_i, s_0) = d_i, \quad d(l_i, s_{j-1}) < d(l_i, s_j), \quad d(s_{j-1}, s_j) = \varepsilon d(l_i, s_{j-1}), \tag{11}$$

for $j = 1, \ldots, \lambda$, where λ is chosen such that $d(l_i, s_\lambda) \leq l < d(l_i, s_{\lambda+1})$. (We used Lemma 4 for justifying the definition of s_0.) In a similar way we define SPs on the segment $\overline{l_i y}$. For the segment $\overline{xy}$ we define the SPs $s_0', \ldots, s_{\lambda'}'$ such that

$$s_0' = a, \ s_{\lambda'}' = b, \ d(l_i, s_{j-1}') < d(l_i, s_j'), \ d(s_{j-1}', s_j') = \varepsilon l \text{ for } j = 1, \ldots, \lambda' - 1,$$

$$\text{and } d(s_{\lambda'-1}', s_{\lambda'}') \leq \varepsilon l.$$

Lemma 5. *The number of SPs on the segments of the triangle t is $O(\ln(l/d)/\varepsilon)$.*

Proof. From (11), $d(l_i, s_j) = (1+\varepsilon)^j d$. Since $d(l_i, s_\lambda) \leq l$, then $(1+\varepsilon)^\lambda d \leq l$ and

$$\lambda \leq \log_{1+\varepsilon}(l/d) = \frac{\ln(l/d)}{\ln(1+\varepsilon)} = O(\ln(l/d)/\varepsilon).$$

Hence each of the segments $\overline{l_i x}$ and $\overline{l_i y}$ contains $O(\ln(l/d)/\varepsilon)$ SPs. The segment $\overline{xy}$ contains $\lceil d(x, y)/(\varepsilon l) \rceil$ SPs, which number is $O(1/\varepsilon)$, since $d(x, y) < 2l$. $\square$

5.2   Description and Analysis of the Algorithm

Next we define a weighted graph $G_\varepsilon = (V_\varepsilon, E_\varepsilon)$ called *approximation graph* with vertex set $\mathrm{Vor}(L) \cup \mathcal{S} \cup \{x, y\}$ and an edge between any pair of vertices corresponding to either points on different edges of the same triangle of $\mathcal{T}$ or to the same new edge. Add also edges joining vertices x and y to the SPs from the triangles containing x and y, respectively. Assign a weight $wt(v, w)$ on each edge (v, w) corresponding to the exposure along the minimum exposure path between v and w in the triangle containing v and w, if v and w belong to different edges, or along the segment containing v and w, otherwise.

Let Q denote the area of the region P and let q denote the minimum distance between any two points of L.

Lemma 6. G_ε *has* $O(n/\varepsilon \ln(Q/q))$ *vertices and* $O(n/\varepsilon^2 \ln^2(Q/q))$ *edges.*

Proof. By Lemma 5 each triangle contains $O(\ln(l/d)/\varepsilon) = O(\ln(\sqrt{Q}/q)/\varepsilon)$ SPs, as P has a bounded aspect ratio. Since $|\mathrm{Vor}(L)| = O(n)$, then the total number of triangles is $O(n)$. The lemma follows. $\square$

Given G_ε, we compute an approximate minimum exposure path between x and y as a shortest path p_ε in G_ε between x and y. That shortest path can be computed using Dijkstra's shortest path algorithm [2] in $O(m_\varepsilon + n_\varepsilon \log n_\varepsilon)$ time, where $n_\varepsilon = |V(G_\varepsilon)|$ and $m_\varepsilon = |E(G_\varepsilon)|$.

Next we analyze how closely p_ε approximates the MEP. For each path p in G_ε let $wt(p)$ denote the sum of the weights of the edges of p.

Lemma 7. *For any* $\varepsilon > 0$ *there exists a path* p *in* G_ε *between vertices* x *and* y *such that* $wt(p) \le (1 + O(\varepsilon/\breve{\alpha})) \exp(\mathrm{MEP}(x, y))$, *where* $\breve{\alpha}$ *is the minimum angle of* $\mathcal{T}$.

Proof. Assume x and y belong to the interior of different triangles of $\mathcal{T}$ and let Δ_x and Δ_y be the triangles containing x and y, respectively. According to Lemma 2, $\mathrm{MEP}(x, y)$ consists of a sequence of subpaths $p_1, \ldots, p_\lambda$ in P, where p_1 and p_λ are MEPs between x or y and a point from Δ_x or Δ_y, respectively, and each of the other paths is either a MEP between points belonging to different edges of a triangle, or is a subsegment of an edge of $\mathcal{T}$. We will construct a sequence $p'_1, \ldots, p'_\lambda$ of paths in G_ε whose concatenation results in a path from x to y and such that $wt(p'_i) \le (1 + O(\varepsilon/\breve{\alpha}))\exp(p_i)$ for any $1 \le i \le \lambda$, which will imply to validity of the lemma. (If x or y is on an edge of the triangulation, then the corresponding paths p'_1 or $p'_{\lambda'}$ will be null paths.)

Choose any $i \in [1, \lambda]$ and consider the case where p_i connects points belonging to different new edges (l, z_1) and (l, z_2) of a triangle Δ, where $l \in L$ (Figure 2). (The proofs for the other cases are similar.) Let h_1 and h_2 be the source and the target of p_i and let h'_1 and h'_2 be the Steiner points on the segments $\overline{h_1 z_1}$ and $\overline{h_2 z_2}$ that are closest to h_1 and h_2, respectively. Denote $d(l_i, h_j) = \eta_j$ and $d(l_i, h'_j) = \eta'_j$ for $j = 1, 2$. By (11) $\eta'_j \le (1+\varepsilon)\eta_j$ for $j = 1, 2$. Define a polar coordinate system with origin l and polar axis $\overline{lz_1}$. Let $\alpha = \angle z_1 l z_2$. Let $p_i = \{(\rho(\theta), \theta) \mid \theta \in [0, \alpha]\}$.

We will define a path $\bar{p}_i = \{(\bar{\rho}(\theta), \theta) \mid \theta \in [0, \alpha]\}$ with source $(h_1', 0)$ and target (h_2', α) that will be a "scaled up" version of p_i. More precisely, our goal is to define a function $k(\theta)$ such that the path defined by the function $\bar{\rho}(\theta) = k(\theta)\rho(\theta)$ will have exposure at most $1 + O(\varepsilon/\breve{\alpha})$ times the exposure of p_i. We will show that it suffices that the following conditions are satisfied:

(i) $\bar{\rho}(0) = \eta_1'$ and $\bar{\rho}(\alpha) = \eta_2'$;

(ii) $\bar{\rho}^{-\mu} = (1 + O(\varepsilon))\rho^{-\mu}$;

(iii) $\sqrt{1 + \left(\dfrac{\bar{\rho}'}{\bar{\rho}}\right)^2} \leq (1 + O(\varepsilon/\breve{\alpha}))\sqrt{1 + \left(\dfrac{\rho'}{\rho}\right)^2}$

We will prove that the function $k(\theta) = \dfrac{\alpha - \theta}{\alpha} \dfrac{\eta_1'}{\eta_1} + \dfrac{\theta}{\alpha} \dfrac{\eta_2'}{\eta_2}$ satisfies those conditions. A direct substitution shows that condition (i) is satisfied. Furthermore,

$$k(\theta) \leq \frac{\alpha - \theta}{\alpha}(1 + \varepsilon) + \frac{\theta}{\alpha}(1 + \varepsilon) = 1 + \varepsilon$$

and hence $\bar{\rho}/\rho \leq 1 + \varepsilon$ and (ii) holds. Property (iii) follows from the previous inequality $k(\theta) \leq 1 + \varepsilon$ and from

$$k'(\theta) = \left(\frac{\eta_2'}{\eta_2} - \frac{\eta_1'}{\eta_1}\right)\frac{1}{\alpha} = O(\varepsilon/\breve{\alpha}).$$

By Properties (ii) and (iii)

$$\exp(\bar{p}_i) = \int_0^\alpha \bar{\rho}^{-\mu}\sqrt{1 + \left(\frac{\bar{\rho}'}{\bar{\rho}}\right)^2}\, d\theta = \int_0^\alpha (1 + O(\varepsilon/\breve{\alpha}))\rho^{-\mu}\sqrt{1 + \left(\frac{\rho'}{\rho}\right)^2}\, d\theta$$

$$= (1 + O(\varepsilon/\breve{\alpha}))exp(p_i).$$

Since by (i) p_i' and $\bar{p}_i$ are paths with the same source and target h_1' and h_2',

$$wt(p_i') = \exp(MEP(h_1', h_2')) < \exp(\bar{p}_i) \leq (1 + O(\varepsilon/\breve{\alpha}))\exp(p_i). \qquad (12)$$

To complete the proof, we add together the inequalities (12) for $i = 1, \ldots, \lambda$. □

Combining Lemma 7 with our analysis of the time complexity, we get the following theorem.

Theorem 1. *Given a convex polygon P of bounded aspect ratio, a sensor field $F = (P, L, S)$ with nonnegative sensibility exponent, two points x and y from P, and any $\varepsilon > 0$, a path p in P between x and y such that $\exp(p) \leq (1 + \varepsilon)\exp(p_{opt})$ can be found in $O(n/\varepsilon^2 \ln^2(\psi)\breve{\alpha}^2)$ time, where $p_{opt} = MEP(x, y, F)$, $n = \max\{|L|, |P|\}$, ψ denotes the ratio of the area of P and the minimum distance between any two points of L, and $\breve{\alpha}$ is the minimum angle of the triangulation of $Vor(L)$.*

Note that although the justification of Theorem 1 is relatively complex, the implementation of the corresponding algorithm requires only running a shortest path algorithm on G_ε.

5.3 Improving the Running Time

The graph G_ε has a relatively high number of edges compared to the number of its vertices. This is due to the fact that in each triangle of $\mathcal{T}$ the number of the edges is roughly proportional to the square of the number of the Steiner points. On the other hand, the set of all shortest paths in $G\varepsilon$ has a structure that allows an efficient implementations of Dijkstra's shortest path algorithm that considers only a fraction of the edges of G_ε.

We will describe the idea of the algorithm BUSHWHACK [13] designed for solving shortest path problems for Euclidean-like distances and show how it can be modified in order to work in our case. The goal is to reduce the number of the edges considered to be roughly proportional to the number of the SPs (within a logarithmic factor). We will only consider here the case $\mu = 0$, where μ is the sensibility exponent from (1). The algorithm for arbitrary values of μ is similar, but needs a more complicated analysis because of the lack of simple analogue of the exposure formula (6).

The BUSHWHACK algorithm is based on Dijkstra's algorithm, which divides the vertices of the graph into two subsets: U, containing vertices to which the exact distances $d_{G_\varepsilon}(x, s)$ from the source x have already been computed, and $V \setminus U$, containing vertices to which approximate distances from x have been assigned based on paths restricted to contain vertices from U only. At each iteration a vertex $s \in V \setminus U$ with a minimum current distance from x is moved to U and the distances to the neighbors of s in $V \setminus U$ are updated.

In order to introduce the BUSHWHACK modification to Dijkstra's algorithm, consider any triangle $\Delta \in \mathcal{T}$. If e is an edge of Δ, we denote by $V(e)$ the set of the vertices of G_ε that correspond to SPs from e. For any two edges e and e' from Δ and vertex v on e that is not on e' we define the set $I(v, e, e')$ consisting of all vertices z from e' such that

$$d_{G_\varepsilon}(x, v) + wt(v, z) \le d_{G_\varepsilon}(x, v') + wt(v', z)$$

for any vertex v' from e. The sets $I(v, e, e')$ can be used to reduce the number of the edges in Δ considered by Dijkstra's algorithm, since for any vertex z from $I(v, e, e')$ there is a shortest paths from x to z that does not contain any vertex from $V(e) \setminus \{v\}$. Hence all edges connecting a vertex from $V(e) \setminus \{v\}$ to a vertex from $I(v, e, e')$ can be ignored in the shortest paths computation.

In fact, the sets $I(v, e, e')$ are dynamic and are updated each time when a new vertex from $V(e)$ is added to U. In order to ensure that these sets can be maintained efficiently, we need to prove that the following two properties hold. Let $\pi(v, w)$ denote the path in P resulting from replacing each edge of the shortest path in G_ε between v and w with the corresponding minimum exposure path and let $d'(v, w)$ denote the exposure of $\pi(v, w)$ (which is also equal to the distance between v and w in G_ε).

Lemma 8. *Let $\pi_1 = \pi(x, y_1)$ and $\pi_2 = \pi(x, y_2)$. Let, for $i = 1, 2$, π_i intersects the edges of a triangle Δ of $\mathcal{T}$ at vertices z_{i1} and z_{i2}, respectively (Figure 1), where all vertices z_{ij}, $1 \le i, j \le 2$, are distinct. Then the segments $\overline{z_{11}z_{12}}$ and $\overline{z_{21}z_{22}}$ do not intersect.*

Lemma 8 implies that each set $I(v, e, e')$ consists of consecutive points on e', i.e., no vertex of $V(e) \setminus I(v, e, e')$ is between two vertices from $I(v, e, e')$ on e'. Hence $I(v, e, e')$ can be identified with an interval (e.g., a pair of points) on e'. The next lemma can be used to compute and maintain such intervals efficiently (in $O(\log |V(e')|)$ time).

A segment s is called *monotonic* [13] with respect to a point z, if the exposure from z to points of s is either monotonically increasing or monotonically decreasing along s.

Lemma 9. *If s is a segment belonging to a line containing the sensor location l_i, then s can be divided into two monotonic segments with respect to any point z in $O(1)$ time.*

Proof. The point z' such that $d(O, z') = d(O, z)$ divides s into monotonic segments, if z is not on s. If z is on s, then s itself is monotonic. □

The proof of an analogue of Lemma 9 for the case where the line containing s does not contain l_i is more complex. Instead of proving such lemma, we notice that we can use other properties to define the set $I(v, e, e')$ in the case where e' is a segment of $V(L)$ and v does not belong to e'. We consider the following two cases for v.

(i) $v = x$. In this case we define $I(v, e, e') = e'$ since $U = \{v\}$.
(ii) $v \neq x$. Note that v can not be either of the endpoints of e as by assumption v is not on e' and v can not be a sensor location l_i as by construction $L \cap S = \emptyset$ (see (11)). Then v is an internal point of e. Denote by v^* the predecessor of v on the shortest path from x to v. That vertex must have already been determined by the algorithm since $v \in U$. Then v is the closest SP to the intersection point between e and a MEP determined by formulas (4)–(5) from v^* to a point on e'. Then $I(v, e, e')$ can be determined as the smallest segment on e' whose endpoints are SPs and which contains the intersection point of e' and the MEP determined by v and v^*.

Further details on the data structures and the analysis of BUSHWHACK algorithm can be found in [13]. We established the following result, which is an improvement of Theorem 1 for the case $\mu = 0$ by a factor of roughly $\Theta(1/\varepsilon)$.

Theorem 2. *Given a polygon P of bounded aspect ratio, a sensor field $F = (P, L, S)$ with zero sensibility exponent, two points x and y from P, and any $\varepsilon > 0$, a path p between x and y in P such that $\exp(p) \leq (1 + \varepsilon) \exp(p_{opt})$ can be found in $O(m \log m)$ time, where $p_{opt} = \text{MEP}(x, y, F)$, $n = \max\{|L|, |P|\}$, $m = O(n/\varepsilon \ln(\psi)\breve{\alpha})$, ψ denotes the ratio of the area of P and the minimum distance between any two points of L, and $\breve{\alpha}$ is the minimum angle of the triangulation of $\text{Vor}(L)$.*

6   Conclusion

In this paper we developed the theoretic framework for designing approximation algorithms for solving minimum exposure path problems for sensor networks.

There are several interesting problems not discussed here that will be subject of our ongoing and future work. These include removing the dependence of the running time of Theorem 2 on $\breve{\alpha}$. Although such a dependence is characteristic for such type of problems, e.g., [3], we will show in the full version of this paper that it can be removed in our case. We also plan to use our MEP algorithms for solving placement problems for sensor networks.

References

1. Abramowitz, M., Stegun, I.A.: Handbook of Mathematical Functions with Formulas, Graphs, and Mathematical Tables. Dover Publications, New York (1964)
2. Ahuja, R.K., Magnanti, T.L., Orlin, J.B.: Network flows: theory, algorithms, and applications. Prentice-Hall, Englewood Cliffs (1993)
3. Aleksandrov, L., Maheshwari, A., Sack, J.-R.: Determining approximate shortest paths on weighted polyhedral surfaces. J. ACM 52(1), 25–53 (2005)
4. Aurenhammer, F.: Voronoi diagrams — surrvey of a fundamental geometric data structure. ACM Computing Surveys 23(3), 345–405 (1991)
5. Cardei, M., Wu, J.: Coverage in wireless sensor networks. In: Ilyas, M., Mahgoub, I. (eds.) Handbook of Sensor Networks, CRC Press, Boca Raton (2004)
6. Clouqueur, T., Phipatanasuphorn, V., Ramanathan, P., Saluja, K.K.: Sensor deployment strategy for detection of targets traversing a region. ACM Mobile Networks and Applications 8(4), 453–461 (2003)
7. Gage, D.W.: Command control for many-robot systems. Unmanned Systems Magazine 10(4), 28–34 (1992)
8. Hata, M.: Empirical formula for propagation loss in land mobile radio services. IEEE Transactions on Vehicular Technology 29(3), 317–325 (1980)
9. Huang, H., Richa, A.W., Segal, M.: Dynamic coverage in ad-hoc sensor networks. Mobile Networks and Applications 10(1-2), 9–17 (2005)
10. Li, X.-Y., Wan, P.-J., Frieder, O.: Coverage in wireless ad hoc sensor networks. IEEE Transactions on Computers 52(6), 753–763 (2003)
11. Megerian, S., Koushanfar, F., Qu, G., Veltri, G., Potkonjak, M.: Exposure in wireless sensor networks: theory and practical solutions. Wireless Networks 8(5), 443–454 (2002)
12. Meguerdichian, S., Koushanfar, F., Potkonjak, M., Srivastava, M.B.: Coverage problems in wireless ad-hoc sensor networks. In: INFOCOM, pp. 1380–1387 (2001)
13. Sun, Z., Reif, J.H.: Bushwhack: An approximation algorithm for minimal paths through pseudo-euclidean spaces. In: Eades, P., Takaoka, T. (eds.) ISAAC 2001. LNCS, vol. 2223, pp. 160–171. Springer, Heidelberg (2001)
14. Veltri, G., Huang, Q., Qu, G., Potkonjak, M.: Minimal and maximal exposure path algorithms for wireless embedded sensor networks. In: SenSys '03: Proceedings of the 1st international conference on Embedded networked sensor systems, pp. 40–50. ACM Press, New York (2003)
15. Weinstock, R.: Calculus of Variations, with Applications to Physics and Engineering. Dover Publications, New York (1974)

Energy Efficient Intrusion Detection in Camera Sensor Networks

Primoz Skraba[1] and Leonidas Guibas[2]

[1] Department of Electrical Engineering
Stanford University, Stanford, CA 94305
primoz@stanford.edu
[2] Department of Computer Science
Stanford University, Stanford, CA 94305
guibas@stanford.edu

Abstract. The problem we address in this paper is how to detect an intruder moving through a polygonal space that is equipped with a camera sensor network. We propose a probabilistic sensor tasking algorithm in which cameras sense the environment independently of one another, thus reducing the communication overhead. Since constant monitoring is prohibitively expensive with complex sensors such as cameras, the amount of sensing done is also minimized. To be effective, a minimum detection probability must be guaranteed by the system over *all* possible paths through the space. The straightforward approach of enumerating all such paths is intractable, since there is generally an infinite number of potential paths. Using a geometric decomposition of the space, we lower-bound the detection probability over all paths using a small number of linear constraints. The camera tasking is computed for set of example layouts and shows large performance gains with our probabilistic scheme over both constant monitoring as well as over a deterministic heuristic.

1 Introduction

Until recently, research in wireless sensor networks (WSN) has focused primarily on low cost, low bandwidth sensors. With dropping costs and advances in imaging technology, there is now increasing interest in camera sensor networks. Several platforms have already been developed for image/video acquisition in a sensor network setting [17, 13]. Cameras provide a higher level of sensor information, but also use more of the limited resources available to a wireless sensor node and so present a new set of challenges. If used continuously, cameras consume too much energy to operate on battery power. While applications such as target tracking usually require constant monitoring, it is unlikely that all areas being monitored will see continuous activity over time. We can achieve significant energy savings if we reduce the amount of sensing a camera node must do, with the goal of first detecting the target(s). Detection can provide a wake-up mechanism for more expensive higher level services such as tracking [16, 2], identity management [24], and occupancy reasoning [27].

J. Aspnes et al. (Eds.): DCOSS 2007, LNCS 4549, pp. 309–323, 2007.

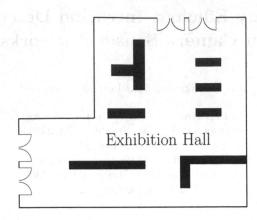

Fig. 1. An exhibition hall with two entrances/exits, allowing for several possible paths through it

In this paper we propose an energy efficient approach to detection in camera sensor networks. Our approach provides two benefits. First, after the initial setup is complete, *no* communication between nodes is required. The amount of sensing done is also minimized, while still providing guarantees on detection quality. Minimizing the use of the cameras is important since they consume as much energy as communication [18]. The algorithm itself is simple: at each time step, a camera independently decides to sense a frame with an assigned probability. These are set so that probabilistic guarantees on detection can be made.

Consider the scenario of tracking people with a camera sensor network in an exhibition hall with the floor plan in Figure 1. In the evening, the building is empty and the network should no longer monitor continuously because there is no one to track. When a person is detected, the network can wake up and begin tracking once again. One possible solution is to have the cameras continuously monitor only the entrances. This either assumes that the cameras have been specifically deployed for the purpose of detection or the cameras which face the entrances will be overused and may soon run out of energy. In this paper, we assume that the cameras are spread out to cover most of the space (as they would be deployed for monitoring or tracking). Surveillance and tracking are canonical applications of camera sensor networks, often aimed at detecting an adverserial intruder. The objective of the intruder is to move between sets of points called sources and destinations. Sources can be thought of as entrances into a space and destinations as secure areas (e.g. a bank vault). Although the intruder model is natural in security applications, the idea of focusing on paths between certain points in space is more general. For example in building monitoring, a destination may be a particular office or all points of a certain distance from the entrance, limiting how far an intruder may travel before being detected.

First we present related work and introduce the models used for the cameras and the space. After introducing the necessary geometric preprocessing of the camera layout, the problem of tasking the cameras subject to a global detection probability constraint is posed in the framework of convex optimization. A deterministic algorithm with a detection probability of 1 is presented for comparison. Finally, analysis of the performance and validation of the algorithms are presented.

2   Related Work

In WSNs, work in tasking sensors has primarily addressed maintaining a minimum level of coverage over the entire space of the network. This is usually referred to as the k-coverage problem. Several centralized and distributed algorithms have been proposed for uniform [1, 15, 12] or differentiated coverage [26]. These all, however, assume a local sensor coverage model[1]. For camera sensor networks, a deployment algorithm which meets coverage constraints over a space was recently proposed in [10]. These types of coverage problems are all closely related the classical Art Gallery Problem and its many variants. For further details, the reader is directed to [21].

   In [20, 25, 19], algorithms for finding maximal exposure and maximal breach paths through a sensor network with local sensors are described. The maximal exposure path is a path which the intruder is exposed to the most sensors. More relevant is the maximal breach path, which is the path of minimal exposure through a sensor network. This work again assumes constant sensing and finds the path through the network which maximally distances itself from all the sensors. The idea of barrier coverage was proposed recently in [14]. Barrier coverage refers to when active sensors form a barrier so that an intruder cannot pass through the network undetected. Results on the probability of random sampling needed to achieve a barrier in the network were given. However, the analysis was done for dense random networks equipped with local sensors and the barriers are static. Energy conservation through limiting the sensing to a small part of the network was also considered in [11, 23]. In [11], the activation pattern follows a user-defined path through the sensor network as a sentry. In [23], the activation pattern is a sweep across the network. Both schemes assume a simple topology and do not handle sensing holes.

   Our work incorporates many similar ideas to the ones mentioned above. However, to the best of our knowledge, there is no prior work on providing energy efficient detection in a non-local camera sensor network.

3   Model

The detection area is modeled as a two dimensional polygonal space. It need not be convex or simple, as seen in the Exhibition Hall example in Figure 1. The space may have occlusions as well as regions not covered by any cameras. Sources and destinations are modeled as points in the space. These may be entrances or

---

[1] A local sensors refers to a sensor that can only detect things close to itself. The circular or Gaussian sensor model fall into this category.

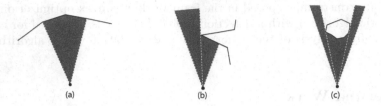

Fig. 2. Examples of visibility regions for cameras: (a) shows the simplest case where two constraint edges suffice, (b) shows an occlusion from one side, where an additional constraint edge is needed and (c) shows an occlusion in the middle of the camera field of view, where two additional constraint edges are needed

"areas of interest." A point may be a source in one setting and a destination in another. The camera's views are modeled by simple two dimensional cones. It is assumed that detection is uniform with probability 1 inside the cone and 0 outside. More complicated models of camera coverage such as a limited depth of view and varying probability of detection can be incorporated with minimal changes to the framework. Time is divided into discrete steps called *frames*. For energy consumption, we assume that taking a shot within a frame by a camera has a fixed cost. This implies that minimizing the energy used by the node to sense is equivalent to minimizing the probability that it senses during a frame. A shortcoming of this model is that it ignores the energy lost when the node is powering the camera up and powering it down. In future work, these costs will be accounted for.

We assume that the cameras know their locations, orientations, and have been properly calibrated. This could be done using structure from motion and automatic registration. The specific problem of camera calibration has been addressed in [8, 9, 22]. Furthermore, we assume that the cameras know the layout of the space in which they are deployed. This could be done by manually uploading a floorplan or learning it automatically with a scanning device [3]. There are no constraints on placement of the cameras or their viewing angles. The locations of sources and destinations along with their corresponding detection constraints are user-defined parameters and assumed to be known.

4   Geometric Preprocessing

The first step is to understand how the cameras cover the space. Beginning with the empty polygonal space, the two edges which make up a camera cone are placed into the space as constraint edges. In the presence of occlusions additional constraint edges are needed to define the boundary of the visibility cone of the camera. Some examples of where additional constraints are needed are shown in Figure 2. To find all the constraint edges, we implement the rotational sweep algorithm described in [10].

The constraint edges of all the camera cones decompose the space into polygonal faces. Each face is characterized by its neighboring faces and the combination

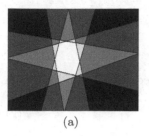
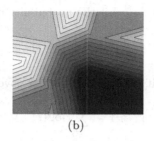
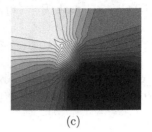

| (a) | (b) | (c) |

Fig. 3. (a) The coverage of 4 cameras facing each other in a rectangular space. The brighter the color the higher the coverage. (b) The distance cost function over the space from a point in the lower righthand corner, where the cost is 1 if the space is covered by a camera and 0 otherwise. Note the distortion due to the uncovered areas in the corners. (c) The cost function where the cost of traveling through the face is proportional to the distance times the number of cameras which see the face.

of cameras which see it. The computation of the arrangement of the visibility cones from the polygonal space and camera constraint edges is a well-studied problem in computational geometry with known efficient solutions [7]. To simplify the implementation, we compute the polygonal faces using the constrained Delaunay triangulation of the space and the camera constraint edges. The faces are reconstructed by joining adjacent triangles which are seen by the same combination of cameras.

From this decomposition, we extract the *spatial adjacency graph* (SAG). The SAG is an undirected graph $G(v, e)$ where each vertex represents a face in the space and two vertices are adjacent if the corresponding faces are adjacent. Two faces are adjacent if an intruder can move between the faces without entering any other face or equivalently, that the faces share a common vertex.

5   Tasking the Cameras

The algorithm tasks the cameras so that whichever path from the source to the destination the intruder chooses, there is high probability that he will be detected. First a simplified intruder motion model is considered. The results are then extended to a more general and realistic motion model.

5.1   Assigning Probabilities in a Graph

Consider any path the intruder can traverse through the space. The continuous path can be mapped to the SAG by marking the corresponding nodes for each face the path crosses. We first assume that at each time step, the intruder may only move one hop in the SAG. Under this simple motion model, each time slot can be considered an independent trial.

Given a path $\mathcal{P} = \{v_1, v_2, \ldots, v_n\}$ the probability of evasion[2] is

$$P_E = \prod_{v_i \in \mathcal{P}} p(v_i), \tag{1}$$

where $p(v_i)$ is the probability that node v_i is not covered. Taking the logarithm of Equation 1, we get

$$\log(P_E) = \sum_{v_i \in \mathcal{P}} \log(p(v_i)). \tag{2}$$

Setting the cost of a node v_i to $\log(p(v_i))$, the cost of path $\mathcal{P}$ is defined as sum of the node costs along the path.

The probability that a node is covered depends on which cameras see the corresponding face and the probability with which they are on. If the set of cameras which cover node v_i is denoted by $\mathcal{C}(v_i)$, node v_i is uncovered if and only if the entire set of cameras $\mathcal{C}(v_i)$ are off. The cost of a node is then given by

$$\log(p(v_i)) = \sum_{c_j \in \mathcal{C}(v_i)} \log(1 - p(c_j)), \tag{3}$$

where $p(c_j)$ is the probability that camera c_j is on. Note that if a node is not covered by any cameras, its cost is 0 and negative otherwise.

Making the change of variables

$$x(c_j) = -\log(1 - p(c_j)),$$
$$\epsilon = -\log(P_E)$$

and substituting into Equation 2, ϵ becomes a linear function of camera weights $x(c_i)$ along a given path. To ensure that P_E is suitably small, we need to ensure that no path of cost less than ϵ exists.

This is a difficult problem because the total cost of the path depends on the individual weights assigned to the cameras. One way to ensure that there is no path of cost less than ϵ is to enumerated all paths between the source and destination. With all the paths as constraints, we can optimize the weight vector $x(c_i)$ over any convex cost function using standard tools from convex optimization [4]. For example, to minimize the maximum amount of time any camera is on, we minimize $||x||_\infty$. Although this is an LP, the number of paths grows exponentially with the number of nodes in G. To reduce the number of constraints, we apply the following lemma:

Lemma 1. *If the cost of a node is set to $\frac{1}{d}$, where d is the shortest distance from the source to the destination through the node, then the total cost of the path cannot be less than 1.*

Proof. Since the length of the shortest path is d and the cost of each along the path step is at least $\frac{1}{d}$, it follows that the cost of any path is at least 1.

---

[2] That is, the probability that the intruder will not be detected.

Using this lower bound, we can provide a constraint for each node individually rather than along paths. If we set the cost of the node v_i to $\frac{\epsilon}{d(v_i)}$ where $d(v_i)$ is path length from v_i to the source plus the path length from v_i to the destination, then by Lemma 1, no path from the source to the destination will have a cost less than ϵ. The constraint for each node becomes

$$\sum_{c_j \in C(v_i)} \log(1 - p(c_j)) \geq \frac{\epsilon}{d(v_i)}, \quad \forall i. \tag{4}$$

The number of constraints is reduced from the number of paths to the number of nodes.

5.2   Assigning Probabilities in Continuous Space

Only assuming an upper bound on the speed of the intruder, we extend the results to the continuous domain. Given the framerates of the cameras, we use the maximum speed of the intruder to convert distance from standard units to frames. For example, if the frame rate of the camera is 15 frames/sec and the maximum speed of the intruder is 3 m/s then the conversion factor is 5 frames/m. The cost of $\mathcal{P}$ is the probability of evasion travelling along that path at the maximum speed. If $\mathcal{P}$ is within the visual field of a camera for a length of n frames and the probability of a camera taking a frame is p, the probability of evasion along $\mathcal{P}$ is $(1 - p)^n$.

To determine the probability of each camera taking a frame, we need to be able to find the distance between two points in a polygonal space. An efficient algorithm is known from computational geometry/robot motion planning. We briefly outline the algorithm, but refer the reader to Chapter 15 of [7] for further details. The algorithm first constructs the *visibility graph* for a space S. The visibility graph consists of nodes representing the vertices of polygonal obstacles in S. An edge connects two vertices if the vertices are visible to one another (i.e. a straight line between the vertices does not intersect any obstacle). The weight of the edge is set as its Euclidean length. To find the shortest distance between two points, we place the points into the visibility graph and add the appropriate edges in the same way (to all visible vertices). Dijkstra's algorithm is used to compute the shortest path through the graph — exactly the shortest path through the space.

There may be regions of space not covered by any cameras which should not contribute to the computed distance as they have a cost of 0. These "holes" act as shortcuts through the space. To account for the holes, the visibility graph must be augmented. In addition to the vertices of the obstacles, there must be a vertex in the visibility graph for each uncovered face. Edges are added if the uncovered face is visible from a vertex. The weight of the edge is the shortest distance from the vertex to the polygon representing the uncovered face. In this augmented visibility graph, the shortest distance returned by Dijkstra will be correct. By comparing Figure 3(a) and Figure 3(b), we see that the distortion introduced by holes can be significant.

The camera probabilities must be set such that no path from the source to the destination has a smaller cost than ϵ. Unlike in the graph case where paths could be enumerated, in the continuous space there are an infinite number of paths. By extending Lemma 1 to the continuous case, we show that one constraint per face is sufficient to guarantee the minimum cost over all paths. The key idea is to find a shortest path *passing through* each face. A path is considered to pass through a face if it intersects or touches any part of the boundary of that face (i.e. it does not have to enter the face). Now we can state the following theorem.

Theorem 1. *For a path through the space from a source to a destination, its length is defined as the distance the path traverses through covered faces. The length does not increase when the path travels through uncovered faces or equivalently each uncovered face is mapped to a point. If the cost of each covered face is set to $\frac{\epsilon}{d}$, where d is the length (as defined above) of the shortest path from the source to the destination passing through the face, then all paths through the space will have a cost of at least ϵ.*

Proof. First note that by definition, any uncovered face has a fixed cost of 0. Therefore, if an uncovered path exists from the source to the destination then the minimum cost from source to destination is 0 and an intruder can traverse the path without being detected with probability 1. Consider a face and the point in the face lying on the shortest path from the source to the destination through the face. In general, neither the point nor the path are unique, however since the value is unique, since it is defined as the minimum. If the shortest total path length is d then we set the cost of the face to $\frac{\epsilon}{d}$. By Lemma 1, all other faces that the path goes through must have equal or greater cost. Therefore, the total cost of the path will be at least ϵ. By similar argument, any other path through the face will have an equal or higher cost, because all sections of the path will have an equal or higher cost.

Since the cost in each face is split between several cameras, all the faces must be sampled. Each face represents a combination of cameras and we do not know a priori which combination of constraints will be active. However, by finding the minimum length path through every face, all possible combinations of cameras that occur in the space are enumerated. If a path of cost lower than ϵ and larger than 0 existed, it would imply that the path crosses a face where the above constraint is not met. Since the constraint for each face is enumerated, this is a contradiction.

5.3   Algorithm

Given a space and camera positions, we find the decomposition of space into faces. Each face is marked with the combination of cameras which see it. Here we only consider the case of one source and one destination but the algorithm extends to multiple sources and destinations. For each face i, we sample the boundary and use the visibility graph algorithm to find the smallest distance from the source to the destination through the face, d_i, by computing the shortest distance to the source and the destination from each point along the boundary of the face. Defining the vector a_i, where $a_{ij} = 1$ if camera j sees face i and 0 otherwise. The optimization problem can be written as

$$\text{minimize} \quad f(x),$$
$$\text{subject to} \quad a_i^T x > \frac{\epsilon}{d_i} \quad \forall i, \text{ and}$$
$$x > 0$$

where $f(x)$ can be any convex function. Then we change $x(c_i)$ back to $p(c_i)$ by

$$p(c_i) = 1 - e^{-x(c_i)} \tag{5}$$

to obtain the vector of camera probabilities.

5.4 Special Case

The special case where we assign one probability to all the cameras can be solved exactly rather than using the constraint at each face. Minimizing this probability is equivalent to minimizing the ∞-norm of the probability vector. The optimal probability can be found if the minimum exposure path with all cameras on is known. The cost of a path through a face is the length of the path through the face multiplied by the number of cameras which see the face, which is the total number of potential frames which the intruder will be exposed to while inside the face. Since each face can have a different cost, the visibility graph cannot be used for computing the minimum exposure path. To find the minimum exposure path, we discretize the space into a grid with 8 neighbors per node. The incoming edge into each node is assigned a cost of the number of cameras which see the node. Then Dijkstra's algorithm is applied to the grid graph to find the minimum cost path from the source to the destination. This gives a good approximation of the true minimum cost. It is simple to show that the discretization error is less than 8% if the source and destination are reasonably far apart compared to the distance between grid points. The distance function with all the cameras on for a simple case is shown in Figure 3(c) where the minimum cost path will clearly need not be the Euclidean shortest path.

The minimum cost is equivalent to the total number of potential frames which will be taken along the path. The required probability is set so that the minimum exposure cost has at least a cost of ϵ. All the camera probabilities are then set to this value. This method only works because the relative weighting of the faces do not change. Changing the relative costs of the faces results in a more difficult version of setting the node weights of the graph in section 5.1, since there is an uncountable number of paths.

6 Deterministic Algorithm

We present a simple heuristic for deterministically detecting intruders as a benchmark to compare the performance of our randomized strategy with. The idea is similar to the algorithm presented in [23]. We try to sweep the space using the cameras while maintaining a barrier between the uncleared or contaminated areas and the destination. By maintaining a barrier, we prevent the intruder from getting closer to the destination and the cleared area from becoming re-contaminated. If we can grow this area, we will eventually clear the entire space.

The deterministic approach has some inherent drawbacks. First, it requires synchronization between the nodes, resulting in communication overhead. Secondly, regions which are not covered by any cameras can never be cleared. This implies that there must always be a separating sweep from the destination to the first hole in coverage, since a potential intruder may hide in it. For the purposes of this comparison, we ignore both issues and compare the deterministic heuristic with the random scheme. The algorithm consists of two parts. First, a schedule of faces which must be turned on at a given time is computed. Then a set of cameras are found which cover the required faces.

To find a set of faces which maintain a barrier and increase the cleared region, we return to the SAG described in Section 4. The sweep begins at the destination, so we mark the appropriate face in the SAG as the choice at time 0. At time 1, all of its adjacent faces are marked to be on. This fulfills both criteria of maintaining a barrier with uncleared regions and expanding the cleared area. In general, if the chosen set of faces at time t is $S(t)$, then $S(t+1)$ will be all the adjacent faces of the faces in $S(t)$ which have not yet been cleared. This is repeated until the source is reached. For the purposes of comparison, we sweep the entire region before repeating the sweep beginning at the initial face.

The second part of the algorithm requires us to choose a set of cameras at each time instance. For each time instance, we must solve the *minimal hitting set problem*. Each face has a set of cameras which see it. The set of faces to be covered at time t form a collection of sets of cameras. We must choose a set of cameras which cover all the sets in the collection. Since, we are solving this problem many times, cameras are chosen using the following method. At $t = 0$, all cameras are assigned a weight of 0. First, all the cameras in the collection are sorted in increasing order according to their weight. From the minimum weight, we find the camera which appears in the most sets (i.e. sees the most faces). This camera is chosen and its weight is increased by 1. It is removed from the list and the procedure is repeated until all the sets are removed. The re-weighting gives preference to cameras which have been chosen fewer times.

7   Simulation Experiments

In this section, we investigate the performance of the algorithm through simulations. The geometric preprocessing was implemented in CGAL [5] and the optimization was done using the CVX [6] package in MATLAB. The four layouts shown in Figure 4(a)-(d) were tested. Layout 1 and 2 are rectangular rooms with differing camera layouts. Layout 1 corresponds to a real camera deployment of 16 webcams. Layout 2 has 14 cameras placed more evenly around the room. Layouts 3 and 4 are larger with multiple occlusions. Although there are more cameras in the latter layouts (31 and 17 respectively), the coverage is not as dense as in the first two cases. The extent of the coverage is shown in Figure 4(e)-(h) where more densely covered areas are lighter in color. For each layout, we compare the results in terms of load balancing between cameras and the total amount of energy spent.

The results summarized in Table 1 were obtained by setting the probability of evasion to 0.01. We compare optimizing the 2-norm and the ∞-norm by bounding the detection probability on individual faces ($||p||_2(1)$; $||p||_\infty||(1)$), the special case of solving for one parameter ($||p||_\infty(2)$) and the deterministic algorithm. The first three columns show the the 2-norm, the 2-norm normalized by the number of cameras, and the ∞-norm of the probability vector respectively. With the probability vector set to the solution, the maximum probability of evasion along the minimum exposure path was computed and is shown in the fourth column of the table. The exception is the deterministic scheme which has a probability of evasion of 0. In all 4 layouts, the deterministic scheme used the cameras much more than any of the probabilistic schemes, illustrating the price of setting P_E to 0.

In Layouts 1 and 2, locally enforcing the minimum detection probability at each face results in overcoverage. The resulting maximum P_E was at least an order of magnitude smaller than the desired 0.01. At the same time the total energy used ($||p||_2$) is at most twice the amount used in the optimal single parameter case. In Layout 3, the performance of the three probabilistic schemes is comparable. The overcoverage is minimal because most of the area is sparsely covered and most of the paths are of similar length. Layout 4 is lies somewhere between. The maximum P_E is still an order of magnitude smaller than the single parameter case, but all other values are comparable.

These results show that the optimization technique provides very good performance giving quite low probabilities of evasion with each camera taking only a small number of frames. Unfortunately, the approximation in the local enforcement of constraints using Theorem 1, makes it difficult to set the true probability of evasion to a desired value. This means that some experimentation is necessary if we do not want overcoverage in the network. The effect of the optimization on coverage be seen in Figure 4(i)-(l). This shows the cost of paths from the source to the destination over the space when the probabilities are set to the 2-norm solution for Layouts 2 and 3. For Layout 2, the coverage becomes much more uniform than in the case of one parameter, while Layout 3 exhibits almost no change.

The notion of load balancing is important because it prevents certain cameras from being overused. A good metric of load balancing is obtained by comparing the average value of the probability vector to its ∞-norm. The average value is given by $||p||_2/N$ in Table 1. The single parameter case obviously has a ratio of 1 in all cases. The other algorithms did nearly as well. For method (1), the maximum was always less than twice the average. Given a particular layout it may not be desirable to achieve perfect load balancing as there may be some cameras which inherently need to do more sensing than others. The deterministic

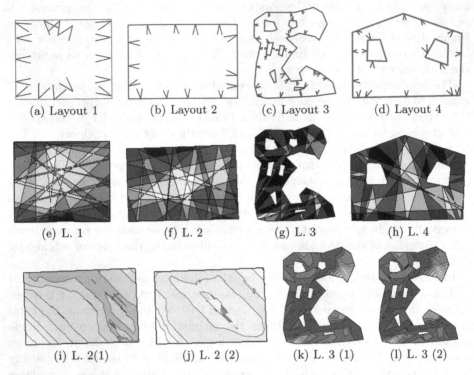

Fig. 4. (a)-(d) 4 example layouts. (e)-(h) Coverage maps for the 4 layouts. The lighter the color of a point, the more cameras cover that point. (i) The minimum total cost from the source to the destination for layout 2 with all the cameras set to one probability. The darkest region is the minimum cost path. (j) The minimum total cost path with the prob. from the approx. optimizaion (k) and (l) The same for layout 3.

algorithm also performed well. However, when there were disjoint faces covered by one camera, it was impossible prevent turning one camera on multiple times.

8   Conclusions and Further Work

We have presented an algorithm for detection where cameras take frames with an assigned probability rather than continuously monitor the space. The camera probabilities can be optimized over a desired convex cost function with relatively few constraints. Once the probabilities are set, no further coordination is required between the nodes. Experimentation shows that although each camera samples infrequently and independently, global guarantees are made on the detection probability.

From the experimentation, we see that locally enforcing the probability of evasion at each face results in overcoverage. Whether it is feasible to efficiently encode the minimum cost over all paths when cameras have variable probabilities is an open question. Furthermore, we assumed that the main energy cost in

Table 1. Results for the 4 layouts. (1) refers to the approximate technique; (2) refers to the optimal ∞-norm value; Det.is the deterministic algorithm

Layout	Method	$\|p\|_2$	$\|p\|_2/N$	$\|p\|_\infty$	Max P_E
1	$\|p\|_2(1)$	0.0608	0.0152	0.0262	0.0005
	$\|p\|_\infty(1)$	0.0785	0.0196	0.0262	0.00002
	$\|p\|_\infty(2)$	0.0320	0.0080	0.008	0.01
	Det.	0.0675	0.2166	0.4545	-
2	$\|p\|_2(1)$	0.0512	0.0137	0.0193	0.001
	$\|p\|_\infty(1)$	0.0646	0.0173	0.0193	0.00012
	$\|p\|_\infty(2)$	0.0344	0.0092	0.0092	0.01
	Det.	1.387	0.3707	0.3722	-
3	$\|p\|_2(1)$	0.1921	0.0345	0.0357	0.0094
	$\|p\|_\infty(1)$	0.1947	0.0353	0.0353	0.0093
	$\|p\|_\infty(2)$	0.1954	0.0351	0.0351	0.01
	Det.	1.400	0.3501	0.3510	-
4	$\|p\|_2(1)$	0.0817	0.0198	0.0228	0.0009
	$\|p\|_\infty(1)$	0.0871	0.0211	0.0228	0.0008
	$\|p\|_\infty(2)$	0.0614	0.0149	0.0149	0.01
	Det.	0.7717	0.1038	0.5729	-

acquiring frames was in actually sampling the frame. Taking the cost of powering on and off into account introduces correlation between the detection probabilities over time complicating the problem significantly.

Acknowledgement. The authors wish to acknowledge support by NSF grants CNS-0435111, CNS-0626151, ARO grant W911NF-06-1-0275, and the DoD Multidisciplinary University Research Initiative (MURI) program administered by ONR under Grant N00014-00-1-0637.

References

[1] Abrams, Z., Goel, A., Plotkin, S.: Set k-cover algorithms for energy efficient monitoring in wireless sensor networks. In: IPSN '04: Proceedings of the third international symposium on Information processing in sensor networks, pp. 424–432. ACM Press, New York (2004)

[2] Arora, A., Dutta, P., Bapat, S., Kulathumani, V., Zhang, H., Naik, V., Mittal, V., Cao, H., Demirbas, M., Gouda, M., Choi, Y., Herman, T., Kulkarni, S., Arumugam, U., Nesterenko, M., Vora, A., Miyashita, M.: A line in the sand: a wireless sensor network for target detection, classification, and tracking. Comput. Networks 46(5), 605–634 (2004)

[3] Biber, P., Fleck, S., Wand, M., Staneke, D., Strasser, W.: First experiences with a mobile platform for flexible 3d model aquisition in indoor and outdoor environments - the waglele. In: 3D-ARCH'2005: 3D Virtual Reconstruction and Visualization of Complex Architectures (2005)

[4] Boyd, S., Vandenberghe, L.: Convex Optimization. Cambridge University Press, Cambridge (2004)

[5] Cgal: Computational geometry algorithms library. http://www.cgal.org
[6] Cvx: Matlab software for disciplined convex programming.
http://http://www.stanford.edu/~boyd/cvx/
[7] de Berg, M., van Kreveld, M., Overmars, M., Schwartzkopf, O.: Computational
Geometry - Algorithms and Applications. Springer, Heidelberg (2000)
[8] Devarajan, D., Radke, R.J., Chung, H.: Distributed metric calibration of ad hoc
camera networks. ACM Trans. Sen. Netw. 2(3), 380–403 (2006)
[9] Devarajan, D., Radke, R.J., Chung, H.: Distributed metric calibration of ad hoc
camera networks. ACM Trans. Sen. Netw. 2(3), 380–403 (2006)
[10] Erdem, U.M., Sclaroff, S.: Automated camera layout to satisfy task-specific and
floor plan-specific coverage requirements. Comput. Vis. Image Underst. 103(3),
156–169 (2006)
[11] Gui, C., Mohapatra, P.: Virtual patrol: a new power conservation design for
surveillance using sensor networks. In: IPSN '05: Proceedings of the 4th international symposium on Information processing in sensor networks, Piscataway,
NJ, USA, p. 33. IEEE Press, New York (2005)
[12] Huang, C.-F., Tseng, Y.-C.: The coverage problem in a wireless sensor network.
In: WSNA '03: Proceedings of the 2nd ACM international conference on Wireless
sensor networks and applications, pp. 115–121. ACM Press, New York (2003)
[13] Kulkarni, P., Ganesan, D., Shenoy, P., Lu, Q.: Senseye: a multi-tier camera sensor
network. In: MULTIMEDIA '05: Proceedings of the 13th annual ACM international conference on Multimedia, pp. 229–238. ACM Press, New York (2005)
[14] Kumar, S., Lai, T.H., Arora, A.: Barrier coverage with wireless sensors. In: Santosh
Kumar, T.H. (ed.) MobiCom '05: Proceedings of the 11th annual international
conference on Mobile computing and networking, pp. 284–298. ACM Press, New
York (2005)
[15] Kumar, S., Lai, T.H., Balogh, J.: On k-coverage in a mostly sleeping sensor network. In: Santosh Kumar, T.H. (ed.) MobiCom '04: Proceedings of the 10th annual
international conference on Mobile computing and networking, pp. 144–158. ACM
Press, New York (2004)
[16] Li, D., Wong, K., Hu, Y., Sayeed, A.: Detection, classification, and tracking of
targets. IEEE Signal Processing Magazine 19(2), 17–30 (2002)
[17] Margi, C.B., Lu, X., Zhang, G., Stanek, G., Manduchi, R., Obraczka, K.:
Meerkats: A power-aware, self-managing wireless camera network for wide area
monitoring. In: Workshop on Distributed Smart Cameras (DSC-06) (October
2006)
[18] Margi, C.B., Petkov, V., Obraczka, K., Manduchi, R.: Characterizing energy consumption in a visual sensor network testbed. In: 2nd International IEEE/Create-
Net Conference on Testbeds and Research Infrastructures for the Development of
Networks and Communities (March 2006)
[19] Megerian, S., Koushanfar, F., Potkonjak, M., Srivastava, M.B.: Worst and best-
case coverage in sensor networks. IEEE Transactions on Mobile Computing 4(1),
84–92 (2005)
[20] Megerian, S., Koushanfar, F., Qu, G., Veltri, G., Potkonjak, M.: Exposure in
wireless sensor networks: theory and practical solutions. Wirel. Netw. 8(5), 443–
454 (2002)
[21] O'Rourke, J.: Art gallery theorems and algorithms. Oxford University Press, Oxford (1987)
[22] Rekletis, I.M., Dudek, G.: Automated calibration of a camera sensor network.
In: IEEE/RSJ International Conference on Intelligent Robots and Systems, pp.
401–406, Edmonton Alberta, Canada (August 2-6, 2005)

[23] Ren, S., Li, Q., Wang, H., Zhang, X.: Design and analysis of wave sensing scheduling protocols for object-tracking applications. In: Prasanna, V.K., Iyengar, S., Spirakis, P.G., Welsh, M. (eds.) DCOSS 2005. LNCS, vol. 3560, Springer, Heidelberg (2005)

[24] Shin, J., Guibas, L., Zhao, F.: A distributed algorithm for managing multi-target identities in wireless ad-hoc sensor networks. In: Zhao, F., Guibas, L.J. (eds.) IPSN 2003. LNCS, vol. 2634, pp. 223–238. Springer, Heidelberg (2003)

[25] Veltri, G., Huang, Q., Qu, G., Potkonjak, M.: Minimal and maximal exposure path algorithms for wireless embedded sensor networks. In: SenSys '03: Proceedings of the 1st international conference on Embedded networked sensor systems, pp. 40–50. ACM Press, New York (2003)

[26] Yan, T., He, T., Stankovic, J.A.: Differentiated surveillance for sensor networks. In: SenSys '03: Proceedings of the 1st international conference on Embedded networked sensor systems, pp. 51–62. ACM Press, New York (2003)

[27] Yang, D., Gonzalez-Banos, H., Guibas, L.: Counting people in crowds with a real-time network of image sensors. In: Proc. IEEE ICCV (2003)

Leveraging Redundancy in Sampling-Interpolation Applications for Sensor Networks

Periklis Liaskovits and Curt Schurgers

University of California San Diego, Electrical and Computer Engineering Department
{pliaskov, curts}@ucsd.edu

Abstract. An important class of sensor network applications aims at estimating the spatiotemporal behavior of a physical phenomenon, such as temperature variations over an area of interest. These networks thereby essentially act as a distributed sampling system. However, unlike in the event detection class of sensor networks, the notion of sensing range is largely meaningless in this case. As a result, existing techniques to exploit sensing redundancy for event detection, which rely on the existence of such sensing range, become unusable. Instead, this paper presents a new method to exploit redundancy for the sampling class of applications, which adaptively selects the smallest set of reporting sensors to act as sampling points. By projecting the sensor space onto an equivalent Hilbert space, this method ensures sufficiently accurate sampling and interpolation, without a priori knowledge of the statistical structure of the physical process. Results are presented using synthetic sensor data and show significant reductions in the number of active sensors.

Keywords: Sensor networks, spatial monitoring, sampling, sensor selection, sensing topology management, energy efficiency, Hilbert space.

1 Introduction

Large scale networks of wireless micro-sensors enjoy increasing popularity in two broad and very distinct classes of applications: event detection and continuous spatiotemporal sampling. In event detection, the goal of the network is to notify the user when particular events take place, such as the presence of an intruder or the start of a fire. A sensor reading such as an elevated temperature indicates for example that a hot object is nearby. A node is often characterized as having a sensing range, which is a measure for the distance over which an event can be detected reliably [1], [2]. The overall goal in terms of network-wide reliable coverage is to ensure that each point in space is covered by a minimum number of sensors [1], [2], [3].

By contrast, in *spatiotemporal sampling* the sensor reading is a single sample, a point value in time and space. The goal of the network is now to create a continuous map of the physical phenomenon, by interpolating the values of the field where no sensor readings exist. As such, there is no concept of a sensing range. Example applications in this class include monitoring humidity and soil decomposition variations in agricultural fields, or building the temperature map inside a warehouse to optimize air-flow. This paper focuses on this sampling-interpolation class of sensor

J. Aspnes et al. (Eds.): DCOSS 2007, LNCS 4549, pp. 324–337, 2007.

network applications and more specifically on how to leverage a dense deployment to increase the overall system lifetime.

It is envisioned that as individual nodes become cheaper, sensor networks will often be over-deployed to compensate for unknown environment and non-exact deployment. As a result, more sensor nodes are present than strictly necessary, which boils down to effective over-sampling in sampling-interpolation type applications. This redundancy can be leveraged to increase the network lifetime, by only having the strictly required set of nodes send their readings, while the others remain in a sleep state. In general, putting nodes in a low-power sleep state has been recognized as one of the most effective means to ensure energy efficient network operation [4].

However, simply finding the minimum required set of nodes is not sufficient by itself. To create a functional *sensing topology management scheme*, multiple such sets should be found, which are as disjoint as possible [1], [5], [6]. At each point in time only one of them is made active while the others are in a low-power sleep state, and this functionality is rotated amongst the sets. Existing work in sensing topology management has focused mainly on the class of event detection applications, which are characterized by a sensing range. Disjoint sets of sensors are found based on a k-coverage criterion, where each point in the field must be covered by at least k sensors to ensure sufficient detection probability [7].

However, k-coverage is unusable for the sampling-interpolation class of sensor network applications, as sensing range is meaningless here. Instead, a different strategy is needed to select sensor sets based on a sampling and reconstruction criterion instead. Such a criterion will rely heavily on the spatiotemporal characteristics of the underlying physical phenomenon, which are typically unknown before deployment, rather than on a property of the sensor. Suitable sensing topology management extends far beyond the application of known sampling theory alone, and indeed provides some unique challenges. First, topology management operates on an already-deployed network. As such, the goal cannot be to find the optimal sampling locations, i.e. positions of sensors, [11] but is restricted to the positions of the already-deployed sensor nodes. Second, it is not sufficient to find just the one minimum set of sensors, but instead the objective is creating multiple equivalent sets to rotate between. Third, the statistical properties of the underlying physical phenomenon to be sensed are typically unknown a priori and therefore have to be learned, preferably with minimal overhead.

To tackle this problem of sensing topology management for sampling-style applications, we will introduce a powerful mathematical tool, namely the use of a *finite-dimensional Hilbert space*. Its unique strength is that it allows a joint representation of both the sensor network topology and the physical process to be sensed, within the same framework. Sensor locations map onto vectors in this Hilbert space, and inner products between them are defined by the correlation structure of the sensed physical process. Consequently, we will be able to prove that the sensing topology management problem is a hard combinatorial problem. The same Hilbert space representation will allow us to devise practical algorithms for creating sets of sensors, while being able to reconstruct the phenomenon within predefined accuracy bounds and without a priori knowledge of the physical process statistics. We will evaluate our approach on synthetic data illustrating that a significant reduction in active sensors can be achieved.

2 Overview and Fundamentals

In this paper, we study the problem of sensing topology management without prior knowledge of the data statistics. We assume a sensor network consisting of N sensor nodes (also referred to as 'nodes' or 'sensors') spread over a 2-D observation area of size F in a uniformly random fashion; there is no redeployment or addition of sensors. Deployed sensors are indexed 1 through N. The setup and definitions we will use are illustrated in Figure 1, and will be detailed in the remainder of this subsection.

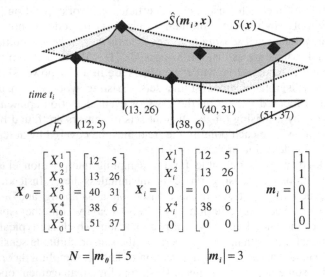

$$X_0 = \begin{bmatrix} X_0^1 \\ X_0^2 \\ X_0^3 \\ X_0^4 \\ X_0^5 \end{bmatrix} = \begin{bmatrix} 12 & 5 \\ 13 & 26 \\ 40 & 31 \\ 38 & 6 \\ 51 & 37 \end{bmatrix} \quad X_i = \begin{bmatrix} X_i^1 \\ X_i^2 \\ 0 \\ X_i^4 \\ 0 \end{bmatrix} = \begin{bmatrix} 12 & 5 \\ 13 & 26 \\ 0 & 0 \\ 38 & 6 \\ 0 & 0 \end{bmatrix} \quad m_i = \begin{bmatrix} 1 \\ 1 \\ 0 \\ 1 \\ 0 \end{bmatrix}$$

$$N = |m_0| = 5 \qquad |m_i| = 3$$

Fig. 1. Example process and reconstruction

At discrete time instants t_i, a subset of the sensors measures the value of a physical quantity of interest (e.g. temperature, humidity). A realization of the spatial process is denoted as $S_i(x)$, where vector x represents 2-D coordinates in the observation field F.

At each time instant, the boolean vector m_i, of length N, indicates if a sensor is active (1) or not (0). The number of non-zero elements of m_i is denoted as $|m_i|$. As a convention, we will reserve subscript i = 0 for the situation where all sensors are active. Therefore, m_0 is the set of all available sensors in the network and $|m_0| = N$. The coordinate matrix X_i consists of the locations of the active sensors, where each of the non-zero rows X_i^k (with $k = 1.. N$) corresponds to one sensor (see also Figure 1). The all-zero rows are never used in what follows, rather they serve to keep indices of sensors the same across sets. As before, by convention X_0 lists the coordinates of all available sensors. Throughput this paper, we assume that sensor locations are known within reasonable accuracy, by running a localization service in the network. In addition, nodes are time synchronized at a coarse level, such that they sample at roughly the same time t_i.

Based on the sensor data reported by the active sensors, a spatial map of the phenomenon is reconstructed. More specifically, we consider point reconstruction that is linear on the measured values. This is an assumption widely employed by existing literature [8], [9], [10], [11], and covers a broad range of interpolation techniques. Formally, the reconstruction $\hat{S}(m_i, x)$ of the phenomenon $S_i(x)$ is written as:

$$\hat{S}(m_i, x) = \sum_{k=1}^{N} g_k(x, X_i) \cdot S_i(X_i^k) \tag{1}$$

Here $S_i(X_i^k)$ is the value of the spatial process at the location X_i^k, or equivalently the sensor value of node k at X_i. The coefficient functions $g_k(x,X_i)$ depend on the specific reconstruction scheme used. To evaluate the performance of this reconstruction, we utilize the distortion metric defined in equation (2). This represents the expected distortion over the field of observation, with respect to all possible realizations of the spatial process:

$$E[D_{m_i}] = \frac{1}{F} \int_F E[(\hat{S}(m_i, x) - S(x))^2]dx \tag{2}$$

The distortion defined in equation (2) is associated with a particular active set of sensors m_i. As explained in the introduction, if multiple disjoint sets can be found, only one of them needs to be active at each point in time. The sensor nodes belonging to the other sets can remain in an energy-efficient sleep state. This is illustrated in Figure 2. In this example, there are three sets, and rotating the sensing functionality between them can roughly result in a three-fold increase in network lifetime.

The overall goal of our sensing topology management scheme is therefore to divide the sensors in as many disjoint sets as possible, while still meeting a desired distortion bound for each set.

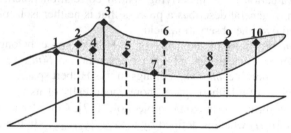

Set 1: nodes {1, 3, 10}   **Set 2:** nodes {2, 5, 6, 8}   **Set 3:** nodes {4, 7, 9}

Fig. 2. Monitoring schedule: rotating sets of active sensors

We will refer to the sequence of sensor selections, i.e. sets, over time as a monitoring schedule. To devise this schedule, we assume that the statistical properties of the process are not known a priori, but instead have to be learned after network deployment. As a result, we propose a two-phase strategy:

1. During the learning phase, all N sensors report their data, where in addition to interpolation itself (i.e. the monitoring application is already operational), the goal is to estimate relevant statistical properties of the process. This learning phase could extend over multiple time instants t_i.

2. During the monitoring phase, only sensors from an active set report and these sets are rotated over time. How such sets are found is the main subject of the rest of this paper.

3 Hilbert Space Representation

To find the sets of active sensors, we will develop a methodology that maps the problem into an equivalent Hilbert space. We observe that the readings $S(X_i^k)$ at a particular sensor k can be viewed as a random variable ($\forall\ i$). Furthermore, the completed span $\{S(X_i^k)\}_{k=1...N}$ of these random variables, i.e. all their possible linear combinations and limits of Cauchy sequences thereof, form a Hilbert space [12]. A Hilbert space is essentially a set of elements, indiscriminately referred to as points or vectors, with appropriate operations defined on them. These operations are linear addition, scalar multiplication, inner product and the norm of an element.

Specifically, the inner product and norm in this space are defined as (the complex conjugate is denoted by *):

$$< S(X_i^k), S(X_i^p) >= E[S(X_i^k) \cdot S^*(X_i^p)] = R(X_i^k, X_i^p) \tag{3}$$

$$\left\| S(X_i^k) \right\|^2 =< S(X_i^k), S(X_i^k) >= E[|\ S(X_i^k)\ |^2] \tag{4}$$

The inner product serves as a measure of similarity between values measured by different sensors, and captures the effect of the locations of the sensors. It is equivalent to evaluation of an underlying spatial correlation function $R(\cdot,\cdot)$ at these locations, which in general describes a process that is neither isotropic nor stationary but only bounded over the observation field.

All linear combinations of elements of a Hilbert space belong to this space. Therefore, as defined in equation (1), the reconstruction $\hat{S}(m_0, \mathbf{x})$ for any point $\mathbf{x}$ (when all sensors report, i.e. $m_i = m_0$) belongs to this Hilbert space.

A useful property of a Hilbert space, more specifically of its inner product, is that of 'approximation through orthogonal projection' [12]. Consider a Hilbert space H_0, and a subspace of it H_1 which is defined by a basis $\{\xi_k\}_{k=1...Q}$. In terms of minimizing the squared norm of the error, the best approximation of an element $\eta \in H_0$ by an element $\eta_1 \in H_1$ will be the orthogonal projection of η onto H_1. In addition, the approximation error induced by this orthogonal projection is given by (5) (assuming that η is not exactly orthogonal to H_1):

$$\min \|\eta - \eta_1\|^2 = \|\eta\|^2 - \frac{(\sum_{k=1}^{Q} |< \eta, \xi_k >|)^2}{\left\| \sum_{k=1}^{Q} < \eta, \xi_k > \cdot \xi_k \right\|^2} \tag{5}$$

As a final remark, we note that the power of the Hilbert space approach is the following: it transforms working with a collection of correlated random variables, into manipulating a set of deterministic vectors where all relevant statistical information is captured in the inner product. It therefore has the potential of being a useful tool in sensor network processing even beyond the specific problem we consider in this paper.

3.1 Primary Subspace

We can now translate the sensor set selection process into this Hilbert space representation. In this case, the Hilbert space, which we denote as H_S, is the space consisting of all random variables $S(x)$ across the whole observation field, i.e, $x \in F$. Each subset of sensors $\{S(X_i^k)\}_{k=1...N}$ defines a subspace H_{Xi}. In our notation of section 3, this corresponds to $\{\xi_k\} = \{S(X_i^k)\}$, $H_1 = H_{Xi}$, $H_0 = H_S$ and η is $S(x)$ for a particular x. Optimal reconstruction as defined in (1) and (2), essentially means that we want to represent each random variable $S(x)$ as a linear combination of vectors $\{S(X_i^k)\}$, optimally with respect to mean squared error. Hence, in view of (4), orthogonal projection provides the best possible performance of a set of sensors $\{S(X_i^k)\}$ in terms of representing all elements of H_S, i.e. the random spatial phenomenon of interest. The expected squared error, which we defined as our distortion metric, can then be evaluated directly through equation (5). However, this requires knowing the inner product of $\eta = S(x)$, $x \in F$, with the sensor vectors. From (3), we see that this is equivalent to knowing the continuous spatial correlation function:

$$< S(x), S(X_i^k) > = E[S(x) \cdot S^*(X_i^k)] = R(x, X_i^k) \qquad (6)$$

Although it is possible to estimate the continuous correlation function [11], this is ultimately a costly and intricate procedure that also does not lend itself to providing distortion guarantees. Instead, we will restrict η to elements of a subspace of H_S, namely that spanned by all deployed sensors H_{X0}. We refer to H_{X0} as the *primary subspace*. It is generated by $\{S(X_0^k)\}_{k=1...N}$ and is therefore of dimension at most N (we have in fact proved that for Poisson sensor deployments it is exactly N, see next section). Essentially, we assume that the initial number of sensors is large enough so that H_{X0} is a close approximation to H_S, or in other words that they can capture enough details of the underlying physical phenomenon in the first place. The distortion metric is thus defined in relation to the maximum information we could extract with our initial deployment. Formally, it means that equation (1) is transformed into:

$$E[D_{m_i}] = \frac{1}{F} \int_F E[(\hat{S}(m_i, x) - \hat{S}(m_0, x))^2] dx \qquad (7)$$

The set $\{S(X_0^k)\}_{k=1...N}$ is a set of vectors spanning the primary subspace, and any reconstruction $\hat{S}(m_i, x)$ is a linear combination of these vectors. Therefore, to evaluate orthogonal projection in equation (5), we ultimately require only the inner products between all sensor vectors. Again, as in (3), these inner products correspond to covariances, which can be estimated as (assuming ergodicity in time):

$$R(X_0^k, X_0^l) \approx \hat{R}_{kl}(X_0^k, X_0^l) = \frac{1}{\Theta - 1} \cdot \sum_{q=1}^{\Theta} S_q(X_0^k) \cdot S_q(X_0^l) \qquad (8)$$

These empirical covariances $\hat{R}_{kl}$ are calculated during the learning phase (see section 2), which requires process realizations from Θ time instants. When put in matrix form, $\hat{R}_{kl}$ can be thought of as an empirical Grammian matrix for projection onto the primary subspace. When estimating distortion in our subsequent algorithms, we will similarly approximate equation (7) with a time average over W instants:

$$\hat{E}[D_{m_i}] = \frac{1}{F} \int_F \frac{1}{W} \cdot \sum_{q=1}^{W} (\hat{S}_q(m_i, x) - \hat{S}_q(m_0, x))^2 dx \qquad (9)$$

4 Sensor Selection

In the Hilbert space framework, introduced in the previous section, each individual sensor can be thought of as a vector in the primary subspace. Consequently, the sensing topology management problem of selecting disjoint sets of sensors translates into creating sets of vectors instead. The goal is to maximize the number of sets that can be found (or equivalently, minimize the average number of vectors in each set), while ensuring that each one can provide a sufficiently accurate reconstruction in terms of expected distortion.

As a first step in tackling this problem, we consider a more basic variant, namely that of finding just a single minimal set: *Given an initial set of sensors, find the minimal subset which yields an interpolated reconstruction that has an expected distortion of at most D_0.*

Finding multiple sets, is a generalization of this problem and hence computationally at least as hard. Furthermore, our algorithm will be built on a good understanding of solutions for the single set selection problem.

The basic single-set problem is related to the issue of sparse signal approximation with general dictionaries, which has been studied in signal processing literature. The term 'dictionary' refers to a set of non-orthogonal vectors used for representation in a Hilbert space, without necessarily forming a basis for that space. A known problem in a V-dimensional Hilbert space is how to select the best vectors out of a redundant dictionary of size $P \geq V$ to approximate an element of that Hilbert space. This requires enumerating all possible subsets of vectors, an operation of which the cost is exponential in P [13], [14]. In the case of a general dictionary, the resulting computation is provably NP-hard [13].

In our scenario, the dimension of the primary subspace H_{X0} is at most N. The set of N vectors that correspond to the initially deployed sensors $\{S(X_0^k)\}_{k=1...N}$ is therefore a redundant dictionary for this space. A dictionary of size N effectively means that the computational cost for optimal sensor selection grows exponentially with the size of the network N. For our particular case we have also proved a stronger result: that for Poisson deployments, the vectors $\{S(X_0^k)\}_{k=1...N}$ are linearly independent on the average. Linear independence means that the dimension of H_{X0} is N, i.e. the dictionary $\{S(X_0^k)\}_{k=1...N}$ is also a basis for the space, rendering optimization over *any* redundant dictionary for this space exponentially hard with the size of the network. More specifically, we have proven the following lemma (proof can be found in a longer journal version of this paper):

Lemma 1: For a deployment where the positions of the sensors form a Poisson point process with constant rate β and the monitored random process $S(x)$ is wide sense stationary, vectors $\{S(X_0^k)\}_{k=1..N}$ are linearly independent on the average.

Since the single-set problem is therefore hard, the same will hold for the extended problem of finding multiple sets. As a result, we have to resort to heuristic algorithms to perform the selection of multiple active sets of sensors. However, the Hilbert space representation provides a powerful framework to build these heuristics, as will be presented in the next subsection.

Finally, we remark that the methods that have been developed to solve the problem of sparse signal approximation with general non-orthogonal dictionaries, such as convex minimization of the ℓ_1 norm or some variant of matching pursuit, are not applicable to our single-set or multiple-set selection problem. The reason is that they deal with approximating deterministic signals that are given beforehand rather than random processes.

5.1 Greedy Algorithm

Based on the expressive power of the Hilbert space representation, we present a class of greedy heuristics to tackle the problem of vector selection.

As an initial step towards identifying a minimal descriptive subset, we consider first the set of all available vectors (sensors). Then, one vector at a time is removed from this set, guided by a score function. It would be appealing to remove the vector whose absence will 'hurt' us the least when trying to describe any element of the primary subspace with the remaining vectors. Since, by definition, all elements of the primary subspace are linear combinations of the initial set of vectors, intuitively, a 'good' vector to remove would be the one that can be best described by a linear combination of the remaining vectors. A measure of this is readily given by equation (5), where $\{\xi_k\}$ are the remaining vectors and η is the candidate vector for removal. The greedy removal thus selects the candidate vector η for which equation (5) is minimized.

The basis of our algorithms is essentially quantifying 'collinearity' or 'orthogonality' between a given candidate vector and an existing set of vectors. This can be done by means of equation (5): if for instance the projection error of a candidate vector onto a set of existing vectors is maximal among all such vectors, then we know that the descriptive power of the set will maximally grow if we add the candidate to it.

The algorithm for finding multiple sets of sensors resulting in adequate reconstruction (or, equivalently, multiple approximate bases for the primary subspace) proceeds as follows. It is not known a priori how many sets can be possibly created. Instead, we start creating the first set by selecting vectors until the distortion criterion is met. Next, the second set will be selected from the remaining vectors, and so forth. Consider, in general, a situation where we are in the process of creating the jth set. At this point, the primary subspace can be considered as being partitioned in three subspaces: (1) the space H_U of vectors in sets 1 through $j-1$; (2) the space H_A of vectors already selected in set j; (3) the space H_R of vectors not yet selected for any of the sets.

Our algorithm considers all candidate vectors η from those not yet belonging to any set. Similar to the greedy removal explained above, for each one of them, it computes the error by orthogonal projection in both of the spaces H_A and H_R (always excluding the vector η):

$$E_A(\eta) = \min\left\|\eta - \sum_\kappa c_k \cdot \xi_\kappa\right\|^2 \quad H_A = span\{\xi_k\}$$

$$E_R(\eta) = \min\left\|\eta - \sum_\kappa d_\kappa \cdot \zeta_\kappa\right\|^2 \quad H_R = span\{\zeta_k\} \tag{10}$$

Based on these two metrics, the 'score' $C(\eta)$ of the vector η is calculated as:

$$C(\eta) = \alpha \cdot E_A(\eta) + (1-\alpha) \cdot E_R(\eta) \tag{11}$$

The vector η with the *maximum* score amongst all candidates is then added to the jth set. The parameter α ($0 \le \alpha \le 1$) allows us to tweak the behavior of the algorithm. Intuitively, the effect of α can be understood as follows:

- If $\alpha = 1$: The vector is added that is 'most orthogonal' to vectors already in H_A, i.e. approximating it by vectors already in the set induces maximal error. This maximally expands the span (i.e. the descriptive value) of the set.
- If $\alpha = 0$: The vector is added that is maximally contracting H_R, i.e. the descriptive value of remaining vectors. It adds the vector that leaves the least amount of information uncaptured.

As a result, the first term in (11), essentially controls how fast the descriptive value of the current set of sensors grows in the primary subspace. On the other hand, the second term partially compensates for situations where the first term favors sensors that are too far apart in the field (and their vectors are likely to be nearly orthogonal), by favoring sensors that also have good descriptive value for the rest of the field.

The detailed set selection algorithm is presented in Figure 3. The strength of this algorithm is that it only utilizes covariance information between vectors instead of over the entire field, through Hilbert space inner products. Note that when the algorithm terminates, there may be some remaining sensors that were not assigned to any active sets (since they could not form a set by themselves that satisfies the distortion target). In this case, they are distributed in a round robin fashion among existing sets in such a way that each set is assigned the sensor that maximally expands it.

Our complete sensing topology management approach is built upon the greedy algorithm of Figure 3. As explained in section 2, it consists of two phases:

1. During the learning phase, all sensors report their values for Θ time instants. This allows us to calculate the empirical covariance matrix $\hat{R}_{kl}$ from (8). The set selection algorithm of Figure 3 takes this matrix as input, and generates the disjoint sets. It estimates distortion through (9) and bounds it by the given target distortion objective D.

2. During the monitoring phase, only one set is activated at a time. The actual scheduling of the different sets can be periodic round robin or based on when the current active set fails.

---

1   **Input:** *vectors* $\{S(X_0^k)\}_{k=1...N}$, $\hat{R}_{kn}$, $\{\hat{S}_q(m_0, x)\}_{q=1...W}$, D

2   **Output:** sets of vectors m_i: $\{S(X_i^k)\}_{k=1...|m_i|}$

3   **begin**

4      $m_0 \leftarrow \bigcup\limits_{k=1}^{N} S(X_0^k)$, $i = 1$

5   **repeat**

6        $m_i \leftarrow \varnothing$

7   **repeat**

8   **if** $(m_i == \varnothing)$ select first available vector into m_i

9   **else**

10   **foreach** candidate vector $S(X)$

11             $k_0 \leftarrow$ index of $S(X)$ in $\bigcup\limits_{k=1}^{N} S(X_0^k)$

12
$$S(X^*) \leftarrow \arg \max_{S(X) \in m_0} \left(\alpha \cdot \left(\hat{R}_{k_0 k_0} - \frac{(\sum\limits_{k \in m_i} (\hat{R}_{k_0 k})^2)^2}{\sum\limits_{k \in m_i, n \in m_i} \hat{R}_{k_0 k} \cdot \hat{R}_{k_0 n} \cdot \hat{R}_{kn}} \right) \right.$$
$$\left. + (1-\alpha) \cdot \left(\hat{R}_{k_0 k_0} - \frac{(\sum\limits_{\substack{k \in m_0 \\ k \neq k_0}} (\hat{R}_{k_0 k})^2)^2}{\sum\limits_{\substack{k \in m_0 \\ k \neq k_0}} \sum\limits_{\substack{n \in m_0 \\ n \neq k_0}} \hat{R}_{k_0 k} \cdot \hat{R}_{k_0 n} \cdot \hat{R}_{kn}} \right) \right)$$

13            $m_i \leftarrow m_i \cup S(X^*)$

14            $m_0 \leftarrow m_0 \setminus S(X^*)$

15   **until** $(\hat{E}[D_{m_i}] \leq D)$

16        $i \leftarrow i + 1$

17   **until** $((m_0 == \varnothing) \text{ or } (\hat{E}[D_{m_0}] > D))$

18   **end**

Fig. 3. Greedy set construction algorithm

5 Evaluation

In order to explore the performance of our solution, we tested it on a wide range of synthetic data. A purely simulated evaluation setting has the major advantage that it gives us access to the ground truth, i.e. the spatial process itself. As a result, we can evaluate distortion of interpolation-based reconstruction with a set of sensors as compared to the actual monitored process. The spatial process was generated by feeding zero mean uniform white noise into a sharp symmetric 2-D low-pass spatial filter. This filter is chosen such that 2-D Nyquist sampling would require a grid of 100 sensors.

For our algorithm, we used a learning phase of $\Theta = 250$ time instants. The reconstruction performance of the final disjoint sets of sensors is evaluated against 250 additional process realizations. Note that these 250 test realizations are only meant to evaluate reconstruction distortion and do not correspond to how many time instants the network physically monitors for; this would normally be much larger. The distortion estimation, given by equation (9), and used in line 15 of Figure 3, requires averaging over W realizations. This is a computationally expensive operation which must be repeated for each sensor added, and choosing W presents a tradeoff between estimation performance and computation cost. We found that $W = 50$ presented a good compromise.

Data interpolation was performed using a standard Delaunay triangulation algorithm provided by Matlab. Interpolation thus achieved was not optimal, but provided reasonable accuracy nonetheless, in the sense that omission of a small number of sensors resulted in very small changes in distortion.

We considered a Poisson-based random deployment with $N = 1000$, 500 and 250 nodes. This allowed us to evaluate the impact of the level of initial over-deployment on our algorithm. For each of the aforementioned values of N, the distortion target D was set to 0.03 (this value was chosen so that the ratio of distortion versus spatial variance of the process was approximately 0.5). Results are reported for $\alpha = 0.5$. Other values of α (e.g. $\alpha = 1$) have also been investigated; in any case, performance did not appear to depend strongly on the specific value chosen. Table 1 shows the number of sets that where obtained for each of the three options in initial number of sensors N.

Table 1. Sets devised by Algorithm 1

N	Set 1	Set 2	Set 3	Set 4	Set 5	Set 6
1000	159	159	158	158	158	208
500	164	163	173	-	-	-
250	125	125	-	-	-	-

As observed in Table 1, for a specific value of N the later sets (higher sequence number) contain more sensors. This is expected as fewer candidate sensors remain at this stage of the algorithm. Across different N, the number of sensors per set is roughly similar, again as is expected, and a larger initial deployment results in more equivalent sets that are found. However, we note that as N decreases initially (e.g. from $N = 1000$ to $N = 500$), slightly more nodes are needed in each set. This can be explained by the fact that with coarser initial sampling, fewer options are available. If N decreases further (e.g. from $N = 500$ to $N = 250$), the set size is reduced, caused by the fact that the primary subspace is a less accurate approximation of H_S. In general, the performance is better the higher the initial value of N, not only in terms of number of sets, but also in terms of distortion (as will be seen in Figure 4).

There is no existing solution to actually compare the performance of our algorithm against. However, we have examined two reasonable alternate approaches, both based on the algorithm of Figure 3, to serve as a basis for comparison:

1. Random selection: At each step of the algorithm, the sensor that is added to the set is selected at random among all remaining ones. Stop when the distortion criterion (line 15 of the algorithm) is satisfied.
2. Distance-based selection: At each step of the algorithm, add the sensor that is furthest away from all sensors currently in the set.

The sets resulting from the random and distance-based selection are shown in Table 2. The distance-based selection yielded fewer sets, with a distortion performance similar or inferior to our algorithm, and is therefore not considered further in this paper. The unsatisfactory performance is caused by the fact that a lot of boundary sensors are chosen in the first set, resulting in more total number of sensors in that set, while at the same time causing a shortage of them in later sets.

Table 2. Sets devised by alternate heuristics

N	Random					Distance-based	
	Set 1	Set 2	Set 3	Set 4	Set 5	Set 1	Set 2
1000	214	204	214	184	184	600	400
500	170	170	160	-	-	330	170
250	125	125	-	-	-	168	82

Random selection performed significantly better. However, for higher values of N (i.e. $N = 1000$), it resulted in fewer sets than our proposed approach, see Table 2. This would result in reduced system lifetime. On the other hand, for the cases in which it selects the same number of sets (i.e. $N = 500$ and $N = 250$), the actual distortion performance is degraded, as will be shown in Figure 4. In addition, our algorithm is generic in terms of the underlying spatial process, which can be non-stationary in space. This is not the case for random selection (or distance-based selection), where uniform spatial characteristics are essentially a best-case scenario.

Figure 4 shows the instantaneous distortion for one run, for each value of N both for our algorithm and random selection. The sets were activated in a sequential manner, as before, with vertical lines indicating points of switching between sets. Realizations up to 50 correspond to the distortion during the learning phase, where all sensors in the network are reporting. This is done to give an idea of the performance of the initial deployment. For $N = 250$ this distortion is very close to the chosen target. This essentially means that the primary subspace is a coarser approximation of the true field, resulting in degraded correlation and distortion estimation performance. This explained the reduced performance of the sets for $N = 250$ in Figure 4. For $N = 500$ and $N = 1000$ the lower initial distortion, however, can be traded off effectively for multiple reporting sets.

We also notice in Figure 4 that, in general, the performance of sets with higher sequence number, i.e. those selected in later stages of the algorithm, is slightly degraded compared to the earlier sets. Since sets are selected sequentially, regions may exist which are crucial for reconstruction and where earlier sets dominate, because 'they were there first'. An implicit assumption of our algorithm is that the space defined by the yet-unselected sensors (i.e. H_R) is adequate to accurately describe the monitored phenomenon. In a network of finite size, this is not always the case. By contrast, for the denser deployments $N = 1000$ and $N = 500$ this limitation is largely alleviated for most sets, as Figure 4 shows. In practical applications, we could therefore designate the first sets as 'working sets', while the later ones serve as 'emergency sets'. Overall, the results shown in Figure 4 indicate that our algorithm succeeds in finding good disjoint sets to meet a specified target distortion. By having only one of them active at each point in time, desired sampling-interpolation performance can still be achieved. In general, by using sets in a sequential activation, the network can remain active for a longer duration of time.

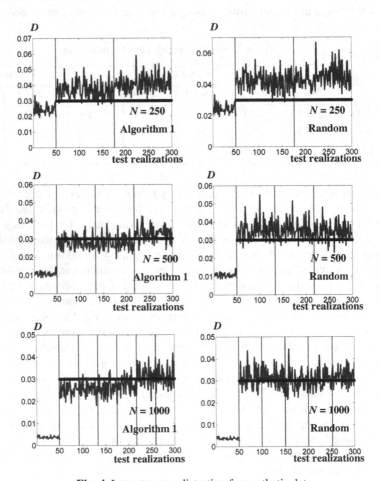

Fig. 4. Instantaneous distortion for synthetic data

6 Conclusion

In this paper, we have tackled the problem of sensing topology management for applications where the user is interested in an interpolated reconstruction of a physical process. In this case, the sensor network essentially behaves as a distributed sampling system, where notions such as sensing range or k-coverage are meaningless. We have proven that an optimal solution is a hard combinatorial problem, and presented an efficient heuristic algorithm to devise disjoint sets of sensors. All these results were obtained by mapping the network onto a Hilbert space representation. This powerful technique transforms a problem dealing with random variables into that of a deterministic vector space where relevant statistical information is captured in the inner product. The Hilbert space representation also enables the application of other important algebraic tools (e.g. sparse principal component analysis) to sensor networks, and therefore might find applications beyond the scope of the specific problem we have tackled in this paper.

References

1. Slijepcevic, S., Potkonjak, M.: Power Efficient Organization of Wireless Sensor Networks. ICC (2001)
2. Cărbunar, B., Grama, A., Vitek, J., Cărbunar, O.: Coverage Preserving Redundancy Elimination in Sensor Networks. SECON (2004)
3. Huang, C.-F., Tseng, Y.-C.: The Coverage Problem in a Wireless Sensor Network. WSNA (2003)
4. Raghunathan, V., Schurgers, C., Park, S., Srivastava, M.B.: Energy-Aware Wireless MicroSensor Networks. IEEE Signal Processing Magazine (March 2002)
5. Koushanfar, F., Taft, N., Potkonjak, M.: Sleeping Coordination for Comprehensive Sensing Using Isotonic Regression and Domatic Partitions. INFOCOM (2006)
6. Liaskovitis, P., Schurgers, C.: A Distortion-Aware Scheduling Approach for Wireless Sensor Networks. In: Gibbons, P.B., Abdelzaher, T., Aspnes, J., Rao, R. (eds.) DCOSS 2006. LNCS, vol. 4026, Springer, Heidelberg (2006)
7. Wang, X., Xing, G., Zhang, Y., Lu, C., Pless, R., Gill, C.: Integrated Coverage and Connectivity Configuration in Wireless Sensor Networks. SenSys (2003)
8. Vuran, M.C., Akyildiz, I.F.: Spatial Correlation-based Collaborative Medium Access Control in Wireless Sensor Networks. In: IEEE/ACM Transactions on Networking, (April 2006)
9. Perillo, M., Ignjatovic, Z., Heinzelman, W.: An Energy Conservation Method for Wireless Sensor Networks Employing a Blue Noise Spatial Sampling Technique. IPSN (2004)
10. Guestrin, C., Bodik, P., Thibaux, R., Paskin, M., Madden, S.: Distributed Regression: An Efficient Framework for Modeling Sensor Network Data. IPSN (2004)
11. Krause, A., Guestrin, C., Gupta, A., Kleinberg, J.: Near Optimal Sensor Placements: Maximizing Information while Minimizing Communication Cost. IPSN (2006)
12. Cramer, H., Leadbetter, M.R.: Stationary and Related Stochastic Processes: Sample Function Properties and Their Applications. Wiley, Chichester (1967)
13. Davis, G., Mallat, S., Avellaneda, M.: Adaptive Greedy Approximations. Journal of Constructive Approximation (1997)
14. Donoho, D.L., Elad, M.: Optimally sparse representation in general (nonorthogonal) dictionaries via l1 minimization. PNAS (2003)

A Fully Polynomial Approximation Algorithm for Collaborative Relaying in Sensor Networks Under Finite Rate Constraints*

Rajgopal Kannan[1], Shuangqing Wei[2], Vasu Chakravarthy[3],
and Murali Rangaswamy[4]

[1] Dept of CS, Louisiana State University, Baton Rouge, LA 70803
[2] Department of ECE, Louisiana State University, Baton Rouge, LA 70803
[3] Air Force Research Laboratories, Wright-Patterson AFB, Dayton, OH 45433
[4] Air Force Research Laboratories, Hanscom AFB, MA 01731
www.csc.lsu.edu/~rkannan, www.ece.lsu.edu/~swei

Abstract. We take an algorithmic approach to a well-known communication channel problem and develop several algorithms for solving it. Specifically, we develop power control algorithms for sensor networks with collaborative relaying under bandwidth constraints, via quantization of finite rate (bandwidth limited) feedback channels. We first consider the power allocation problem under collaborative relaying where the tradeoff between minimizing ones own energy expenditure and the energy for relaying is considered under the constraints of packet outage probability and bandwidth constrained (finite rate) feedback. Then we develop bandwidth constrained quantization algorithms (due to the finite rate feedback) that seek the optimal way of quantizing channel quality and power values in order to minimize the total average transmission power and satisfy the given probability of outage. We develop two kinds of quantization protocols and associated quantization algorithms. For separate source-relay quantization, we reduce the problem to the well-known k-median problem [1] on line graphs and show a a simple $O((K_J)^2 N)$ polynomial time algorithm, where $\log_2 K_J$ is the quantization bandwidth and N is the size of the discretized parameter space. For joint quantization, we first develop a simple 2-factor approximation of complexity $O(K_J N + N \log N)$. Then, for $\epsilon > 0$, we develop a fully polynomial approximation scheme (FPAS) that approximates the optimal quantization cost to within an $1 + \epsilon$-factor. The running time of the FPAS is polynomial in $1/\epsilon$, size of the input N and also $\ln F$, where F is the maximum available transmit power.

1 Introduction

Energy efficiency is an important consideration in wireless sensor networks. One technique for minimizing transmission energy in a cluster is collaborative

---

* This work was supported by NSF grants IIS-0329738, ITR-0312632 and by AFRL under contract #F33615-02-D-1283 (sub #05-2D1005.001). The opinions expressed herein are those of the individual authors and independent of the sponsoring agencies.

J. Aspnes et al. (Eds.): DCOSS 2007, LNCS 4549, pp. 338–353, 2007.

relaying. Nodes can select partners to act as relays for forwarding their data to the clusterhead or sink. Relaying exploits cooperative diversity, the fact that sometimes the relay-clusterhead channel quality is significantly better than the direct source-clusterhead. Thus if the relay is able to receive and decode the source message, even if there are errors (packet outage) between the source and clusterhead, the relay can correctly transmit the packet to the destination.

Cooperative diversities under relaying can be exploited to further improve reliability and energy efficiency by using Channel State Information (CSI) [2,3]. Communication channel quality is estimated and fed-back to the nodes in order to decide the metrics of relaying. The preceding cited works share a common feature in that they assume a set of relay nodes is already selected and the issue is to determine power allocations across all transmitting nodes without considering data originating from relay nodes themselves. No consideration is given toward the relay's own needs other than its function as a relay.

In our model, we assume that source sensor and relay sensor both have their own data to transmit to the clusterhead along with an individual quality of service requirement, e.g. outage probability P_{out} as a good approximation for frame error rate (FER). We do not consider partner selection protocols but assume a relay has been apriori selected. Each node divides its entire energy budget into two parts. One is for transmitting its own data, the other is devoted to relaying information. As a partner relationship is established between two nodes such that each of them helps the other forward/relay information, we are interested in a fundamental energy tradeoff question: What power allocation policy should be adopted by each node in order to minimize its own total energy consumption while meeting the outage probability constraints and complying with its obligation as a relay.

In [2,3], perfect CSI at each node is assumed available to the source and relay nodes. However perfect CSI can only be available under the assumption of unlimited feedback channel capacity in order for the receiver to transmit back the measurements to its transmitter without any error. Adaptive signaling under the finite rate feedback constraint has attracted considerable attentions lately because of its more practical implications compared with the perfect CSI assumption [4] (and references therein).

Not much work has been done yet for adaptive signaling schemes in sensor networks with relay channels under the finite rate feedback constraint. In [5], the power control problem is tackled for relay channels with finite rate feedback. However, only amplify-and-forward relaying is considered, in which the issue of availability of CSI for the source-relay link is relatively easier to address than the decode-and-forward case. In addition, the majority of work in the literature on finite rate feedback problems approach the resulting quantization problems directly by finding out the optimal quantization regions of fading vectors, as well as associated power allocation functions [4].

In this paper, we take an algorithmic approach to collaborative relaying under finite rate feedback by using the technique of discretization of variables (in our case channel fading coefficients). We first briefly present results obtained in [6] where we optimize the total average power expenditure of both relay and source nodes under the assumption that nodes have *perfect* CSI in a network of two transmitting sensors and one clusterhead. Based on the power control strategies developed in [6] for decode-and-forward relaying, we develop bandwidth constrained quantization algorithms (due to the finite rate feedback) that seek the optimal way of quantizing channel quality and power values in order to minimize the total average transmission power and satisfy the given probability of outage.

We develop two kinds of quantization protocols and associated quantization algorithms. First we consider separate source and receiver quantization, where the clusterhead splits its available quantization bandwidth for feedback, independently between the source and relay node. We reduce this quantization problem to the well-known k-median problem [1] on line graphs and show a a simple $O(NK_J(K_J + \log N))$ polynomial time algorithm, where $\log K_J$ is the quantization bandwidth and N is the size of the discretized parameter space. Then we consider joint quantization. Here the base station can exploit the joint probability distributions of source and relay channels and power values and use the entire quantization bandwidth to jointly feedback both the source and relay. Unfortunately, the joint quantization problem is NP-hard by reduction from the k-median problem, which has itself been a subject of study for several decades (problem ND51 in [1]). Therefore, we develop a simple 2-factor approximation of complexity $O(N(K_J + \log N))$. Then, for $\epsilon > 0$, we develop a fully polynomial approximation scheme (FPAS) that approximates the optimal quantization cost to within an $1 + \epsilon$-factor. The running time of the FPAS is polynomial in $1/\epsilon$, size of the input N and also $\ln F$, where F is the maximum available transmit power.

The paper is organized as follows. We first present the system model in Section 2. Power control strategies with perfect CSI are then provided in Section 3. When finite rate feedback constraint is imposed, the independent and joint quantization algorithms for source and relay nodes are given in the next two Sections.

2   System Model

To illustrate the major idea of power control across relay nodes, we first consider a simple model in which there are two nodes N_1 and N_2 transmitting to a common receiver N_D with help from each other. Narrow-band quasi-static fading channel is assumed, where channel fading coefficients remain fixed during the transmission of a whole packet, but are independent from node to node. The complex channel coefficient $h_{i,j}$ captures the effects of both pathloss and the quasi-static fading on transmissions from node N_i to node N_j, where $i \in \{1, 2\}$, and $j \in \{2, 1, D\}$. Statistically, $h_{i,j}$ are modeled as zero mean, mutually independent proper complex Gaussian random variables with variances: $E|h_{i,j}|^2 = 2\sigma_{i,j}^2$. We first assume a non-causal system model in which amplitudes $|h_{i,j}|$ are available to all transmitters and receivers at the beginning of transmissions. In a quasi-static fading channel, CSI can be obtained by exploiting training sequences sent by transmitters.

Consider a time-division (TD) multiple access scheme in which an entire time period is divided into 4 slots [7, Fig. 2]. A repetition coding-based decode-and-forward strategy (R-DF)is assumed at N_j, $j = 1, 2$, where relay node transmits the same codeword as what source sends if its decoding is successful. The cooperative communication protocol can be described as follows: Based on the available CSI, N_1 can determine whether relaying from N_2 is needed or not, as explained in the power control algorithms below. If such collaboration is sought, N_1 transmits as a source to N_D in the first slot and then in the second slot N_2 forwards its decoded messages to the destination. If N_2 is not asked for relaying, N_1 transmits in the first 2 slots of on its own. Over the last two slots, N_1 and N_2 exchange their roles as a source and relay.

The mathematical characterization of the whole process is:

$$Y_{1,D}[k] = h_{1,D}S_1[k] + W_{1,D}[k], Y_{1,2}[k] = h_{1,2}S_1[k] + W_{1,2}[k]$$

for $k \in [0, N/4]$; and

$$Y_{2,R}[k] = h_{2,D}\tilde{S}_1[k] + W_{2,R}[k]$$

for $k \in (N/4, N/2]$, if relay N_2 is needed and decoding is successful. The figure N is the total number of degrees of freedom available over the entire transmission period, and $W_{i,j}$ are independent complex white Gaussian noise with two-sided power spectral density $\mathcal{N}_0 = 1$. For R-DF schemes, $\tilde{S}_j[k]$ are scaled versions of the transmitted Gaussian codewords $S_j[k]$. Over the last two slots, similar models can be set up for node 2 based on symmetry over $k \in (N/2, N]$.

Given CSI on $|h_{i,j}|$, transmission powers over various periods are denoted as: $E|S_1[k]|^2 = P_{1,D}$, $k \in [0, N/4]$ and $E|\tilde{S}_1[k]|^2 = P_{2,R}$, $k \in (N/4, N/2]$ if N_2 is needed and decoding is successful; $E|S_1[k]|^2 = P_{1,D}$, $k \in [0, N/2]$ and $E|\tilde{S}_1[k]|^2 = 0$, $k \in [0, N/2]$, if N_2 is not needed. Similarly, we define $E|S_2[k]|^2 = P_{2,D}$, $k \in (N/2, \frac{3}{4}N)$ and $E|\tilde{S}_2[k]|^2 = P_{1,R}$, $k \in (\frac{3}{4}N, N]$ if N_1 is needed and decoding is successful; $E|S_2[k]|^2 = P_{2,D}$, $k \in (N/2, N]$ and $E|\tilde{S}_2[k]|^2 = 0$, $k \in (N/2, N]$ if N_1 is not needed.

3   Total Energy Minimization for Collaborative Relaying with Perfect CSI

Under the constraint that each sensor node has an outage probability no greater than $P_{j,out}$, i.e. $\Pr[I_j < R_j] \leq P_{j,out}$, where I_j is the mutual information of the overall link for transmitting node $j \in \{1, 2\}$'s information, our objective is to investigate power control policies under which the total energy of these two nodes is minimized in a complete collaborative manner. This **Collaborative Relaying** problem can be formulated as below:

$$\min \sum_{j=1}^{2} E[P_{j,D} + P_{j,R}], \text{ subject to } \Pr[I_k < R_k] \leq P_{k,out}, \text{ for } k = 1, 2. \quad (1)$$

Under the collaborative relaying approach, the optimal power allocation policy $[P_{i,D}, P_{j,R}]$ to solving problem (1) can be characterized by the following Theorem [6].

Theorem 1. *The optimal power allocation vector $[P_{i,D}, P_{j,R}]$ depends on channel strength ratios captured by $|h_{i,D}|/|h_{j,D}|$ and $|h_{i,D}|/|h_{i,j}|$ for $i \neq j$ and $i, j \in \{1, 2\}$. The resulting solutions are:*
$P_{i,D} = \hat{P}_{i,D}, P_{j,R} = \hat{P}_{j,R}$ *if $h_{i,j}$ are in the set*

$$A_i = \left\{ |h_{i,j}| : \frac{|h_{i,D}|^2}{|h_{i,j}|^2} < \frac{2}{2^{R_i} + 1} \ and \ \frac{|h_{i,D}|^2}{|h_{i,j}|^2} + \frac{|h_{i,D}|^2}{|h_{j,D}|^2} \left(1 - \frac{|h_{i,D}|^2}{|h_{i,j}|^2} \right) \le \frac{2}{2^{R_i} + 1} \right\}$$
(2)

and $\hat{P}_{i,D} + \hat{P}_{j,R} \le s_i^*$. Otherwise if $h_{i,j} \in A_i^c$, the complimentary set of A_i, and $2\tilde{P}_{i,D} \le s_i^*$, the solution is $P_{i,D} = \tilde{P}_{i,D}, P_{j,R} = 0$. For all other cases, transmission powers are all set to zero $P_{i,D} = P_{j,R} = 0$. Transmission power functions are defined as follows:

$$\tilde{P}_{i,D} \triangleq (2^{R_i} - 1)(|h_{i,D}|^2), \ \hat{P}_{i,D} \triangleq (2^{2R_i} - 1)(|h_{i,j}|^2), \ \hat{P}_{i,R} \triangleq \frac{2^{2R_j} - 1}{|h_{i,D}|^2} \left(1 - \frac{|h_{j,D}|^2}{|h_{j,i}|^2} \right).$$
(3)

The thresholds $s_i^*, i = 1, 2$ are determined by solving the following equations to meet outage probability constraints:

$$1 - P_{i,out} = Pr\left\{ 2\tilde{P}_{i,D} < s_i^*, \ for \ \left(\frac{|h_{i,D}|^2}{|h_{i,j}|^2}, \frac{|h_{i,D}|^2}{|h_{j,D}|^2} \right) \in A_i^c \right\}$$

$$+ Pr\left\{ \hat{P}_{i,D} + \hat{P}_{j,R} < s_i^*, \ for \ \left(\frac{|h_{i,D}|^2}{|h_{i,j}|^2}, \frac{|h_{i,D}|^2}{|h_{j,D}|^2} \right) \in A_i \right\}.$$

4   Optimal Quantization for Optimal Collaborative Relaying

In the previous section, we assumed the availability of perfect CSI at each sensor node in order to develop an optimal power control algorithm for collaborative relaying. However, in reality, perfect CSI is not possible since bandwidth limitations prevent the full exchange of precise channel information between the source, relay and base-station[1]. This motivates the idea of developing power control algorithms for sensor networks with relaying under bandwidth constraints, specifically via quantization of finite rate (bandwidth limited) feedback channels.

In this paper, we develop optimal quantization algorithms for optimal sensor relaying by selecting appropriate quantization parameters and quantized values.

---

[1] Imperfect CSI can also arise due to measurement errors and the time lag between channel state measurements and actual transmission. In this paper, we do not consider measurement errors and also assume slowly time-varying channel parameters.

Quantized information received at the source and relay nodes is then mapped to corresponding transmit powers. The overall objective of the quantization algorithm is to minimize the expected sum of source and relay transmit powers, as in the previous section. For the quantization algorithms, we need to consider the power consumed by source and relay to satisfy the outage probability of the source only during the first two mini-slots (the first half of the collaborative relaying process). The algorithm can then be separately applied to develop quantization for the source-relay pairs during the second half of the collaborative process (when source and relay switch roles).

4.1 Quantization Protocol

The proposed quantization algorithms are associated with a specific protocol for exchanging quantized information between the participants. We describe our protocol below. Quantized information is exchanged between the participants in four sequential steps as follows: In the first step, prior to data transmission, the source node broadcasts a training sequence to the base station as well as the relay. The clusterhead/basestation uses the training sequence to determine $h_{1,D}$ while the relay node simultaneously determines $h_{1,2}$. In the second step, the relay node broadcasts another training sequence along with the quantized value of measured $h_{1,2}$ using the quantization algorithm QRB (described subsequently) to the clusterhead and the source. This is used by the clusterhead to determine $h_{2,D}$. At this point, the clusterhead has perfect $h_{1,D}$ and $h_{2,D}$ measurements and quantized $h_{1,2}$, while the relay and source have measured and quantized $h_{1,2}$ values, respectively. Next, in the third step of the quantization protocol, using either joint or separate quantization algorithms (described subsequently) the station broadcasts quantized values to the source and relay. This value is sufficient for the source and relay to determine their respective transmit power levels for data transmission and relaying.

All algorithms can be implemented at all three nodes, so each node is aware of the mapping from quantization to power levels without separate information. Also each node is aware of the mapping for the other nodes. We also note as a characteristic of the algorithms that power values are not quantized through rounding, rather a set of feasible transmit power values is derived and there is a mapping from channel space to this power space.

4.2 Preliminaries

We develop the proposed quantization algorithms by discretizing the parameter space. For notational simplicity, let h denote any of the channel fading parameters $h_{1,2}$, $h_{1,D}$ and $h_{2,D}$. Let $\gamma > 0$ be an (arbitrary) discretization unit such that the range of each h is divided into M discrete and contiguous intervals $I_j = [h_j, h_{j+1})$, where $h_j = j\gamma$ and $j = 0, 1, \ldots M - 1$. Each interval is of length γ, except the last interval $[h_{M-1}, \infty)$, which extends to infinity. The channel fading variables h are exponentially distributed and hence we can choose as a design parameter a maximum value, after which h is very small.

First, assume that the range of the source-relay fading coefficient is restricted, i.e., it is known that $h_{1,2} \in [h_a, h_b)$, where $h_a = k\gamma$, $h_b = l\gamma$, $l > k$. Now consider the discretized $\{h_{1,D}, h_{2,D}\}$ space as divided into $N = M^2$ blocks each of dimension $\gamma \times \gamma$. Let $b_{u,t}$ denote the $(u, t)^{th}$ block in this space and let $H_{u,t}$ be the apriori probability that the $h_{1,D}, h_{2,D}$ channel fading coefficients fall into $b_{u,t}$ where $H_{u,t} = Pr.\{u\gamma \leq h_{1,D} < (u+1)\gamma\} \cdot Pr.\{t\gamma \leq h_{2,D} < (t+1)\gamma\}$.

Also let $P_{u,t} = (P_{u,t}^s, P_{u,t}^r)$ denote the minimum (source,relay) transmit power vector such that data can be collaboratively transmitted without outage if channel quality falls anywhere within block b_j. We define,

$$\left(P_{u,t}^s = \max\{P_{1,D}\},\ P_{u,t}^r = \max\{P_{2,D}\}\right) \forall (h_{1,D}, h_{2,D}) \in b_{u,t},\ \forall h_{1,2} \in [h_a, h_b) \tag{4}$$

where $P_{1,D}$ and $P_{2,D}$ are obtained using Theorem 1. Note that $P_{u,t}^s = 0$ ($P_{u,t}^r = 0$, resp.) if the block is one of those for which we require outage (non-cooperation from the relay, resp.) i.e the channel configuration corresponding to the given block falls under the threshold s_1^* (s_2^*, resp.). $H_{u,t}$ and $P_{u,t}$ can be obtained in $O(1)$ time for each block $b_{u,t}$.

Consider the N blocks in the discretized $h_{1,D}, h_{2,D}$ space. We state that a block $b_{i,j}$ s-covers (r-covers, resp.) block $b_{k,l}$ if $P_{i,j}^s \geq P_{k,l}^s$ ($P_{i,j}^r \geq P_{k,l}^r$, resp.). Consider a block that is s-covered as well as r-covered. If the source transmits at the source power of the s-covering block and the relay transmits at the relay power of the r-covering block, then we are guaranteed there will be no outage if the realized (actual) channel fading coefficients happen to fall within the covered block. Note that if the source and relay powers of a block are both zero, then we want the block to be in outage and there is no need to cover the block.

5 QBS and QBR: Independent Basestation-Source and Basestation-Relay Quantization Algorithms

We assume the total downlink quantization bandwidth (from clusterhead to source and relay) is k_J, i.e. the clusterhead has k_J bits available to transmit the results of quantizations QBS and QBR to the source and relay. Under independent quantization, the clusterhead, after measuring the exact $h_{1,D}, h_{2,D}$ values, transmits independent quantization information to the source and relay, such that the realized block (under measured h values) will be s-covered by the corresponding source power and and r-covered by the corresponding relay power.

Let k_s and k_r denote the choices for separate quantization bandwidths to source and relay respectively, where $k_s + k_r = k_J$. Let $K_J = 2^{k_J}$, $K_s = 2^{k_s}$ and $K_r = 2^{k_r}$.

The cost of the optimal independent quantization scheme OptIQ is given by

$$\text{Cost}_{\text{OptIQ}} = \min_{k_s + k_r = k_J} (QBS(k_s) + QBR(k_r)) \tag{5}$$

We show that optimal independent quantization algorithms can be obtained through simple reductions from the k-median problem, whose running time is polynomial in the discretization parameter N. As we show below, the running time of $QBS(k_s)$ and $QBR(k_r)$ are $O(NK_s + N \log N)$ and $O(NK_r + N \log N)$ respectively. Thus from Eq.5, the cost of the optimal independent quantization algorithm is $O(NK_J(K_J + \log N))$. We describe $QBS(k_s)$ and $QBR(k_r)$ below.

5.1   Algorithm $QBS(k_s)$

First, QRB quantizes $h_{1,2}$ and this encoded value is sent to the base station, which must implement either joint or separate quantization. Thus the quantization of the $\{h_{1,D}, h_{2,D}\}$ space is conditioned on the received quantized value of $h_{1,2}$, i.e. for every code of $h_{1,2}$, there is a particular quantization in the $\{h_{1,D}, h_{2,D}\}$ space. This quantization must be designed to minimize the source and relay power consumption. Since the optimal result also depends on the quantization of $h_{1,2}$, we must find the optimal quantization of $h_{1,2}$ for which the optimal quantization of the $\{h_{1,D}, h_{2,D}\}$ gives the minimal power consumption. QRB achieves this optimal recursive quantization as described in the last section.

Let $K_s = 2^{k_s}$, i.e. the $\{h_{1,D}, h_{2,D}\}$ space must be encoded by the clusterhead into K_s levels, given the restricted $h_{1,2}$ space. The objective of algorithm QBS is to find a set F_s of K_s blocks (equivalently K_s power levels) such that all blocks are s-covered and the expected transmission power of the source required to satisfy the outage probability over the entire $\{h_{1,D}, h_{2,D}\}$ space and restricted $h_{1,2}$ channel space is minimized. QBS can be expressed as the following minimization problem:

$$QBS : \underset{F_s}{\operatorname{argmin}}\{\sum_{u,t} H_{u,t} \underset{b_{i,j} \in F_s | b_{i,j} \, s-\text{covers} b_{u,t}}{\min} P_{i,j}^s\} \tag{6}$$

We can now relate QBS to the k-median problem. The general k-median problem on a graph G can be formulated as finding the optimal set F of vertices (medians) that satisfies

$$\text{kcost}_G = \underset{F}{\operatorname{argmin}}\{\sum_{u \in G} w_u \min_{v \in F} d_{uv}\} \tag{7}$$

where $|F| \leq k$, w_u is the weight of vertex u and d_{uv} is the minimum distance between u, v in G. While the k-median problem is known to be NP-hard in the general case, (ref. problem ND51 in [1]), it is solvable in polynomial time for trees [8,9,10] and lines (paths) [11,12,13,14]. In this case, QBS can be easily reduced to an instance of k-median on paths by using the fact that the s-cover relationship is transitive.

The reduction is as follows: Sort the N blocks in non-decreasing order of source power $P_{u,t}^s$. Construct the directed path G^* whose vertices are the elements of the sorted list in order. The directed edge cost between adjacent vertices $v_i = b_{u,t}$ and $v_{i+1} = b_{k,l}$ is set to $c_{i,i+1} = P_{k,l}^s - P_{u,t}^s$ while vertex v_i is assigned a weight $w_{v_i} = H_{u,t}$. After running the k-median algorithm on G^* (with $k = K_s$), the source power values of the k selected median nodes are mapped to the K_s quantization levels under QBS i.e the qth quantization level corresponds to the power value of the qth vertex in the k-median solution. Since QBS is implemented at both the source and clusterhead, the power-level to quantization mapping is apriori available to the source and it can transmit at the appropriate level when the the quantized level is fed-back by the clusterhead.

The cost of quantization algorithm QBS is obtained as:

$$\text{cost}_{QBS} = \text{kcost}_{G^*} + \sum_{u,t} H_{u,t} P_{u,t}^s \qquad (8)$$

Since s-cover is transitive, the block represented by each vertex in G^*, s-covers all the blocks represented by vertices to its left. If v_i is selected to be one of the k-medians, then its contribution towards being a median for v_j is $(P_i^s - P_j^s)H_j$ while its contribution to being an s-cover for v_j is $P_i^s H_j$. For any set of k-medians from G^*, the difference in cost from QBS is the constant $\sum_j P_j^s H_j$ and thus there is a one-to-one correspondence between the optimal solution to k-median on G^* and the optimal quantization QBS. Thus we have,

Theorem 2. *QBS is an optimal quantization of the source channel.*

We note that k-median on a path can be implemented in $O(kn)$ time [8] Hence the time complexity of QBS is $O(NK_s + N\log N)$. A slightly weaker $O(K_s N^2)$ algorithm for QBS, based on dynamic programming, is presented in [15].

5.2   Algorithm $QBR(k_r)$

Algorithm $QBR(k_r)$ is identical to $QBS(k_s)$ with s-cover replaced by r-cover and all P^s values replaced by P^r.

6   Joint Source/Relay Quantization

We now consider the case when the clusterhead devotes the entire downlink quantization bandwidth to jointly quantize the source and relay transmit powers. Intuitively, this approach should prove more efficient in terms of total power minimization as the clusterhead can consider quantization over the joint probability distribution of transmit powers and channel fadings, as opposed to treating them independently. Unfortunately the related optimization problem is no longer polynomial. It can be shown that joint quantization is NP-hard by reduction from the general k-median problem. Therefore we consider bounded approximations.

The k-median problem has been the subject of study for several decades. There has been much work on developing efficient heuristics and approximation algorithms [16,8], particularly on trees and line graphs, as cited earlier. For some more general cases, a constant factor approximation was presented in [16] for graphs with a Euclidean distance metric (a 6-factor approximation). Here, we first present a simple 2-factor approximation that exploits the much simpler structure of joint quantization (as opposed to general k-median) and is easy to implement. Then we develop a $(1 + \epsilon)$-FPAS for joint quantization that can approximate the quantized total power to within an arbitrarily close ϵ factor.

For both algorithms, we assume that the total downlink bandwidth for joint quantization is k_J, with $K_J = 2^{k_J}$. As before, the relay first transmits a quantized value corresponding to a range of $h_{1,2}$. Thus both algorithms quantize the $\{h_{1,D}, h_{2,D}\}$ space into K_J values, given the restricted $h_{1,2}$ space. Each quantized value corresponds to a (source,relay) power level pair. After measuring channel quality, the clusterhead broadcasts the corresponding quantized value to the source and relay nodes. Subsequent data transmission is accomplished using the corresponding source and relay power levels. Note that the nodes are each aware of the others power requirements since the algorithm is implemented at both nodes. This is necessary since if the relay power is 0 (no relaying), the source can transmit at the required level during both slots.

6.1   2-Factor Approximation for Joint Quantization

Consider an arbitrary block $b_{u,t}$. For notational simplicity, we drop the dual subscripts u, t and use b_j to denote the block. The minimum total power required to transmit this block without outage is given by $P_j = P_j^s + P_j^r$. H_j denotes the source-clusterhead and relay-clusterhead channel fading coefficient probability of b_j. We use P and H to denote this set of minimum total powers (per block) and channel fading probabilities over all blocks.

$QJ1$, the 2-approximation algorithm for joint quantization of source and relay powers is defined as follows: For each block b_j, replace (P_j^s, P_j^r) with $(P_j^{s'}, P_j^{r'}) = (\max(P_j^s, P_j^r), \max(P_j^s, P_j^r))$. Let $P_j' = P_j^{s'} + P_j^{r'}$ and sort the blocks in nondecreasing order of P_j'. Construct the line graph $G^*(P')$ on the vertices corresponding to this sorted list, similar to algorithm QBS, and run the k-median algorithm (with $k = K_J$) on G^*. Let F_J (with $|F_J| = K_J$) denote the subset of blocks corresponding to vertices returned by the k-median algorithm. Each block in F_J corresponds to a quantization level q, $0 \leq q \leq K_J - 1$. The corresponding source and relay transmit powers are $\max(P_i^s, P_i^r)$, where b_i is the block corresponding to quantization level q. In this case, the source and relay transmit powers are identical, thus they will transmit at the same power when a given quantization level is fedback from the clusterhead.

Theorem 3. *QJ1 is a 2-approximation to the optimal joint quantization algorithm.*

Proof. Let cost_{QJ^*} denote the cost of the optimal joint quantization algorithm for the given set of blocks. QJ^* finds a subset of K_J blocks such that all N blocks in the set are s- as well as r-covered by the source and relay power values represented by these K_J blocks and the average total power of the blocks minimally meets the outage probability requirements. Let $\text{cost}_{G^*(P)}$ denote the cost of the optimal k-median algorithm (with $k = K_J$) on directed line graph G^* using power values P and constructed as in the previous section.

We first note that $\text{cost}_{QJ^*} \geq \text{kcost}_{G^*(P)} + \sum_j H_j P_j$. Clearly, equality is met when $P_j^r = 0$ for all blocks. Further, every solution to QJ^* is a solution to k-median on $G^*(P)$. If b_j was a selected block in QJ^*, then $P_j > P_i$ for all blocks

b_i that are simultaneously s-covered and r- covered by b_j. Thus in $G^*(P)$, vertex v_j would be to the right of all such vertices v_i. v_j can therefore be a median for these vertices. However the converse is not true and $G^*(P)$ need not correspond to a feasible quantization. The block corresponding to a median in G^* need not be a solution to QJ^*, since $P_j > P_i$ does not imply that b_j can simultaneously s-cover and r-cover b_i. Hence cost_{QJ^*} is larger than the right hand side in these cases.

Let cost_P denote the sum $\text{kcost}_{G^*(P)} + \sum_j H_j P_j$ for any set of powers P. Now consider the system of blocks with (P_j^s, P_j^r) replaced with $(P_j^{s''}, P_j^{r''}) = \left(P_j^s + P_j^r, P_j^s + P_j^r\right)$. Let $P_j'' = P_j^{s''} + P_j^{r''}$ i.e $P_j'' = 2P_j$. Clearly, $\text{cost}_{P''} = 2\text{cost}_P \leq 2\text{cost}_{QJ^*}$, from the discussion above. Note that every solution to k-median on $G^*(P'')$ can be converted to a lower cost feasible solution for k-median on $G^*(P')$ since $P_j'' \geq P_j'$ for all blocks b_j. Thus $\text{cost}_{P'} \leq \text{cost}_{P''}$. Putting the two observations together, we get $\text{cost}_{P'} \leq 2\text{cost}_{QJ^*}$ and hence $QJ1$ is a 2-approximation.

6.2   Fully Polynomial Approximation Scheme

For the $(1 + \epsilon)$-FPAS (labeled QJ2), we transform the problem from the $(h_{1,D}, h_{2,D})$ channel space to a covering problem in the 2-dimensional power space as follows: Each block b_t in the $(h_{1,D}, h_{2,D})$ channel space is characterized by the vector (P_t^s, P_t^r) in the power space, $1 \leq t \leq N$. Let $P^s = \{P_t^s\}_t$ and $P^r = \{P_t^r\}_t$ represent the set of source and relay powers. Without loss of generality, we assume that $|P^s| = |P^r| = N$. We construct an $N \times N$ grid of cells $C = (P^s \times P^r)$, where cell c_{ij} represents source power $P_i^s \in P^s$ and relay power $P_j^r \in P^r$, $1 \leq i, j \leq N$. As before, the total power of c_{ij} is represented by $P_{ij} = P_i^s + P_j^r$ while H_{ij} denotes the channel probability of c_{ij}, where $H_{ij} = H_k$ if c_{ij} corresponds to some block b_k, $1 \leq k \leq N$. Note that c_{ij} need not correspond to an actual block and in this case $H_{ij} = 0$.

We define s- and r-covering as before. For this problem we are interested only in joint s- and r-covering. The cells jointly covered by c_{ij} are defined by the rectangle with left bottom endpoint at the origin and top right corner at (P_i^s, P_j^r). However for the algorithm, we prefer to express the joint covering relationship as a directed graph G with $2N - 1$ levels numbered from 2 to $2N$. Level $2N$ consists of only one node c_{NN} with incoming edges from parents $c_{N-1,N}$ and $c_{N,N-1}$ in level $2N-2$. In general, node c_{ij} is located in level $i + j$ and has two outgoing edges to its two children in level $i + j + 1$ ($c_{i+1,j}$ and $c_{i,j+1}$) and two incoming edges from its two children in level $i + j - 1$ ($c_{i-1,j}$ and $c_{i,j-1}$). The nodes in level $i + j$ are listed left to right in the order $c_{1,i+j-1}, \ldots, c_{i+j-1,1}$.

We use $\overline{H}_{ij}$ to represent the cumulative channel probability of all cells that are jointly covered by c_{ij}, where $\overline{H}_{ij} = \sum_{k=1}^{i} \sum_{l=1}^{j} H_{kl}$. Henceforth, we drop the dual subscript and use v_t to refer a generic node c_{ij} in G. We will also slightly abuse the notation and let $\overline{H}_a$ denote the set of nodes covered by a as well as the cumulative channel probability of these nodes. Thus for example, $\overline{H}_a \backslash \overline{H}_b$

denotes the nodes covered by a and not b as well as the cumulative value of their channel probabilities.

It can be seen that solving the k-median problem on directed graph G will also lead to a solution to the joint quantization problem. Recent results show that k-median can be solved in polynomial time on a directed tree [10]. However G is not a directed tree (removing the directions on edges leads to cycles) and this result cannot be applied. Instead we are able to develop an FPAS for this problem.

Let r-set $L_r = (v_1, v_2, \ldots, v_r)$ denote an ordered list of r nodes from G. The nodes in an r-set are ordered by increasing levels. For nodes in the same level, we impose a left to right ordering. Note that the ordering ensures $P_1 \leq P_2 \leq \ldots \leq P_r$. Let $QC(L_r)$ denote the quantization cost if all nodes from L_r (and only L_r) were chosen to represent quantization power levels. The total power required for transmitting each cell in the $(h_{1,D}, h_{2,D})$ space is the power level of its nearest ancestor in G belonging to L_r. Thus we have

$$QC(L_r) = \sum_{i=1}^{r} P_i \left(\overline{H}_i \setminus \left(\bigcup_{k=1}^{i-1} \overline{H}_k \right) \right) \tag{9}$$

Define $\overline{H}(L_r) = \bigcup_{k=1}^{r} \overline{H}_k$. $\overline{H}(L_r)$ represents the cumulative channel probability of nodes covered by L_r. Let $S_r^T = \{L_r^1, L_r^2, \ldots, L_r^T\}$ denote an ordered list of T distinct r-sets arranged in non-decreasing order of cost $QC(L_r^i)$, $1 \leq i \leq T$. Each r-set represents a potential sub-solution (with r levels) to the overall K_J level quantization problem. However, we would like to reduce the number of potential sub-solutions without losing essential information. Thus we prune the list by retaining only those potential solutions with a specific channel quality property.

Let $\delta > 0$ be an arbitrary parameter. The operation $\text{Prune}_{r,\delta}(S_r^T)$ returns the reduced list $S_r^{n(r)}$ of size $n(r)$ and is defined as follows:

Algorithm $Prune_{r,\delta}(S_r^k)$
---

1. Initialize $i \longleftarrow 1$, $j \longleftarrow 1$, $S_r^{n(r)} \longleftarrow \phi$.
2. $tempH \longleftarrow \overline{H}(L_r^i)$.
3. **While** $(QC(L_r^j) \leq (1+\delta)QC(L_r^i))$ **and** $(j \leq k)$
   if $(\overline{H}(L_r^j) \geq tempH)$
   $\{tempH \longleftarrow \overline{H}(L_r^j); x \longleftarrow j; j++\}$
   Endif
   Endwhile
4. $S \longleftarrow S \bigcup L_r^x$.
5. **While** $(QC(L_r^j) \leq (1+\delta)QC(L_r^x))$ **and** $(\overline{H}(L_r^j) \leq \overline{H}(L_r^x))$ $\{j++\}$
6. $i \longleftarrow j$. **If** $j \leq k$ Go to Step 2.

Lemma 1. *For every $L_r^j \in S_r^T$, there exists an $L_r^x \in S_r^{n(r)}$, such that either ($QC(L_r^j)/(1+\delta) \leq QC(L_r^x)$ or $(1+\delta)QC(L_r^j) \geq QC(L_r^x))$ and $\overline{H}(L_r^x) > \overline{H}(L_r^j)$.*

Proof. Step 3 ensures that a representative L_r^x is found that has the highest $\overline{H}$ among all L_r^j's with $j \leq x$ and $QC(L_r^x) \leq (1 + \delta)QC(L_r^j)$. Once L_r^x is found, step 5 ensures that we keep eliminating all L_r^j's within a δ-neighborhood of L_r^x that have smaller $\overline{H}$ values, i.e $QC(L_r^j) \leq (1 + \delta)QC(L_r^x)$ and $\overline{H}(L_r^j) \leq \overline{H}(L_r^x)$.

Lemma 1 indicates a key requirement for the overall algorithm. By selecting the particular r-set with maximum $\overline{H}$ within each δ-neighborhood, we are minimizing the future cost of covering similar costing r-sets while potentially paying a factor of $(1 + \delta)$ extra current cost.

We now define the key iterative step to be used in algorithm QJ2. Consider an arbitrary r-set $L_r = (v_1, v_2, \ldots, v_r)$. Let v_r correspond to actual node $u_j \in G$. We define the operation Create_{r+1} that creates new $r+1$ sets from L_r by considering $L_r \bigcup u_k, \forall k, k = j + 1, j + 2 \ldots$. New nodes are considered in increasing order as per our ordering convention. Thus the last node to be considered corresponds to cell c_{NN}. let $R = N^2$, the size of graph G. Then for each L_r, we create $R - j$ new $r + 1$ sets.

$QC(L_{r+1})$ can be calculated in $O(R)$ time as follows: Assume all nodes covered by L_r are marked. Then $\overline{H}_{u_k} \backslash \overline{H}_{L_r}$ can be calculated and marked by breadth-first traversal of G starting from u_k in the reverse direction of arrows.

Finally, $QC(L_{K_J})$ is created by adding to each list in $QC(L_{K_J-1})$ the lowest possible node in G such that all nodes are covered.

Algorithm QJ2 is now defined below. We assume some arbitrary node u_i as the first member of the K_J quantization and proceed as follows:

Algorithm QJ2(u_i, ϵ)
---

1. $L_1^1 \longleftarrow (u_i)$, $S_1^1 \longleftarrow (L_1^1)$, $n(1) \longleftarrow 1$, $\delta \longleftarrow \frac{\epsilon}{2K_J}$.
2. **For** $r = 1$ to $K_J - 1$
   - (a) $S_{r+1}^T \longleftarrow \text{Create}_{r+1}(S_r^{n(r)})$;
   - (b) Sort S_{r+1}^T by Quantization Costs $QC(L_{r+1}^i)$'s ;
   - (c) $S_{r+1}^{n(r+1)} \longleftarrow \text{Prune}_{r+1,\delta}(S_{r+1}^T)$;
3. $\text{Cost(QJ2)} \longleftarrow QC(L_{K_J}^1)$. Return $L_{K_J}^1$.

The minimum cost algorithm is given by $QJ2 = \min_i QJ2(u_i)$. We now analyze the complexity and correctness of QJ2.

Theorem 4. *For $0 < \epsilon < 1$, $QJ2(\epsilon)$ is a FPAS for the joint quantization problem.*

Proof. We need to show that (a) the solution returned is within a factor of $1 + \epsilon$ of the optimal solution and (b) the running time is polynomial in $1/\epsilon$.

For the first part, we have to show that our policy of selecting the r-set with the largest $\overline{H}$ within a δ-neighborhood is not suboptimal, i.e it does not create solutions with a cost that exceeds a $1 + \epsilon$ factor of the optimal solution. Assume some L_{r-1}^q is optimal for the inductive hypothesis. Some L_1^1 is certainly optimal as we run R instances of the algorithm starting at each node. Let L_r^x be the r-set chosen during the rth stage of pruning and let L_r^y be the optimal choice in the same δ-neighborhood, which was not chosen because of L_r^x. Let v_x and v_y be the two nodes that created the respective r-sets from L_{r-1}^q.

Let $A = \overline{H}_{v_x} \backslash \overline{H}_{L_{r-1}^q}$ and $B = \overline{H}_{v_y} \backslash \overline{H}_{L_{r-1}^q}$ be the marginal $\overline{H}$ contributions of the nodes. AP_x and BP_y are the marginal costs of adding v_x and v_y respectively to L_{r-1}^q. Also let $A_1 = A \bigcup \overline{H}_{L_{r-1}^q}$ and $B_1 = B \bigcup \overline{H}_{L_{r-1}^q}$

Since L_r^x was chosen over L_r^y, we know from lemma 1 that $A_1 \geq B_1$ and also $AP_x \leq (1 + \delta)BP_y$. Now let u_t be a node that is added during step $r + 1$. We consider two cases:

First, let u_t be the terminal node, i.e after u_t all nodes in G are covered. The cost of $L_r^x \bigcup u_t$ and $L_r^y \bigcup u_t$ are given by

$$C_1 = QC(L_{r-1}^q) + AP_x + P_t(\overline{H}_t \backslash A_1) \tag{10}$$

$$C_2 = QC(L_{r-1}^q) + BP_y + P_t(\overline{H}_t \backslash B_1), \tag{11}$$

respectively. ¿From the above observations on A, B, A_1, B_1, we have $\overline{H}_t \backslash A_1 \leq \overline{H}_t \backslash B_1$ and thus $C_1 \leq (1 + \delta)C_2 < (1 + \epsilon)C_2$ since $\delta = \epsilon/2K_J$.

Suppose u_t is a non-terminal node. We argue that there always exists another vertex u_w to be added in future whose cost will be within a $1 + \epsilon$ factor by going with L_r^x instead of L_r^y. Suppose now $\overline{H}_t \backslash A > \overline{H}_t \backslash B$ even though $A > B$. Thus u_t has a larger overlap with B. Clearly the Quantization Cost of $L_r^y \bigcup u_t$ can be unboundedly smaller than the cost of $L_r^x \bigcup u_t$. Now consider another additional node u_w that is added later than u_t such that A and B are both covered. u_w exists since u_t is non-terminal and A and B have to be covered before the algorithm terminates. By definition, $P_w \geq P_t$. Consider the costs of $L_r^x \bigcup u_t \bigcup u_w$ and $L_r^y \bigcup u_t \bigcup u_w$ given by

$$C_1 = QC(L_{r-1}^q) + AP_x + P_t(\overline{H}_t \backslash A_1) + P_w(\overline{H}_w \backslash (\overline{H}_t \bigcup A_1)) \tag{12}$$

$$C_2 = QC(L_{r-1}^q) + BP_y + P_t(\overline{H}_t \backslash B_1) + P_w(\overline{H}_w \backslash (\overline{H}_t \bigcup B_1)), \tag{13}$$

respectively. Now using the fact that $P_w \geq P_t$ and $A_1 \geq B_1$, we can see that again $C_1 \leq (1 + \delta)C_2 < (1 + \epsilon)C_2$ as desired. Hence we have shown that choosing the L_r^x representative as defined in the algorithm is not suboptimal by larger than a $1 + \delta$ factor at each stage.

Since at each stage we are no more than a $1 + \delta$ factor from the optimal, after K_J stages, we will be within a factor

$$(1 + \delta)^{K_J} = (1 + \frac{\epsilon}{2K_J})^{K_J} \leq (1 + \epsilon) \tag{14}$$

For the second part, we need to show that the running time is polynomial in $1/\epsilon$. Algorithm Prune takes clearly takes time $O(T)$ assuming all costs and $\overline{H}$ values are known. Let $F = QC(L_r^T)$, i.e F is the maximum quantization cost in S_r^T. Note that $F \leq P_{max}$, the maximum allowed transmission power for source and relay. This is usually imposed as a practical limitation. Now the size $n(r)$ of the pruned list can be determined as follows: Let $S_r^{n(r)} = \{L_r^1, \ldots, L_r^{n(r)}\}$.

By lemma 1, we have $QC(L_r^i) > (1+\delta)QC(L_r^{i-1}$ and $\overline{H}(L_r^i) > \overline{H}(L_r^{i-1}$. Since successive elements in $S_r^{n(r)}$ differ by at least a $(1+\delta)$ factor, we get

$$n(r) \leq 2 + \log_{1+\delta} F = 2 + \frac{\ln F}{\ln 1 + \delta} \leq 2 + \frac{2\ln F}{\delta} = O\left(\frac{K_J \ln F}{\epsilon}\right) \quad (15)$$

The time complexity of $QJ2(\epsilon)$ can now be determined as follows. The $Create_{r+1}$ operation of step 2 takes $O(|n(r)|R^2) = O(K_J \ln FR^2/\epsilon)$ time since $O(R)$ nodes are separately added to each existing list and each addition takes $O(R)$. The size T of the new $r+1$-list S_{r+1}^T is $O(|n(r)|R)$ and so sorting takes $O(|n(r)|R \log(|n(r)|R))$ time. Finally, the pruning operation takes $O(|n(r)|R)$ time. Hence the overall complexity is dominated by the first step which is $O(K_J \ln FR^2/\epsilon)$. Since the algorithm is called $O(R)$ times (one for each u_i), the total complexity is $O(K_J \ln FR^3/\epsilon)$ which is polynomial in $1/\epsilon$, K_J and F.

7   Relay to Clusterhead Quantization Algorithm QRB

Finally, we describe the Quantization between relay and base station. Let k_b be the relay to clusterhead quantization bandwidth and $K_b = 2^{k_b}$. If the clusterhead is a more powerful node than an ordinary sensory, then $K_b << \{K_s, K_r\}$. Let $QRB(k, t\gamma)$ represent the optimal cost of quantizing the range of $h_{1,2}$ represented by $0 < h_{1,2} < t\gamma$ into k levels, $1 \leq t \leq M$. In the case of independent clusterhead to source/relay quantization, QRB can be specified by the following dynamic program.

$$QRB_{k,t\gamma} = \min_{1 \leq r < t} \{QRB_{k-1,r\gamma} + \text{OptIQ}_{K_J}^{r\gamma,t\gamma}\} \quad (16)$$

where the second term is a call to the independent quantization algorithm with restricted $h_{1,2}$ and total bandwidth parameters as described. The boundary conditions are evaluated at $QRB_{1,t\gamma_1}$ for $1 \leq t \leq N_1$, using the recursive calls and the fact that $QRB_{0,t\gamma_1} = 0$. QRB is calculated in a bottom-up manner with increasing t and k.

8   Conclusions

In this paper, we address the problem of developing power control algorithms for sensor networks with collaborative relaying under bandwidth constraints, via quantization of finite rate feedback channels. We develop a system model using channel fading parameters as the metric and are able to develop power control policies that minimize aggregate source and relay power. Perfect Channel State Information is not available due to bandwidth constraints and thus we focused on developing quantization algorithms. We develop two quantization protocols: independent quantization and joint quantization of source and relay channels by the clusterhead. Our proposed quantization problem is related to the k-median problem. Independent quantization can be reduced to k-median on line graphs and hence easily solved in polynomial time. However joint quantization is NP-hard and therefore we are forced to develop approximations. We show an easy to implement 2-factor approximation and then develop a Fully Polynomial Approximation Scheme that can approach the optimal to within a $(1 + \epsilon)$ factor. In future work, we will work on further simplification of the FPAS.

References

1. Garey, M.R., Johnson, D.S.: Computers and Intractability: A Guide to the Theory of NP-completeness. Freeman, San Francisco (1979)
2. Lin, Z., Erkip, E., Stefanov, A.: Cooperative regions and partner choice in coded cooperative systems. IEEE Transactions on Communications 4(54), 760 (2006)
3. Nosratinia, A., Hunter, T.E.: Grouping and partnership selection in cooperative wireless networks. IEEE J. Select. Areas Commun. 25(2), 1–10 (2007)
4. Love, D.J.L., Heath Jr, R.W., Strohmer, T.: Grassmannian beamforming for multiple-input multiple-output wireless systems. IEEE Trans. Inform. Theory 49(10), 2735–2747 (2003)
5. Ahmed, N., Khojastepour, M.A., Sabharwal, A., Aazhang, B.: Outage minimization with limited feedback for the fading relay channel. IEEE Trans. Commun. 54(4), 659–669 (2006)
6. Wei, S., Kannan, R.: Strategic versus collaborative power control in relay fading channels. In: IEEE International Symposium on Information Theory (ISIT), Seattle (July 2006)
7. Laneman, J., Tse, D., Wornel, G.: Cooperative diversity in wireless networks: efficient protocols and outage behavior. IEEE Trans. Inform. Theory 50, 3062–3080 (2004)
8. Kariv, O., Hakimi, S.L.: An algorithmic approach to network location problems. Part II: The p-medians. SIAM J. Appl. Math. 37, 539–560 (1979)
9. Tamir, A.: An o(pn) algorithm for p-median and related problems on tree graphsi. Operation Research Letters 19, 59–64 (1996)
10. Benkoczi, R., Bhattacharya, B., Chrobak, M.L.L.: Faster algorithms for k-medians in trees. Extended Abstract.
11. Hassin, R., Tamir, A.: Improved complexity bounds for location problems on the real line. Operation Research Letters 10, 395–402 (1991)
12. Auletta, V., Parente, D., Persiano, G.: Placing resources on a growing line. J. Algorithms 26(1), 87–100 (1998)
13. Li, B., Golin, M.J., Italiano, G.F., Deng, X.: On the optimal placement of web proxies in the internet. In: Proc. of IEEE INFOCOM (1999)
14. Woeginger, G.: Monge strikes again: optimal placement of web proxies in the internet. Operation Research Letters 27, 93–96 (2000)
15. Kannan, R., Wei, S., Deng, G., Chakravarthy, V., Rangaswamy, M.: Energy efficient relaying via channel quantization in wireless networks. In: 41st Annual Conference on Information Sciences and Systems (CISS 07), JHU (March 2007)
16. Charikar, M., Guha, S., Tardos, E., Shmoys, D.B.: A constant-factor approximation algorithm for the k -median problem (extended abstract). In: ACM Symposium on Theory of Computing, pp. 1–10 (1999)

A Connectivity Based Partition Approach for Node Scheduling in Sensor Networks

Yong Ding, Chen Wang, and Li Xiao

Department of Computer Science and Engineering,
Michigan State University
{dingyong, wangchen, lxiao}@cse.msu.edu

Abstract. This paper presents a Connectivity based Partition Approach (CPA) to reduce the energy consumption of a sensor network by sleep scheduling among sensor nodes. CPA partitions sensors into groups such that a connected backbone network can be maintained by keeping only one arbitrary node from each group in active status while putting others to sleep. Nodes within each group swap between active and sleeping status occasionally to balance the energy consumption. Unlike previous approaches that partition nodes geographically, CPA is based on the measured connectivity between pairwise nodes and does not depend on nodes' locations. In this paper, we formulate node scheduling as a constrained optimal graph partition problem, and propose CPA as a distributed heuristic partition algorithm. CPA can ensure k-vertex connectivity of the backbone network for its partition so as to achieve the trade-off between saving energy and preserving network communication quality. Moreover, simulation results show that CPA outperforms other approaches in complex environments where the ideal radio propagation model does not hold.

1 Introduction

Wireless sensor networks consist of a large number of small battery powered nodes that need to operate in unattended status for months. In order to sustain sensors to run for a long period of time with limited energy capacity, it is critical to save energy in sensor operations. Since wireless communication consumes the majority of energy among all the sensors' activities, reducing power consumption in communication is the most effective approach to prolong sensors' lifetime. Two strategies are usually used to minimize energy dissipation in sensor communication: i) adjust the radio transmission power of each node; or ii) schedule the wireless interfaces of sensor nodes to rotate between active and sleeping status.

Several approaches have been proposed to reduce the energy consumption of a sensor network by minimizing sensors' transmission power while maintaining the network connectivity [1] [2] [3]. However, the major energy of a sensor network is often consumed by idle listening instead of packet transmission and reception under light traffic or in a dense network. It has been broadly observed that the energy consumption of a wireless interface cannot be ignored even when it is in

J. Aspnes et al. (Eds.): DCOSS 2007, LNCS 4549, pp. 354–367, 2007.
© Springer-Verlag Berlin Heidelberg 2007

the listening mode [4] [5]. Therefore, energy can be further saved by reducing the time spent in idle listening of sensor nodes.

In this paper, we adopt the sleep scheduling approach to reduce the energy consumption without causing dramatic data delivery delay in a dense sensor network. Since only a small portion of the sensors are involved in packet transmission and reception in a dense sensor network where broadcast is not frequently initiated, it will be most effective to save energy by turning off the wireless interfaces of those redundant sensors that only operate in listening status. Therefore, we can divide the sensor nodes into groups such that nodes in each group are equivalent with regard to data delivery. At each time, one node is selected from each group to operate in active radio mode (listening, transmitting, receiving), while other nodes put themselves into sleeping mode by turning off their wireless interfaces. No matter which node is selected from each group, all the active nodes need to form a connected backbone network. In addition, the roles of active nodes and sleeping nodes need to be swapped once a while to balance the power consumption among all the nodes, which prolongs the network's lifetime.

GAF [6] [7] partitions the nodes based on their geographic locations. It divides the deployed area into multiple equal-size squared cells so that nodes in the same cell form a group. By assuming an ideal radio propagation model and choosing appropriate side length of cells according to the radio transmission range, it ensures that a connected backbone network can be formed as long as at least one node in each cell remains in active mode. However, this geographic partition suffers from several drawbacks besides its dependence on the localization infrastructure. GAF uses fixed cells in its partition, which consequently guarantees a backbone network with vertex connectivity of 4. Thus, it lacks the flexibility to provide different partitions that can ensure different connectivity levels of the backbone networks. Moreover, GAF depends on the assumption of an ideal radio propagation model, which does not always hold due to radio's irregular transmission pattern and multipath effect. Thus, it is often invalid to judge the connectivity between pairwise nodes simply based on the distance calculated from their locations.

Motivated by these limitations of GAF, we propose a Connectivity based Partition Approach (CPA), which divides nodes based on their measured connectivity instead of guessing connectivity by their positions. In comparison with GAF, our approach has more flexibility in that it can generate partitions while ensuring k-connectivity of the backbone network. In addition, as CPA is based on measured connectivity, it can ensure the connectivity between nodes of neighboring groups in the partition even under an un-ideal radio propagation model, which makes it more adaptive to complex environments compared with GAF. To illustrate the basic idea of CPA, we introduce the motivation and give a formal problem description in Section 2. The detailed description of the algorithm is discussed in Section 3. In Section 4, we evaluate our proposed approach by comparing it with the GAF approach. Previous studies are summarized in Section 5 and we conclude our work in Section 6.

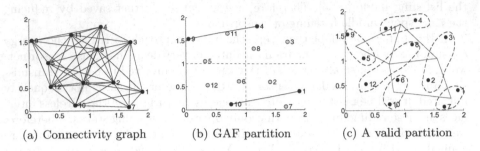

(a) Connectivity graph   (b) GAF partition   (c) A valid partition

Fig. 1. GAF under Irregular Radio Propagation Models

2   Overview

In this section, we first analyze several limitations of GAF that motivate us to partition the nodes based on their measured connectivity rather than their locations. After that, a formal description is given for the problem of partitioning nodes based on their connectivity graph in a large sensor network. Our solution is given in the next section.

2.1   Motivation

Although turning off nodes can reduce the energy consumption dramatically, the change of network graph property may affect the communication performance and therefore incur more power dissipation. Vertex connectivity is a useful metric to evaluate the communication quality of the backbone network with regard to node failure and congestion. GAF actually provides a 4-connected backbone network for a large sensor network. It guarantees that each active node is connected with at least four active nodes in its four neighboring grids respectively in the backbone network. However, GAF lacks the flexibility to provide backbone networks with different vertex connectivity under different requirements. If the nodes are relatively more robust and the traffic rate is not high, a backbone network with lower connectivity is desired to achieve more energy saving by maintaining fewer active nodes. On the other hand, a backbone network with higher connectivity can cope with higher node failure and traffic rate. Therefore, a more flexible algorithm is desired to partition the nodes into groups of appropriate size.

Another problem with GAF is that it may not work well under irregular radio propagation models. To illustrate this, we use DOI (Degree of Irregularity) [8] as a radio propagation model. This model assumes an upper and lower bound on signal propagation range. The parameter DOI is defined as the maximum radio range variation per unit degree change in the direction of radio propagation. The DOI model used in our example is shown in Fig. 6(b). The upper bound is the maximum radio transmission radius R, the lower bound is half of the upper bound, and DOI is set to 0.1.

We deploy 12 sensor nodes with maximum radio transmission radius of $\sqrt{5}$ uniformly into a 2×2 area as shown in Fig. 1(a). Each edge between pairwise sensors represents a symmetric link between them based on the DOI model. In Fig. 1(b), the deployed area is divided into 2×2 grids, each of which owns 3 nodes according to GAF. However, there is a possibility that the backbone graph is disconnected. As the case shown in Fig. 1(b), nodes 1, 4, 9, and 10 are selected from each grid to become active nodes, but they form a disconnected backbone network. The partition of GAF is invalid because the connectivity between nodes in neighboring grids is no longer ensured under the irregular radio model. On the other hand, Fig. 1(c) shows a valid partition. It consists of 6 groups where the nodes in each group are mutually connected. Each edge between two groups means that any node from one group is connected with any node in the other group. It is obvious that the backbone formed by selecting one node from each group is always connected. This partition is valid because it is based on the measured connectivity between nodes instead of guessing connectivity from nodes' locations.

In this paper, we will study how to flexibly partition the nodes based on their measured connectivity. Before presenting our algorithm, a formal problem description is given in the following.

2.2   Problem Formulation

To reduce the energy consumption of communication in sensor networks, we can divide sensor nodes into groups such that only one node in each group keeps active at each snapshot while others are put into sleeping mode. The partition must satisfy the following constraints:

- Any node is within one hop away from all the other nodes in the same group. Under such a constraint, each node can be covered by the communication backbone, that is, each node is either in the backbone network if it is an active node or directly connected to the backbone network otherwise.
- The backbone network formed by active nodes at each snapshot must satisfy some connectivity properties such that it does not suffer significant loss of communication quality as compared with the original network.
- The analysis in [9] shows that for those sensor applications where data are collected by a sink, the sensors closer to the sink always deplete their energy faster under uniform distribution of nodes, no matter what sleep scheduling is used. However, some mobility assisted approaches, such as [10] and [11], can help achieve uniform energy consumption in sensor networks. Therefore, in order to better evaluate the sleep scheduling algorithm, we assume the uniform energy consumption for sensor nodes in this paper. In order for all the groups to remain alive together as long as possible, the energy needs to be evenly distributed among groups.
- A smaller number of groups is preferred without degrading the communication quality of the original network, because more energy conservation can be achieved by decreasing the number of active nodes at each time.

By referring to some terms in graph partition problems [12], we can formalize the problem as below.

Let G(V, E) be the connectivity graph of the original sensor network, where each vertex in V corresponds to a sensor node and each edge in E represents a symmetric communication link between the two nodes.

Definition 1. We can partition G into k parts A_1, A_2, ..., A_k, where A_i is completely adjacent. In other words, G is partitioned into k cliques. We can encode this partition by a symmetric k-by-k matrix M in which the diagonal entry $M_{i,i}$ is 1 representing each part as a clique, and off-diagonal entry $M_{i,j}$ is 0, 1, or 2, if A_i and A_j are completely non-adjacent, have arbitrary connections, or are completely adjacent, respectively. This partition is called an *M-partition*.

Let G' denote the backbone graph formed by selecting one arbitrary active node from each part of the M-Partition. If $M_{i,j}$ is 2, then the active nodes from A_i and A_j are guaranteed to be connected; if $M_{i,j}$ is 0, they are disconnected; otherwise, they will be either connected or disconnected.

Definition 2. Let H be the graph induced by an M-partition P in which each part in P is mapped to a vertex in H and each 2-value entry in P is mapped to the corresponding edge in H. We call H the *2-induced graph* of P.

We are interested in H because it reflects the minimum connectivity property of G' at each snapshot. In H, if two vertices are adjacent, any arbitrary node in the corresponding part of one vertex is connected with any arbitrary node in the corresponding part of the other vertex. In other words, E(H) is a subset of E(G'). As a result, suppose $\kappa(H) = k$, then we have $\kappa(G') \geq k$, where κ denotes the vertex connectivity of the corresponding graph.

Let l be a label on V of the original network graph G(V, E) where $l(v)$ ($v \in V$) is the energy amount in the sensor node v, then the total energy of the sensor network is $E_{total} = \sum_{v \in V} l(v)$. We can also derive another label g on the M-partition P of G, which represents the total energy amount in each part, that is, $g(A) = \sum_{v \in A} l(v)$ for each $A \in P$.

Problem Description. Given a graph G(V, E), which represents the original sensor network, and a label l on V, which represents the energy in each sensor node, find a smallest-size M-partition P* of G such that i) $\kappa(H) \geq k$, where H is the 2-induced graph of P* and k is the minimum vertex connectivity required by the backbone network. ii) $(1 - \delta)\frac{E_{total}}{N} \leq g(A_i) \leq (1 + \delta)\frac{E_{total}}{N}$ for each $A_i \in$ P*, where N is the cardinality of P* and $0 \leq \delta < 1$ is the unbalanced factor.

As each part in the partition P* is a clique, each node is one hop away from all the other nodes in the same part. The connectivity property of the backbone network can be guaranteed by the first constraint of the problem, and the balanced energy distribution can be satisfied by the second constraint. Moreover, the optimization nature of this problem requires the most efficient partition in energy-saving.

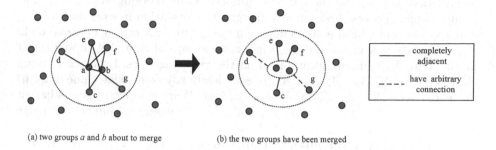

(a) two groups a and b about to merge   (b) the two groups have been merged

Fig. 2. CPA Group Merging Process

3 CPA Design

The formulated problem is NP hard [12]. In this section, we propose a Connectivity based Partition Approach (CPA) to approximate a good partition for this problem, which is a heuristic distributed algorithm, where only local computation is involved. CPA is a distributed iterative process. It starts from the initial partition where each node forms a unique group. CPA continuously merges two groups into a larger one until further merging will break the constraints of the problem.

In CPA, there are two kinds of nodes in each group: ordinary nodes and a head node. Each kind of node maintains its node ID and associated group information including its group ID, IDs of other group members, and ID of the head node in its group. One head node is selected from each group to maintain some additional information on the connectivity between its group and the neighboring groups in the current M-partition. Let $N_l(A_i)$ be the set of neighboring groups that are connected with group A_i through l-value edges in the current M-partition, i.e., $N_l(A_i) = \{A_j \mid M_{i,j} = l\}$. Thus, each head node of group i will store $N_1(A_i)$ and $N_2(A_i)$, which are the set of neighboring groups having arbitrary connection with group i and completely adjacent with group i respectively.

CPA starts from the initial partition of one node in each group. Let A_i denote the group formed by node v_i. Consequently, v_i acts as the head node and stores group connectivity information $N_1(A_i)$ and $N_2(A_i)$, where $N_2(A_i)$ is the set of groups A_j whose node v_j is connected with node v_i, and $N_1(A_i)$ is empty because any two groups are either completely adjacent or completely non-adjacent in the initial partition. CPA goes through a group merging process iteratively before it gets to the final partition.

3.1 Group Merging

In the group merging process, head nodes of each two completely adjacent groups exchange group connectivity information to decide whether their groups should merge. Only completely adjacent groups can merge so that the new group is also a clique. Suppose A_i and A_j are two completely adjacent groups in the current

M-partition. Let A_{ij} be the new group obtained by merging A_i and A_j. The group merging process first updates the group information in each node of A_{ij}, and keeps only one head node to maintain the group connectivity information in A_{ij}. Then, the new group and the neighboring groups of A_i and A_j update their group connectivity information based on the following rules. i) For each group $A' \in N_2(A_i) \cap N_2(A_j)$, A_{ij} and A' are completely adjacent (edge value of 2). ii) For each group $A' \in N_0(A_i) \cap N_0(A_j)$, A_{ij} and A' are completely non-adjacent (edge value of 0). iii) Otherwise, A_{ij} and A' have arbitrary connections (edge value of 1). Therefore, an updated M-partition is formed. Fig. 2 illustrates the process of merging two groups into a larger group. In Fig. 2(a), a and b are two completely adjacent groups, that is, any node in a is connected with any node in b. When the two groups merge as shown in Fig. 2(b), as groups c, e, f are completely adjacent to both a and b, each of them is completely adjacent to the new merged group. On the other hand, groups d and g only have arbitrary connection with the new group, which are illustrated by the dashed lines in the figure, because d is not completely adjacent with b and g is not completely adjacent with a.

Contentions may occur when multiple neighboring groups want to merge simultaneously. We resolve this by imposing a randomized backoff delay on the time when the two groups announce their willingness to merge. If no contention is observed at the end of the delay, the two groups about to merge will announce their decision to all of their neighboring groups. Otherwise, they will reevaluate the backoff delay based on the updates from other group merges.

The goodness of the final partition depends on the sequence of group merge. We consider several factors for deciding which two groups are preferred to be merged first in the current partition in order to arrive at a good partition eventually. These factors can be reflected as a utility function in the randomized backoff delay so that higher priority groups will announce their intentions to merge with a shorter time of delay.

– For any two groups A_i and A_j in the current partition, let $P = |N_2(A_i) \cap N_2(A_j)|$ and $Q = |N_2(A_i) \cup N_2(A_j)|$, then $C = P/Q$ indicates the level of equivalence between A_i and A_j. The two groups with higher C value will be given higher priority in the group merging process. Specifically, at the starting phase of the algorithm, where each node constitutes a single group, nodes with exactly the same set of neighbors will be merged first, because these nodes are exactly equivalent with regard to data routing.
– Let $g(A_i)$ denote the energy in group A_i. We want the total energy to be evenly distributed in each group so as to maximize the network's lifetime. For any two groups A_i and A_j, let $D = [g(A_i) + g(A_j)]/E_{total}$ where E_{total} is the total energy of all the sensor nodes in the network, then we will give pairwise groups with lower D value higher priority in the group merging process.

Therefore, each two completely adjacent groups can be assigned with a utility value $U = k_1(1 - C) + k_2 D$, where k_1 and k_2 are coefficients. The backoff delay for each pair of groups is set to be proportional to $U + R$, where R is a random value uniformly distributed in $[0, 1]$, which is used to resolve contentions among pairwise groups with the same utility value. As a result, the appropriate assignment of backoff delay enables pairwise groups with lower utility value to merge first as well as resolving contentions in the group merging process.

In M-partition, we refer "2-degree of A_i" to the number of groups that are completely adjacent to A_i. Each group keeps track of its 2-degree value during the group merging process, which is used to decide whether pairwise groups should be merged. If a group merge may cause the 2-degree of some group to drop below k, then these two groups will give up their intention to merge. The group merging process will be terminated when no groups can be merged. By Penrose's theorem [13], we can guarantee the k-connectivity of the backbone graph by ensuring the minimum degree of k in the 2-induced graph of M-partition when the sensor network can be modeled as a random geometric graph. Under the irregular radio propagation model where the sensor network cannot be concisely modeled as a random geometric graph, we will show by simulation that a k-connected backbone network can still be formed with high probability.

3.2   Load Balancing Energy Usage in Groups

As all the nodes in the network are equally important, running a node in active status until its energy is depleted is not an appropriate energy usage strategy. In order to prolong the lifetime of each node, the nodes in each group need to switch between active and sleeping status periodically so that all nodes remain alive together for as long as possible.

Assume all the nodes in each group can be synchronized. In order to elect an active node, each group member broadcasts a message to the whole group stating its willingness to become active. Each node waits for a certain time delay before its announcement. The earliest announcement will suppress the others so that the corresponding node will become the active node in the group. The time delay for each node is set to be inversely proportional to its residual energy. Therefore, the node with maximum residual energy will be selected. Then, the selected active node informs other nodes of the time it will remain active, after which all the nodes need to reselect an active node again within the group.

4   Performance Evaluation

We evaluate our schemes in a 10×10 square area where 500 sensors are uniformly deployed. The simulation is based on the energy consumption model observed in [4], that is, the ratio of energy consumed in listening, receiving, and transmitting status is 1:1.2:1.7. The initial energy level of each node is set to 500, which means the node will remain alive for 500 units of time in the listening status. According to the assumption in *Section 2.2* that the energy consumption is uniform over all the sensor nodes with the mobility assisted approaches helping collect data, we simulate the energy consumption in the sensor network by an equivalent scenario. In each time slice, we randomly select 20 traffic nodes, which send and receive packets between each other. In addition, we use load balanced energy aware routing [14] in the backbone network. One slight modification we make is that we use the total residual energy of the group to denote the residual energy of the corresponding active node in the load balanced route decision.

We perform the simulation in MATLAB. The energy consumption is calculated based on the changing status of each node and the energy consumption ratio for each status. In our evaluation of the sensor network's lifetime, we do not take the energy consumed in the partition process into consideration, because it runs only once at the deployment phase of the sensor network so that it only consumes a trivial portion of the network's total energy. In this section, we evaluate CPA in comparison with GAF under both the ideal radio transmission model and the irregular radio transmission models.

Table 1. Partitions of GAF and CPA

Partition Approach	CPA (min-deg=2)	CPA (min-deg=3)	CPA (min-deg=4)	GAF	CPA (min-deg=5)	CPA (min-deg=6)
Number of Groups	71	84	91	100	106	116
Average Group Size	7.0	6.0	5.5	5.0	4.7	4.3
Standard Deviation of Group Size	0.92	0.77	0.79	0.73	0.67	0.63

4.1   Under Ideal Radio Propagation Model

In the ideal radio propagation model, the radio transmission range is the same in different directions. In our simulation, the radio transmission radius R is set to $\sqrt{5}$. GAF uses square cells with length of $R/\sqrt{5} = 1$ to partition the deployed area, thus all the nodes are divided into 100 groups. We also run CPA, which is based on the connectivity between nodes instead of their locations under the same experiment setting. CPA is executed with different values for parameter *mindeg*, which controls the minimum degree of the 2-induced graph of the final partition. CPA guarantees that the backbone network generated based on this partition will be *mindeg*-connected.

The partition results are shown in Table 1. For CPA, the number of groups increases with the parameter *mindeg*, because more active nodes are needed each time in order to ensure higher connectivity of the backbone network. GAF ensures that each node is connected with at least four nodes in its four neighboring cells in the backbone network. Therefore, we can regard GAF as comparable to CPA(*mindeg*=4). As shown in the table, CPA(*mindeg*=4) partitions the nodes into 91 groups, which is fewer than GAF. This indicates that CPA can identify redundant nodes more sufficiently than GAF. As discussed in previous sections, it is preferable to have the total energy evenly distributed in the groups so as to prevent the early death of some groups, which may disrupt the connectivity of the backbone network. We assume that each node has the same initial energy when deployed, so the standard deviation of group size can be an indication

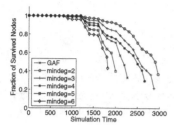
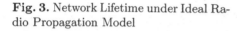

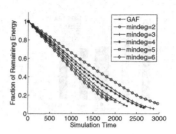

Fig. 3. Network Lifetime under Ideal Radio Propagation Model

Fig. 4. Energy Consumption under Ideal Radio Propagation Model

of how evenly the total energy is distributed in the groups. Table 1 shows the standard deviation of group size for each partition result. It shows that our approach can divide energy as evenly in groups as GAF under random uniform distribution of nodes.

We then evaluate the lifetime of the network when the partition results are applied. Each group keeps only one active node each time to form a backbone network. The nodes in each group balance the energy usage by reselecting the active node periodically, which may change the topology of the backbone network. We refer to lifetime as the time when the backbone network formed by active nodes turns out to be disconnected. If no sleeping schedule scheme is used, that is, all nodes keep active until death, the network lifetime will be less than 500. Fig. 3 illustrates the network lifetime where different partitions are adopted. As shown in the figure, the fraction of surviving nodes decreases with time, and the simulation stops when the backbone network becomes disconnected. The partition of CPA($mindeg$=2) achieves the longest network lifetime (around 3000), because it keeps the fewest number of active nodes each time. The network lifetime decreases for partitions of CPA with higher $mindeg$ values which can, however, ensure better connectivity of the backbone network. We can also observe from the figure that the partition of CPA($mindeg$=4) has longer network lifetime than the partition of GAF, even though they ensure the same level of connectivity. Fig. 4 illustrates the energy consumption of the whole network with regard to time. The energy consumption rate is relatively constant because the traffic nodes generate traffic at a constant speed and the number of active nodes remains constant each time as well.

Although lowering node density reduces the energy consumption of the network, the topology change may affect the network's communication quality. For example, if a packet goes through a much longer path in the backbone network than it does in the original network, longer data transfer delay will be experienced. Fig. 5 shows the ratio of the average routing path length in the backbone network and the original network for different partitions. The ratio decreases for CPA with higher $mindeg$ because its partition ensures higher connectivity. As can be seen, the ratio is quite low (below 1.3) for all the partitions listed in

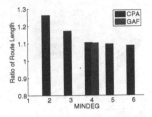

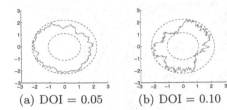

(a) DOI = 0.05   (b) DOI = 0.10

Fig. 5. Comparison of the Impact on Route Length under GAF and CPA

Fig. 6. Irregular Radio Propagation Model

the figure, which means the node scheduling based on these partitions does not dramatically increase packet delivery delay.

4.2   Under Irregular Radio Propagation Model

In order to evaluate the performance of CPA in comparison with GAF under complex environments, we choose the DOI model [8]. This model assumes an upper and lower bound on the signal propagation range. The parameter DOI is defined as the maximum radio range variation per unit degree change in the direction of radio propagation. In our simulation of radio irregularity, we set the upper bound to $\sqrt{5}$ and lower bound to half of upper bound. Fig. 6 shows two examples of radio propagation range in different directions where the DOI value is set to 0.05 and 0.10 respectively. The higher the DOI value, the more irregularity in the radio propagation range.

GAF cannot adapt to different levels of irregularity in the radio propagation model, because it is based on the sensors' locations and consequently cannot detect the irregularity level. Unlike GAF, CPA partitions sensors based on their measured connectivity, which enables it to obtain appropriate partition sizes under different levels of radio irregularity. Fig.7 shows the partition sizes of CPA with different *mindeg* under irregular radio with different DOI values. We can observe from the figure that the partition size increases with the DOI value. As the communication between sensors is more seriously influenced by higher irregularity in the radio propagation model, more active nodes are needed to maintain the same level of connectivity in the backbone network, leading to larger partition size.

We perform simulations to study the network lifetime under GAF and CPA (*mindeg* = 2 or 4). For GAF, we use the same partition for different DOI values, that is, 100 groups with cell length of 1, because GAF is unaware of the radio irregularity. We run the simulation multiple times for each partition, and the comparison of average lifetime is shown in Fig. 8. Our simulation finds that the lifetime for GAF is not stable through repeated simulations. As the connectivity between neighboring cells is no longer guaranteed by GAF under the irregular radio propagation model, there is a possibility that the backbone network formed

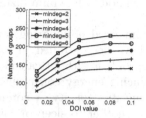

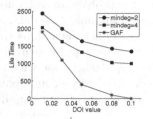

Fig. 7. Partition Sizes under Irregular Radio Propagation Model

Fig. 8. Network Lifetime under Irregular Radio Propagation Model

by randomly selecting an active node from each group is disconnected. The higher the radio irregularity, the higher the probability of a disconnected backbone network. As illustrated in the figure, when DOI is close to 0.1, the GAF partition cannot even work. In contrast, CPA works well under different conditions. The lifetime for CPA decreases with the radio irregularity level, because more active nodes are needed to maintain the same connectivity of the backbone network, and thus more energy is consumed per time unit. On the other hand, GAF does work if it divides the deployed area by cells with side length of 0.5 so that any nodes in two neighboring cells are within $\sqrt{5}/2$, the lower bound of radio transmission range in DOI model. However, this results in 400 groups with an average group size of 1.25. Apparently, this partition can only prolong the network's lifetime for a very small portion.

5   Related Work

Energy in sensor networks can be saved by adjusting the radio transmit power of each node. Several topology control algorithms [1] [2] [3] have been proposed to reduce energy consumption by selecting adequate node transmit power while maintaining network connectivity.

[15] [16] [17] select the set of active nodes for routing purposes based on the idea of approximating a minimum connected dominating set (MCDS). [18] further discusses how to balance energy dissipation in the cluster heads of the CDS. Deb and Nath [19] proposed a node scheduling approach that can adapt to the trade-off between energy conservation and data delivery quality. Although CDS approaches save energy by decreasing the number of active nodes, they are not efficient at balancing energy consumption among nodes so as to maximize the network lifetime.

To reduce the energy waste in idle listening, duty cycling has been proposed in [20] and [21], where the wireless interface of each node follows a periodic cycle of active/sleep states. Although duty cycling is energy-efficient, it increases the delay of data delivery, because the intermediate node has to wait for the next-hop node to wake up to receive the packet. [22] analyzes the bounds of data delivery delay by using completely decentralized duty cycling. In [23], the authors formulate the problem of assigning duty cycle to each node while minimizing the end-to-end communication delay.

Node scheduling algorithms that maintain a connected dominating set and balance energy usage by switching node status have been studied in [4] and [6]. These approaches cope with the idle listening problem without causing dramatical data delivery delay. Span [4] aims at reducing energy consumption of a wireless network without significantly diminishing its capacity or connectivity. In Span, each node makes a local decision on whether to sleep or join the backbone as a coordinator by periodically checking the status of its neighbors. Unlike Span, GAF [6] divides nodes into groups such that a communication backbone is formed by selecting an arbitrary active node from each group while keeping others in sleeping mode. Compared with Span, GAF imposes less overhead on switching node status, because only nodes within each group need to communicate with each other for load balance purposes.

Our proposed CPA schedules nodes based on partitioning. Different from GAF, CPA is based on the measured connectivity between nodes instead of their locations. Besides preserving GAF's advantage in efficient load balancing of energy dissipation, it aims at ensuring k-vertex connectivity of the backbone network, and better adaptivity for unideal radio propagation. CEC [7] divides nodes into clusters based on measured connectivity similarly, but it cannot efficiently switch node status within each group like GAF and CPA. Instead, it needs to re-form clusters to balance energy consumption among nodes.

6   Conclusion

In this paper, we propose to partition the nodes based on their measured connectivity instead of geographic locations. We formulate it as a constrained optimal graph partition problem, and present CPA, a distributed heuristic algorithm, to approximate a good partition. CPA outperforms other partition approaches in two aspects. First, CPA can guarantee k-vertex connectivity of the backbone network under ideal radio propagation models, which balances the trade-off between saving energy and preserving the network's communication quality. In addition, simulation results show that CPA can also ensure k-vertex connectivity of the backbone network with high probability under irregular radio propagation models. Therefore, CPA has better adaptivity to complex environments.

Acknowledgement

This work was supported in part by the US National Science Foundation under grants CCF-0514078, CNS-0549006, and CNS 0551464.

References

1. Ramanathan, R., Hain, R.: Topology control of multihop wireless networks using transmit power adjustment. In: INFOCOM (2000)
2. Wattenhofer, R., Li, L., Bahl, P., Wang, Y.-M.: Distributed topology control for wireless multihop ad-hoc networks. In: INFOCOM (2001)

3. Li, N., Hou, J.C.: Flss: A fault-tolerant topology control algorithm for wireless networks. In: MobiCom (2004)
4. Chen, B., J. K., B. H., M. R.: Span: An energy-efficient coordination algorithm for topology maintenance in ad hoc wireless networks. In: Mobicom (2001)
5. Stemm, M., Katz, R.H.: Measuring and reducing energy consumption of network interfaces in hand-held devices. In: IEICE Transactions on Communications (1997)
6. Xu, Y., Heidemann, J., Estrin, D.: Geography-informed Energy Conservation for Ad Hoc Routing. In: MobiCom (2001)
7. Xu, Y., et al.: Topology control protocols to conserve energy in wireless ad hoc networks. tech. rep. (2003)
8. He, T., Huang, C., Blum, B.M., Stankovic, J.A., Abdelzaher, T.F.: Range-free localization schemes in large scale sensor networks. In: MobiCom (2003)
9. Subramanian, R., Fekri, F.: Sleeping scheduling and lifetime maximization in sensor networks: Fundamental limits and optimal solutions. In: IPSN (2006)
10. Luo, J., Hubaux, J.-P.: Joint mobility and routing for lifetime elongation in wireless sensor networks. In: InfoCom (2005)
11. Wang, W., Srinivasan, V., Chua, K.-C.: Using mobile relays to prolong the lifetime of wireless sensor networks. In: MobiCom (2005)
12. Feder, T., Hell, P., Klein, S., Motwani, R.: Complexity of graph partition problems. In: ACM STOC (1999)
13. Penrose, M.D.: On k-connectivity for a geometric random graph. In: Wiley Random Structures and Algorithms (1999)
14. Sankar, A., Liu, Z.: Maximum lifetime routing in wireless ad-hoc networks. In: INFOCOM (2004)
15. Das, B., Bharghavan, V.: Routing in ad-hoc networks using minimum connected dominating sets. In: ICC (1997)
16. Guha, S., Khuller, S.: Approximation algorithms for connected dominating sets. In: European Symposium on Algorithms (1996)
17. Banerjee, S., Khuller, S.: A clustering scheme for hierarchical control in multi-hop wireless networks. In: INFOCOM (2001)
18. Wu, J., Dai, F., Gao, M., Stojmenovic, I.: On calculating power-aware connected dominating sets for efficient routing in ad hoc wireless networks. Journal of Communications and Networks (2002)
19. Deb, B., Nath, B.: On the node-scheduling approach to topology control in ad hoc networks. In: MobiHoc (2005)
20. Ye, W., Heidemann, J., Estrin, D.: An energy-efficient mac protocol for wireless sensor networks. In: InfoCom (2002)
21. van Dam, T., Langendoen, K.: An adaptive energy-efficient mac protocol for wireless sensor networks. In: Sensys (2003)
22. Dousse, O., Mannersalo, P., Thiran, P.: Latency of wireless sensor networks with uncoordinated power saving mechanisms. In: MobiHoc (2004)
23. Gang Lu, B.K.A.G., Sadagopan, N.: Delay efficient sleep scheduling in wireless sensor networks. In: InfoCom (2005)

Energy-Efficient Data Acquisition Using a Distributed and Self-organizing Scheduling Algorithm for Wireless Sensor Networks

Supriyo Chatterjea[1], Tim Nieberg[2], Yang Zhang[1], and Paul Havinga[1]

[1] Department of Computer Science, University of Twente,
P.O. Box 217 7500AE, Enschede, The Netherlands
{supriyo,zhangy,havinga}@cs.utwente.nl
[2] Research Institute for Discrete Mathematics, University of Bonn,
Lennestr. 2, 53113 Bonn, Germany
nieberg@or.uni-bonn.de

Abstract. Wireless sensor networks are often densely deployed for environmental monitoring applications. Collecting raw data from these networks can lead to excessive energy consumption. Thus using the spatial and temporal correlations that exist between adjacent nodes we appoint a few as representative nodes that perform in-network aggregation. This reduces the total number of transmissions. Our distributed scheduling algorithm autonomously assigns a particular node to perform aggregation and reassigns schedules when network topology changes. These topology changes are detected using cross-layer information from the underlying MAC layer. We also present theoretical performance estimates and upper bounds of our algorithm and evaluate it by implementing the algorithm on actual sensor nodes, demonstrating an energy-saving of up to 80% compared to raw data collection.

1 Introduction

Densely deployed wireless sensor networks (WSNs) allow environmental monitoring at extremely high spatial and temporal resolutions. However, extracting the raw data from such networks can have problems, e.g. batteries may get drained rapidly due to excessive operation of the transceiver or data quality may deteriorate due to dropped packets caused by network congestion (Figure 1(a)).

To solve the above problems, we exploit the high degree of spatial correlation that exists between the sensor readings of adjacent nodes in a densely deployed network. Thus, instead of every node transmitting individual readings, we appoint a subset of nodes, referred to as *correlating nodes* that transmit the messages representative of all the remaining nodes at any given point in time. Every correlating node initially transmits information to the sink, indicating the correlation of its readings with its adjacent neighbors. Subsequently, it continues transmitting its own readings until a change in correlation is detected, in which case, it transmits an updated correlation message. The sink then *estimates* the

J. Aspnes et al. (Eds.): DCOSS 2007, LNCS 4549, pp. 368–385, 2007.

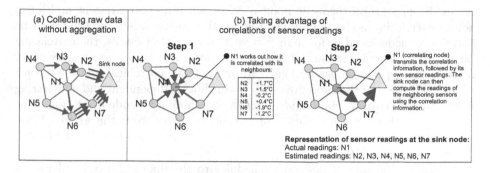

Fig. 1. Advantage of using correlation information (b) instead of transmitting raw data (a)

Table 1. List of contributions

No.	Contribution
1	We present a distributed scheduling algorithm that enables every node to autonomously choose schedules based only on locally available information.
2	Although we assume our network is static, we prove that our algorithm possesses self-stabilizing properties that allow it to recover within a finite time regardless of any disturbances in the network such as topology changes or communication errors. We present theoretical upper bounds for message transmissions and network stabilization times when topology changes occur.
3	We illustrate how our algorithm takes advantage of cross-layer information to improve energy-efficiency and adapt quickly to topology changes.
4	We present theoretical performance estimates and upper bounds for the performance of our algorithm. We evaluate the algorithm by presenting results based on an implementation on actual sensor nodes and present results indicating energy-savings of up to 80% compared to raw data collection.

readings of the adjacent neighbors of the correlating node by combining the current readings of the correlating node with the previously transmitted correlation information as shown in Figure 1(b). We present a completely distributed and self-organizing scheduling algorithm that (i) prevents two adjacent nodes acting as correlating nodes simultaneously, (ii) increases the robustness and accuracy of the readings by giving every node a chance at some point to act as a correlating node. This ensures that no node is always represented only by estimated readings. Our contributions are stated in Table 1.

2   An Overview of the $\mathcal{DOSA}$ Approach

The primary objective of $\mathcal{DOSA}$ is to help decide *when* a particular node should act as a correlating node and thus represent the sensor readings of its 1st order neighbors. During the correlating node's schedule, the node initially transmits correlation information to the sink node followed by its own sensor readings. All the 1st order neighbors *do not* transmit their sensor readings to the sink during this period. Since $\mathcal{DOSA}$ is intended to solve a scheduling problem, we make use of a distributed graph coloring algorithm to assign schedules to individual nodes. From a graph theoretic point of view, since no two adjacent nodes can act as correlating nodes simultaneously, all the nodes chosen by $\mathcal{DOSA}$ to be correlating nodes need to form an *independent set*. Additionally, the correlating nodes for a particular instant of time need to form a *dominating set* since every non-correlating node must be joined to at least one correlating node by some edge. Also note that the subset of nodes that is both independent and dominating is known as a *maximal independent set*. A maximal independent set cannot be extended further by the addition of any other nodes from the graph. These requirements help define the constraints outlined in Section 5 that $\mathcal{DOSA}$ follows in order to perform its task.

To hasten the rate of assigning schedules to the nodes, $\mathcal{DOSA}$ utilizes the information provided by the underlying MAC protocol, LMAC [8,12]. Instead of coloring all the nodes from scratch, $\mathcal{DOSA}$ meets its requirements by building up on the colors already assigned by LMAC. An added advantage of this form of cross-layer optimization is that a lesser number of messages need to be transmitted for all the schedules to be assigned properly as we make use of information that already exists. Furthermore, $\mathcal{DOSA}$'s dependence on LMAC makes it more reactive to changes in topology as any changes in neighborhood detected by LMAC are immediately filtered to $\mathcal{DOSA}$. As the operation of $\mathcal{DOSA}$ is completely dependent on LMAC, we first give a brief overview of LMAC and then proceed to present the operation of $\mathcal{DOSA}$.

3   LMAC: A Lightweight Medium Access Control Protocol

LMAC is a TDMA-based medium access control protocol designed for WSNs. Time in LMAC is divided into frames, each of which is further divided into a fixed number of time slots. Every node chooses its own slot using a distributed algorithm that uses only locally available information. A node is allowed to pick any slot as long as it is not owned by any other node within its two-hop neighborhood.

A time slot consists of two sections, the Control Message (CM) and the Data Message (DM). The CM, which contains control information and has a fixed length, is broadcast by a node to its neighbors during its own time slot once *every* frame irrespective of whether the node has any data to send. The CM contains a table which indicates the slots that are occupied by itself and its one-hop neighbors and other control information. Every node maintains a *Neighbor Table* that stores the information about its one-hop neighbors, e.g. ID, occupied slot, number of hops to sink node, etc. Thus a node can automatically work out its degree from its Neighbor Table. Occupied slots are marked with a 1 where as unoccupied ones are marked with a 0. A node joining the network first listens out for the CMs of all its neighbors and then picks one of the slots that is marked as unoccupied by performing an OR-operation. The DM contains higher layer protocol messages. The length of the DM can vary depending on the amount of data that a node needs to send. It does however, have a maximum length.

4   Preliminaries for Self-stabilization

The self-stabilization approach is essential for $\mathcal{DOSA}$ to initialize during start-up and recover from topology changes due to addition and removal of nodes and also to formalize the self-organizing properties of the algorithm. It also allows a system that has entered an illegal state (due to occurrence of transient faults) to converge back to a legitimate state within a finite time and with no external intervention. We now look at the preliminaries of self-stabilization and refer the reader to [5] for more details on the subject.

All nodes in the network are assumed to have unique IDs and knowledge of their adjacent neighbors. Each node has a state specified by its local variables. The state of the entire system called the *global state* or *configuration* is the union of the local states of all the nodes. The objective of the system is to reach a desirable global final state called a *legitimate state*. The state of a system can either be *legitimate* or *illegitimate*. We use $\mathcal{S}$ to denote the set of all possible states. We denote the set of all legitimate states by $\mathcal{L}$ such that $\mathcal{L} \subseteq \mathcal{S}$. We define $\mathcal{R} \in \mathcal{S} \times \mathcal{S}$ such that $(s_i, s_j) \in \mathcal{R}$. An execution of e is a maximal sequence of states, $e = s_i, s_{i+1}, ... s_j$ such that $\forall i \geq 1, s_i \in \mathcal{S}$, and s_i is reached from s_{i-1} by executing a particular rule. A system can be considered to be self-stabilizing if the following two conditions hold:

- **Closure:** If $s \in \mathcal{L}$ and $s \rightarrow s'$ then $s' \in \mathcal{L}$. Therefore the closure property means that when a system is in a legitimate state, the following state is always a legitimate state as well regardless of the rule executed.
- **Convergence:** Starting from any configuration $s \in \mathcal{S}$, every execution reaches $\mathcal{L}$ within a finite number of transitions.

5 $\mathcal{DOSA}$: A Distributed and Self-organizing Scheduling Algorithm

$\mathcal{DOSA}$ uses a distributed graph coloring approach to decide when a particular node should be a correlating node. Every color owned by a node represents a particular frame of time during which a node is required to act as a correlating node. In conventional graph coloring approaches, colors are assigned to vertices such that adjacent vertices are assigned different colors and the number of colors used is minimized. While $\mathcal{DOSA}$'s graph coloring approach also ensures that adjacent nodes in the network do not own the same colors it differs in the sense that each node is allowed to own *multiple* colors, i.e. a node can have multiple schedules. Moreover, the number of colors used in $\mathcal{DOSA}$ is fixed and is equal to the number of slots that are assigned to an LMAC frame.

Before we proceed, we first state certain definitions that are used through out the rest of this paper. We model the network topology as an undirected graph G where $G = (V, E)$. V represents the vertices or nodes in the network while two nodes are connected by an edge in E if they are within radio transmission range of each other. $\mathbf{K}$ represents the set of colors used to color *all* the nodes. So $|\mathbf{K}|$

is equal to the number of slots per frame in LMAC. Also, we denote the *closed* neighborhood of a node $v \in V$ by $\Gamma(v)$ i.e. $\Gamma(v) := \{u \in V | (u,v) \in E\} \cup \{v\}$.

Using the graph-theoretic distance $d_G(u,v)$, that denotes the number of edges on a shortest path in G between vertices u and v, we can define the rth neighborhood of v as $\Gamma_r(v) := \{u \in V | d_G(u,v) \leq r\}$. Similarly, we define the *open* neighborhood of a node v by $\Gamma'(v)$ where, $\Gamma'(v) := \{u \in V | (u,v) \in E\}$. Given that $\Gamma'(v)$ denotes the *open* neighborhood of node v, we refer to C_v as the set of colors owned by node v. Then for C_v it holds that $0 < |C_v| < (|\mathbf{K}| - |\Gamma'(v)|)$.

Given that a *node-induced subgraph* is a subset of the nodes of a graph G together with edges whose endpoints are both in this subset, we define a *component* as a node induced subgraph of a subset of nodes. Furthermore, we call two components independent if they are not connected by an edge.

Before describing the details of the operation of $\mathcal{DOSA}$, we first state the constraints derived from the requirements stated in Section 2, which define its behavior. The following two constraints must be met when two nodes u and v are adjacent to each other:

Constraint 1: $C_v \cap C_u = \emptyset$
In other words, two adjacent nodes cannot own the same colors. This is because two adjacent nodes should not be assigned as correlating nodes in the same time instant.

Constraint 2: $C_{\Gamma(v)} = \mathbf{K}$
*All colors should be present within the one-hop neighborhood of node v, i.e. if node v does not own a particular color itself, the color **must** be present in one of its neighboring nodes that is one hop away. This ensures that every node's readings will be represented at the sink node for every time instant either directly or through a correlated reading.*

Lemma 1. *The combination of constraints 1 and 2 ensures that at any time slot, c_i, all nodes owning the color c_i, which correspond to that time slot, form a maximal independent set on G.*

Proof. At any time instant according to Constraint 1, two adjacent nodes will never own the color c_i, thus resulting in an independent set I. Constraint 2 ensures that in the closed neighborhood of every node $v \in V$, every color is present. This clearly results in a maximal independent set.

5.1   Dependency of $\mathcal{DOSA}$ on LMAC

As mentioned in Section 3, LMAC assigns a slot to every node in the network. $\mathcal{DOSA}$ begins its distributed coloring scheme by considering the initial slot assignment phase in LMAC as an input. Slot assignments in LMAC correspond to partial color assignments in $\mathcal{DOSA}$. Thus while LMAC assigns every node with a single color, $\mathcal{DOSA}$ assigns the remaining colors that ensure the adherence to the constraints 1 and 2 given in the previous section. We can then state that, $C_v = C_{v_{LMAC}} \cup C_{v_{DOSA}}$, where $C_{v_{LMAC}}$ refers to the color corresponding to the LMAC slot owned by node v and $C_{v_{DOSA}}$ refers to the colors assigned to node v by $\mathcal{DOSA}$. Similarly, the colors owned by the nodes adjacent to node v, $C_{\Gamma'(v)}$, are also made up of LMAC and $\mathcal{DOSA}$ colors. Thus we can state, $C_{\Gamma'(v)} = C_{\Gamma'(v)_{LMAC}} \cup C_{\Gamma'(v)_{DOSA}}$. The dependency of $\mathcal{DOSA}$ on LMAC allows nodes to adapt autonomously and immediately to changes in network topology. For example, the addition or removal of a node results in the change being reflected in the LMAC Neighbor Tables of all other neighboring nodes within range. $\mathcal{DOSA}$ detects changes in LMAC's Neighbor Table and performs a re-assignment of schedules if any of the neighboring nodes do not meet the constraints mentioned above. Utilizing such cross-layer information from LMAC ensures that $\mathcal{DOSA}$ does not spend additional resources trying to detect topology changes itself.

5.2   General Operation of $\mathcal{DOSA}$

$\mathcal{DOSA}$ uses a *greedy* approach to assign colors to nodes. Coloring is performed using two types of colors: *LMAC Colors* and *$\mathcal{DOSA}$ Colors*. LMAC Colors refer to the colors that have been assigned by LMAC - due to the slot assignment. $\mathcal{DOSA}$ Colors refer to the additional colors that are assigned by $\mathcal{DOSA}$ to ensure that constraints 1 and 2 are met. This occurs *after* the LMAC colors have been assigned. $\mathcal{DOSA}$ does not have any control over the LMAC Color of a node as it depends purely on the slot assignment performed by LMAC. In fact, such control is also not required. Therefore, in the following, we refer to $\mathcal{DOSA}$ Colors simply as colors unless otherwise indicated.

Colors are acquired based on a calculated priority. A node computes its priority within its one-hop neighborhood based on its degree and node ID. The higher the degree of a node, the higher its priority. If two neighboring nodes have the same degree, priority is calculated based on the unique node ID; the node with the larger node ID will have the higher priority.

Once all nodes have acquired their LMAC slots, a *BeginSecondPhase* message is injected into the network through the sink node requesting the nodes to begin the $\mathcal{DOSA}$ coloring phase. At this stage, every node receiving the *BeginSecondPhase* message only has an LMAC Color and does not satisfy the constraints mentioned earlier. Thus these nodes mark themselves as Unsatisfied. A node only attains the Satisfied status when it satisfies the two constraints mentioned in Section 5. Upon receiving the *BeginSecondPhase* message, a node broadcasts the *NodeStatus* message. This message contains information about the node's status (i.e. Satisfied/Unsatisfied) and the list of colors owned. The *ColorsOwned* field is a string of $|\mathbf{K}|$ bits where every color owned by a node is marked with a 1. The rest of the bits are marked with a 0. Initially, a node only marks its own LMAC Color as 1 due to the initial LMAC slot assignment. A neighboring node that receives the *NodeStatus* message then performs coloring using $\mathcal{DOSA}$ as outlined in Algorithm 1. Note that the *NodeStatus* message is the only message that is used for the operation of $\mathcal{DOSA}$.

Algorithm 1. $\mathcal{DOSA}$ - Normal Initialization

Input: NodeStatusMSG(SatisfiedStatus(TRUE/FALSE), ColoursOwned)
Output: NodeStatusMSG(SatisfiedStatus(TRUE), ColoursOwned)/ NIL
1: UPDATE(LocalInfoTable, v)
2: **if** LocalInfoTable contains entries from ALL adjacent nodes **then**
3:     **if** SatisfiedStatus(v)=$FALSE$ **then**
4:         Compute PRIORITY(v)
5:         **if** PRIORITY(v)=Highest **then**
6:             $C_v \leftarrow \mathbf{K} \backslash C_{\Gamma'(v)}$
7:             ColorsOwned$\leftarrow C_v$
8:             SatisfiedStatus$\leftarrow TRUE$
9:             UPDATE(LocalInfoTable, v)
10:            BROADCAST NodeStatusMSG(Degree, SatisfiedStatus, ColoursOwned)
11:        **end if**
12:    **end if**

13: **end if**

We now briefly describe the operation of $\mathcal{DOSA}$ outlined in Algorithm 1. Upon receiving a *NodeStatus* message, a node first updates its *LocalInfoTable* (Line 1). This table stores all the information contained in the *NodeStatus* messages that are received from all the adjacent nodes. Once a node receives *NodeStatus* messages from *all* its immediate neighbors (Line 2), and if its status is Unsatisfied(Line 3), the node proceeds to compute its priority. PRIORITY computes the priority of a node *only* among its unsatisfied neighbors (Line 4), i.e. as time progresses and more nodes attain the Satisfied status, PRIORITY needs to consider a smaller number of neighboring nodes. The highest priority is given to the node with the largest degree among its adjacent Unsatisfied neighbors. If more than one node has the same degree, then the highest priority is given to the Unsatisfied node with the largest NodeID.

The node that has the highest priority among all its immediate unsatisfied neighbors, acquires *all* the colors that are not owned by any of its adjacent neighbors (Line 7). As the node has then satisfied both constraints of $\mathcal{DOSA}$, it switches to the Satisfied state, updates its own *LocalInfoTable* and informs all its neighbors through a broadcast operation (Lines 8-10). Note that this technique corresponds to a highest degree greedy approach.

Figure 2 provides a step-by-step example of how the $\mathcal{DOSA}$ algorithm assigns colors to the nodes in a network. We make the assumption in the example that LMAC uses 16 slots.

Correctness of $\mathcal{DOSA}$. In this section we illustrate how $\mathcal{DOSA}$ is able to successfully carry out initialization within a finite time given any arbitrary network. We initially assume that no transmission errors occur throughout the initialization phase but subsequently describe how such issues are handled in Section 5.2.

In order for $\mathcal{DOSA}$ to operate properly, it is absolutely imperative that every node always has up-to-date state information of its immediate neighbors. If a node n experiences a certain change in state (e.g. change from Satisfied to Unsatisfied) and fails to inform an adjacent neighbor of the change, this neighbor node might execute certain inappropriate steps based on its outdated state

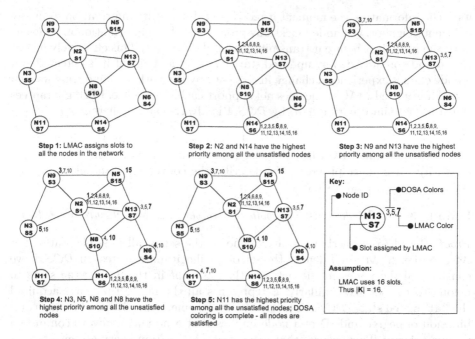

Fig. 2. A step-by-step example of how $\mathcal{DOSA}$ colors are assigned

information of n. This error may prevent $\mathcal{DOSA}$ from stabilizing within a finite time. Thus it is essential for $\mathcal{DOSA}$ to possess the *cache coherence* property [7].

Let each node $v \in V$ in the sensor network have a variable, C_v indicating the colors owned by node v. For each $(u, v) \in E$, let u have a variable $\Diamond_u C_v$ which denotes a cached version of C_v. We can call a system *cache coherent* if $\forall u, v : (u, v) \in E : \Diamond_u C_v = C_v$ [7]. This means that whenever v assigns a value to C_v, node v also broadcasts the new value to all its neighbors. The moment a node u receives an updated value of C_v, it instantaneously (and atomically) updates $\Diamond_u C_v$.

If we consider the operation of LMAC alone, the cache coherency property does not hold. Let us consider the case where two adjacent nodes v and u own the slots i and j respectively where $j > i$. Suppose v first broadcasts its updated state information to u during its own slot i. Now consider the case where the state of v changes in slot l where $i < l < j$. In this case, v will be unable to broadcast its newly updated status to u as the earliest time when it can transmit will be in slot $i + n$ where n is the number of slots in a single frame, i.e. v would have to wait one entire frame. This delay in transmission prevents the cache coherence property from existing. Nevertheless, for $\mathcal{DOSA}$ we have the following lemma:

Lemma 2. *Assuming no errors occur, nodes executing the $\mathcal{DOSA}$ algorithm on top of the LMAC protocol are all cache coherent.*

Proof. In order to ensure cache coherence, $\mathcal{DOSA}$ carries out *pre-transmission state information processing* or PSIP. PSIP ensures that while a node updates its cache information the moment it receives updated state information from any adjacent neighbor, the node blocks any processing of the information in its cache until the point just before it transmits during its own slot. This effectively means that a node broadcasts any updated state change the moment it is detected and a node cannot experience a change in state at any time other than during its own slot. Thus while LMAC alone does not support cache coherence, PSIP guarantees that the state information used by $\mathcal{DOSA}$ is always cache coherent.

There are a few properties that $\mathcal{DOSA}$ possesses that ensure that it stabilizes within a finite time: (i)Cache coherence (Shown in Lemma 2), (ii)Closure property, (iii)Convergence property. We describe the convergence and closure properties in greater detail below.

Lemma 3. $\mathcal{DOSA}$ *demonstrates both the convergence and closure properties.*

Proof. Recall from Section 4 that $\mathcal{S}$ denotes the set of all possible states. Let $\mathcal{M} \in \mathcal{S}$ (i.e. $\mathcal{S}\backslash\mathcal{M} = \mathcal{L}$) denote the set of all illegitimate states. In $\mathcal{DOSA}$, we consider all the nodes in the network that are *not* in the Satisfied state to belong to the set $\mathcal{M}$. Similarly, $\mathcal{L}$ represents all the nodes that have acquired the Satisfied state. $\mathcal{DOSA}$'s prioritization scheme, which is based on the combination of degree and ID of a node implies that a node can always compute a unique priority. This ensures that as long as $|\mathcal{M}| > 0$, in every atomic step, at least one node is enabled and thus attains the Satisfied state, i.e. if $n \in \mathcal{M}$, $|\mathcal{M}| = i$ and $|\mathcal{L}| = j$ in step r, then at step $r + 1$, $n \in \mathcal{L}$, $|\mathcal{M}| = i - k$ and $|\mathcal{L}| = j + k$ where $k > 0$. Thus over a finite number of steps, all nodes in $\mathcal{M}$ eventually converge towards $\mathcal{L}$.

Furthermore, as we assume that no communication errors or topology changes occur during the initialization process, a node that acquires the Satisfied state, remains in that state forever, regardless of the messages received. This is synonymous to the closure property.

Lemma 4. *Assuming no transmission errors or topology changes occur, given that d is the number of nodes in G'_{max}, which is the largest independent component in G, the time taken for all nodes in G to attain the Satisfied state, t_s (in frames) in $\mathcal{DOSA}$ during the initialization is such that $d + 1 \leq t_s \leq 2d - 1$.*

Lemma 5. *During the initialization of DOSA, every node in the network transmits a total of 3 messages.*

We refer the reader to [2] for the proof of the above two lemmas.

Handling Message Corruption. Up to now, we have assumed that all communication is error free. However, $\mathcal{DOSA}$ does take certain steps to ensure that it continues to operate normally even when transmission errors or topology changes occur. Due to lack of space, we refer the reader to [2] for further details.

6   Performance of $\mathcal{DOSA}$

The effectiveness of $\mathcal{DOSA}$ can be evaluated by observing the number of correlating nodes at any point of time and comparing it against the case of raw data collection where every node will be involved in transmitting raw sensor readings. The cardinality of the maximal independent set can vary greatly depending on the set of chosen nodes. This results in varying degrees of energy efficiency since a larger cardinality means lower efficiency as compared to raw data collection.

This then leads us to the following question: *Given a particular graph, what is the maximum cardinality of the maximal independent set formed by $\mathcal{DOSA}$?* This would give us an estimation or bound on the worst case performance of $\mathcal{DOSA}$. Since computing the maximum maximal independent set of a given graph is NP-hard [4], we take a "covering" approach to give a bound on the worst case performance of $\mathcal{DOSA}$.

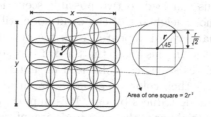

Fig. 3. Estimating the cardinality of the maximum maximal independent set generated by $\mathcal{DOSA}$

Lemma 6. *The worst case performance of $\mathcal{DOSA}$ can be guaranteed to result in a savings of at least $\left(\frac{2nr^2}{xy} - 1\right) \times 100\%$ compared to raw data collection when n nodes are uniformly distributed in an area of dimensions $x \times y$ and every node has a circular transmission radius of r.*

Proof. Let us divide the area $x \times y$ into m squares where, $m = \frac{xy}{2r^2}$. Since the nodes are assumed to be randomly distributed, we may reasonably assume that nodes are present in all m squares, Figure 3. Note that this results in a worst-case estimation. Furthermore, we assume that exactly one node in every square forms part of a maximal independent set. We immediately see that it is not possible to have more than one node which is part of the maximal independent set in a single square as these 'extra' nodes would be in range of the first chosen node. Thus this consequently implies that the cardinality of the maximal independent set would be m. It would be impossible to increase the size any further by adding any more nodes. We can then conclude that the maximum cardinality of the maximal independent set created by $\mathcal{DOSA}$ is m. Thus the percentage savings of $\mathcal{DOSA}$ compared to the collection of raw data would then be, $\frac{n-m}{m} \times 100$. This can then be simplified to $\left(\frac{2nr^2}{xy} - 1\right) \times 100\%$.

Network density, μ can be defined as follows: $\mu = \frac{n\pi r^2}{xy}$. Using the above equations, we can then state $|I| \leq \frac{n\pi}{2\mu}$ where I is any independent set also including the one computed by $\mathcal{DOSA}$. However, network density is approximately equal to average connectivity, i.e. $\frac{n\pi}{2\mu} \approx \frac{n\pi}{2(\rho-1)}$ where ρ is the average connectivity. This result is used to plot the graph in Figure 5(a) which estimates the cardinality of $\mathcal{DOSA}$ as the average connectivity is varied.

6.1   Coping with a Dead Node

$\mathcal{DOSA}$ ensures that a node is able to reorganize the scheduling algorithm within a finite time autonomously the moment a neighboring node disappears from the network by retrieving cross-layer information from the underlying LMAC protocol, i.e. the death of a node triggers an update in the LMAC Neighbor Table.

The death of a node leads to the disappearance of the colors that were owned by the dead node. This can lead to two possible scenarios. Firstly, it may be possible that one or more neighbors of the dead node still satisfy constraints 1 and 2 as the colors that have disappeared with the dead node are also present in its neighboring nodes. In this case, the Satisfied neighboring nodes continue to maintain their existing schedules and do not transmit any messages. Note however, that while their color assignments are invariant, the degree of the neighbors of the dead node reduces by one. It is important that nodes that are one hop away from the neighbor of the dead node are informed about this change of degree as this information would be required in case any schedules need to be reassigned in the future due to certain network perturbations. However, as our design takes advantage of cross-layer information from LMAC, explicit message transmissions are not required to relay information regarding a change of degree of a node. This information is instead automatically disseminated through the periodic broadcast of the CM section of the LMAC protocol. Recall that the Occupied Slot list in the CM section can also be used to deduce the degree of a node.

In the second scenario, the death of a node may result in one or more neighboring nodes ending up with certain missing colors. As these nodes no longer satisfy constraints 1 and 2, the nodes switch to the Unsatisfied state and broadcast this change in status to their immediate one-hop neighborhood. A node then waits for one frame to see if any of the neighboring nodes are also in the Unsatisfied state. After waiting one frame, the node with the missing color(s) acquires all the colors it lacks if it has the highest priority among all the unsatisfied nodes. This whole process is described in Algorithm 2. If a node lacks a color but does not have the highest priority, it continues to wait until all its higher priority unsatisfied neighbors have become satisfied. In other words the node continues to execute Algorithm 1 every time it receives a *NodeStatus* message until it finally acquires the Satisfied state.

In order to explain the timing bounds of $\mathcal{DOSA}$ when a node dies, we use the same argument as in the proof of Lemma 4. We can extend this lemma in Lemma 7. (Refer to [2] for the proofs of Lemmas 7 - 9.)

Algorithm 2. $\mathcal{DOSA}$ - Coping with the loss of a node

---

Input: LMAC Neighbor Table indicates at least one missing node
Output: NodeStatusMSG(SatisfiedStatus(FALSE & TRUE), ColoursOwned)/NIL
 1: UPDATE(LocalInfoTable, v)
 2: **if** MissingColours(v) = $TRUE$ (i.e. SatisfiedStatus(v)=$FALSE$) **then**
 3:    BROADCAST NodeStatusMSG(Degree,SatisfiedStatus(FALSE), ColoursOwned)
 4:    WAIT one frame
 5:    Compute PRIORITY(v)
 6:    **if** Priority(v)=Highest **then**
 7:       $C_v \leftarrow \mathbf{K} \backslash C_{\Gamma'(v)}$
 8:       ColorsOwned$\leftarrow C_v$
 9:       SatisfiedStatus$\leftarrow TRUE$
10:       UPDATE(LocalInfoTable, v)
11:       BROADCAST NodeStatusMSG(Degree, SatisfiedStatus(TRUE), ColoursOwned)
12:    **end if**
13: **end if**

Lemma 7. *When a node v with x neighbors dies, the maximum time taken for all nodes to converge towards the* Satisfied *state is $x + 1$ frames where $x \le |K| - 1$.*

Lemma 8. *When a node v with x neighbors dies, the maximum possible number of messages that may be transmitted is $2x$ where $x \le |K| - 1$.*

Lemma 9. *When a node v dies, only its first order neighbors may be affected, i.e. may switch from the* Satisfied *to the* Unsatisfied *state.*

6.2 Coping with a New Node

When a node dies, $\mathcal{DOSA}$ executes one fixed set of steps to ensure that the scheduling scheme stabilizes within a finite time. In the node addition operation however, the set of steps taken by $\mathcal{DOSA}$ depends on the events that occur when a new node v is added to the network. For example, the node v may detect an LMAC collision or may cause colliding or missing colors in neighboring nodes or may even cause a combination of these events. Different permutations and combinations of these events can cause the network to react in a multitude of ways. This makes it impractical to analyze the performance bounds of every particular sequence of events that causes the network to react in a certain manner. Instead, we categorize all the permutations and combinations of events in terms of how far the network disturbance propagates when a node v is added to the network.

For example, depending on the combination of events, nodes that are either 2 or 3 hops away from node v may become Unsatisfied. (Refer to [2] for a detailed description of all possible events.) However, the addition of a node does not cause a *domino* effect in $\mathcal{DOSA}$ as explained in the following lemma. We refer the reader to [2] for the proof.

Lemma 10. *When a node v is added, all nodes beyond the 3rd order neighborhood of v can be **guaranteed** to be unaffected (i.e. they remain in the* Satisfied *state).*

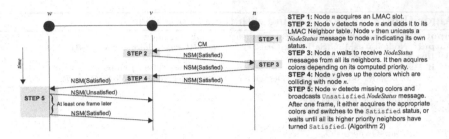

Fig. 4. Timing diagram for addition of a new node, n (Node v is adjacent to n and node w is 2 hops from n)

Algorithm 3. $\mathcal{DOSA}$ - Coping with a new node

Input: NodeStatusMSG(SatisfiedStatus(TRUE), ColoursOwned)
Output: NodeStatusMSG(SatisfiedStatus(TRUE), ColoursOwned)
1: UPDATE(LocalInfoTable,n)
2: **if** LocalInfoTable contains entries from ALL adjacent nodes **then**
3:     Compute PRIORITY(n)
4:     **if** PRIORITY(n)=Highest **then**
5:         $C_n \leftarrow \mathbf{K} \backslash C_{\Gamma'_{LMAC}(n)}$
6:     **else**
7:         $C_n \leftarrow \mathbf{K} \backslash C_{\Gamma'(n)}$
8:     **end if**
9:     UPDATE(LocalInfoTable,n)
10:     BROADCAST NodeStatusMSG(SatisfiedStatus(TRUE), ColoursOwned)
11: **end if**

While theoretically the addition of a node can cause a network disturbance to propagate 3 hops, our earlier simulation results (based on 10,000 node additions over 100 randomly deployed topologies) indicate that in around 92% of the cases, the network disturbance is restricted to *within* the second order neighborhood of the newly added node [2]. In 8% of the cases, none of the neighbors are affected. Third order neighbors are only affected in less than 1% of the cases.

Irrespective of the sequence of events that occur, $\mathcal{DOSA}$ always executes a few common steps when a new node joins the network. The next set of steps depends on how far the network disturbance will propagate. We first explain the initial common steps below.

When a new node, n is added to the network, LMAC ensures that the node occupies a slot that is not used by any other nodes within 2 hops of n, Figure 4, Step 1. Node n then begins broadcasting its CM section. Neighboring nodes then detect the new node and add its entry into their LMAC Neighbor Table, Figure 4, Step 2. We explain the remaining steps taken by $\mathcal{DOSA}$ by referring to a neighboring node of node n as node v. It is also explained in Algorithm 3.

The moment v, which is already in the Satisfied state, detects a new node, n, it unicasts a *NodeStatus* message to node n, Figure 4, Step 2. Node n then waits to receive *NodeStatus* messages from all its adjacent neighbors, Figure 4,

Table 2. Upper bounds for time and message transmission when a node is added

	No effect (Group 1, 8%)	1st order (Group 2, 46%)
		Color collision
Event type	-	Color collision
Max. time (Frames)	≤ 1	≤ 1
Max Msgs Tx	$= \|\Gamma'_1(v)\| + 1$	$\leq 2\|\Gamma'_1(v)\| + 1$

	2nd order	
	Color collision (Group 3, <1%)	Missing color (Group 4, 46%)
Event type	Color collision (Group 3, <1%)	Missing color (Group 4, 46%)
Max. time (Frames)	≤ 2	$2 + \|\Gamma'_2 \setminus \Gamma'_1\|$
Max Msgs Tx	$\leq 2\|\Gamma'_1(v)\| + 1 + \|\Gamma'_2 \setminus \Gamma'_1\|$	$\leq 2\|\Gamma'_1(v)\| + 1 + 2\|\Gamma'_2 \setminus \Gamma'_1\|$

	3rd order
	Missing color (Group 5, <1%)
Event type	Missing color (Group 5, <1%)
Max. time (Frames)	$3 + \|\Gamma'_3 \setminus \Gamma'_2\|$
Max Msgs Tx	$\leq 2\|\Gamma'_1(v)\| + 1 + 2\|\Gamma'_2 \setminus \Gamma'_1\| + 2\|\Gamma'_3 \setminus \Gamma'_2\|$

Step 3. By this stage, n would know about the existence of all its adjacent neighbors as any existing LMAC collisions would have already been resolved.

From this point onwards, the actions taken by $\mathcal{DOSA}$ are dependent on the sequence of events that occur. Once n has received *NodeStatus* messages from all its adjacent neighbors, it checks if it has the highest priority within its immediate neighborhood. If n finds that it has the highest priority, it acquires *all* colors *except* the LMAC colors of the adjacent neighboring nodes. This helps to ensure that over time, even if the network topology changes, the cardinality of the maximal independent set continues to be low.

However, if node n realizes that it does not have the highest priority, it simply acquires all the colors that it is presently lacking. As node n has now satisfied constraints 1 and 2, it broadcasts a *NodeStatus* message indicating that it is Satisfied.

At this stage, a neighboring node v, that receives the *NodeStatus* message from n may detect that some of its colors are colliding with n. This would mean that Constraint 1 is not being met. Thus node v gives up the colliding colors, attains the Satisfied state, updates its own *LocalInfoTable* and informs all its neighbors through a broadcast operation, Figure 4, Step 4.

As v has given up certain colors, it could be possible that a node w, that is adjacent to v but not to n (i.e. w is 2 hops away from n), may become Unsatisfied (due to the *NodeStatus* message transmitted in Step 4 of Figure 4). Node w can then resolve the situation by executing Algorithm 2 which allows it to recover when certain colors are found to be missing, Figure 4, Step 5.

Next we present the upper bounds of $\mathcal{DOSA}$ in terms of time taken to stabilize the network and number of message transmissions when a new node is added. Since the addition of a node can result in the occurrence of several events, we breakdown the analysis into 5 possible groups, based on the depth of propagation of the network disturbance as shown in Table 2. Note that these 5 groups encompass all the possible sequence of events that can happen due to the addition of a new node, e.g. colliding LMAC slots, new node acquiring the highest priority, etc.

We refrain from explaining the derivations for the theoretical upper bound times for network stabilization shown in Table 2 as they have been derived using the same arguments presented earlier in Lemma 4. However, in order to present

Table 3. Rules describing number of message transmissions

No.	Description of rule
1	When a node that has already acquired $\mathcal{DOSA}$ colors detects a new neighbor node, v, it unicasts one *NodeStatus* message to node v.
2	A new node, v broadcasts one *NodeStatus* message once it has acquired its LMAC slot, resolved all LMAC collisions amongst its neighbors and has received *NodeStatus* messages from all its neighbors.
3	A node, that acquires a new LMAC color that is not listed in its existing list of $\mathcal{DOSA}$ colors broadcasts one *NodeStatus* message.
4	A node that experiences a color collision event transmits one *NodeStatus* message.
5	A node that experiences a missing color event transmits two *NodeStatus* messages - the first to indicate that the node is Unsatisfied due to the missing color(s) and the second to indicate the node is Satisfied after it has acquired the appropriate colors.

a more concise explanation, we present the theoretical upper bounds for the number of message transmissions using a set of 5 rules listed in Table 3. We refer the reader to [2] for an example on how these rules can be used to work out the upper bounds in Table 2.

7   Details of Implementation and Results

We evaluate the performance of $\mathcal{DOSA}$ by implementing the algorithm on 25 Ambient 2.0 μNodes [1]. While the footprint of LMAC and the AmbientRT operating system [1] is 2782 bytes, $\mathcal{DOSA}$ only takes up 869 bytes. All 25 nodes including the sink are within transmission range of each other, i.e. they form a complete graph. However, testing $\mathcal{DOSA}$ ideally requires a multihop network. As it is impractical to test a large number of different multihop topologies, we generate the topologies randomly and broadcast this information from the sink to the entire network. On receiving this information, each node uses it to create a virtual set of neighbors which is obviously a subset of the actual neighbors. This results in a multihop topology. While LMAC uses the actual neighbors to perform slot allocation, $\mathcal{DOSA}$ uses the virtual set of neighbors to operate. We generate topologies for 5 different average connectivities ranging from 5 to 9. For each average connectivity, we create 50 random topologies and each topology consists of 25 nodes. Thus we have 250 topologies in total.

Our tests are designed to investigate the performance of $\mathcal{DOSA}$ in terms of (i) energy savings compared to collecting raw data and (ii) network stabilization times and message transmissions when a node is removed or added to the network. We have already computed the upper bounds for stabilization times and message transmissions. We now perform experiments to investigate how often and to what proportion these upper bounds are reached in reality.

Figure 5(a) shows that even at high cardinalities, i.e. low average connectivity the number of correlating nodes is not greater than 32% which results in 68% energy savings when compared to collecting raw data. As the average connectivity of the network increases, the cardinality reduces producing savings of up to 80% when the connectivity rises to 9. The difference between the estimate and the actual results can be attributed to boundary effects.

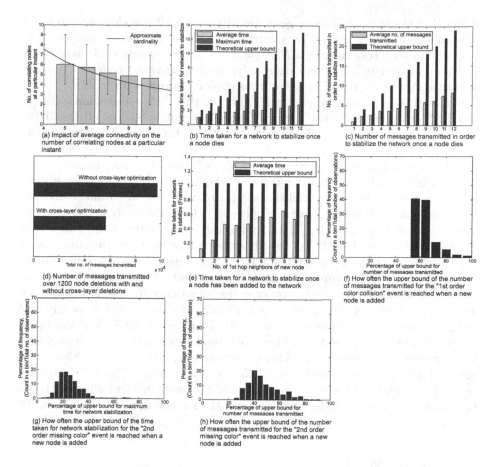

Fig. 5. Implementation results of $\mathcal{DOSA}$

According to Figure 5(b) the average stabilization times, maximum stabilization times and the theoretical upper bound increase with the number of neighbors of the dead node. However, the rate of increase of the average and maximum times decreases as the number of neighbors of the dead node increases. This is because the probability of having a large number of nodes arranged in an increasing manner (refer to [2] for more details) reduces as the number of neighbors increases. Thus a higher density network may not necessarily mean a longer recovery time when a node is removed. In fact, according to the results, when a node with 12 neighbors dies, the average recovery time is around 23% of the theoretical upper bound.

As explained earlier, in the worst case, when a node dies causing all its neighbors to become `Unsatisfied`, every single neighbor will need to transmit two messages (as indicated in Lemma 8). However, in random network topologies, as shown in Figure 5(c) the average number of messages transmitted when a node dies is only around 30% of the maximum theoretical upper bound. Figure 5(d) shows that the number of messages transmitted over 1200 node deletions is significantly lower when cross-layer information is used. Instead of each neighbor of the dead node transmitting a *NodeStatus* message, regardless of their status, use of cross-layer information causes only the Unsatisfied nodes to transmit messages leading to overall message transmission savings of up to 42%.

Our earlier simulation results in [2] indicate that "1st order color collision" and "2nd order missing color" events (Table 2) occur in 92% of all the 10,000 simulation runs while "No effect" occurs in 8% of the cases. The other events occur in less than 1% of the cases. We present the implementation results only for the more significant "1st order color collision" and "2nd order missing color" events. Figure 5(e) shows that regardless of the number of 1st order neighbors a new node has, the network stabilization time remains within 1 frame coinciding with the bounds stated in Table 2. Figure 5(f) shows that only in 1% of all 1st order color collision cases the number of messages transmitted when a new node is added was around 90-100% of the upper bound for message transmissions. In around 80% of the cases the number of messages transmitted was around 50-70% of the upper bound. Figure 5(g) shows that around 92% of time, the network stabilization time when the 2nd order nodes experience a missing color event, was less than 40% of the upper bound. Figure 5(h) shows that in nearly 90% of cases, the number of messages transmitted was less than 70% of the upper bound. Notice that the results in Figure 5(f) tend closer to the upper bound than those presented in Figure 5(h). This is because while the results in Figure 5(f) only require the 1st order nodes to be affected, the results in Figure 5(h) involve both the 1st and 2nd order nodes. Naturally the probability of affecting nodes both in the 1st and 2nd order nodes is lower than affecting nodes in only the 1st order.

8   Related Work

Extensive literature can be found describing various energy-efficient data extraction techniques for WSNs. The approach presented in [13] takes advantage of spatially correlated sensor readings but it is primarily designed for event-based queries. They also assume that individual nodes are location aware which is not required in $\mathcal{DOSA}$. Unlike $\mathcal{DOSA}$ which is designed for multihop networks, the strategies proposed in [9,6] are not scalable as they require all nodes in the WSN to be in direct transmission range of the base station. Ken [3] takes advantage of spatial and temporal correlations and works in a multihop environment, but does not mention any details of how to cope with topology changes. Though PAQ [11] takes advantage of spatial correlations between adjacent nodes, non-clusterheads are always represented at the sink only by estimated readings unlike $\mathcal{DOSA}$ which gives every node a chance to act as a correlating node. SAF [10] reduces energy consumption by sending trends instead of individual readings. However, it forms clusters off-line and thus fails to take advantage of correlations between adjacent nodes. In the event of a sudden trend change, SAF requires all the affected nodes to transmit model updates to the sink while $\mathcal{DOSA}$ causes only the affected correlating nodes to transmit a model update. SAF and PAQ also do not take advantage of any cross-layer optimizations using the MAC or mention any theoretical bounds on possible energy savings.

9 Conclusion and Future Work

We have presented a distributed and self-organizing scheduling algorithm for energy-efficient data acquisition, that takes advantage of spatial correlations of sensor readings of adjacent nodes and cross-layer information from the MAC protocol to result in up to 80% energy savings when compared to conventional raw data collection. We present theoretical performance bounds and also show results based on an actual implementation on sensor nodes to support our claim. We are currently working on strategies to identify correlations and keep correlation models updated.

References

1. Ambient systems (2006) http://www.ambient-systems.net/ambient/index.htm
2. Chatterjea, S., Nieberg, T., Meratnia, N., Havinga, P.J.M.: A distributed and self-organizing scheduling algorithm for energy-efficient data aggregation in wireless sensor networks. Technical Report TR-CTIT-07-10, Enschede (February 2007)
3. Chu, D., Deshpande, A., Hellerstein, J.M., Hong, W.: Approximate data collection in sensor networks using probabilistic models. In: ICDE, p. 48 (2006)
4. Crescenzi, P., Kann, V.: A compendium of np optimization problems: Maximum independent set (2005)
   http://www.nada.kth.se/viggo/wwwcompendium/node34.html
5. Dolev, S.: Self-Stabilization. MIT Press, Cambridge (2000)
6. Heinzelman, W.B., Chandrakasan, A.P., Balakrishnan, H.: An application-specific protocol architecture for wireless microsensor networks. Wireless Communications, IEEE Transactions on 1(4), 660–670 (2002)
7. Herman, T.: Models of self-stabilization and sensor networks. In: IWDC 2003. LNCS, vol. 2918, pp. 205–214. Springer, Heidelberg (2003)
8. Hoesel, L.v. Havinga, P.: A lightweight medium access protocol (lmac) for wireless sensor networks: Reducing preamble transmissions and transceiver state switches. In: INSS, Tokyo, Japan (June 2004)
9. Liu, C., Wu, K., Pei, J.: An energy efficient data collection framework for wireless sensor networks by exploiting spatiotemporal correlation. IEEE Transactions on Parallel and Distributed Systems (To appear)
10. Tulone, D., Madden, S.: An energy-efficient querying framework in sensor networks for detecting node similarities. In: MSWiM, pp. 191–300 (2006)
11. Tulone, D., Madden, S.: Paq: Time series forecasting for approximate query answering in sensor networks. In: Römer, K., Karl, H., Mattern, F. (eds.) EWSN 2006. LNCS, vol. 3868, pp. 21–37. Springer, Heidelberg (2006)
12. van Hoesel, L.F.W., Havinga, P.J.M.: Design aspects of an energy-efficient, lightweight medium access control protocol for wireless sensor networks. Technical Report TR-CTIT-06-47, Enschede (July 2006)
13. Vuran, M.C., Akan, B., Akyildiz, I.F.: Spatio-temporal correlation: theory and applications for wireless sensor networks. Comput. Networks 45(3), 245–259 (2004)

An Adaptive Scheduling Protocol for Multi-scale Sensor Network Architecture*

Santashil PalChaudhuri[1] and David B. Johnson[2]

[1] Aruba Networks, Sunnyvale, CA
[2] Department of Computer Science, Rice University, Houston, TX

Abstract. In self-organizing networks of battery-powered wireless sensors that can sense, process, and communicate, energy is the most crucial and scarce resource. However, since sensor network applications generally exhibit specific limited behaviors, there is both a need and an opportunity for adapting the network architecture to match the application in order to optimize resource utilization. Many applications–such as large-scale collaborative sensing, distributed signal processing, and distributed data assimilation–require sensor data to be available at multiple resolutions, or allow fidelity to be traded-off for energy efficiency. We believe that cross-layering and application-specific adaptability are the primary design principles needed to build sensor networking protocols. In previous work, we proposed an adaptive cross-layered self-organizing hierarchical data service under COMPASS architecture, that enables multi-scale collaboration and communication. In this paper we propose a time division multiplexing medium scheduling protocol tailored for this hierarchical data service, to take advantage of the communication and routing characteristics to achieve close to optimal latency and energy usage. We present an analytical proof on the bounds achieved by the protocol and analyze the performance via detailed simulations.

1 Introduction

Sensor networking has emerged as a promising tool for monitoring and actuating the devices of the physical world. It employs self-organizing networks of battery-powered wireless sensors that can sense, process, and communicate. Such networks can be rapidly deployed at low cost, enabling large-scale, on-demand monitoring and tracking over a wide area. The sensors can be deployed where it is difficult or resource-intensive to monitor the environment otherwise. Typical deployment examples are natural or man-made crises like severe weather, wild-fires, earthquakes, volcanic activities, chemical, biological or nuclear agents, and structural and habitat monitoring. In general, the sensors measure physical quantities at different spatial and temporal levels, supporting in-network signal processing and data assimilation. Queried results are next communicated with one

* This work was supported in part by NSF under grants CNS-0520280, CNS-0435425, CNS-0338856, and CNS-0325971; and by a gift from Schlumberger. The views and conclusions contained here are those of the authors and should not be interpreted as necessarily representing the official policies or endorsements, either express or implied, of NSF, Schlumberger, Aruba Networks, Rice University, or the U.S. Government or any of its agencies.

J. Aspnes et al. (Eds.): DCOSS 2007, LNCS 4549, pp. 386–403, 2007.

or more data sinks. Finally, based on the prevailing result, the sink nodes perform the desired computation to monitor and actuate the physical devices.

In this paper, we propose the design of an adaptive cross-layered scheduling service for a hierarchical network architecture providing for multi-scale collaboration. Energy and communication bandwidth are two of the most crucial and scarce resources in a sensor network. Also, sensor network applications generally exhibit application specific characteristics. Consequently, there is both a need and an opportunity for adapting the network architecture to match the applications. The objective is to minimize the resource consumption while extending the life of the network. There are specific requirements and limitations of sensor networks, which make their architecture and protocols both challenging and significantly different from that of traditional network architectures.

1.1 Motivating Application

The motivation to undertake this research came from the industrial need of sensor network architecture for real life applications. Deployment of sensor networks to monitor the ultra-clean environment for Intel fabrication plant [1] is a glaring example. The fabrication plants use some of the most delicate equipments to produce hundreds of thousands of chips daily. The fab process requires an ultra-clean environment with controlled temperature and humidity. This critical environment is maintained by a complicated cooling system composed of a diversity of pumps and coolers, amongst other machines, with multiple moving components. The breakdown of any of this equipment has a critical impact on the production line and results in significant losses. Today, Intel technicians manually carry around sensors to monitor abnormal vibrations in the equipment, which can detect worn bearings, failing compressors, and defective chillers. This vibration data is fed to an application which uses Fourier analysis to compare against expected profiles. When the analysis detects a variation in vibration outside the normal range, the affected equipment is scheduled for maintenance during normal down time.

Automation of this process makes it less dependent on human frailties—fatigue, oversight, and error—as well as enables more frequent monitoring for failures with rapid deterioration characteristics. Recently, Intel deployed a sensor network for gathering requisite operational data in an efficient way.

Real life deployment of proposed scheme demands robustness and optimization in many critical areas. Challenges in such a system include clustering together nodes of interest, providing communication paradigms for ease of application development, and efficient scheduling of communication between the battery-powered nodes. Optimizations possible include the following: (1) Temperature and vibration are monitored at different scales, and the reading of coarse granularity is reported frequently. Finer granularity readings of different scales are reported less frequently to conserve energy. For uni-dimensional data, like temperature or humidity, simple operations like computing the maximum or average suffice in the hierarchy of sensors. For multi-dimensional data, like vibration data, more complex compression algorithms can be utilized to reduce the data for communication. (2) In case of anomalous data, finer resolution sensor readings are necessary to localize the area from which the anomaly was generated. Automatic alarms, with exact location and nature of the problem, are triggered for the operator, who then schedules immediate maintenance of the equipment.

1.2   COMPASS Overview

Many applications, such as large-scale collaborative sensing, distributed signal processing, and distributed data assimilation require the sensor data to be available at multiple resolutions, or allow fidelity to be traded-off for energy efficiency. The COMPASS architecture [2] we proposed enables scalability, localization and resolution-tuning, as well as provides communication abstractions to simplify application design. A few illustrations of applications that take advantage of this multi-scale approach include the following:

- *Multi-resolution Data Extraction*: A monitoring application requires variable resolution of sampled data. It requires higher resolution during operation at specific times of day or during periods of large variation, but lower resolution at other times. Also, a finite energy budget might be specified, and the resolution has to conform to this budget while providing the best possible resolution.
- *Requisite Resolution Drilling*: An application may demand relatively higher resolution data from a specific region. For example, the specific region might be a high security area thereby requiring closer monitoring, or the region might be deemed to be of greater interest because of the inherent nature of low-resolution data emanating from it. The specific region might actually have a fire, or may be generated due to malfunctioning sensors generating anomalous data.

The sensed data are the measured quantities governed by laws of physics. Consequently, there exists considerable temporal and spatial locality of measured data. For example, a sensor measuring temperature will have a high correlation with that of nearby sensors, as well as with its values sensed at earlier times. A simple hierarchical structure supporting multi-resolution will not be able to exploit such correlation effectively. A network hierarchy aligned to the communication flow can lead to operational efficiency in respect of volume and frequency of data to be transmitted. In this background we proposed a *self-organizing, self-adapting* sensor network architecture based on the specific requirements of an application.

The proposed architecture takes advantage of the fact that sensor network applications exhibit specific sets of behavior, compared to completely generalized network applications. The communication pattern in respect of source-destination pairs, as well as duration and periodicity, is known *a priori* in many applications. As opposed to multi-hop forwarding in ad hoc networks, the fusion of forwarded data in a sensor network opens the possibility of reduced communication, achieving higher efficiency of energy management. Hence, there exists opportunities to adapt the network protocols to meet the application requirements. Such adaptations require cross-layer optimizations in the networking stack, which goes beyond the strict protocol layering.

2   Adaptive Scheduling Service

This paper presents the design and evaluation of a scheduling protocol tailored for a multi-scale architecture. This energy efficient scheme enforces a tight bound on latency. It exploits the characteristics of sensor networks—periodic communication, limited communication abstractions, and known fusion function properties. To the best of our knowledge, scheduling based on information dependency of this type has not been attempted earlier. This scheduling takes into account the data flow as well as the aggregation performed at each hop. The scheduling protocol is adaptive: it is driven by the application demand and optimized for energy usage. Since no one single scheduling protocol is well-suited for all sensor applications [3], the network services need to adapt to the application-specific requirements. We design the scheduling protocol for data monitoring applications like that in the Intel fabrication plant noted earlier. Although the scheduling protocol has been optimized for such an application, it can support any generalized communication paradigm.

In the following sections, we present the background followed by a survey the related work. Next, we provide the design of the protocol, analyze its bounds, and report the simulation results.

2.1   Design Principles

In this section, we explore the design space for the scheduling protocol. The scheduling is a way to allow contending nodes to share a common channel by allocating non-conflicting time slots. The alternate to such a scheduling approach is contention-based access.

In shared-medium networks, one of the fundamental tasks of a medium access control (MAC) protocol is to avoid collisions between two interfering nodes. The protocol allocates the channel to the nodes efficiently, so that each node can communicate within a bounded waiting time and with as little overhead as possible. Some of the important attributes for traditional MAC protocols are fairness, latency, throughput, and bandwidth utilization. In contrast, the important attributes of a MAC protocol for WASN are energy efficiency and scalability towards size and topology change. The major sources of energy waste as elaborated by Ye et al. [4] are:

- **Collisions:** A collision results in corruption of a packet and subsequent retransmission, leading to increased energy consumption and latency.
- **Idle listening and overhearing:** Listening for packets addressed to this node or destined for other nodes wastes energy. Idle listening consumes significant energy comparable to actually receiving a packet.
- **Control packet overhead:** Increased control overhead consumes extra energy for transmitting and receiving these control packets.

The prior MAC protocols proposed [5,4] for WASN have identified and addressed many of the WASN *environment* requirements. However, the inherent advantages that can be derived from the specific characteristics of WASN have not been fully exploited. The following properties of WASN applications can be exploited for the MAC protocol design:

- Many sensor network applications have communication requirements that are *periodic* and known beforehand such as collecting temperature statistics at regular intervals. This periodic nature can be used to schedule the medium access by the nodes and thus minimize collisions. This can also aid the radio interfaces sleep/wake-up decisions and thereby decrease the idle listening and overhearing.
- A contention-based medium access scheme will also be necessary to support *event-driven* applications, such as intrusion or fire detection. The forwarding node can be woken up in time to process event-driven data by making it application-aware. Real-time constraints can be communicated from the application to adapt the MAC protocol to meet its deadlines.
- Frequently, sensor network applications are used for data gathering or monitoring, which implies that those communication flows are destined towards the sink. This single-destination communication characteristic can be taken advantage of in the MAC protocol to build efficient transmission schedules.
- In-network processing is used in sensor networks to reduce communication requirements. Knowledge of the aggregation characteristics can help determine a bound on the traffic requirements of each node.

The MAC protocol should have two modes to support the two different communication requirements of sensor applications, namely periodic and event-driven. The need to support these two kinds of modes was recently proposed in the IEEE standard for low-power sensors [6]. The relative proportion of the two modes in a frame can be dynamically determined by the applications, depending on the application's current needs.

Periodic Contention-Free Period: Medium access in this mode is of the class of Spatial Time Division Multiple Access (STDMA) [7] protocols. The application's deterministic traffic distribution during the periodic communication can be used to compute an efficient slot allocation policy. The length of each slot should be large enough to send a complete data packet of fixed a size.

Event-Driven Contention Access Period: During this mode, a sensor will be in sleep-mode except when necessary to communicate. The mode is based on IEEE 802.11 protocol [8] with carrier sense and RTS-CTS.

3   Related Work

There is a significant body of research that addresses the problem of providing access to the medium for next-hop communication. These protocols can be categorized as contention-based and schedule-based.

3.1   Contention-Based Protocols

IEEE 802.11 [8] is the best-known example of a contention-based MAC protocol. It uses a carrier sense multiple access technique combined with a handshake mechanism to access the channel and avoid collisions. A key limitation of IEEE 802.11 is that the nodes stay in idle mode for a long time, and hence this protocol is not suitable for energy-constrained nodes. A simple Power Save Mode (PSM) is specified in order to reduce the energy consumption. Nodes are time-synchronized and awake at the beginning of a certain multiple of a beacon interval (called the Listening Interval and depends on the client configuration). It stays awake for a certain fixed period, and during this time, packets are exchanged to inform the nodes of buffered traffic. The nodes for which traffic is destined will remain awake after the fixed interval to receive the packets.

El-Hoiydi [9] proposed a low-level carrier sense technique that effectively duty cycles the radio. The basic idea is to shift the cost of data transfer from the receiver to the transmitter by increasing the length of the preamble. This leads to greater energy efficiency, as the number of receivers is more numerous than the number of transmitters in a broadcast network. In a recent implementation of this idea in the B-MAC procotol [10] as part of TinyOS, this preamble length is exposed as a parameter to the upper layers, so the application can select the optimal trade-off between energy savings and performance.

To reduce the overhead of periodic waking up and listening for incoming packets, there have been a number of proposals addressing this issue. The PAMAS protocol [11] was one of the first power-saving protocols that allows nodes to sleep when there is ongoing transmission in the neighborhood, leading to increased energy savings. S-MAC [4] adds a fixed slot structure and does duty cycling within each slot. At the beginning of each slot, nodes wake up and contend for the channel. The duty cycle can be directly controlled by the application for energy-performance trade-off, but the duty cycle must be fixed *a priori*. T-MAC [5] introduced an adaptive duty-cycle to automatically adjust to traffic fluctuations inherent in many applications.

All of the above techniques are useful for energy optimization when the traffic pattern is unknown. However, for situations when traffic pattern and routing path are known *a priori*, it makes sense to take advantage of this information for additional optimization.

3.2 Schedule-Based Protocols

Schedule-based MAC protocols are attractive because they are collision-free and there is no idle-listening. In this class of protocols, a slot is allocated for one node per neighborhood uniquely. This collision-free slot is used by that node for transmissions to any or all of it's neighbors. Thus two nodes cannot be assigned the same slot if one station is within the range of the other, or if two stations have common neighbors. The objective of these schedule-based protocols is to allow communication without interference, while maximizing the number of parallel transmissions.

Over the years, a substantial number of algorithms [12,7,13] have been published in an effort to solve the collision-free scheduling problem in multi-hop wireless networks. As the optimal scheduling is an NP-hard problem, all of these algorithms provide approximate solutions. These mostly centralized protocols are not optimized for sensor networks that have several unique characteristics compared to typical multi-hop ad hoc networks.

Bluetooth scheduling [14] is a closely related field of work. It enables formation of multi-hop networks using a combination of piconets, called scatternets. Each piconet has a master and up to seven slaves. Bluetooth does not use a global slot synchronization mechanism. Each link is individually synchronized by a reference provided by the master in the link. Therefore, for multi-hop forwarding, slots are wasted as nodes need to switch time references. Various techniques have been proposed to address this time reference alignment problem. However, global synchronization is frequently required by the applications themselves in sensor networks, so it is reasonable to assume the existence of such a mechanism. Also, the master-slave structure in bluetooth is only one-hop and hence cannot be directly applicable to schedule the multi-hop trees formed by our proposed multi-scale architecture.

For multi-hop sensor networks, the TRAMA protocol [15] is one of the first proposals and is based on the NAMA protocol [16]. The nodes periodically awake to exchange broadcasts to learn the two-hop neighborhood. Using this knowledge, nodes reserve slots in the future for backlogged traffic. The protocol uses a hash function for detecting priority among contending neighbors. This leads to priority chaining whereby nodes can get higher priority in one neighborhood and lower in another, causing inefficient scheduling. TRAMA tries to schedule packets hop-by-hop instead of flows, leading to higher end-to-end latency.

Sichitiu's work [17] is the closest to ours with regard to the fact that it tries to schedule flows end-to-end. It dynamically sends a *SETUP* packet to schedule new flows. If this packet is received without collision by the receiver, this slot is scheduled for this pair of neighbors. This receiver then uses the same mechanism for forwarding the packet over the next hop towards the data sink, thereby creating a scheduled flow for the whole path. This approach is similar to the one we use for scheduling flows in our inter-cluster scheduling. However, our intra-cluster protocol takes advantage of the fact that sensed data are processed at each hop in many typical sensor network deployments, leading to better compression and lesser number of slots necessary. In Sichitiu's protocol, each flow is individually scheduled, leading to higher latency in data sensing applications.

3.3   Graph Coloring

The scheduling problem can be directly mapped to the graph coloring problem. Minimum graph coloring is a well-known NP-hard problem and has been researched for several decades in order to find solutions closest to the optimal. In the graph, a valid distance-2 coloring assigns different colors to any pair of nodes between which there is a path length of at most 2; this is very similar to producing a conflict-free schedule. A good background work based on this is provided by Marathe et al. [18]. However, graph coloring corresponds only to the 2-hop conflict-free property necessary for scheduling. In our scheduling protocol, we have additional constraints on slot allocation due to which previous graph coloring techniques cannot be directly applied. These constraints are elaborated in the next section.

4   Medium Access Scheduling in a Hierarchy

This section describes the design of the medium access protocol for a hierarchically clustered network. This medium access protocol is specifically designed for supporting the communication abstractions supported in this architecture. There is a Conflict Free Period (CFP) for periodic *a priori* known traffic, and a Contention-based Access Period

(CAP) for nodes to send event-driven traffic, and to introduce new periodic traffic. A TDMA-style protocol is designed for the CFP, and we leverage the existing literature on energy-efficient CSMA protocols [19,4,5] for the CAP.

A time frame in the protocol is shown in Figure 1. Interval L_1 interval corresponds to communication amongst nodes within the first-level cluster. A near optimal protocol proposed to schedule intra-cluster communication is the main contribution in this chapter, and is elaborated in Section 4.1. Interval L_i for $i > 1$, is the time interval during which communication takes place between cluster-heads at level i. The intervals for communicating among different levels are separated so as minimize the perturbation at higher levels when localized change takes place in lower-level schedules. The total length of this CFP period is kept as small as possible for two reasons: (1) the CFP period gives the lower bound on the real-time responsiveness of the system; and (2) the nodes need to stay awake only during part the CFP period.

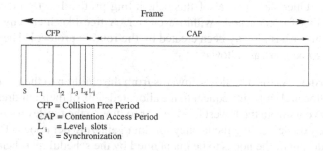

CFP = Collision Free Period
CAP = Contention Access Period
L_i = Level$_i$ slots
S   = Synchronization

Fig. 1. Frame diagram

We make the following assumptions in the design of the protocol:

– Wireless links used for sending or receiving packets are bi-directional. This assumption is typical for most MAC protocols.
– The interference range is the same as the range for transmission and reception. The interference range depends on a variety of factors and cannot be represented easily as a function of the transmission range. Most collision-free scheduling work on the approximation of equality between transmission and interference range.
  This assumption can be relaxed using techniques such as RID [20], where instead of 2-hop communication to reach interfering nodes, multi-hop forwarding is used to reach the interfering nodes. My theory of conflict-free slot allocation proposed in this design requires the ability to ascertain and communicate with interfering nodes. Thus, our protocol can be adapted to use RID in future, as RID provides these.
– A clock synchronization facility [21] exists.

This new scheduling protocol is designed specifically for multi-scale data retrieval applications, where data flows from the sensors to the sink. This is the common case usually encountered in a typical sensor network application.

Algorithm 1. Centralized Algorithm for Intra-Cluster Scheduling

---

Require: k slots necessary per node known; Depends on fusion function and number of children

1: Traverse tree in Depth First order from cluster root
2: Finish order of the visit forms a topological sort to give a partial order
3: For each visited node, assign a minimum possible slot number s, with the following constraints
4: **repeat**
5:         $s >$ max (slot number allocated to any child)
6:         $s \neq$ slot already allocated to any node within 2 hops
7: **until** k consecutive slots are found

4.1 Intra-cluster Scheduling

Protocol. The chief design goal of this scheduling protocol is to allocate spatially conflict-free slots for each node within a cluster. A tree-like hierarchy within each lowest-level cluster has already been created by the routing protocol. The other design goals of this protocol are as follows:

1. **Partial order:** As the data flows upwards from the children to their parent, the parents are allocated slots subsequent to the allocated slots of their children. This property allows data from the lowest level sensor to flow to the top-level sink within one frame, thereby decreasing the latency for data gathering per epoch. This induces a partial ordering of the nodes to be maintained by the scheduling scheme.
2. **Contiguous allocation:** The slots allocated to the children of a node should be close together, such that the parent can wake up for one contiguous time to receive data packets from its children, and can then switch off the radio to save power. As switching between on and off state consumes time and energy, contiguous allocation is beneficial. This entails a scheduling policy in which the slots of siblings needs to be grouped together.
3. **Fusion characteristics:** The fusion function characteristics at each node are known *a priori*, such as through the Information Exchange Service (IES) [22]. Fusion function characteristics are analyzed by the scheduling protocol to determine the number of slots to be allocated at each level. For example, for a fusion function such as average sensed value, the parent needs to average its own value with the value received from each of its children and then send it upwards in a single slot. The number of slots to be allocated to a node per frame also depends on the application-determined periodic rate, which can also be known through the IES. Therefore, a node might have multiple slots allocated, as deemed necessary by the fusion function.

We first describe a base centralized scheduling scheme, that allocates conflict-free slots and later generalize it to a distributed scheduling scheme. The centralized scheme is shown in Algorithm 1, Within the lowest level cluster, a cluster-head coordinates the traffic out of that cluster. It does a Depth-First Search (DFS) of the tree rooted at itself, and sorts the nodes according to the finish order (of a DFS visit) to give a post-order traversal of the nodes within the cluster. The post-order traversal has the desirable property that the parent has a finish order immediately after the finish order of its children in the tree. The slot times are assigned according to that given by the topological sort, thereby providing the desired partial order. This algorithm produces a collision-free slot allocation within a single cluster. However, this centralized scheme has the following three problems: (1) It assumes a complete graph knowledge. (2) It leads to a very inefficient slot allocation, with no concurrent communication at all. This produces a large CFP, which is detrimental for real-time response. This is shown in Figure 2, where 7 slots are used for 7 nodes. In the figure, the lines show the parent-child relationship in the cluster and the dotted lines show that that nodes at the two ends of that line are within range of each other. (3) Inter-cluster interference is not accounted for.

We next describe a distributed token-passing protocol to compact the slot allocation and meet all the other design goals. This algorithm is formally described in Algorithm 3. The algorithm requires each node to know its 2-hop neighborhood, the cluster it belongs to, and its parent and children. The cluster root generates a token packet and passes the token around in DFS order. The node holding the token allocates a slot for itself when the token comes back to it after traversing all its children. The allocated slot is put inside the token, and hence the token contains the allocated slots of all the nodes it has traversed. As the siblings need to be allocated slots as close to each other as possible, a slot is chosen according to this condition. There is a trade-off possible between choosing closer slots to minimize the number of transitions, or choosing the minimum possible slot to minimize the total number of slots needed for CFP.

Choosing the next child to visit in the DFS can be random (which we call the Base algorithm) or based on some graph property. Based on observation from Figure 4 and experimental comparison with a few other techniques, we order the children based on their degree and choose the next child to visit based on that order. The other techniques we compared with are: ordering based on the number of children each node has, and ordering based on the number of nodes in the subtree rooted at the node. The intuitive reason is to allocate the lesser slots to the nodes in a dense area, thereby leading to less conflict allocating slots in the less dense areas. We call this version of the algorithm the Degree algorithm.

Analysis. This problem can be mapped to a graph coloring problem as follows. The network is taken as a graph (G), and a graph G^2 is produced. G^2 has the same vertex set as G, and all the pairs of vertices which are 2-edge reachable in G have an edge in G^2. Thus, G^2 gives the interference graph similar to our node scheduling problem. Coloring a graph is an NP-hard problem even in a centralized setting, and many approximate coloring solutions have been proposed in the graph theory literature. However, due to the constraints on slot allocation induced by the partial ordering, none of these solutions can be directly applied for our scheduling scenario. The graph being that of a wireless network, it is an Unit Disk Graph (UDG) having special properties. In a UDG, a link between two nodes exists if the distance between the nodes is less than the radio range. The *competitive ratio* is the ratio between the chromaticity given by the approximate algorithm and the optimal solution.

Algorithm 2. Distributed Protocol for Intra-Cluster Scheduling

---

Require: k slots necessary per node known; Depends on fusion function and number of children
Require: Each node knows its 2-hop neighborhood
Require: Each node knows the cluster to which it belongs
Require: Each node knows its parent and its children
 1: For all token packets overheard, a node remembers slots its neighbor is sending or receiving
 2: **while** All children of a node not visited **do**
 3:     Send token to one of the children not visited. The child to visit next next depends on the heuristic chosen.
 4:     Wait until token comes back
 5:     Then remember the slot allocation of the child
 6: **end while**
 7: All nodes in the subtree rooted at the node has been visited
 8: For each visited node, assign a possible slot number s, with the following constraints
 9: **repeat**
10:         $s >$ maximum (slot number allocated to any child)
11:         $s \neq$ slot already allocated in token packet
12:         $s \neq$ slot overheard being allocated by another node in 1-hop neighborhood
13:         Assign the k slots close to slots already allocated by it's siblings
14: **until** k consecutive slots are found
15: Append the node id, its chosen sending slot, and its send mode (parent or all) to the token
16: Append the node id and receiving slots to the token
17: Send token to its parent with flag stating SUBTREE_DONE
18: **while** Token not received from parent with flag SUBTREE_DONE **do**
19:     Wait
20: **end while**
21: Broadcast the final token packet locally

Lower bound on optimality: We show using a sub-graph, which we believe is the worst-case for our algorithm, that the bound of this competitive ratio is 1.2 for our Base algorithm. Figure 3 and Figure 4 show the nodes in a cluster labled A through H. They form a tree hierarchy denoted via the lines. The dotted lines show two nodes within range of each other. The numbers beside the nodes shows the slot that the node has been allocated. Comparing the worst case slot allocation in Figure 3 with the optimal slot allocation for the same graph in Figure 4, we get the 1.2 ratio. So, if the adversary chooses the graph as well as the ordering of DFS visits in the base algorithm, our protocol cannot expect an optimality lesser than that value.

Upper bound on optimality: Employing the properties of an UDG, an upper bound of 6 has been also established for the competitive ratio for the base algorithm. This is based on the observation that there can be at most 5 neighbors of a node which are not connected to each other. If in a naive approach our algorithm gives different color to all of its neighbors, it can at most be 6 times as bad, as there will be a clique in the size of degree 6.

This small constant bound on the competitive ratio is a good indication of the optimality of our proposed solution. The time and cost for the scheduling protocol are both $O(N)$ where N is the number of nodes in the smallest level cluster. The clustering

---

Algorithm 3. Distributed Protocol for Intra-Cluster Scheduling

---

Require: k slots necessary per node known; Depends on fusion function and number of children
Require: Each node knows its 2-hop neighborhood
Require: Each node knows the cluster to which it belongs
Require: Each node knows its parent and its children
1: For all token packets overheard, a node remembers slots its neighbor is sending or receiving
2: **while** All children of a node not visited **do**
3:     Send token to one of the children not visited. The child to visit next next depends on the heuristic chosen.
4:     Wait until token comes back
5:     Then remember the slot allocation of the child
6: **end while**
7: All nodes in the subtree rooted at the node has been visited
8: For each visited node, assign a possible slot number s, with the following constraints
9: **repeat**
10:         $s >$ maximum (slot number allocated to any child)
11:         $s \neq$ slot already allocated in token packet
12:         $s \neq$ slot overheard being allocated by another node in 1-hop neighborhood
13:         Assign the k slots close to slots already allocated by it's siblings
14: **until** k consecutive slots are found
15: Append the node id, its chosen sending slot, and its send mode (parent or all) to the token
16: Append the node id and receiving slots to the token
17: Send token to its parent with flag stating SUBTREE_DONE
18: **while** Token not received from parent with flag SUBTREE_DONE **do**
19:     Wait
20: **end while**
21: Broadcast the final token packet locally

---

hierarchy ensures that the number of nodes in a cluster is constant, subject to the constant node density. So, the algorithm is both constant in time and cost with the number of nodes in the network.

We have also proved the upper and lower bounds on the number of slots used in the graph.

Theorem 1. The upper bound on the number of slots used is $\Delta^2 + h$, where Δ is the maximum node degree and h is the height of the tree.

Proof. Δ^2 implies the 2-hop neighborhood of a node. As We allocate the nodes in conflict-free 2-hop neighborhood, the whole neighborhood can always be colored using the same number of colors as there are nodes. Hence, the whole graph can be colored using the number of colors in the maximum degree of any node. For uniform random distribution of nodes, we take Δ as the average degree.

The h factor comes in as one slot increases for each level of the tree. The parent needs to be strictly one slot more than its children.

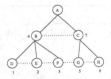 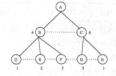 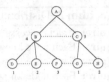

Fig. 2. Centralized slot allocation in a subgraph

Fig. 3. Possible slot allocation
in a worst-case subgraph

Fig. 4. Optimal slot allocation
in a worst-case subgraph

Theorem 2. The lower bound on the number of slots used is $\Delta + h$, where Δ is the maximum node degree and h is the height of the tree.

Proof. As a 2-hop neighborhood is to be made conflict-free of every node, the range of connectivity of nodes becomes $2 \times r$, where r is the range of the node. So, within a circle of r radius, all nodes are within 2-hops of each other and hence form a clique—a fully connected graph. So, atleast Δ colors are needed to color this clique.

As before, the h factor comes in as one slot increases for each level of the tree. The parent needs to be strictly one slot more than its children.

4.2   Inter-cluster Scheduling

Inter-cluster scheduling occurs after the data has reached the cluster-heads at the lowest level cluster. Each level of communication is separated into time, and hence does not conflict with each other. This is shown in Figure 1, where the CFP is divided into multiple periods for each level.

For inter-cluster communication between peers, the medium access protocol utilizes knowledge of the routing path. Beacons transmitted by cluster-heads at all levels of the hierarchy by the routing protocol specifies the routing path for communication between peers at different levels. A packet reserving the channel is sent for the periodic traffic along the routing path specified by the beacons. Each node in the path allocates slots in a conflict-free fashion such that two different peer-paths do not conflict in time, while ensuring maximum parallelism in the communication paths.

5   Evaluation

To evaluate the performance of this proposed protocol, we measure the latency of data gathering and associated energy savings. As we utilize extra cross-layer information, which no previous approach has used before, we compare against the optimal solution.

We used a simulator to randomly generate graphs, construct a tree in the way described in the routing protocol, and then implement the scheduling protocol. The simulator helped in the validation and experimentation with a number of different heuristics to select the best one. Each of the points was generated as an average of 50 runs.

Finding the optimal solution for scheduling was the primary reason we used a graph simulator, as using a traditional simulator with full simulation of physical and medium

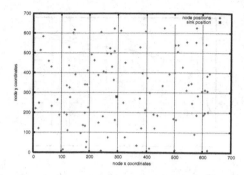

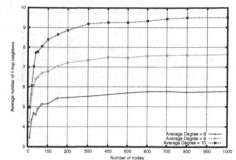

Fig. 5. Uniform random distribution of 100 nodes with average degree of 8

Fig. 6. Number of 1-hop neighbors with number of nodes, for different degree

access later would have been too slow to generate the optimal solution. We implemented simulated annealing technique to get the optimal solution needing exponential complexity. Up to 10 million runs were necessary for finding each point on the optimal graphs.

The data gathering latency is directly related to the total number of slots needed to cover the entire network × the duration of each slot. Hence, we show the number of total slots as a measure of latency. The energy savings come from the amount of time (or slots) each node has to remain in active mode to receive or transmit data or idle. This is also calculated as the average number of slots each node has to remain awake.

Figure 5 shows a random uniform distribution of 100 nodes with an average degree of 8, and range of 100. This corresponds to a square with side as 626.6. The node closest to the center is chosen as the sink node. All graphs have the number of nodes varying from 10 to 1000, and average degree of 6, 8 and 10.

As the analytical upper and lower bounds depend on the 1-hop and 2-hop neighbors, the neighborhood degrees are plotted with increasing number of nodes. Figure 6 shows the plot for degree of 6, 8 and 10. The plots approach those values with increasing number of nodes, but are lesser because of edge effects which lower the average.

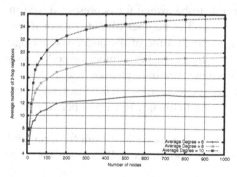

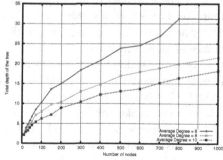

Fig. 7. Number of 2-hop neighbors with number of nodes, for different degree

Fig. 8. Height of tree with number of nodes, for different degree

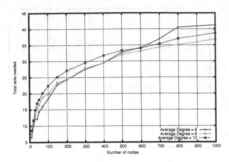 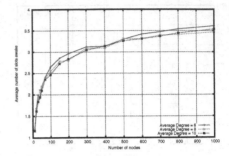

Fig. 9. Latency with number of nodes, for different degree

Fig. 10. Average number of slots in which each node is awake, for different degree

Figure 7 is similarly a plot of the number of 2-hop neighbors for a degree of 6, 8 and 10. The number of nodes is proportional to the area for a given uniform density. If one takes the ratio of the area of two circles, with twice the radius in one from the other, it will have a value of 4. But, in the graph the value is about 2.5 times the 1-hop neighbor value. Other than edge effects, another factor which reduces it is the density of nodes. The ratio of 4 will be reached only in a very dense topology where a 2-hop route covers all the area of a circle with twice the radius.

The analytical bounds also depends on the depth of the tree formed. Hence, the depth of the tree formed for increasing number of nodes is shown in Figure 8. The height increases logarithmically as has been analytically shown before. The curve for degree 6 leads to poor connectivity and frequently forms disconnected graph, specially at larger number of nodes. It also leads to large depths due to long circuitous paths because of lack of density. That is the reason for the depth curve not being smooth like the curves for other degrees. For 500 nodes, average degree of 10 leads to height of about 12. Figure 9 shows the latency for gathering data from every sensor up to the sink node. It is given in terms of slots, where each slot takes the duration equivalent to the transmission of the largest data packet. The number of slots necessary increases logarithmically with n, thereby scaling to a large number. For 500 nodes in a cluster, about 33 slots are needed for degree equal to 10.

The average number of slots each node needs to be awake for is presented next in Figure 10. For 500 nodes, this value is about 3.2. The number of slots idle or transmitting or receiving is inversely proportional to the lifetime of the node. The average number of slots awake increases very slowly with increasing number of nodes, also pointing to the scalability of the system. Assuming 500 byte payload and 100 Kbps radio, approximate slot size would be 5 ms. So, on an average each node needs to be awake for $5 \times 3.2 = 16$ ms. So, if data monitoring frequency is 1 event per second, a node has to be awake only 1.6% of the time.

Figure 11 shows the analytical bounds proved earlier, as well as result from running experiments to find the optimal solution. The upper bound as well as lower bound is strictly met by the optimal graph.

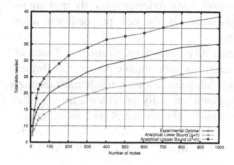

Fig. 11. Analytical bounds on optimality

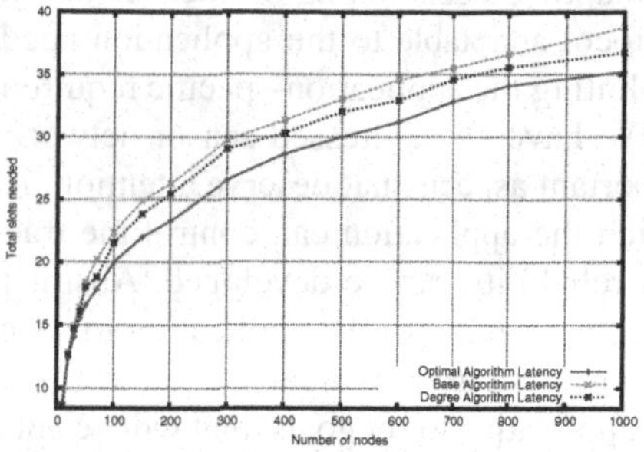

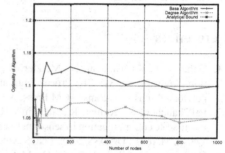

Fig. 12. Comparison of the latency for different algorithms

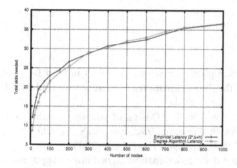

Fig. 13. Empirical and experimental latency

Fig. 14. Normalized optimality of algorithm

Figure 12 compares the latency of the base and the degree algorithm with that of the optimal. Normalized optimality if shown in Figure 14. The optimality reduces and stabilizes at 1.05 for Degree algorithm, and 1.1 for Base Algorithm. We analytically found a bound of 1.2 for the base algorithm on a worst-case graph. A value of $2 \times \Delta + h$ is empirically plotted and it is seen that it closely follows the experimental optimal for the degree algorithm. This total number of slots gives the chromatic number of the graph, and there does not exist any polynomial solution to this. However, this empirically observed formula seems to match closely with the experimental results.

These results show that the scheduling protocol is very close to optimal scheduling in terms of total latency of data gathering, as well as in terms of energy efficiency.

6   Conclusions and Future Work

Apart from enabling multi-resolution collaboration, a clustering hierarchy allows the network to scale to a very large number of sensors. Our architecture design adapts to the communication and collaboration requirements of the application, reducing communication energy and bandwidth usage. This work highlights the need for and possibility of adapting a scheduling protocol to suit the application requirements. By making the protocol adaptable to the application needs dynamically, the new protocol will allow exploiting the application-specific requirements.

We have not addressed sensor network reliability or QoS requirements. These are important aspects that deserve attention. Abstractions with tunable parameters through which the application can control the trade-off between resource usage and accuracy and reliability can be developed. Abstract Regions currently provides a tuning interface, but it requires the application to specify low-level parameters such as number of retransmissions. The tunable parameter needs to be at a higher level, which will enable the application to set goals, that will be automatically translated by the networking layer into the low-level parameters.

Making the routing and scheduling adapt without application aid can be another future direction of work. Instead of explicit communication from the application as is done in this paper, information flow can be learned. Thereafter, the adaption is done based on this learning.

References

1. Chhabra, J., Kushalnagar, N., Metzler, B., Sampson, A.: Sensor networks in intel fabrication plants. In: SenSys '04: Proceedings of the 2nd international conference on Embedded networked sensor systems, ACM Press, New York (2004)
2. PalChaudhuri, S., Kumar, R., Baraniuk, R., Johnson, D.: Design of adaptive overlays for multi-scale communication in sensor networks. In: Prasanna, V.K., Iyengar, S., Spirakis, P.G., Welsh, M. (eds.) DCOSS 2005. LNCS, vol. 3560, Springer, Heidelberg (2005)
3. Heidemann, J., Silva, F., Estrin, D.: Matching data dissemination algorithms to application requirements. In: SenSys '03: Proceedings of the 1st international conference on Embedded networked sensor systems, pp. 218–229. ACM Press, New York (2003)
4. Ye, W., Heidemann, J., Estrin, D.: An energy-efficient mac protocol for wireless sensor networks. In: Proceedings of the IEEE Infocom, New York, NY, USA, USC/Information Sciences Institute, IEEE, pp. 1567–1576 (June 2002)
5. van Dam, T., Langendoen, K.: An adaptive energy-efficient mac protocol for wireless sensor networks. In: Proceedings of the first international conference on Embedded networked sensor systems, pp. 171–180. ACM Press, New York (2003)
6. IEEE Computer Society LAN MAN Standards Committee: Wireless Medium Access Control (MAC) and Physical Layer (PHY) Specifications for Low-Rate Wireless Personal Area Networks (LR-WPANs), IEEE Std 802.15.4. The Institute of Electrical and Electronics Engineers, New York (2003)
7. Nelson, R., Kleinrock, L.: Spatial-TDMA: A collision-free multihop channel access control. IEEE Transactions on Computers 33, 934–944 (1985)
8. IEEE Computer Society LAN MAN Standards Committee: Wireless LAN Medium Access Control (MAC) and Physical Layer (PHY) Specifications, IEEE Std 802.11-1997. The Institute of Electrical and Electronics Engineers, New York (1997)
9. El-Hoiydi, A.: Aloha with Preamble Sampling for Sporadic Traffic in Ad-hoc Wireless Sensor. In: Proceedings of IEEE International Conference on Communications (ICC), New York, USA (2002)
10. Polastre, J., Hill, J., Culler, D.: Versatile low power media access for wireless sensor networks. In: SenSys '04: Proceedings of the 2nd international conference on Embedded networked sensor systems, pp. 95–107. ACM Press, New York (2004)

11. Singh, S., Raghavendra, C.: PAMAS: Power Aware Multi-Access Protocol with Signalling for Ad Hoc Networks. SIGCOMM Computer Communication Review, vol. 28(3) (July 1998)
12. Chou, A.M., Li, V.: Slot allocation strategies for TDMA protocols in multihop packet radio networks. In: Proceedings of INFOCOM 1992, pp. 710–716 (1992)
13. Chlamtac, I., Farago, A.: Making transmission schedules immune to topology changes in multi-hop packet radio networks. IEEE/ACM Trans. Netw. 2(1), 23–29 (1994)
14. Salonidis, T., Tassiulas, L.: Asynchronous TDMA ad hoc networks: Scheduling and Performance. In: Proceedings of European Transactions in Telecommunications (ETT) (2004)
15. Rajendran, V., Obraczka, K., Garcia-Luna-Aceves, J.J.: Energy-efficient collision-free medium access control for wireless sensor networks. In: Sensys '03: Proceedings of the first international conference on Embedded networked sensor systems, pp. 181–192. ACM Press, New York (2003)
16. Bao, L., Garcia-Luna-Aceves, J.J.: A new approach to channel access scheduling for ad hoc networks. In: MobiCom '01: Proceedings of the 7th annual international conference on Mobile computing and networking, pp. 210–221. ACM Press, New York (2001)
17. Sichitiu, M.: Cross-Layer Scheduling for Power Efficiency in Wireless Sensor Networks. In: Proceedings of INFOCOM 2004 (2004)
18. Marathe, M., Breu, H., Ravi, H., Rosenkrantz, D.: Simple heuristics for unit disk graphs. Networks 25, 59–68 (1995)
19. Woo, A., Culler, D.E.: A transmission control scheme for media access in sensor networks. In Mobile Computing and Networking, pp. 221–235 (2001)
20. Zhou, G., He, T., Stankovic, J.A., Abdelzaher, T.F.: Rid: Radio interference detection in wireless sensor networks. In: Proceedings of IEEE Infocom (2005)
21. PalChaudhuri, S., Saha, A., Johnson, D.B.: Adaptive clock synchronization in sensor networks. In: Proceeding of the Information Processing in Sensor Networks(IPSN), Berkeley, CA (April 2004)
22. Kumar, R., PalChaudhuri, S., Ramachandran, U.: System support for cross-layering in sensor network stack. In: Proceedings of the International Conference on Mobile Ad Hoc and Sensor Networks, Hong Kong, China (December 2006)

Minimum-Energy Broadcast with Few Senders*

Stefan Funke, Sören Laue, and Rouven Naujoks

Max-Planck-Institut für Informatik,
Stuhlsatzenhausweg 85, 66123 Saarbrücken, Germany
{funke,soeren,naujoks}@mpi-inf.mpg.de

Abstract. Broadcasting a message from a given source node to all other nodes is a fundamental task during the operation of a wireless network. In many application scenarios the network nodes have only a limited energy supply, hence minimizing the energy consumption of any communication task prolongs the lifetime of the network. During a broadcast operation using intermediate nodes to relay messages within the network might decrease the overall energy consumption since the cost of transmitting a message grows super-linearly with the distance. On the other hand using too many intermediate nodes during a broadcast operation increases both latency as well as the chances that some transmission could not properly received (e.g. due to interference).

In this paper we consider a constrained broadcast operation, where a source node wants to send a message to all other nodes in the network but at most k nodes are allowed to participate actively, i.e. transmit the message. Restricting the number of transmitting nodes helps in reducing interference, latency and increasing reliability of the broadcast operation, of course at the cost of a slightly higher energy consumption. For the case of network nodes embedded in the Euclidean plane we provide a $(1 + \epsilon)$-approximation algorithm which runs in time linear in n and polynomial in $1/\epsilon$ but with an exponential dependence on k. As an alternative we therefore also provide an $O(1)$-approximation whose running time is linear in n and polynomial in k. The existence of a $(1 + \epsilon)$-approximation algorithm is in stark contrast to the unconstrained broadcast problem where even in the Euclidean plane no algorithm with approximation factor better than 6 is known so far.

1   Introduction

In contrast to wired or cellular networks, ad hoc wireless networks a priori are unstructured in a sense that they lack a predetermined interconnectivity. An ad hoc wireless network is built of a set of radio stations P, each of which consists of a receiver as well as a transmission unit. A radio station v can send a message by setting its *transmission range* $r(v)$ and then by starting the transmission process. All other radio stations at distance at most $r(v)$ from v will be able to receive

---

* This work was supported by the Max Planck Center for Visual Computing and Communication (MPC-VCC) funded by the German Federal Ministry of Education and Research (FKZ 01IMC01).

J. Aspnes et al. (Eds.): DCOSS 2007, LNCS 4549, pp. 404–416, 2007.

v's message (we are ignoring interference for now). For transmitting a message across a transmission range $r(v)$, the power consumption of v's transmission unit is proportional to $r(v)^\alpha$, where α is the *transmission power gradient*. In the idealistic setting of empty space, $\alpha = 2$, but it may vary from 2 to more than 6 depending on the environment conditions of the location of the network. Given some transmission range assignment $r : P \to \mathbb{R}_{\geq 0}$ for all nodes in the network, we can derive the so-called *communication graph* $G^{(r)} := G(P, E)$. $G(P, E)$ is a directed graph with vertex set P which has a directed edge (p, q) iff $r(p) \geq |pq|$, where $|pq|$ denotes the Euclidean distance between p and q. The cost of the transmission range assignment r is

$$\text{cost}(r) := \sum_{v \in P} r(v)^\alpha$$

Numerous optimization problems can now be considered by looking for the minimum cost transmission range assignment r such that the respective communication graph G satisfies some property Π, see [5] for an overview. One classic property Π is defined as follows: *Given a specific source node s we want the communication graph G to contain a directed spanning tree rooted at s.* This problem is called the *energy-minimal broadcast problem (EMBC)*, since the respective transmission range assignment allows the source node s to distribute a message over the whole network at minimum total energy cost.

It is known, that if the points are located in the Euclidean plane, the minimum spanning tree (MST) of the point set induces a transmission range assignment which has cost at most a factor of 6 above the optimum solution [1]. On the other hand, there are point sets where the MST-based solution is a factor 6 worse than the optimal solution, so the bound of 6 is tight [15].

One problem that is particularly prominent for the MST–based solution is the fact that in the resulting transmission range assignment a very large fraction of the network nodes are transmitting (i.e. have non-zero transmission range). In the MST-based range assignment, at least $n/6$ nodes are actually senders during the broadcast operation (since the maximum degree of the minimum spanning tree of a set of points in the Euclidean plane is bounded by 6). This raises several critical issues: (a) The more network nodes are transmitting in the process of one broadcast operation, the more likely it is that some nodes in the network experience interference due to several nearby nodes transmitting at the same time (unless special precautions are taken that interference does not occur). (b) Every retransmission of a message implies a certain delay which is necessary to setup the transmission unit etc; that is, the more senders are involved in the broadcast operation, the higher the latency. This effect is even amplified by the previous problem if due to interference messages have to be resent. (c) Network nodes are not 100% reliable; if for example the probability for a network node to operate properly is 99.9%, the probability for a network broadcast to fail, i.e. not all nodes receiving the message, is $1 - 0.999^{(n/6)}$, which for a network of $n = 3000$ nodes is around 40%! This suggests to look for broadcast operations in the network that use only very few sending nodes. Of course, this comes at the cost of an increased total power consumption, but the behaviour with respect to the critical issues (a) to (c) can be drastically improved.

In this paper we suggest the following restricted broadcast operation: *Given a specific source node s we want to find a transmission range assignment r of minimum total cost such that the respective communication graph G contains a directed spanning tree rooted at s* **and at most *k* nodes have a non-zero transmission range assigned**. We call this problem the *k-set energy-minimal broadcast (k-SEMBC)* problem. Allowing only a small number *k* of sending nodes during the broadcast operation has several advantages: (a) The *k* transmissions can be easily scheduled in *k* different time slots, hence avoiding any interference at all. (b) The latency is obviously bounded by $O(k)$. (c) In the above scenario the probability of a broadcast operation to fail is $1 - 0.999^k$, which e.g. for $k = 10$ is 1%.

Of course, the best behaviour in terms of interference, latency and reliability can be achieved by having the source node *s* transmit its message directly to all nodes in the network. Assume w.l.o.g. that by scaling the maximum distance of another node in the network to *s* is 1, this operation would cost one unit of power. Consider the scenario in Figure 1; here we have the *n* radio stations equally distributed on a segment of length 1. As mentioned above, the direct broadcast from *s* incurs a power cost of $1^\alpha = 1$. In case of an unbounded number of allowed sending nodes, essentially every node just forwards the message to the next node to the right. The total power consumption of this broadcast (which involves $n - 1$ sending nodes) is $(\frac{1}{n-1})^\alpha \cdot (n - 1) = (\frac{1}{n-1})^{\alpha-1}$. That is the power savings compared to the direct transmission is a factor of about $n^{\alpha-1}$, using $n-1$ sending nodes, though, which implies the above issues w.r.t. interference, latency, and reliability. If, on the other hand, we allow only *k* sending nodes, we could in the ideal case select $k - 1$ stations at about equal distance from each other on the segment and incur a cost of $(\frac{1}{k})^\alpha \cdot k = (\frac{1}{k})^{\alpha-1}$. This simple example illustrates the (maximum) potential energy savings due to the use of intermediate stations when compared to a direct transmission from the source node *s*. In practice, though, the experienced energy savings are far from such high factors, and also the loss in energy efficiency when allowing only *k* stations to send compared to the unrestricted case is far less pronounced. In general, the advantage of using intermediate stations for a broadcast operation is greater if the nodes are distributed along 1-dimensional patterns and curves; for a very dense uniform distribution of the nodes e.g. within square, the direct transmission from the source is almost optimal (for $\alpha = 2$, for $\alpha > 2$ the gain of using intermediate stations grows). Unfortunately, the MST-based algorithm will always create a transmission range assignment where at least $n/6$ nodes are sending, even if there exist equally good or even better assignments with few sending nodes.

1.1   Our Contribution

In this paper we consider the *k*-set minimum energy broadcast problem from an analytical point of view. We show that somewhat surprisingly for any network of *n* radio stations there exists a subset *S* of the stations whose size is *independent*

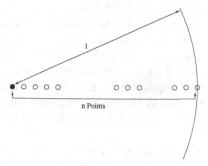

Fig. 1. Energy savings by using intermediate radio stations

of n and which preserves all the important characteristics of S with respect to a energy efficient k-set broadcast. We call S a *synopsis* or *core-set* of the network topology w.r.t. the k-SEMBC. We will show that using a synopsis of size $|S| = O((k/\epsilon)^2)$, any solution of the k-SEMBC for S translates to a solution for k-SEMBC for the original set P at a cost at most a $(1+\epsilon)$ factor away and vice versa. Since the size of this synopsis is independent of the network size, we can run even a brute-force algorithm to compute the optimum k-set broadcast. The running time of this algorithm is linear in n but still exponential in k. So we also present an $O(1)$-approximation algorithm whose running time is linear in n but polynomial in k. We want to emphasize that the focus of this paper is to examine the fundamental structure of the k-SEMBC problem rather than provide practical algorithms for direct use in a wireless network. Though we believe that with some engineering effort variants of our algorithms can be made practical, we have not conducted simulations yet to show practicability of our approaches.

1.2 Related Work

The EMBC problem is known to be $\mathcal{NP}$-hard ([6,5]), For arbitrary, non-metric distance functions the problem can also not be approximated better than a log-factor unless $\mathcal{P} = \mathcal{NP}$ [14]. For the Euclidean setting in the plane, [6] and [15] have shown a lower bound of 6 for the approximation ratio of the MST-based solution. In a sequence of papers the upper bound for this solution was lowered from in several steps (e.g. [6,15,8]) to finally match its lower bound of 6 (Ambühl, [1]). While all these papers focused on analytical worst-case bounds for the algorithm performance, simulation studies e.g. in [7] show that the actual performance in "real-world" networks is much better. There has also been work on more restricted broadcast operations in the spirit of k-SEMBC. In [2] the authors consider a *bounded-hop* broadcast operation where the resulting communication graph has to contain a spanning tree rooted at the source node s of depth at most h. They show how to compute an optimal h-hop broadcast range assignment for $h = 2$ in time $O(n^7)$. For $h > 2$ they show how to obtain a $(1 + \epsilon)$-approximation in time $O(n^{O(\mu)})$ where $\mu = (h^2/\epsilon)^{2^h}$, that is, their running time is triply exponential in the number of hops h and this shows up in the exponent of n. In [9] Funke and Laue show how to obtain a $(1 + \epsilon)$ approximation for the h-hop broadcast problem in time doubly exponential in h. Their approach is also based on a synopsis of the network, but in contrast to this paper they require a synopsis S that has size exponential in h. We note that bounded-hop broadcasts address the issue of latency since a message will arrive at any network node after at most h intermediate stations, still the reliability and interference problems remain as potentially very many network nodes might actively participate in the broadcast. General surveys of algorithmic range assignment problems can be found in [5,16,12]. Closely related in particular to the $O(1)$-approximation algorithm that we will present is the work by Bilò et al. [4]. They consider the problem of covering a set of n points in the plane using at most k disks such that the sum of the areas of the disks is minimized. They provide a $(1 + \epsilon)$-approximation to this problem in time $O(n^{\alpha^2/\epsilon^2})$. They do not address the problem of enforcing connectivity which is part of the k-SEMBC problem.

1.3   Outline

Section 2 recaps a known complexity result for the the unconstrained broadcast problem EMBC and sketches a simple folklore-brute-force algorithm to solve the k-SEMBC problem. Section 3 contains the core contributions of our paper; we show how to extract a small synopsis of the network topology (Section 3.1) and how to use that to obtain a $(1 + \epsilon)$-approximation algorithm. In Section 3.2 we show how a faster algorithm obtains an $O(1)$-approximation. Finally, in Section 4 we point out directions for future research.

2   Preliminaries

The unconstrained broadcast problem EMBC is known to be $\mathcal{NP}$-hard and for non-metric distance functions even not well approximable ([6,14]. Since the unconstrained broadcast problem is a special case of the k-set broadcast problem with $k = n$ these hardness results carry over to the k-set broadcast problem, if k is not treated as a constant. If k is regarded a constant, the problem can be solved in polynomial time as we will see in the following.

2.1   A Naive, Brute-Force Algorithm

The k-set broadcast problem can be solved in a brute force manner. Essentially, one can try out all $\binom{n}{k-1}$ different subsets for the $k-1$ active senders apart from the source s. For each of those (and the source node s), one then assigns all possible $n-1$ ranges. In total we have then $O(n^{k-1}(n-1)^k) = O(n^{2k})$ potential power assignments. For each of those we can check in $O(n^2)$ time whether it is a valid k-set broadcast.

That is, overall we have the following corollary:

Corollary 1. *For n points we can compute the optimal k-set broadcast in time* $O(n^{2k+2})$.

For most practical applications, we expect k to be a small constant, but unfortunately not small enough that this naive algorithm can be applied to networks of not too small size (e.g. several thousand nodes). In the following Section we lower our expectations and aim for *approximate* solutions to the k-set broadcast problem. This allows for more efficient algorithms as we will see.

3 Algorithms

3.1 Small Synopsis of the Network Topology

We say a range assignment r is *valid* if the induced communication graph $G^{(r)}$ contains a directed spanning tree reaching all nodes $p \in P$ and rooted at s, and at most k nodes have non-zero transmission range assigned; otherwise we call r *invalid*.

Definition 1. *Let P be a set of n points, $s \in P$ a designated source node. Consider another set S of points (not necessarily a subset of P). If for any valid range assignment $r : P \to \mathbb{R}_{\geq 0}$ there exists a valid range assignment $r' : S \to \mathbb{R}_{\geq 0}$ such that $\mathrm{cost}(r') \leq (1 + \epsilon) \cdot \mathrm{cost}(r)$ and for any valid range assignment $r' : S \to \mathbb{R}_{\geq 0}$ there exists a valid range assignment $r : P \to \mathbb{R}_{\geq 0}$ such that $\mathrm{cost}(r) \leq (1 + \epsilon) \cdot \mathrm{cost}(r')$ then S is called $(1 + \epsilon)$-synopsis for (P, s).*

A $(1+\epsilon)$-synopsis for a problem instance (P, s) can hence be viewed as a problem sketch of the original problem. If we can show that a $(1+\epsilon)$-synopsis of small size (independent of n) exists, solving the k-SEMBC problem on this problem sketch immediately leads to an $(1 + \epsilon)^2$-approximate solution to the original problem. The former can be even done using a brute force algorithm. Of course, the transformation from range assignment r' for the synopsis S to range assignment r for the input point set P has to be practical in order to derive a solution for the original problem. We will see that this can be done in linear time.

The definition of a synopsis can be seen as a generalization of core-sets defined in previous papers. For example, the term core-set has been defined for k-median [11] or minimum enclosing disk [13]. However, in the case of the k-SEMBC problem we have to consider two more issues. The first is feasibility. While any solution to the k-median problem is feasible not every solution is feasible for the k-SEMBC problem. The second issue is monotonicity. For the problem of the smallest enclosing disk the optimal solution does not increase if we remove points from the input. We do not have this property here. An optimal solution can increase or decrease if we remove input points. Hence, the above definition of a $(1 + \epsilon)$-synopsis can be seen as a generalization of core-sets to any optimization problem.

We will now show that we can find a small synopsis to the original problem. We assume that the maximum distance from the source node s to another node is 1.

Lemma 1. *For any k-SEMBC instance there exists a $(1+\epsilon)^\alpha$-synopsis of size* $O(\frac{k^2}{\epsilon^2})$.

Proof. We place a grid of grid width $\Delta = \frac{1}{\sqrt{2}}\frac{\epsilon}{k}$ on the plane. Notice, that the grid has to cover an area of radius 1 around the source only because the furthest distance from node s to any other node is 1. Hence its size is $O(\frac{k^2}{\epsilon^2})$. Now we assign each point in P to its closest grid point. Let S be the set of grid points that had at least one point from P snapped to it.

It remains to show that S is indeed a $(1+\epsilon)^\alpha$-synopsis. We can transform any given valid range assignment r for P into a valid range assignment r' for S. We define the range assignment r' for S as

$$r'(p') = \max_{p \text{ was snapped to } p'} r(p) + \sqrt{2}\Delta.$$

Since each point p is at most $\frac{1}{\sqrt{2}}\Delta$ away from its closest grid point p' we certainly have a valid range assignment for S. It is easy to see that the cost of r' for S is not much larger than the cost of r for P. We have:

$$\sum_{p' \in S} (r'(p'))^\alpha = \sum_{p' \in S} (\max_{p \text{ was snapped to } p'} r(p) + \sqrt{2}\Delta)^\alpha$$

$$\leq \sum_{p' \in S} (\max_{p \text{ was snapped to } p'} r(p) + \frac{\epsilon}{k})^\alpha$$

$$\leq \sum_{p \in P} (r(p) + \frac{\epsilon}{k})^\alpha.$$

The relative error satisfies

$$\frac{\text{cost}(r')}{\text{cost}(r)} \leq \frac{\sum_{p \in P}(r(p) + \frac{\epsilon}{k})^\alpha}{\sum_{p \in P}(r(p))^\alpha}.$$

Notice, that $\sum_{p \in P} r(p) \geq 1$ and r is positive for at most k points p. Hence, the above expression is maximized when $r(p) = \frac{1}{k}$ for all points p that are assigned a positive value. Thus

$$\frac{\text{cost}(r')}{\text{cost}(r)} \leq \frac{k \cdot (\frac{1}{k} + \frac{\epsilon}{k})^\alpha}{k \cdot (\frac{1}{k})^\alpha} = (1+\epsilon)^\alpha.$$

On the other hand we can transform any given valid range assignment r' for S into a valid range assignment r for P as follows. We select for each grid point $g \in S$ one representative g_P from P that was snapped to it. For the grid point to which s (the source) was snapped we select s as the representative. If we define the range assignment r for P as $r(g_P) = r'(g) + \sqrt{2}\Delta$ and $r(p) = 0$ if p does not belong to the chosen representatives, then r is a valid range assignment for P because every point is moved by the snapping by at most $\Delta/\sqrt{2}$. Using the same reasoning as above we can show that $\text{cost}(r) \leq (1+\epsilon)^\alpha \text{cost}(r')$. Hence, we have shown that S is indeed a $(1+\epsilon)^\alpha$-synopsis.

Once we have solved the k-SEMBC problem for the $(1 + \epsilon)^\alpha$-synopsis S we can easily transform the obtained solution to a $(1 + \epsilon)^{2\alpha}$-approximate solution to the original problem. Let us now concentrate on solving the k-SEMBC problem for the synopsis S. Since we were able to reduce the problem size to a constant independent of n, we can employ a brute-force strategy to compute an optimal solution for the reduced problem (S, s).

When looking for an optimal, energy-minimal solution for S, it is obvious that each node needs to consider only $|S|$ different ranges. Hence, naively there are at most $\left(\frac{k^2}{\epsilon^2} \right) \cdot \left(\frac{k^2}{\epsilon^2} \right)^k$ different range assignments to consider at all. We enumerate all these assignments and for each of them we check whether the range assignment is valid; this can be done in time $|S|$. Of all the valid range assignments we return the one of minimal cost.

Assuming the floor function a $(1 + \epsilon)^\alpha$-synopsis S for an instance of the k-SEMBC problem for a set of n radio nodes in the plane can be constructed in linear time. Hence we obtain the following theorem:

Theorem 1. *A $(1 + \epsilon)^{2\alpha}$-approximate solution to the k-SEMBC problem on n points in the plane can be computed in time $O(n + \left(\frac{k}{\epsilon} \right)^{4k})$.*

A simple observation allows us to improve the running time slightly. Since eventually we are only interested in an approximate solution to the problem, we are also happy with only approximating the optimum solution for the synopsis S. Such an approximation for S can be found more efficiently by not considering all possible at most $|S|$ ranges for each grid point. Instead we consider as admissible ranges only 0 and $\frac{\epsilon}{k} \cdot (1 + \epsilon)^i$ for $i \geq 0$. That is, the number of different ranges a node can attain is at most $1 + \log_{1+\epsilon} \frac{k}{\epsilon} \leq \frac{2}{\epsilon} \cdot \log \frac{k}{\epsilon}$ for $\epsilon \leq 1$. This comes at a cost of a $(1 + \epsilon)$ factor by which each individual assigned range might exceed the optimum. The running time of the algorithm improves, though, which leads to our main result in this section:

Theorem 2. *A $(1 + \epsilon)^{3\alpha}$-approximate solution to the k-SEMBC problem on n points in the plane can be computed in time $O\left(n + \left(\frac{k^2 \log \frac{k}{\epsilon}}{\epsilon^3} \right)^k \right)$.*

Obviously, a $(1 + \psi)$-approximate solution can be obtained by choosing $\epsilon = \theta(\psi/\alpha)$.

3.2 Faster $O(1)$-Approximations

We now show how to compute a constant approximation for the k-set broadcast problem. The idea is to first cluster the points into k clusters. Then we ensure connectivity of these point sets by increasing their cluster sizes. As clustering we define the *k-disk cover problem*:

Definition 2 (k-disk cover problem (k-DCP)). *Given a set P of n points in the Euclidean plane $\mathbb{R}^2$, find a subset $C \subseteq P$ of cardinality at most k and* *radii $r_v \geq 0$ associated with each element $v \in C$ such that $\sum_{v \in C} r_v^\alpha$ is minimized and all points in P are covered by the disks $D_v^{r_v} := \{x \in \mathbb{R}^2 \mid |xv| \leq r_v\}$.*

Given a k-disk cover $D := (C, (r_v)_{v \in C})$ for P with center points C and radii r_v, we associate with D a range assignment r_D on P as follows:

$$\forall v \in P: \ r_D(v) := \begin{cases} r_v & \text{if } v \in C \\ 0 & \text{otherwise} \end{cases}$$

By D_i we denote a disk in D. Note that the k-DCP with the additional constraint that the communication graph $G^{(r_D)}$ is connected is exactly the k-set broadcast problem and that an instance of one problem is an instance of the other. Thus, the cost of an optimal solution for an instance I of the k-DCP is a lower bound on I for the k-set broadcast problem. Unfortunately k-DCP is NP-hard (see [4]) but it admits a PTAS as shown in [4] by Bilò et al. A direct consequence of their results is:

Corollary 2. *There exists an algorithm for the k-DCP that computes $(1 + \epsilon)$-approximate solutions in time $n^{(\frac{\alpha}{\epsilon})^d}$ for a constant d.*

By setting ϵ to 1 we obtain a 2-approximation algorithm for the k-DCP that runs in time $n^{c'_\alpha}$ for a constant c'_α. Note that their algorithm can easily be modified such that the source s is the center of one of the disks.

Our approximation algorithm works as follows: First we compute an approximate k-disk cover $D := (C, (r_v)_{v \in C})$ over P. Then we determine for the center points in C an approximate broadcast with range assignment r_B by using an MST based algorithm (see [1]) that has an approximation guarantee of 6. Now we construct a range assignment r_A for P in the following way:

$$\forall v \in P: \ r_A(v) := \begin{cases} \max\{r_v, r_B(v)\} & \text{if } v \in C \\ 0 & \text{otherwise} \end{cases}$$

Note that $G^{(r_A)}$ is connected and therefor induces a valid k-set broadcast since only k stations are sending. We still have to show that we have computed an approximate solution:

Theorem 3. $\text{cost}(r_A) \leq 36 c_\alpha \cdot \text{cost}(r_{opt})$, *where r_{opt} is the range assignment of an optimal k-set broadcast and c_α is a constant depending only on α.*

Proof. The proof idea is as follows: Assuming knowledge about an optimal range assignment r_{opt} for the k-set broadcast we transform the range assignment r_D into r'_D such that

a) r'_D is a valid k-set broadcast
b) the sending nodes in r'_D are exactly the center points of D and
c) r'_D is a constant factor approximation of r_{opt}.

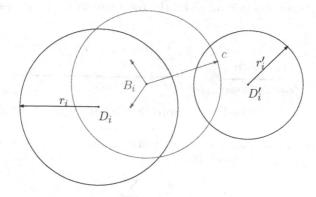

Fig. 2. Proof illustration for the constant factor approximation algorithm

If we know that such a broadcast r'_D exists, we can simply compute an optimal broadcast r_B over the center points of D. Then we know that $\text{cost}(r_B) \leq \text{cost}(r'_D)$ and r_B must also be a constant factor approximation of r_{opt}.

Consider now the communication tree T which is defined as a subtree of $G^{(r_{opt})}$ spanning P. The idea of the construction of r'_D is to replace the inner nodes of T (i.e. the sending stations of r_{opt}) by increasing the radii of the disks in C appropriately so that r'_D is valid.

We increase the nonzero values of r_D in the following way: with each of the inner nodes B_i of T we associate arbitrarily one disk D_i in which B_i is contained. Note that there must be at least one such disk for each B_i since the disks cover P. We now update r_D in a breath first search manner on T starting from source node s (see figure 2):

Given an inner node B_i of T if all children of B_i in T lie in the associated disk D_i then all of them can be reached from node C_i without increasing r_i. The interesting case is if there are children of B_i that are not contained in D_i but contained in a disk D'_i whose center is not covered so far. Assume that there is exactly one such child c. We then set the radius of D_i to $r_i + r'_i + r_{opt}(B_i)$. If there is more than one such child, let c be the one that maximizes r'_i so that each child of B_i and the centers of the disks in which the children of B_i are contained in can be reached by D_i. Note that it can happen that two different inner nodes B_i and B_j are associated with the same disk D_k, so that D_k is updated more than once in the process. In such a case we update D_k only if r_k is increasing. Now let us assume that for a disk D_i the last update involved disk D'_k then we call disk D'_k the target disk of D'_k.

By induction $G^{(r_D)}$ is connected after these updates. furthermore note that the sending stations are still exactly the center points of D. Let $D^* \subseteq D$ be the set of disks that are updated and let B_i be the node in T in the update step for disk $D_i \in D^*$. Summing over all disks, the total cost of the broadcast is therefor bounded by:

$$\underbrace{\sum_{D_i \in D \setminus D^*} r_i^\alpha}_{\leq \mathrm{cost}(D) \leq 2\,\mathrm{cost}(r_{opt})} + \underbrace{\sum_{D_i \in D^*} (r_i + r_i' + r_{opt}(B_i))^\alpha}_{(**)}$$

Before we bound the second term, note that a disk appears as a target disk only once in the process of updating the disk radii since once its center point is covered it is never considered as a target disk again. Thus each r_i' in the above sum can also appear only once. Thus,

$$(**) \leq c_\alpha \Big[\sum_{D_i \in D^*} r_i^\alpha + \sum_{D_i \in D^*} r_i'^\alpha + $$

$$\sum_{D_i \in D^*} r_{opt}(B_i)^\alpha \Big]$$

$$\leq c_\alpha \Big[2 \underbrace{\sum_{D_i \in D} r_i^\alpha}_{\leq 4\,\mathrm{cost}(r_{opt})} + \underbrace{\sum_{D_i \in D} r_{opt}(B_i)^\alpha}_{=\mathrm{cost}(r_{opt})} \Big]$$

$$\leq 5 \cdot c_\alpha \cdot \mathrm{cost}(r_{opt})$$

where the constant c_α can be bounded by $3^{\alpha-1}$. Thus there exists a broadcast over the center points C with total cost upper bounded by

$$2\,\mathrm{cost}(r_{opt}) + 5 \cdot c_\alpha \cdot \mathrm{cost}(r_{opt})$$
$$\leq 6 \cdot c_\alpha \cdot \mathrm{cost}(r_{opt})$$

Since we use a 6-approximate broadcast, the algorithm has an approximation ratio of $36c_\alpha$.

Theorem 4. *There exists a constant factor approximation algorithm for the k-set broadcast problem over n points in the Euclidean plane that runs in $O(n^{c_\alpha})$.*

The theorem can be further improved by using the results of the previous section. By setting ϵ to 1 we obtain a synopsis of size k^2. Using theorem 4 we obtain directly a constant factor approximation algorithm whose running time is only linear in n and polynomial in k:

Theorem 5. *There exists a constant factor approximation algorithm for the k-set broadcast problem over n points in the Euclidean plane that runs in time linear in n and polynomial in k, i.e. in $O(n + k^{2 \cdot c_\alpha'})$.*

4   Future Work

4.1   Simple, Distributed Algorithms with Good Performance

In the introduction we have pointed out why we believe that a broadcast operation with a bounded number of senders can be of great benefit for the efficient operation of a wireless network. The main part of the paper considered the *k*-SEMBC problem from a purely analytical point of view, though. The most important research direction to follow in the near future is to design simple and *distributed* algorithms for the *k*-SEMBC problem. One possible idea could be to construct distributedly a low-weight *k*-disk-cover of the network and then connect the components in some way such that the overall cost does not increase by too much – very much in the spirit of our $O(1)$-approximation algorithm. We also plan to look at the *k*-SEMBC problem from a more empirical point of view. Several heuristic solutions could be thought of and examined using extensive simulations on different network deployments.

4.2   Extension to Metrics of Bounded Doubling Dimension

While actual network deployments often are in a planar setting, the experienced metric for several reasons is typically not exactly of the Euclidean type (due to obstacles, interference etc.), but often in some sense 'close' as it still retains some correlation between geographic distance and distance in the metric. One way to measure similarity between metrics is the so called *doubling dimension* [10]. It remains to examine whether our algorithms also work on metrics of bounded doubling dimension.

4.3 $(1 + \epsilon)$-Approximation with Running Time Polynomial in *k* (and $1/\epsilon$) ?

One major drawback of the $(1 + \epsilon)$-approximation algorithm presented is that it still requires time exponential in the number of senders *k*. It is unclear whether this exponential dependence could be removed. One idea might be to use the approach by Arora ([3]) based on a shifted dissection which was already used successfully to obtain fast approximation algorithms for geometric problems. This might induce an exponential dependence on $1/\epsilon$, though.

References

1. Ambühl, C.: An optimal bound for the mst algorithm to compute energy-efficient broadcast trees in wireless networks. In: Caires, L., Italiano, G.F., Monteiro, L., Palamidessi, C., Yung, M. (eds.) ICALP 2005. LNCS, vol. 3580, Springer, Heidelberg (2005)
2. Ambühl, C., Clementi, A.E.F., Di Ianni, M., Lev-Tov, N., Monti, A., Peleg, D., Rossi, G., Silvestri, R.: Efficient algorithms for low-energy bounded-hop broadcast in ad-hoc wireless networks. In: Diekert, V., Habib, M. (eds.) STACS 2004. LNCS, vol. 2996, Springer, Heidelberg (2004)
3. Arora, S.: Approximation schemes for geometric np-hard problems: A survey. In: Hariharan, R., Mukund, M., Vinay, V. (eds.) FST TCS 2001: Foundations of Software Technology and Theoretical Computer Science. LNCS, vol. 2245, Springer, Heidelberg (2001)

4. Bilò, V., Caragiannis, I., Kaklamanis, C., Kanellopoulos, P.: Geometric clustering to minimize the sum of cluster sizes. In: Brodal, G.S., Leonardi, S. (eds.) ESA 2005. LNCS, vol. 3669, Springer, Heidelberg (2005)
5. Clementi, A., Huiban, G., Penna, P., Rossi, G., Verhoeven, Y.: Some recent theoretical advances and open questions on energy consumption in ad-hoc wireless networks. In: Proc. 3rd Workshop on Approximation and Randomization Algorithms in Communication Networks (ARACNE), pp. 23–38 (2002)
6. Clementi, A.E.F., Crescenzi, P., Penna, P., Rossi, G., Vocca, P.: On the complexity of computing minimum energy consumption broadcast subgraphs. In: Ferreira, A., Reichel, H. (eds.) STACS 2001. LNCS, vol. 2010, pp. 121–131. Springer, Heidelberg (2001)
7. Clementi, A.E.F., Huiban, G., Rossi, G., Verhoeven, Y.C., Penna, P.: On the approximation ratio of the mst-based heuristic for the energy-efficient broadcast problem in static ad-hoc radio networks. In : IPDPS, p. 222 (2003)
8. Flammini, M., Navarra, A., Klasing, R., Pérennes, S.: Improved approximation results for the minimum energy broadcasting problem. In: DIALM-POMC, pp. 85–91 (2004)
9. Funke, S., Laue, S.: Bounded-hop energy-efficient broadcast in low-dimensional metrics via coresets. In: Thomas, W., Weil, P. (eds.) STACS 2007. LNCS, vol. 4393, pp. 272–283. Springer, Heidelberg (2007)
10. Gupta, A., Krauthgamer, R., Lee, J.R.: Bounded geometries, fractals, and low-distortion embeddings. In: FOCS, pp. 534–543 (2003)
11. Har-Peled, S., Mazumdar, S.: Coresets for k-means and k-median clustering and their applications. In: STOC, pp. 291–300 (2004)
12. Kirousis, L.M., Kranakis, E., Krizanc, D., Pelc, A.: Power consumption in packet radio networks. Theor. Comput. Sci. 243(1-2), 289–305 (2000)
13. Kumar, P., Mitchell, J.B., Yildirim, E.A.: Approximate minimum enclosing balls in high dimensions using core-sets. J. Exp. Algorithmics, vol. 8(1.1) (2003)
14. Guha, S., Khuller, S.: Improved methods for approximating node weighted steiner trees and connected dominating sets. Information and Computation 150, 57–74 (1999)
15. Wan, P.-J., Calinescu, G., Li, X., Frieder, O.: Minimum-energy broadcast routing in static ad hoc wireless networks. In: INFOCOM, pp. 1162–1171 (2001)
16. Wieselthier, J.E., Nguyen, G.D., Ephremides, A.: On the construction of energy-efficient broadcast and multicast trees in wireless networks. In: INFOCOM, pp. 585–594 (2000)

Author Index

Lecture Notes in Computer Science

For information about Vols. 1–4450

please contact your bookseller or Springer

Vol. 4500: N. Streitz, A. Kameas, I. Mavrommati (Eds.), The Disappearing Computer. XVIII, 304 pages. 2007.

Vol. 4499: Y.Q. Shi (Ed.), Transactions on Data Hiding and Multimedia Security II. IX, 117 pages. 2007.

Vol. 4497: S.B. Cooper, B. Löwe, A. Sorbi (Eds.), Computation and Logic in the Real World. XVIII, 826 pages. 2007.

Vol. 4496: N.T. Nguyen, A. Grzech, R.J. Howlett, L.C. Jain (Eds.), Agent and Multi-Agent Systems: Technologies and Applications. XXI, 1046 pages. 2007. (Sublibrary LNAI).

Vol. 4495: J. Krogstie, A. Opdahl, G. Sindre (Eds.), Advanced Information Systems Engineering. XVI, 606 pages. 2007.

Vol. 4494: H. Jin, O.F. Rana, Y. Pan, V.K. Prasanna (Eds.), Algorithms and Architectures for Parallel Processing. XIV, 508 pages. 2007.

Vol. 4493: D. Liu, S. Fei, Z. Hou, H. Zhang, C. Sun (Eds.), Advances in Neural Networks – ISNN 2007, Part III. XXVI, 1215 pages. 2007.

Vol. 4492: D. Liu, S. Fei, Z. Hou, H. Zhang, C. Sun (Eds.), Advances in Neural Networks – ISNN 2007, Part II. XXVII, 1321 pages. 2007.

Vol. 4491: D. Liu, S. Fei, Z.-G. Hou, H. Zhang, C. Sun (Eds.), Advances in Neural Networks – ISNN 2007, Part I. LIV, 1365 pages. 2007.

Vol. 4490: Y. Shi, G.D. van Albada, J. Dongarra, P.M.A. Sloot (Eds.), Computational Science – ICCS 2007, Part IV. XXXVII, 1211 pages. 2007.

Vol. 4489: Y. Shi, G.D. van Albada, J. Dongarra, P.M.A. Sloot (Eds.), Computational Science – ICCS 2007, Part III. XXXVII, 1257 pages. 2007.

Vol. 4488: Y. Shi, G.D. van Albada, J. Dongarra, P.M.A. Sloot (Eds.), Computational Science – ICCS 2007, Part II. XXXV, 1251 pages. 2007.

Vol. 4487: Y. Shi, G.D. van Albada, J. Dongarra, P.M.A. Sloot (Eds.), Computational Science – ICCS 2007, Part I. LXXXI, 1275 pages. 2007.

Vol. 4486: M. Bernardo, J. Hillston (Eds.), Formal Methods for Performance Evaluation. VII, 469 pages. 2007.

Vol. 4485: F. Sgallari, A. Murli, N. Paragios (Eds.), Scale Space and Variational Methods in Computer Vision. XV, 931 pages. 2007.

Vol. 4484: J.-Y. Cai, S.B. Cooper, H. Zhu (Eds.), Theory and Applications of Models of Computation. XIII, 772 pages. 2007.

Vol. 4483: C. Baral, G. Brewka, J. Schlipf (Eds.), Logic Programming and Nonmonotonic Reasoning. IX, 327 pages. 2007. (Sublibrary LNAI).

Vol. 4482: A. An, J. Stefanowski, S. Ramanna, C.J. Butz, W. Pedrycz, G. Wang (Eds.), Rough Sets, Fuzzy Sets, Data Mining and Granular Computing. XIV, 585 pages. 2007. (Sublibrary LNAI).

Vol. 4481: J. Yao, P. Lingras, W.-Z. Wu, M. Szczuka, N.J. Cercone, D. Ślęzak (Eds.), Rough Sets and Knowledge Technology. XIV, 576 pages. 2007. (Sublibrary LNAI).

Vol. 4480: A. LaMarca, M. Langheinrich, K.N. Truong (Eds.), Pervasive Computing. XIII, 369 pages. 2007.

Vol. 4479: I.F. Akyildiz, R. Sivakumar, E. Ekici, J.C.d. Oliveira, J. McNair (Eds.), NETWORKING 2007. Ad Hoc and Sensor Networks, Wireless Networks, Next Generation Internet. XXVII, 1252 pages. 2007.

Vol. 4478: J. Martí, J.M. Benedí, A.M. Mendonça, J. Serrat (Eds.), Pattern Recognition and Image Analysis, Part II. XXVII, 657 pages. 2007.

Vol. 4477: J. Martí, J.M. Benedí, A.M. Mendonça, J. Serrat (Eds.), Pattern Recognition and Image Analysis, Part I. XXVII, 625 pages. 2007.

Vol. 4476: V. Gorodetsky, C. Zhang, V.A. Skormin, L. Cao (Eds.), Autonomous Intelligent Systems: Multi-Agents and Data Mining. XIII, 323 pages. 2007. (Sublibrary LNAI).

Vol. 4475: P. Crescenzi, G. Prencipe, G. Pucci (Eds.), Fun with Algorithms. X, 273 pages. 2007.

Vol. 4474: G. Prencipe, S. Zaks (Eds.), Structural Information and Communication Complexity. XI, 342 pages. 2007.

Vol. 4472: M. Haindl, J. Kittler, F. Roli (Eds.), Multiple Classifier Systems. XI, 524 pages. 2007.

Vol. 4471: P. Cesar, K. Chorianopoulos, J.F. Jensen (Eds.), Interactive TV: a Shared Experience. XIII, 236 pages. 2007.

Vol. 4470: Q. Wang, D. Pfahl, D.M. Raffo (Eds.), Software Process Dynamics and Agility. XI, 346 pages. 2007.

Vol. 4469: K.-C. Hui, Z. Pan, R.C.-k. Chung, C.C.L. Wang, X. Jin, S. Göbel, E.C.-L. Li (Eds.), Technologies for E-Learning and Digital Entertainment. XVIII, 974 pages. 2007.

Vol. 4468: M.M. Bonsangue, E.B. Johnsen (Eds.), Formal Methods for Open Object-Based Distributed Systems. X, 317 pages. 2007.

Vol. 4467: A.L. Murphy, J. Vitek (Eds.), Coordination Models and Languages. X, 325 pages. 2007.

Vol. 4466: F.B. Sachse, G. Seemann (Eds.), Functional Imaging and Modeling of the Heart. XV, 486 pages. 2007.

Vol. 4465: T. Chahed, B. Tuffin (Eds.), Network Control and Optimization. XIII, 305 pages. 2007.

Vol. 4464: E. Dawson, D.S. Wong (Eds.), Information Security Practice and Experience. XIII, 361 pages. 2007.

Vol. 4463: I. Măndoiu, A. Zelikovsky (Eds.), Bioinformatics Research and Applications. XV, 653 pages. 2007. (Sublibrary LNBI).

Vol. 4462: D. Sauveron, K. Markantonakis, A. Bilas, J.-J. Quisquater (Eds.), Information Security Theory and Practices. XII, 255 pages. 2007.

Vol. 4459: C. Cérin, K.-C. Li (Eds.), Advances in Grid and Pervasive Computing. XVI, 759 pages. 2007.

Vol. 4453: T. Speed, H. Huang (Eds.), Research in Computational Molecular Biology. XVI, 550 pages. 2007. (Sublibrary LNBI).

Vol. 4452: M. Fasli, O. Shehory (Eds.), Agent-Mediated Electronic Commerce. VIII, 249 pages. 2007. (Sublibrary LNAI).

Vol. 4451: T.S. Huang, A. Nijholt, M. Pantic, A. Pentland (Eds.), Artifical Intelligence for Human Computing. XVI, 359 pages. 2007. (Sublibrary LNAI).